# Lecture Notes in Computer Science 11305

*Commenced Publication in 1973*
Founding and Former Series Editors:
Gerhard Goos, Juris Hartmanis, and Jan van Leeuwen

More information about this series at http://www.springer.com/series/7407

Long Cheng · Andrew Chi Sing Leung
Seiichi Ozawa (Eds.)

# Neural
# Information Processing

25th International Conference, ICONIP 2018
Siem Reap, Cambodia, December 13–16, 2018
Proceedings, Part V

 Springer

*Editors*
Long Cheng (iD)
The Chinese Academy of Sciences
Beijing, China

Seiichi Ozawa
Kobe University
Kobe, Japan

Andrew Chi Sing Leung
City University of Hong Kong
Kowloon, Hong Kong SAR, China

ISSN 0302-9743          ISSN 1611-3349 (electronic)
Lecture Notes in Computer Science
ISBN 978-3-030-04220-2          ISBN 978-3-030-04221-9 (eBook)
https://doi.org/10.1007/978-3-030-04221-9

Library of Congress Control Number: 2018960916

LNCS Sublibrary: SL1 – Theoretical Computer Science and General Issues

This Springer imprint is published by the registered company Springer Nature Switzerland AG
The registered company address is: Gewerbestrasse 11, 6330 Cham, Switzerland

# Preface

The 25th International Conference on Neural Information Processing (ICONIP 2018), the annual conference of the Asia Pacific Neural Network Society (APNNS), was held in Siem Reap, Cambodia, during December 13–16, 2018. The ICONIP conference series started in 1994 in Seoul, which has now become a well-established and high-quality conference on neural networks around the world. Siem Reap is a gateway to Angkor Wat, which is one of the most important archaeological sites in Southeast Asia, the largest religious monument in the world. All participants of ICONIP 2018 had a technically rewarding experience as well as a memorable stay in this great city.

In recent years, the neural network has been significantly advanced with the great developments in neuroscience, computer science, cognitive science, and engineering. Many novel neural information processing techniques have been proposed as the solutions to complex, networked, and information-rich intelligent systems. To disseminate new findings, ICONIP 2018 provided a high-level international forum for scientists, engineers, and educators to present the state of the art of research and applications in all fields regarding neural networks.

With the growing popularity of neural networks in recent years, we have witnessed an increase in the number of submissions and in the quality of submissions. ICONIP 2018 received 575 submissions from 51 countries and regions across six continents. Based on a rigorous peer-review process, where each submission was reviewed by at least three experts, a total of 401 high-quality papers were selected for publication in the prestigious Springer series of *Lecture Notes in Computer Science*. The selected papers cover a wide range of subjects that address the emerging topics of theoretical research, empirical studies, and applications of neural information processing techniques across different domains.

In addition to the contributed papers, the ICONIP 2018 technical program also featured three plenary talks and two invited talks delivered by world-renowned scholars: Prof. Masashi Sugiyama (University of Tokyo and RIKEN Center for Advanced Intelligence Project), Prof. Marios M. Polycarpou (University of Cyprus), Prof. Qing-Long Han (Swinburne University of Technology), Prof. Cesare Alippi (Polytechnic of Milan), and Nikola K. Kasabov (Auckland University of Technology).

We would like to extend our sincere gratitude to all members of the ICONIP 2018 Advisory Committee for their support, the APNNS Governing Board for their guidance, the International Neural Network Society and Japanese Neural Network Society for their technical co-sponsorship, and all members of the Organizing Committee for all their great effort and time in organizing such an event. We would also like to take this opportunity to thank all the Technical Program Committee members and reviewers for their professional reviews that guaranteed the high quality of the conference proceedings. Furthermore, we would like to thank the publisher, Springer, for their sponsorship and cooperation in publishing the conference proceedings in seven volumes of *Lecture Notes in Computer Science*. Finally, we would like to thank all the speakers, authors, reviewers, volunteers, and participants for their contribution and support in making ICONIP 2018 a successful event.

October 2018

Jun Wang
Long Cheng
Andrew Chi Sing Leung
Seiichi Ozawa

# ICONIP 2018 Organization

## General Chair

Jun Wang      City University of Hong Kong,
Hong Kong SAR, China

## Advisory Chairs

Akira Hirose      University of Tokyo, Tokyo, Japan
Soo-Young Lee      Korea Advanced Institute of Science and Technology,
South Korea
Derong Liu      Institute of Automation, Chinese Academy of Sciences,
China
Nikhil R. Pal      Indian Statistics Institute, India

## Program Chairs

Long Cheng      Institute of Automation, Chinese Academy of Sciences,
China
Andrew C. S. Leung      City University of Hong Kong, Hong Kong SAR,
China
Seiichi Ozawa      Kobe University, Japan

## Special Sessions Chairs

Shukai Duan      Southwest University, China
Kazushi Ikeda      Nara Institute of Science and Technology, Japan
Qinglai Wei      Institute of Automation, Chinese Academy of Sciences,
China
Hiroshi Yamakawa      Dwango Co. Ltd., Japan
Zhihui Zhan      South China University of Technology, China

## Tutorial Chairs

Hiroaki Gomi      NTT Communication Science Laboratories, Japan
Takashi Morie      Kyushu Institute of Technology, Japan
Kay Chen Tan      City University of Hong Kong, Hong Kong SAR,
China
Dongbin Zhao      Institute of Automation, Chinese Academy of Sciences,
China

## Publicity Chairs

Zeng-Guang Hou                  Institute of Automation, Chinese Academy of Sciences,
                                    China
Tingwen Huang                   Texas A&M University at Qatar, Qatar
Chia-Feng Juang                 National Chung-Hsing University, Taiwan
Tomohiro Shibata                Kyushu Institute of Technology, Japan

## Publication Chairs

Xinyi Le                        Shanghai Jiao Tong University, China
Sitian Qin                      Harbin Institute of Technology Weihai, China
Zheng Yan                       University Technology Sydney, Australia
Shaofu Yang                     Southeast University, China

## Registration Chairs

Shenshen Gu                     Shanghai University, China
Qingshan Liu                    Southeast University, China
Ka Chun Wong                    City University of Hong Kong,
                                    Hong Kong SAR, China

## Conference Secretariat

Ying Qu                         Dalian University of Technology, China

## Program Committee

Hussein Abbass                  University of New South Wales at Canberra, Australia
Choon Ki Ahn                    Korea University, South Korea
Igor Aizenberg                  Texas A&M University at Texarkana, USA
Shotaro Akaho                   National Institute of Advanced Industrial Science
                                    and Technology, Japan
Abdulrazak Alhababi             UNIMAS, Malaysia
Cecilio Angulo                  Universitat Politècnica de Catalunya, Spain
Sabri Arik                      Istanbul University, Turkey
Mubasher Baig                   National University of Computer and Emerging
                                    Sciences Lahore, India
Sang-Woo Ban                    Dongguk University, South Korea
Tao Ban                         National Institute of Information and Communications
                                    Technology, Japan
Boris Bačić                     Auckland University of Technology, New Zealand
Xu Bin                          Northwestern Polytechnical University, China
David Bong                      Universiti Malaysia Sarawak, Malaysia
Salim Bouzerdoum                University of Wollongong, Australia
Ivo Bukovsky                    Czech Technical University, Czech Republic

| | |
|---|---|
| Ke-Cai Cao | Nanjing University of Posts and Telecommunications, China |
| Elisa Capecci | Auckland University of Technology, New Zealand |
| Rapeeporn Chamchong | Mahasarakham University, Thailand |
| Jonathan Chan | King Mongkut's University of Technology Thonburi, Thailand |
| Rosa Chan | City University of Hong Kong, Hong Kong SAR, China |
| Guoqing Chao | East China Normal University, China |
| He Chen | Nankai University, China |
| Mou Chen | Nanjing University of Aeronautics and Astronautics, China |
| Qiong Chen | South China University of Technology, China |
| Wei-Neng Chen | Sun Yat-Sen University, China |
| Xiaofeng Chen | Chongqing Jiaotong University, China |
| Ziran Chen | Bohai University, China |
| Jian Cheng | Chinese Academy of Sciences, China |
| Long Cheng | Chinese Academy of Sciences, China |
| Wu Chengwei | Bohai University, China |
| Zheru Chi | The Hong Kong Polytechnic University, SAR China |
| Sung-Bae Cho | Yonsei University, South Korea |
| Heeyoul Choi | Handong Global University, South Korea |
| Hyunsoek Choi | Kyungpook National University, South Korea |
| Supannada Chotipant | King Mongkut's Institute of Technology Ladkrabang, Thailand |
| Fengyu Cong | Dalian University of Technology, China |
| Jose Alfredo Ferreira Costa | Federal University of Rio Grande do Norte, Brazil |
| Ruxandra Liana Costea | Polytechnic University of Bucharest, Romania |
| Jean-Francois Couchot | University of Franche-Comté, France |
| Raphaël Couturier | University of Bourgogne Franche-Comté, France |
| Jisheng Dai | Jiangsu University, China |
| Justin Dauwels | Massachusetts Institute of Technology, USA |
| Dehua Zhang | Chinese Academy of Sciences, China |
| Mingcong Deng | Tokyo University of Agriculture and Technology, Japan |
| Zhaohong Deng | Jiangnan University, China |
| Jing Dong | Chinese Academy of Sciences, China |
| Qiulei Dong | Chinese Academy of Sciences, China |
| Kenji Doya | Okinawa Institute of Science and Technology, Japan |
| El-Sayed El-Alfy | King Fahd University of Petroleum and Minerals, Saudi Arabia |
| Mark Elshaw | Nottingham Trent International College, UK |
| Peter Erdi | Kalamazoo College, USA |
| Josafath Israel Espinosa Ramos | Auckland University of Technology, New Zealand |
| Issam Falih | Paris 13 University, France |

| | |
|---|---|
| Bo Fan | Zhejiang University, China |
| Yunsheng Fan | Dalian Maritime University, China |
| Hao Fang | Beijing Institute of Technology, China |
| Jinchao Feng | Beijing University of Technology, China |
| Francesco Ferracuti | Università Politecnica delle Marche, Italy |
| Chun Che Fung | Murdoch University, Australia |
| Wai-Keung Fung | Robert Gordon University, UK |
| Tetsuo Furukawa | Kyushu Institute of Technology, Japan |
| Hao Gao | Nanjing University of Posts and Telecommunications, China |
| Yabin Gao | Harbin Institute of Technology, China |
| Yongsheng Gao | Griffith University, Australia |
| Tom Gedeon | Australian National University, Australia |
| Ong Sing Goh | Universiti Teknikal Malaysia Melaka, Malaysia |
| Iqbal Gondal | Federation University Australia, Australia |
| Yue-Jiao Gong | Sun Yat-sen University, China |
| Shenshen Gu | Shanghai University, China |
| Chengan Guo | Dalian University of Technology, China |
| Ping Guo | Beijing Normal University, China |
| Shanqing Guo | Shandong University, China |
| Xiang-Gui Guo | University of Science and Technology Beijing, China |
| Zhishan Guo | University of Central Florida, USA |
| Christophe Guyeux | University of Franche-Comte, France |
| Masafumi Hagiwara | Keio University, Japan |
| Saman Halgamuge | The University of Melbourne, Australia |
| Tomoki Hamagami | Yokohama National University, Japan |
| Cheol Han | Korea University at Sejong, South Korea |
| Min Han | Dalian University of Technology, China |
| Takako Hashimoto | Chiba University of Commerce, Japan |
| Toshiharu Hatanaka | Osaka University, Japan |
| Wei He | University of Science and Technology Beijing, China |
| Xing He | Southwest University, China |
| Xiuyu He | University of Science and Technology Beijing, China |
| Akira Hirose | The University of Tokyo, Japan |
| Daniel Ho | City University of Hong Kong, Hong Kong SAR, China |
| Katsuhiro Honda | Osaka Prefecture University, Japan |
| Hongyi Li | Bohai University, China |
| Kazuhiro Hotta | Meijo University, Japan |
| Jin Hu | Chongqing Jiaotong University, China |
| Jinglu Hu | Waseda University, Japan |
| Xiaofang Hu | Southwest University, China |
| Xiaolin Hu | Tsinghua University, China |
| He Huang | Soochow University, China |
| Kaizhu Huang | Xi'an Jiaotong-Liverpool University, China |
| Long-Ting Huang | Wuhan University of Technology, China |

| | |
|---|---|
| Panfeng Huang | Northwestern Polytechnical University, China |
| Tingwen Huang | Texas A&M University, USA |
| Hitoshi Iima | Kyoto Institute of Technology, Japan |
| Kazushi Ikeda | Nara Institute of Science and Technology, Japan |
| Hayashi Isao | Kansai University, Japan |
| Teijiro Isokawa | University of Hyogo, Japan |
| Piyasak Jeatrakul | Mae Fah Luang University, Thailand |
| Jin-Tsong Jeng | National Formosa University, Taiwan |
| Sungmoon Jeong | Kyungpook National University Hospital, South Korea |
| Danchi Jiang | University of Tasmania, Australia |
| Min Jiang | Xiamen University, China |
| Yizhang Jiang | Jiangnan University, China |
| Xuguo Jiao | Zhejiang University, China |
| Keisuke Kameyama | University of Tsukuba, Japan |
| Shunshoku Kanae | Junshin Gakuen University, Japan |
| Hamid Reza Karimi | Politecnico di Milano, Italy |
| Nikola Kasabov | Auckland University of Technology, New Zealand |
| Abbas Khosravi | Deakin University, Australia |
| Rhee Man Kil | Sungkyunkwan University, South Korea |
| Daeeun Kim | Yonsei University, South Korea |
| Sangwook Kim | Kobe University, Japan |
| Lai Kin | Tunku Abdul Rahman University, Malaysia |
| Irwin King | The Chinese University of Hong Kong, Hong Kong SAR, China |
| Yasuharu Koike | Tokyo Institute of Technology, Japan |
| Ven Jyn Kok | National University of Malaysia, Malaysia |
| Ghosh Kuntal | Indian Statistical Institute, India |
| Shuichi Kurogi | Kyushu Institute of Technology, Japan |
| Susumu Kuroyanagi | Nagoya Institute of Technology, Japan |
| James Kwok | The Hong Kong University of Science and Technology, SAR China |
| Edmund Lai | Auckland University of Technology, New Zealand |
| Kittichai Lavangnananda | King Mongkut's University of Technology Thonburi, Thailand |
| Xinyi Le | Shanghai Jiao Tong University, China |
| Minho Lee | Kyungpook National University, South Korea |
| Nung Kion Lee | University Malaysia Sarawak, Malaysia |
| Andrew C. S. Leung | City University of Hong Kong, Hong Kong SAR, China |
| Baoquan Li | Tianjin Polytechnic University, China |
| Chengdong Li | Shandong Jianzhu University, China |
| Chuandong Li | Southwest University, China |
| Dazi Li | Beijing University of Chemical Technology, China |
| Li Li | Tsinghua University, China |
| Shengquan Li | Yangzhou University, China |

| | |
|---|---|
| Ya Li | Institute of Automation, Chinese Academy of Sciences, China |
| Yanan Li | University of Sussex, UK |
| Yongming Li | Liaoning University of Technology, China |
| Yuankai Li | Universitat of Science and Technology of China, China |
| Jie Lian | Dalian University of Technology, China |
| Hualou Liang | Drexel University, USA |
| Jinling Liang | Southeast University, China |
| Xiao Liang | Nankai University, China |
| Alan Wee-Chung Liew | Griffith University, Australia |
| Honghai Liu | University of Portsmouth, UK |
| Huaping Liu | Tsinghua University, China |
| Huawen Liu | University of Texas at San Antonio, USA |
| Jing Liu | Chinese Academy of Sciences, China |
| Ju Liu | Shandong University, China |
| Qingshan Liu | Huazhong University of Science and Technology, China |
| Weifeng Liu | China University of Petroleum, China |
| Weiqiang Liu | Nanjing University of Aeronautics and Astronautics, China |
| Dome Lohpetch | King Mongkut's University of Technology North Bangoko, Thailand |
| Hongtao Lu | Shanghai Jiao Tong University, China |
| Wenlian Lu | Fudan University, China |
| Yao Lu | Beijing Institute of Technology, China |
| Jinwen Ma | Peking University, China |
| Qianli Ma | South China University of Technology, China |
| Sanparith Marukatat | Thailand's National Electronics and Computer Technology Center, Thailand |
| Tomasz Maszczyk | Nanyang Technological University, Singapore |
| Basarab Matei | LIPN Paris Nord University, France |
| Takashi Matsubara | Kobe University, Japan |
| Nobuyuki Matsui | University of Hyogo, Japan |
| P. Meesad | King Mongkut's University of Technology North Bangkok, Thailand |
| Gaofeng Meng | Chinese Academy of Sciences, China |
| Daisuke Miyamoto | University of Tokyo, Japan |
| Kazuteru Miyazaki | National Institution for Academic Degrees and Quality Enhancement of Higher Education, Japan |
| Seiji Miyoshi | Kansai University, Japan |
| J. Manuel Moreno | Universitat Politècnica de Catalunya, Spain |
| Naoki Mori | Osaka Prefecture University, Japan |
| Yoshitaka Morimura | Kyoto University, Japan |
| Chaoxu Mu | Tianjin University, China |
| Kazuyuki Murase | University of Fukui, Japan |
| Jun Nishii | Yamaguchi University, Japan |

| | |
|---|---|
| Haruhiko Nishimura | University of Hyogo, Japan |
| Grozavu Nistor | Paris 13 University, France |
| Yamaguchi Nobuhiko | Saga University, Japan |
| Stavros Ntalampiras | University of Milan, Italy |
| Takashi Omori | Tamagawa University, Japan |
| Toshiaki Omori | Kobe University, Japan |
| Seiichi Ozawa | Kobe University, Japan |
| Yingnan Pan | Northeastern University, China |
| Yunpeng Pan | JD Research Labs, China |
| Lie Meng Pang | Universiti Malaysia Sarawak, Malaysia |
| Shaoning Pang | Unitec Institute of Technology, New Zealand |
| Hyeyoung Park | Kyungpook National University, South Korea |
| Hyung-Min Park | Sogang University, South Korea |
| Seong-Bae Park | Kyungpook National University, South Korea |
| Kitsuchart Pasupa | King Mongkut's Institute of Technology Ladkrabang, Thailand |
| Yong Peng | Hangzhou Dianzi University, China |
| Somnuk Phon-Amnuaisuk | Universiti Teknologi Brunei, Brunei |
| Lukas Pichl | International Christian University, Japan |
| Geong Sen Poh | National University of Singapore, Singapore |
| Mahardhika Pratama | Nanyang Technological University, Singapore |
| Emanuele Principi | Università Politecnica elle Marche, Italy |
| Dianwei Qian | North China Electric Power University, China |
| Jiahu Qin | University of Science and Technology of China, China |
| Sitian Qin | Harbin Institute of Technology at Weihai, China |
| Mallipeddi Rammohan | Nanyang Technological University, Singapore |
| Yazhou Ren | University of Science and Technology of China, China |
| Ko Sakai | University of Tsukuba, Japan |
| Shunji Satoh | The University of Electro-Communications, Japan |
| Gerald Schaefer | Loughborough University, UK |
| Sachin Sen | Unitec Institute of Technology, New Zealand |
| Hamid Sharifzadeh | Unitec Institute of Technology, New Zealand |
| Nabin Sharma | University of Technology Sydney, Australia |
| Yin Sheng | Huazhong University of Science and Technology, China |
| Jin Shi | Nanjing University, China |
| Yuhui Shi | Southern University of Science and Technology, China |
| Hayaru Shouno | The University of Electro-Communications, Japan |
| Ferdous Sohel | Murdoch University, Australia |
| Jungsuk Song | Korea Institute of Science and Technology Information, South Korea |
| Andreas Stafylopatis | National Technical University of Athens, Greece |
| Jérémie Sublime | ISEP, France |
| Ponnuthurai Suganthan | Nanyang Technological University, Singapore |
| Fuchun Sun | Tsinghua University, China |
| Ning Sun | Nankai University, China |

| | |
|---|---|
| Norikazu Takahashi | Okayama University, Japan |
| Ken Takiyama | Tokyo University of Agriculture and Technology, Japan |
| Tomoya Tamei | Kobe University, Japan |
| Hakaru Tamukoh | Kyushu Institute of Technology, Japan |
| Choo Jun Tan | Wawasan Open University, Malaysia |
| Shing Chiang Tan | Multimedia University, Malaysia |
| Ying Tan | Peking University, China |
| Gouhei Tanaka | The University of Tokyo, Japan |
| Ke Tang | Southern University of Science and Technology, China |
| Xiao-Yu Tang | Zhejiang University, China |
| Yang Tang | East China University of Science and Technology, China |
| Qing Tao | Chinese Academy of Sciences, China |
| Katsumi Tateno | Kyushu Institute of Technology, Japan |
| Keiji Tatsumi | Osaka University, Japan |
| Kai Meng Tay | Universiti Malaysia Sarawak, Malaysia |
| Chee Siong Teh | Universiti Malaysia Sarawak, Malaysia |
| Andrew Teoh | Yonsei University, South Korea |
| Arit Thammano | King Mongkut's Institute of Technology Ladkrabang, Thailand |
| Christos Tjortjis | International Hellenic University, Greece |
| Shibata Tomohiro | Kyushu Institute of Technology, Japan |
| Seiki Ubukata | Osaka Prefecture University, Japan |
| Eiji Uchino | Yamaguchi University, Japan |
| Wataru Uemura | Ryukoku University, Japan |
| Michel Verleysen | Universite catholique de Louvain, Belgium |
| Brijesh Verma | Central Queensland University, Australia |
| Hiroaki Wagatsuma | Kyushu Institute of Technology, Japan |
| Nobuhiko Wagatsuma | Tokyo Denki University, Japan |
| Feng Wan | University of Macau, SAR China |
| Bin Wang | University of Jinan, China |
| Dianhui Wang | La Trobe University, Australia |
| Jing Wang | Beijing University of Chemical Technology, China |
| Jun-Wei Wang | University of Science and Technology Beijing, China |
| Junmin Wang | Beijing Institute of Technology, China |
| Lei Wang | Beihang University, China |
| Lidan Wang | Southwest University, China |
| Lipo Wang | Nanyang Technological University, Singapore |
| Qiu-Feng Wang | Xi'an Jiaotong-Liverpool University, China |
| Sheng Wang | Henan University, China |
| Bunthit Watanapa | King Mongkut's University of Technology, Thailand |
| Saowaluk Watanapa | Thammasat University, Thailand |
| Qinglai Wei | Chinese Academy of Sciences, China |
| Wei Wei | Beijing Technology and Business University, China |
| Yantao Wei | Central China Normal University, China |

| | |
|---|---|
| Guanghui Wen | Southeast University, China |
| Zhengqi Wen | Chinese Academy of Sciences, China |
| Hau San Wong | City University of Hong Kong, Hong Kong SAR, China |
| Kevin Wong | Murdoch University, Australia |
| P. K. Wong | University of Macau, SAR China |
| Kuntpong Woraratpanya | King Mongkut's Institute of Technology Chaokuntaharn Ladkrabang, Thailand |
| Dongrui Wu | Huazhong University of Science and Technology, China |
| Si Wu | Beijing Normal University, China |
| Si Wu | South China University of Technology, China |
| Zhengguang Wu | Zhejiang University, China |
| Tao Xiang | Chongqing University, China |
| Chao Xu | Zhejiang University, China |
| Zenglin Xu | University of Science and Technology of China, China |
| Zhaowen Xu | Zhejiang University, China |
| Tetsuya Yagi | Osaka University, Japan |
| Toshiyuki Yamane | IBM, Japan |
| Koichiro Yamauchi | Chubu University, Japan |
| Xiaohui Yan | Nanjing University of Aeronautics and Astronautics, China |
| Zheng Yan | University of Technology Sydney, Australia |
| Jinfu Yang | Beijing University of Technology, China |
| Jun Yang | Southeast University, China |
| Minghao Yang | Chinese Academy of Sciences, China |
| Qinmin Yang | Zhejiang University, China |
| Shaofu Yang | Southeast University, China |
| Xiong Yang | Tianjin University, China |
| Yang Yang | Nanjing University of Posts and Telecommunications, China |
| Yin Yang | Hamad Bin Khalifa University, Qatar |
| Yiyu Yao | University of Regina, Canada |
| Jianqiang Yi | Chinese Academy of Sciences, China |
| Chengpu Yu | Beijing Institute of Technology, China |
| Wen Yu | CINVESTAV, Mexico |
| Wenwu Yu | Southeast University, China |
| Zhaoyuan Yu | Nanjing Normal University, China |
| Xiaodong Yue | Shanghai University, China |
| Dan Zhang | Zhejiang University, China |
| Jie Zhang | Newcastle University, UK |
| Liqing Zhang | Shanghai Jiao Tong University, China |
| Nian Zhang | University of the District of Columbia, USA |
| Tengfei Zhang | Nanjing University of Posts and Telecommunications, China |
| Tianzhu Zhang | Chinese Academy of Sciences, China |

| | |
|---|---|
| Ying Zhang | Shandong University, China |
| Zhao Zhang | Soochow University, China |
| Zhaoxiang Zhang | Chinese Academy of Sciences, China |
| Dongbin Zhao | Chinese Academy of Sciences, China |
| Qiangfu Zhao | University of Aizu, Japan |
| Zhijia Zhao | Guangzhou University, China |
| Jinghui Zhong | South China University of Technology, China |
| Qi Zhou | University of Portsmouth, UK |
| Xiaojun Zhou | Central South University, China |
| Yingjiang Zhou | Nanjing University of Posts and Telecommunications, China |
| Haijiang Zhu | Beijing University of Chemical Technology, China |
| Hu Zhu | Nanjing University of Posts and Telecommunications, China |
| Lei Zhu | Unitec Institute of Technology, New Zealand |
| Pengefei Zhu | Tianjin University, China |
| Yue Zhu | Nanjing University, China |
| Zongyu Zuo | Beihang University, China |

# Contents – Part V

## Pattern Recognition

## Word, Text and Document Processing

# Prediction

# Predicting Degree of Relevance of Pathway Markers from Gene Expression Data: A PSO Based Approach

Pratik Dutta[✉], Sriparna Saha, and Agni Besh Chauhan

Department of Computer Science and Engineering,
Indian Institute of Technology Patna, Patna, India
{pratik.pcs16,sriparna,agni.cs13}@iitp.ac.in

**Abstract.** In functional genomics, a pathway is defined as a set of genes which exhibit similar biological activities. Given a microarray expression data, the corresponding pathway information can be extracted with the use of some public databases. All member genes of a given pathway may not be equally relevant in estimating the activity of that pathway. Some genes can participate adequately in the given pathway, some may have low-associations. Existing literature has either considered all the genes wholly or discarded some genes completely in estimating the corresponding pathway-activity. Inspired by this, the current work reports about an automated approach to measure the degree of relevance of a given gene in predicting the pathway-activity. As a large search space has to be dragged, the exploration properties of particle swarm optimization are utilized in the current context. Particles of the PSO represent different scores of relevance for the member genes of different pathways. In order to deal with the relevance-score, the popular $t$-score which is widely used in measuring the pathway-activity is expanded in the name of weighted $t$-score. The proposed PSO-based weighted framework is then evaluated on three gene expression data sets. In order to show the supremacy of the proposed method, top 50% pathway markers are selected for each data set and the quality of these measures is checked after performing 10-fold cross-validation with respect to different quality measures. The results are further validated using biological significance tests.

**Keywords:** Particle swarm optimization · Pathway activity
Gene markers · Weighted $t$-score · Degree of relevance

## 1 Introduction

### 1.1 Background

With the enhancement of biotechnology, microarray technology becomes a leading method for measuring the activity levels of genome-wide expression profiles [4,5,22]. Analyzing these profiles helps to identify gene pathways and important biomarkers that help in improving diagnosis, prognosis, and treatment of

© Springer Nature Switzerland AG 2018
L. Cheng et al. (Eds.): ICONIP 2018, LNCS 11305, pp. 3–14, 2018.
https://doi.org/10.1007/978-3-030-04221-9_1

the disease [9]. Among all genes of microarry profiles, some genes have different expression values across different samples or time points. These differentially expressed genes [21] are called biomarkers or gene markers which act as a strong candidate for the pathogenic role. The selection of biomarkers is still a challenging problem due to the inherent noise and high dimensional microarray data [2,15]. Moreover, gene markers are selected independently, though they share same functional attributes. This further increases the redundancy of the system and may lead to decrease the overall classification performance.

## 1.2  Motivation and Methodology

To alleviate this problem, several studies have proposed the use of pathway-based markers, instead of individual gene-markers [13,17]. In this pathway based marker, the pathway activity is inferred by summarizing the expression values of its member genes. It is shown by different literature surveys [11,19] that pathway-marker helps in better understanding biological insights into the underlying physiological mechanisms. In [14], automatic identification of relevant genes is posed as an optimization problem where some subset of genes are selected to take part in the pathway activities. As an objective function t-score [20] is utilized which is calculated using the subset of genes which are present in a particular solution. The search capability of particle swarm optimization (PSO) [18] is explored to identify the best gene subset which can lead to obtain optimized value of t-score. Furthermore, these methods act as better classification tools than traditional gene-marker based classifiers.

But most of the recent works assumed that either a particular gene can take part in the pathway activity or it can not be considered while determining the $t$-score corresponding to a pathway. Instead of ignoring a gene entirely, quantifying the importance of a gene in regulating the activity of a pathway would be more imperative.

The current work expands this idea and develops a PSO based automated approach to suitably identify the weight of importance of a particular gene in a given pathway. For this very purpose several modifications are integrated in the current framework. The particles of the PSO based framework are now some real-valued strings with values ranging between 0 and 1. In order to capture the goodness of the weight-combination, the existing $t$-score is also extended to grab the hypothesis that each member gene of the particle is participated to infer pathway activity with some weight value. The modified version of $t$-score (described in Sect. 3.2) is used as the objective function in the proposed optimization framework.

This proposed method is applied on three real life datasets (described in Sect. 2) and compared to five different existing algorithms namely binary particle swarm optimization (BPSO) [10], mean [7], median [7], log-likelihood ratio (LLR) [19], and condition-responsive genes (CORGs) [12]. This comparative approach establishes the efficacy of the proposed algorithm with respect to five performance metrics namely, *sensitivity, specificity, accuracy, Fscore* and *AUC*.

This comparative study delineates that the overall accuracy of identifying informative pathways of the proposed method is better than all other methods. To prove that the resultant pathway markers are biological relevant, a biological relevance test is also conducted.

The rest of the paper is structured as follows. Section 2 provides the idea about datasets. Section 3 demonstrates how to infer the pathway activity and the proposed particle swarm optimization technique [18]. The experimental results and comparison of different algorithms are finally summarized in Sect. 4. Finally Sect. 5 concludes the paper.

## 2   Datasets

In this proposed approach, three real life gene expression datasets are used for the purpose of the experiments. These datasets are publicly available from: www. biolab.si/supp/bi-cancer/projections/info/.

- **Prostate:** In this dataset, two types of samples are present; one class is prostate tumours and another is prostate tissue not containing tumours. Here the number of prostate tumour samples is 52 and the number of non-tumour (normal) samples is 52. This gene expression profile consists of 12,533 genes and 102 numbers of samples.
- **GSE1577 (Lymphoma Leukaemia):** This dataset is constructed by 15434 genes with 19 samples. Among these 19 samples, nine are T-cell lymphoblastic lymphoma (T-LL) and remaining ten are T-cell acute lymphoblastic leukaemia (T-ALL).
- **GSE412 (Child-ALL):** The childhood ALL dataset (GSE412) contains 110 childhood acute lymphoblastic leukemia samples. Among these 110 samples, 50 samples are of type "before therapy" and remaining 60 are of type "after therapy". This gene expression profile dataset has 8280 genes.

## 3   Proposed Method

In this section, the proposed weighted gene selection technique for a given pathway is described in detail. The assumption is that all the genes responsible for anticipating the functionality of a given pathway may not be equally responsible in doing the same. The quantification of relevance of different genes in participating in a given pathway to estimate its functionality is requisite. The appropriate relevance combination can be determined automatically using the search capability of particle swarm optimization (PSO). In order to calculate the fitness function of the particle, a new real version weighted $t$-score is used.

Figure 1 illustrates the flow diagram of the proposed method. The detailed description of important steps of the proposed algorithm is provided in the following subsections.

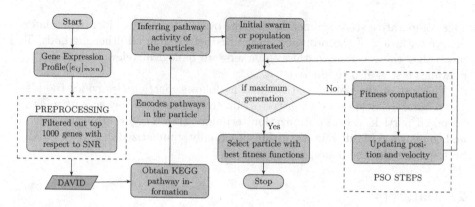

**Fig. 1.** Flowchart of our proposed single-objective PSO-based optimization framework.

### 3.1 Pre-processing of Microarray Dataset

In the existing publicly available gene expression datasets, expression values of all the genes are not uniformly distributed over some specific range. Hence, for identifying the differentially expressed genes, we need to use a quality measure that is platform independent. Biological signal to noise ratio (SNR) is one of the quality measures that is platform independent. Genes with the higher values of SNR indicate that the genes are more significant (differentially expressed) over a large range of values. Here, genes are sorted in descending order of signal to noise ratio (SNR) values and top 1000 genes are filtered out for further data analysis.

### 3.2 Calculation of Weighted *t*-score

In our proposed approach, we have designed a new version of *t*-score that is used as a fitness function of our PSO approach. Generally t-score indicates the testing ability to differentiate the cumulative expressions of constituent genes for a given pathway. Actually, it is the difference between central tendency and variation or dispersion exists from the average value of the data points. Therefore, the t-score test statistics is computed as:

$$t(\alpha) = \frac{\mu_g^1 - \mu_g^2}{\sqrt{\frac{\sigma_g^1}{N_1} + \frac{\sigma_g^2}{N_2}}} \tag{1}$$

Where $\alpha$ is the pathway activity level of the given pathway $\mathcal{P}$. Here $\mu_g^i$ and $\sigma_g^i$ are the mean and the standard deviation of the gene $g$ for class $i \mid i \in \{1, 2\}$, respectively. $N_i$ denotes the number of samples in two different classes. Here $\mu_g^i$ and $\sigma_g^i$ are described by

$$\mu_g^i = \frac{\sum_{k=1}^{N_i} e_{kg}}{N_i} \quad \text{and} \quad \sigma_g^i = \sqrt{\frac{1}{N_i - 1} \sum_{k=1}^{N_i} (e_{kg} - \mu_g^i)^2} \tag{2}$$

Where $e_{kg}$ is the expression value of the $g^{th}$ gene (sample) for $k^{th}$ condition. Normally $t$-score is used as the discriminative power checker i.e., higher t-score indicates higher ability of differentiation.

Normally it is assumed that during the calculation of pathway activity, genes belonging to the pathway either actively participate or not participate in estimating the pathway activity. But in the weighted version of $t$-score calculation, all the genes participate in pathway activity calculation with some weights. Hence in our approach, modified weighted mean $((\mu_g^i)_w)$ and weighted standard deviation of the sample (gene) $g$ for class $i$ are defined as follows:

$$(\mu_g^i)_w = \frac{\sum_{k=1}^{N_i} (e_{kg}) * w_k}{\sum_{k=1}^{N_i} w_k} \tag{3}$$

and

$$(\sigma_g^i)_w = \sqrt{\frac{N_i}{N_i - 1} \frac{\sum_{k=1}^{N_i} (e_{kg} - \mu_g^i)^2 * w_k}{\sum_{k=1}^{N_i} w_k}} \tag{4}$$

Hence the proposed weighted $t$-score of the our method is calculated as

$$t_w(\alpha) = \frac{(\mu_g^1)_w - (\mu_g^2)_w}{\sqrt{\frac{(\sigma_g^1)_w}{N_1} + \frac{(\sigma_g^2)_w}{N_2}}} \tag{5}$$

### 3.3 Inferring Pathway Activity

In this section, a brief description of inferring pathway activity is summarized in steps.

- The preprocessed 1000 genes (described in Sect. 3.1) are fed to DAVID [8], a pathway database, to obtain the pathway information of these genes. DAVID returns the KEGG pathway information along with the corresponding genes.
- Now each pathway obtained from the DAVID, contains a set of gene IDs. The gene expression values of the member genes of a particular pathway $\mathcal{P}_k$ are gathered to infer pathway activity.
- Then a gene expression data matrix is created considering the member genes of each pathway $(\mathcal{P}_k)$. Now PSO is applied on this data matrix. Each particle of PSO encodes only the indices of the member genes.
- A portion of particle $(\mathfrak{P}_k)$ consisting of two pathways $(\mathcal{P}_1, \mathcal{P}_2)$ is illustrated in Fig. 2. The value of each cell in the particle represents the degree of relevance of the particular gene in inferring pathway activity.
- Now the pathway activity of a particular pathway $(\mathcal{P}_i)$ of the particle is calculated by dividing weighted sample wise sum by sum of the weights i.e., $\alpha(\mathcal{P}_i) = \frac{\sum_{i=1}^{G_i} \sum_{j=1}^{N_i} (e_{ji}) * w_i}{\sum_{i=1}^{G_i} w_i}$. Here $G_i$ is the number of the genes of the pathway $\mathcal{P}_i$ in the particle $\mathfrak{P}_k$ and $N_i$ is the number of samples of each gene. The denominator is used to stabilize the variance of the mean.

– Then the weighted $t$-score of the pathway activity is calculated by Equation-5. The main objective of the proposed method is to maximize the weighted $t$-score of each particle in each generation. Particle with the maximum weighted t-score and the corresponding pathway activities are considered as the final solution of our proposed method.

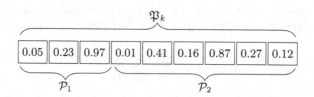

**Fig. 2.** Particle encoding technique.

### 3.4 Proposed PSO Based Approach

**Particle Encoding.** Generally in PSO, the population is called swarm and the swarm consists of $m$ number of candidate solutions or particles. Each of the particles has $n$ number of cells where each cell represents degree of importance of a gene responsible for inferring a pathway. In our approach, if $n$ number of pathways are selected after applying the steps of Sect. 3.1 (which are responsible for further analysis), then the genes responsible of those $n$ pathways are encoded in a particle. All the pathways are not of the same length. In this real version of PSO, each cell of the particle has some real value. This real value quantifies the relevance of the particular gene in inferring the activity of the given pathway.

**Initialization.** In this phase, the value of each cell is randomly initialized by a real value between 0 and 1. After the initial particles of the swarm are generated, the fitness value of each particle is calculated. Then the velocity of each cell of the particle is initialized to zero. Different steps of PSO are executed for 300 times. The swarm size of the proposed PSO based approach is 25. Other inputs of the proposed method are weighting factors c1 and c2 which are cognitive and social parameters, respectively. These weighting factors are set to 2 as traditionally used in the existing literature [18].

**Fitness Computation.** The main objective of the proposed particle swarm optimization (PSO) technique is to maximize the average t-score produced by different pathways. The real value of each cell of the particle represents the activity of the gene in computing the fitness value. Here the value of each cell lies between 0 and 1. The value for a gene near to 1 indicates that the gene takes part more actively in inferring pathway activity. The weighted $t$-score of the corresponding particle is calculated using Eq. 5.

If the particle $\mathfrak{P}$ has $n$ number of pathways, i.e., $\mathfrak{P} = (\mathcal{P}_1, \mathcal{P}_2, \mathcal{P}_3 \ldots \mathcal{P}_n)$, the fitness function of the particle is the mean of the weighted $t$-score of the corresponding $n$ pathways. Hence the mathematical equation of the fitness function is

$$fitness = \frac{\sum_{i=1}^{n} t_w(\alpha(\mathcal{P}_i))}{n} \tag{6}$$

Here $t_w(\alpha(\mathcal{P}_i))$ is the weighted $t$-score of the pathway activity $\alpha$ inferred from pathway $\mathcal{P}_i$.

**Updating Position and Velocity.** Initially, the position of each gene in a particle is initialized to a weight value present in the corresponding cell of the particle. The velocity of each gene is initialized to zero. Then in each iteration, the position and velocity of the genes are updated. The main rule to update the velocity and position is to keep track of position and velocity history. The PSO process monitors the best position obtained so far in the history. The best position is also called $pb$ or local best. The best position among all the particles is called $gb$ or global best. The position and velocity of the particle are updated according to the following equations

$$v_{ij}(t+1) = w * v_{ij}(t) + c_1 * r_1 * (pb_{ij}(t) - x_{ij}(t)) + c_2 * r_2 * (gb_{ij}(t) - x_{ij}(t)) \tag{7}$$

$$S(v_{ij}(t+1)) = \frac{1}{(1 + e^{(v_{ij}(t+1))})} \text{ and } x_{ij}(t+1) = \begin{cases} 0, & \text{if } r_3 < S(v_{ij}(t+1)) \\ 1, & \text{otherwise} \end{cases} \tag{8}$$

Here $t$, $i$ and $j$ represent the time stamp, the particle and the position, respectively. The velocity of any particular timestamp depends on the previous timestamp velocity i.e., $v_{ij}(t+1)$ is acquired on the basis of $v_{ij}(t)$. Then new position $x_{ij}(t+1)$ is obtained depending on the value of $S(v_{ij}(t+1))$, which is calculated by the Eq. 8. Here $r_1$, $r_2$ and $r_3$ are the random numbers and $c_1$, $c_2$ are constants which are set to 2. The values of $r_1$ and $r_2$ range between 0 and 1. The inertia weight($w$) is calculated by the following equation $w = 1.1 - \frac{gbest}{pbest}$.

## 4    Experimental Results

In this section, the results illustrate that the proposed method is superior to different existing works with respect to different performance measures. This delineates the efficacy of the proposed method in identifying robust pathway activity. The proposed method is applied on three real life datasets (described in Sect. 2). The performance of the proposed PSO based technique is compared with five existing methods namely binary particle swarm optimization (BPSO) [10], Mean [7], Median [7], log-likelihood ratio(LLR) [19], and condition-responsive genes(CORGs) [12] with respect to five performance metrics. These five performance metrics are *sensitivity* [16], *specificity* [16], *accuracy* [16], *Fscore* [16] and *AUC* [16].

## 4.1  Comparative Methods

To establish the superiority of the proposed method, the performance of the proposed weighted PSO (wPSO) based technique is compared with several existing optimization techniques having different complexity levels. The comparing methods are general particle swarm optimization (BPSO) [10], traditional statistical methods i.e., Mean [7] and Median [7], pathway-marker based classification techniques like LLR [19] and CORGs [12]. In BPSO [10], a particular gene either actively participated for inferring pathway activity or not. The second and third comparing approaches (Mean and Median) are based on the basic statistical methods and do not take care of any biological or functional relevance of genes during classification. The fourth one (LLR) [19] is based on probabilistic inference of pathway activities rather than individual gene markers. In this case, the pathway activity is inferred by combining the log-likelihood ratios (LLR) [19] of the constituent genes. Motivated by its reliable and accurate classification capacity than other traditional techniques, in this current study LLR is used as one of the comparing approaches. The last one (CORGs) [12] is a pathway activity based classification technique where the pathway activity is inferred from the gene expression levels of its condition-responsive genes (CORGs). Here a greedy search is used to identify the member genes of a pathway and the set of genes corresponding to maximum pathway activity are represented as CORGs.

The proposed method is compared to the above mentioned algorithms in terms of five performance metrics i.e. *sensitivity, specificity, accuracy, Fscore* and *AUC*. For the input gene expression profiles, three real life datasets(described in Sect. 2) are used. The steps for calculating the above metrics are as follows:

1. The final solution of the proposed optimization(wPSO) technique reports the weight combination of set of genes present in different pathways encoded in a particle. The pathway activities are calculated using weighted values present in the solution.
2. Then the $p$-value [6] for each pathway activity is calculated using Wilcoxon Ranksum method [3]. Now the pathways are sorted in ascending order in the basis of the $p$-values.
3. From these sorted pathways, top 50% are extracted for further processing. These pathways are indicated as pathway markers.
4. Then all the genes are present in the pathway markers are collected and the corresponding expression values are gathered. This constitutes a new gene expression dataset. This is fed to SVM classifier. As in this proposed approach, gene datasets contain two types of samples i.e. normal and tumour, SVM acts as a binary classifier [1].
5. After the binary classification, the process of 10-fold cross validation is executed to calculate the above mentioned five performance metrics.

The corresponding *sensitivity, specificity, accuracy, Fscore* and *AUC* values are shown in Table 1. From Table 1, it is clearly evident that the proposed method is more efficient than other comparative algorithms with respect to above

**Table 1.** Comparative study of our proposed framework with several existing techniques in terms of different metrics

| Datasets | Algorithms | Sensitivity | Specificity | Accuracy | F-score | AUC |
|----------|-----------|-------------|-------------|----------|---------|-----|
| Prostate | **Proposed** | **0.9600** | **0.9808** | **0.9706** | **0.9697** | **0.9770** |
|          | BPSO | 0.8846 | 0.9000 | 0.8922 | 0.8932 | 0.9677 |
|          | CORGS | 0.8846 | 0.8800 | 0.8824 | 0.8846 | 0.9315 |
|          | LLR | 0.9038 | 0.8600 | 0.8824 | 0.8868 | 0.9400 |
|          | Mean | 0.9038 | 0.8200 | 0.8627 | 0.8704 | 0.9054 |
|          | Median | 0.9038 | 0.7800 | 0.8431 | 0.8545 | 0.9092 |
| LL | **Proposed** | **1.0000** | **1.0000** | **1.0000** | **1.0000** | **1.0000** |
|    | BPSO | 0.9000 | 1.0000 | 0.9474 | 0.9474 | 1.0000 |
|    | CORGS | 0.9312 | 0.9898 | 0.9474 | 0.9434 | 0.9889 |
|    | LLR | 0.8000 | 1.0000 | 0.8947 | 0.8889 | 1.0000 |
|    | Mean | 0.8801 | 0.8896 | 0.8766 | 0.9000 | 0.9769 |
|    | Median | 0.8714 | 0.8753 | 0.7263 | 0.6897 | 0.9348 |
| Child | **Proposed** | **0.8500** | **0.8000** | **0.8273** | **0.8430** | **0.8600** |
|       | BPSO | 0.7600 | 0.7333 | 0.7455 | 0.7308 | 0.9034 |
|       | CORGS | 0.7600 | 0.7000 | 0.7273 | 0.7170 | 0.8887 |
|       | LLR | 0.7600 | 0.7176 | 0.7364 | 0.7238 | 0.9023 |
|       | Mean | 0.7501 | 0.7000 | 0.7273 | 0.7170 | 0.8960 |
|       | Median | 0.7500 | 0.7167 | 0.7364 | 0.7238 | 0.8953 |

mentioned five performance metrics. Table 1 also illuminates the effectiveness of the proposed method with respect to other comparative methods for all three real life datasets. The proposed method shows improvements of 6.22%, 8.98%, 8.78%, 8.56% and 0.96% with respect to *sensitivity, specificity, accuracy, F score* and *AUC*, respectively, for the Prostrate dataset in comparison to the other best performing systems. Similarly, for the Lymphoma Leukaemia (LL) dataset, the proposed method shows improvements of 11.11%, 5.52% and 5.52% with respect to *sensitivity, accuracy,* and *AUC*, respectively. For the Child-all dataset, though the *AUC* of the proposed method is slightly less than few existing algorithms, other four metrics *sensitivity, specificity, accuracy* and *F score* the proposed method are 11.84%, 9.09%, 10.9% and 15.35% higher than the existing methods.

After analyzing the outcomes of different approaches across all datasets, it can be seemingly concluded that the proposed method outperforms other comparative methods.

## 4.2 Discussion of Results

In this paper, we have proposed a novel weighted particle swarm optimization (wPSO) based technique to identify biologically relevant pathways. In this method, the pathway activity is inferred from the weighted $t$-score measure proposed in this article. The novelty of the proposed method lies in predicting the proper value of relevance for a given gene in inferring the corresponding pathway. None of the genes are completely discarded as done in the previous approaches, rather each of them is automatically assigned a real-value between 0 and 1 which indicates the degree of its suitability in assessing the corresponding pathway activity. As the genes are randomly initialized with some real values, so each gene has participated to infer the pathway activity. For that reason, the proposed method can generate various possibilities of pathways and among them, pathways with maximum weighted $t$-scores are selected for the final solution. As $t$-score basically calculates the correlation among the samples(genes), higher value of $t$-score related to any pathway indicates that the candidate genes corresponding to that pathway are strongly functionally correlated. Hence the proposed method can capture genes which have more discriminative power due to their consistent expression values. Furthermore, the proposed method fully utilizes the available discriminative information of all the member genes rather some of them as done in [14]. The degree of relevance of each gene is determined automatically with the aim to increase the corresponding pathway activity. Therefore the proposed approach can prudently identify biological significant pathways. As Table 1 demonstrates that the proposed method can significantly improve the overall classification, this method can lead to the construction of more reproducible classifiers.

## 4.3    Biological Relevance

In this section, the biological relevance of resultant pathway markers is explored in terms of disease-gene association. In this regard, the top 50% pathway markers

**Table 2.** Disease associated with the resultant pathway markers

| Disease | Gene symbol (# PMID) |
|---|---|
| Prostatic neoplasms | ERG(37), GSTP1(28), BCL2(22), IGFBP3(18), IGF1(14), GSTM1(12), TNFSF10(11), CAV1(11), PLAU(8) |
| Prostate carcinoma | ERG(245), BCL2(120), ERBB3(96), FKBP4(69), RASGRF1(48), HNRNPK(44) PBX1(35), PRKCA(28), RPL19(25), AKR1A1(23), ITPR1(18) |
| Malignant neoplasm of prostate | ERG(267), BCL2(117), IRAK1(108), ERBB3(67), FKBP4(57), RASGRF1(50), PBX1(34), RPL19(27), AKR1A1(23), ITPR1(18) |
| Carcinoma ALL (LL) | KRAS(107), VEGFA(73), BCL2(52), CCND1(43), MMP9(35), PIK3CB(31), CCND1(30), VEGFA(27), MDM2(24), BCL2(23) |
| Leukemia (LL) | BCL2(115), VEGFA(35), PIK3CB(20), MDM2(17), CCND1(16), RB1(11), MMP9(8), EPAS1(7), MAPK14(7), KRAS(7) |
| Lymphoma (LL) | BCL2(312), CCND1(96), VEGFA(20), MDM2(18), PIK3CB(10), NXT1(8), MMP9(8), RB1(7), SKP2(5) |
| Childhood all | ITGB3(2), TGFB1(2), FZR1(1), ABL1(1), MET(1), ATM(1), ERBB3(1), RB1(0) |
| Leukemia | MYC(284), TGFB1(121), CALM1(48), IKBKG(44), ABL1(24), HSP90AB1(20), AURKA(20), SMAD5(20), BCR(17), RAF1(11) |
| Acute lymphocytic leukemia | MYC(235), ABL1(217), AURKA(84), BCR(17), TMSB4X(11), EPAS1(9), FAS(7), BAX(7), RB1(6), ERCC2(5), PLK1(5) |
| Lymphoma | MYC(1), TGFB1(7), CALM1(3), PIAS2(1), TGFBR1(1), AURKA(2), ABL1(7), BCR(15), RAF1(2), HSPA1L(1) |

are searched in disease-gene association database (http://www.disgenet.org/) for associated disease. This database basically returns the number of Pubmed citations for the disease-gene association. In Table 2, a part of the disease-gene association record is enumerated.

The first column of the table contains the disease names, the second column represents the corresponding gene symbols with the number of Pubmed citations in parentheses. The higher value in the parenthesis represents that the particular gene is largely cited by Pubmed. The obtained result is a strong evidence to corroborate that the member genes of the pathway are very much responsible for different biochemical process or disease of the living cell.

## 5   Conclusions

Pathway-based marker finding plays a key role in disease diagnosis and its classification. In the current work the problem of automatic determination of appropriate relevance of a member gene participated in a pathway is posed as an optimization problem. This is further solved after employing the search capability of particle swarm optimization. The experimental results evidently illustrate the potency of the proposed approach in detecting appropriate pathway markers.

Future work includes extension of the proposed framework using multiobjective optimization based architecture. Development of some new scoring mechanisms to judge the quality of a pathway activity could also be a future research direction.

## References

1. Alon, U., et al.: Broad patterns of gene expression revealed by clustering analysis of tumor and normal colon tissues probed by oligonucleotide arrays. Proc. Natl. Acad. Sci. **96**(12), 6745–6750 (1999)
2. Bandyopadhyay, S., Mallik, S., Mukhopadhyay, A.: A survey and comparative study of statistical tests for identifying differential expression from microarray data. IEEE/ACM Trans. Comput. Biol. Bioinform. **11**(1), 95–115 (2014)
3. Deng, L., Pei, J., Ma, J., Lee, D.L.: A rank sum test method for informative gene discovery. In: Proceedings of the Tenth ACM SIGKDD International Conference on Knowledge Discovery and Data Mining, pp. 410–419. ACM (2004)
4. Ding, C., Peng, H.: Minimum redundancy feature selection from microarray gene expression data. J. Bioinform. Comput. Biol. **3**(02), 185–205 (2005)
5. Dutta, P., Saha, S.: Fusion of expression values and protein interaction information using multi-objective optimization for improving gene clustering. Comput. Biol. Med. **89**, 31–43 (2017)
6. Ghosh, A., Dhara, B.C., De, R.K.: Selection of genes mediating certain cancers, using a neuro-fuzzy approach. Neurocomputing **133**, 122–140 (2014)
7. Guo, Z., et al.: Towards precise classification of cancers based on robust gene functional expression profiles. BMC Bioinform. **6**(1), 58 (2005)
8. Huang, D.W., Sherman, B.T., Lempicki, R.A.: Systematic and integrative analysis of large gene lists using david bioinformatics resources. Nat. Protoc. **4**(1), 44 (2009)

9. Huang, T., Chen, L., Cai, Y.D., Chou, K.C.: Classification and analysis of regulatory pathways using graph property, biochemical and physicochemical property, and functional property. PLOS ONE **6**(9), 1–11 (2011). https://doi.org/10.1371/journal.pone.0025297

10. Kennedy, J.: Particle swarm optimization. In: Encyclopedia of Machine Learning, pp. 760–766. Springer, Heidelberg (2011)

11. Khunlertgit, N., Yoon, B.J.: Identification of robust pathway markers for cancer through rank-based pathway activity inference. Adv. Bioinform. **2013** (2013)

12. Lee, E., Chuang, H.Y., Kim, J.W., Ideker, T., Lee, D.: Inferring pathway activity toward precise disease classification. PLoS Comput. Biol. **4**(11), e1000217 (2008)

13. Ma, S., Kosorok, M.R.: Identification of differential gene pathways with principal component analysis. Bioinformatics **25**(7), 882–889 (2009)

14. Mandal, M., Mondal, J., Mukhopadhyay, A.: A PSO-based approach for pathway marker identification from gene expression data. IEEE Trans. Nanobiosci. **14**(6), 591–597 (2015)

15. Mandal, M., Mukhopadhyay, A.: A graph-theoretic approach for identifying non-redundant and relevant gene markers from microarray data using multiobjective binary PSO. PloS ONE **9**(3), e90949 (2014)

16. Mukhopadhyay, A., Mandal, M.: Identifying non-redundant gene markers from microarray data: a multiobjective variable length PSO-based approach. IEEE/ACM Trans. Comput. Biol. Bioinform. (TCBB) **11**(6), 1170–1183 (2014)

17. Pang, H., Zhao, H.: Building pathway clusters from random forests classification using class votes. BMC Bioinform, **9**(1), 87 (2008)

18. Parsopoulos, K.E.: Particle Swarm Optimization and Intelligence: Advances and Applications: Advances and Applications. IGI Global, Hershey (2010)

19. Su, J., Yoon, B.J., Dougherty, E.R.: Accurate and reliable cancer classification based on probabilistic inference of pathway activity. PloS ONE **4**(12), e8161 (2009)

20. Wang, K., Li, M., Bucan, M.: Pathway-based approaches for analysis of genomewide association studies. Am. J. Hum. Genet. **81**(6), 1278–1283 (2007)

21. Wang, Y., Makedon, F.S., Ford, J.C., Pearlman, J.: HykGene: a hybrid approach for selecting marker genes for phenotype classification using microarray gene expression data. Bioinformatics **21**(8), 1530–1537 (2004)

22. Yang, K., Cai, Z., Li, J., Lin, G.: A stable gene selection in microarray data analysis. BMC Bioinform. **7**, 228 (2006). https://doi.org/10.1186/1471-2105-7-228

# Prediction of Taxi Demand Based on ConvLSTM Neural Network

Pengcheng Li[(⊠)], Min Sun, and Mingzhou Pang

School of Electrical Information and Electrical Engineering,
Shanghai Jiao Tong University, Shanghai, People's Republic of China
{princeli,msun,pangmz}@sjtu.edu.cn

**Abstract.** As an important part of the urban public transport system, taxi has been the essential transport option for city residents. The research on the prediction and analysis of taxi demand based on the taxi GPS data is one of the hot topics in transport recently, which is of great importance to increase the incomes of taxi drivers, reduce the time and distances of vacant driving and improve the quality of taxi operation and management. In this paper, we aim to predict the taxi demand based on the ConvLSTM network, which is able to deal with the spatial structural information effectively by the convolutional operation inside the LSTM cell. We also use the LSTM network in our experiment to implement the same prediction task. Then we compare the prediction performances of these two models. The results show that the ConvLSTM network outperforms LSTM network in predicting the taxi demand. Due to the ability of handling spatial information more accurately, the ConvLSTM can be used in many spatio-temporal sequence forecasting problems.

**Keywords:** ConvLSTM · LSTM · Taxi demand

## 1 Introduction

In modern cities, taxi has been the essential transport option for city residents in their daily life. However, sometimes a driver has to spend a lot of time searching for the next passenger, and the passengers often complain about the inconvenience of taking a taxi due to the long wait-time. Therefore, the prediction and analysis of taxi demand throughout a city is of great importance in solving this kind of problem, which can lead to the effective taxi dispatching, help the drivers to improve their incomes and reduce the wait-time for passengers. In recent years, with the rapid development of information technology and data science, more data resources are available and computable than before. Historical taxi trips have been widely used in many tasks, such as urban traffic congestion estimation and prediction [1], the understanding of taxi service strategies [2], and taxi operation optimization [3].

However, there are a few researches on the prediction of taxi demand based on historical taxi trips. Zhao *et al.* [4] define a maximum predictability for the taxi demand and prove that the taxi demand is highly predictable. Besides, they validate their theory by implementing three prediction algorithms. Li *et al.* [5] propose an ARIMA-based model to predict the spatial-temporal variation of picking up passengers in a densely

L. Cheng et al. (Eds.): ICONIP 2018, LNCS 11305, pp. 15–25, 2018.
https://doi.org/10.1007/978-3-030-04221-9_2

populated region, the predicted results can be used to help taxi drivers to find the next passenger. Kong *et al.* [6] propose a time-location-relationship (TLR) combined taxi service recommendation model, which not only predicts the number of potential passengers in the subregions, but also recommends some suitable areas with the large taxi demand. The above-mentioned methods mainly focus on predicting the taxi demand by analyzing the historical taxi data, such as the amount of pick-up passengers. Actually, many factors may exert an influence on the prediction of taxi demand, for example, weather, date, traffic condition and so on. In addition, this is also a time series forecasting problem, the long-term dependencies exist in the different time periods. We need a better method to deal with the long-term dependency and to improve the prediction accuracy. With the recent advances of the deep neural networks, Long Short Term Memory (LSTM) [7] networks, as a special kind of Recurrent Neural Network (RNN), have been proved to be effective in sequence learning problems. LSTM is capable of learning long-term dependencies by using some gating mechanisms to store information for future use. Nowadays, LSTMs have been widely used in many applications such as speech recognition, time series predictions and language processing. Xu *et al.* [8] propose a real-time method for predicting taxi demands based on LSTM network and achieve a good prediction accuracy. Although the LSTM network is powerful for handling temporal correlation, it involves much redundancy for the spatial data. The input data of the general stacked LSTM layers is one-dimensional, but it often tends to contain multiple variables, so the input data has to be a single long vector, leading to a great deal of weights and considerable computational cost.

In this paper, in order to predict the taxi demand more accurately, we introduce Convolutional LSTM (ConvLSTM) [9], one of the variants of LSTM. ConvLSTM was designed to embed the convolutional operation inside the LSTM cell to deal with the spatial data more effectively, which had been proved to be better for some general spatio-temporal sequence forecasting problem. Therefore, we apply the ConvLSTM to the problem of predicting taxi demands based on the historical taxi data, and we also adopt the traditional LSTM network as the general method in our experiment. Then we compare the different prediction performances of these two models. The experimental results show that the ConvLSTM network model, which considers the spatio-temporal property more precisely, outperforms the traditional LSTM network for the prediction of taxi demand.

## 2    Models

### 2.1    LSTM

Long Short Term Memory (LSTM) network was introduced by Hochreiter and Schmidhuber in 1997, and was popularized and refined by many people in their following work. The LSTM architecture consists of a set of recurrently connected subnets, known as memory blocks [10]. Each block contains one or several memory cells and three gates - input gate, output gate and forget gate. These gates can regulate the cell state by adding or removing the information. The forget gate tend to decide what kind of information we will throw away from the cell state. The input gate can decide the values we need to update, and after updating the old cell state into a new one, the output gate decides what information we are going to output. In general, the gating mechanisms are able to allow the LSTM cells to store and update the information over long periods of time. The equations of the gates (input gate, output gate and forget gate) in LSTM are as follows:

$$i_t = \sigma(W_{xi}x_t + W_{hi}h_{t-1} + W_{ci}c_{t-1} + b_i) \tag{1}$$

$$f_t = \sigma(W_{xf}x_t + W_{hf}h_{t-1} + W_{cf}c_{t-1} + b_f) \tag{2}$$

$$c_t = f_t c_{t-1} + i_t \tanh(W_{xc}x_t + W_{hc}h_{t-1} + b_c) \tag{3}$$

$$o_t = \sigma(W_{xo}x_t + W_{ho}h_{t-1} + W_{co}c_t + b_o) \tag{4}$$

$$h_t = o_t \tanh(c_t) \tag{5}$$

Where $i$, $f$ and $o$ refer respectively to the input gate, forget gate and output gate, $c$ refers to the memory cell, $W$ is the weight matrix, $x_t$ is the input data at time $t$, $h_{t-1}$ is the hidden output at time $t-1$, $c_t$ is the cell state at time $t$, the activation function of the gates is denoted $\sigma$, and $b$ is the bias value. The structure of stacked LSTMs is shown in Fig. 1.

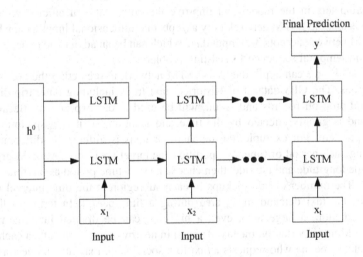

**Fig. 1.** The structure of stacked LSTMs model

## 2.2 ConvLSTM

ConvLSTM was proposed by Shi *et al.* in 2015 [9], and it is devised to learn the spatial information effectively in the dataset. The input data of the LSTM is one-dimensional, but the real input data may consist of multiple variables, and sometimes there exists a spatial correlation between these variables. When the LSTM network is applied to solve the problems related to the time series, the input data need to be a single long vector, so the spatial information can not be taken into consideration accurately.

Different from the traditional LSTM network structures, the ConvLSTM network has the convolutional structures in both the input-to-state and state-to-state transitions [9]. The equations of the gates (input gate, output gate and forget gate) in ConvLSTM are as follows:

$$i_t = \sigma(W_{xi} * x_t + W_{hi} * h_{t-1} + W_{ci}c_{t-1} + b_i) \tag{6}$$

$$f_t = \sigma(W_{xf} * x_t + W_{hf} * h_{t-1} + W_{cf}c_{t-1} + b_f) \tag{7}$$

$$c_t = f_t c_{t-1} + i_t \tanh(W_{xc} * x_t + W_{hc} * h_{t-1} + b_c) \tag{8}$$

$$o_t = \sigma(W_{xo} * x_t + W_{ho} * h_{t-1} + W_{co}c_t + b_o) \tag{9}$$

$$h_t = o_t \tanh(c_t) \tag{10}$$

As we can see, the difference between equations in the two kinds of LSTM networks is that the Hadamard product between the weight $W$ and the input data $x_t$, the hidden output $h_{t-1}$ in each gate is replaced with the convolution operator (*). By doing so, the ConvLSTM network captures underlying spatial features by convolution operations in multiple-dimensional data [11]. In addition, the convolutional layer is a reasonable substitute for the fully connected layer in the network, which reduces the number of weight parameters in the model and improve the computational efficiency. Another difference is that the LSTM network only accepts one-dimensional input data, while the ConvLSTM network accepts 3-D input data, which can be an advantage in dealing with many spatio-temporal sequence forecasting problems.

In this study, we can apply the ConvLSTM network in predicting the taxi demand in the regions. The GPS dataset of Historical taxi trips, including date, trip distance, pick-up and drop-off records and so on, will be used in our experiment. Because the taxi demand is greatly affected by the time and location, so it is a spatio-temporal sequence problem. For example, the demand for taxis is different in different social areas during each period of time. We partition the target area into a M × M grid map based on the longitude and latitude, then choose a small time period as the time-step of the model. The number of taxi pickups in each subregion at the time interval will be regarded as the taxi demand in an area during a time period. In terms of the taxi demand prediction, a large area, even a big city, can be divided into many small subregions. More importantly, the taxi demand in nearby regions may affect each other. For example, someone who requests a taxi to a social area, may also request a taxi to return the previous location after several hours. Therefore, we need a method which is able to handle the spatial information. The ConvLSTM network contains a convolution operator which has the powerful ability to capture the spatial structural information. The structure of our model using ConvLSTM is shown in Fig. 2. The input data of the model at one time-step is three-dimensional as M × M × 1. M refers to the side length of our grid map and the number 1 is the amount of channels. The input shape is similar to the data shape of gray images in the process of image processing. We also need to decide the time steps $t$, then we can predict the taxi demand in the next time-step based on the historical taxi dataset in the past several time steps. The output of this model is the taxi demands of all subregions at time step $t + 1$. More details will be discussed in the experiment part. The structure of ConvLSTM is shown in Fig. 2.

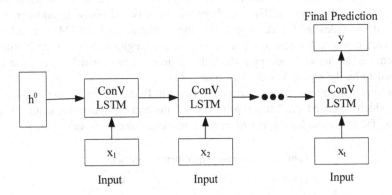

**Fig. 2.** The structure of ConvLSTM model

## 3    Experiment

### 3.1    Data

The historical taxi trip dataset used in our experiment is New York City taxi trip dataset [12], which was distributed by NYC government for research purposes. The dataset contains the taxi trip records from January 2013 through December 2013. Each taxi trip consists of several records, such as hack license, vendor id, pickup date/time, dropoff date/time, passenger count, trip time and so on. Limited by the hardware condition and computational power, we choose a part of the whole dataset from January 2013 through April 2013, and we use the taxi trip data related to our experiment, including pickup date/time, dropoff date/time and latitude/longitude coordinates for the pickup and dropoff locations. We select 80% of the data for training our model and keep the remaining 20% for validation.

### 3.2    Model Setup

In our experiment, we choose a geographical area with the latitude ranging from 40.750° to 40.765° and the longitude ranging from −73.996° to −73.978°, then we partition this area into equal small ones according to latitude and longitude, which means that each subregion possesses the same size. In order to analyse the influence of the number of subregions in predicting the taxi demands, we will divide our experimental area into 9 (3 × 3) subregions and 25 (5 × 5) subregions respectively, then evaluate the predicted results on each model. We use the ConvLSTM network and LSTM network to implement the prediction task, and compare their performances on the basis of performance metrics. The time-step length is 60 min in our study, and every one week data is used as a sequence, then the sequence length will be 168

(24 × 7). That means we utilize the historical taxi demand data of the past 168 time intervals to predict the number of passengers in the next time period. Therefore, the taxi demands of all subregions in the next time-step is the output of our models. In terms of the input data, the ConvLSTM network and the LSTM network have a different shape, which is one of the major differences between these two models. In addition to the number of samples and time steps, the input features of LSTM network is a $1 \times N$ vector, where $N$ refers to the number of the subregions. $N$ is 9 or 25 in our test. The ConvLSTM network accept a three-dimensional tensor, which means the input data need to be the shape $3 \times 3 \times 1$ or $5 \times 5 \times 1$.

Our implementations of these two models are in Python 3.5 with the assistance of Keras, which is based on TensorFlow, one of the most popular backends in deep learning. Table 1 includes the list of major parameters in our experiments.

**Table 1.** Important experimental parameters.

| Parameter | Value |
|---|---|
| Number of subregions | 9/25 |
| Time-step length | 60 min |
| Sequence length | 168 |
| Number of hidden layers (LSTM) | 2 |
| Number of nodes in each hidden layer | 9–18 |
| Number of channels (ConvLSTM) | 1 |
| Number of filters (ConvLSTM) | 9 |

### 3.3   Performance Metrics

In order to evaluate the performance of our prediction model, we adopt two kinds of prediction error metrics: Root Mean Square Error (RMSE) [13] and Symmetric Mean Absolute Percentage Error (SMAPE) [14]. The SMAPE is an alternative to Mean Absolute Percentage Error (MAPE) when there are zero or near-zero demand for items [15]. In contrast to the original formula defined by Armstrong in 1985, we use the currently accepted version of SMAPE without the factor 0.5 in denominator. The RMSE and SMAPE in region $i$ over time periods $[1 - T]$ are as follows:

$$RMSE_i = \sqrt{\frac{1}{T} \sum_{t=1}^{T} (P_{i,t} - P_{i,t}^*)^2} \tag{11}$$

$$SMAPE_i = \frac{1}{T} \sum_{t=1}^{T} \frac{\left| P_{i,t} - P_{i,t}^* \right|}{P_{i,t} + P_{i,t}^* + k} \tag{12}$$

Here $P_{i,t}$ refers to the real taxi demand in region $i$ at time-step $t$, and $P_{i,t}^*$ refers to the predicted taxi demand. In order to avoid division by zero when both $P_{i,t}$ and $P_{i,t}^*$ are zero, the small number $k$ is added to Eq. 11 [14]. The RMSE and SMAPE of all regions at time-step $t$ would be:

$$RMSE_t = \sqrt{\frac{1}{N}\sum_{i=1}^{N}(P_{i,t} - P_{i,t}^*)^2} \qquad (13)$$

$$SMAPE_t = \frac{1}{N}\sum_{i=1}^{N}\frac{\left|P_{i,t} - P_{i,t}^*\right|}{P_{i,t} + P_{i,t}^* + k} \qquad . \qquad (14)$$

The total number of subregions in our experiment is denoted by $N$. The Eq. 13 and Eq. 14 can be used to evaluate the prediction performance over the whole area.

## 4  Results

### 4.1  Spatiotemporal Feature

We use a portion of the New York City taxi trip dataset to analyse the spatiotemporal feature of the taxi demands, which is of great importance in our experiment. We utilize the records of taxi pickups to find the spatial distribution pattern of taxi demands throughout NYC, which is visualized with the software ArcGIS. We also analyse the taxi demands at different time periods in the same location. The taxi demand distribution on January 2, 2013 throughout NYC is shown in Fig. 3, and the taxi demands during different periods of time at JFK airport is shown in Fig. 4.

As we can see, affected by the multiple factors, such as geography, social property, economic development and so on, there is an obvious difference about the taxi demands in the areas. The majority of the taxi demands are in the Manhattan, while few people request the taxis in Staten Island. In terms of the temporal feature, as one of the most busiest airport in the United States, JFK airport possesses the different taxi pickups at different time periods, it is not easy for taxi drivers to find the passengers at 3:00 am.

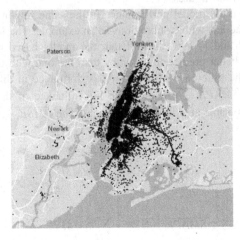

**Fig. 3.** The distribution of taxi pickups throughout NYC on January 2, 2013

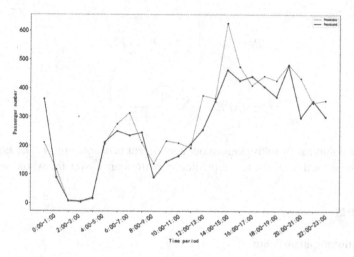

**Fig. 4.** The passenger number of taxi pickups at JFK airport at time periods

## 4.2    Prediction Results

In our experiment, we use the predicted results of 168 time steps to analyse the prediction accuracy. When the number of subregions is 9 (3 × 3), we name these small areas with numbers from 1 to 9, which is the same for 25 subregions. The $RMSE_t$ and $SMAPE_t$ are shown in Figs. 5 and 6. The time steps is 24 in the figures, LSTM_9_Regions denotes the result of the model based on LSTM network with 9 subregions, while ConvLSTM_25_Regions denotes the result of the model based on the ConvLSTM network with 25 subregions. As shown in Fig. 5, the $RMSE_t$ share the similar pattern, with more subregions in the model, the $RMSE_t$ tend to be lower. While in Fig. 6, we find that the model which includes more small areas has the higher $SMAPE_i$. However, it is obvious that the ConvLSTM network outperforms LSTM network as shown in both figures.

In order to analyse the model performance in the same region at all time periods simply, we take the models with 9 subregions as an example. The $RMSE_i$ and $SMAPE_i$ are shown in Figs. 7 and 8. We can find that the model based on the ConvLSTM network also has the better prediction accuracy than the one with the LSTM network, though there exists the difference in prediction performance in different regions. We use $\overline{RMSE - N}$ and $\overline{SMAPE - N}$ to evaluate the influence of the number of small regions divided by the same area in our experiment. The $\overline{RMSE - N}$ and $\overline{SMAPE - N}$ are defined as follows:

$$\overline{RMSE - N} = \frac{1}{N} \sum_{i=1}^{N} RMSE_i \tag{15}$$

$$\overline{SMAPE - N} = \frac{1}{N} \sum_{i=1}^{N} SMAPE_i \tag{16}$$

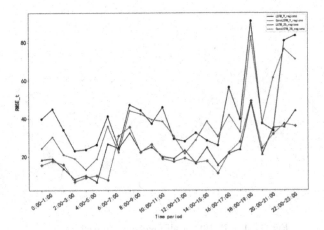

**Fig. 5.** The RMSE of all subregions at each time-step

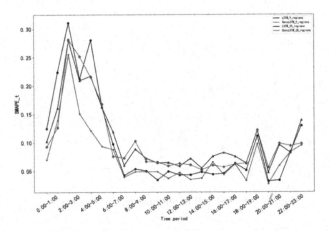

**Fig. 6.** The SMAPE of all subregions at each time-step

Where $N$ refers to the number of subregions. The results of different models are shown in Table 2.

Table 2 shows that the ConvLSTM network has the better prediction performance than the LSTM network, in terms of the size of each subregion, we find that the model involving more subregions tends to be less accurate in predicting the taxi demand, which means we need to partition the experimental area into the small regions with a proper size.

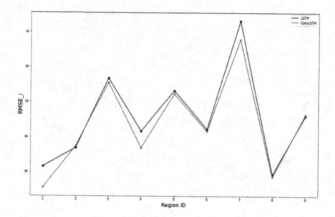

**Fig. 7.** The RMSE of all time steps in each region

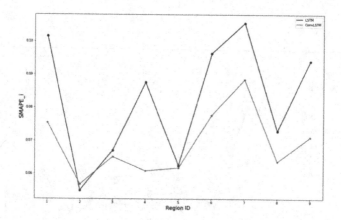

**Fig. 8.** The SMAPE of all time steps in each region

**Table 2.** Model performance metrics

| Model | $RMSE - N$ | $SMAPE - N$ |
|---|---|---|
| LSTM (9 regions) | 45.72 | 0.083 |
| ConvLSTM (9 regions) | 43.64 | 0.069 |
| LSTM (25 regions) | 59.26 | 0.261 |
| ConvLSTM (25 regions) | 57.74 | 0.238 |

## 5   Conclusion and Future Work

In this paper, based on the New York City taxi trip dataset, we analyse the spatiotemporal feature of taxi demand, then apply the ConvLSTM network and LSTM network to predict the taxi pickups in divided subregions respectively, the experimental results show that the prediction accuracy of the ConvLSTM network is better than that of the LSTM network, which proves that the ConvLSTM network is able to capture the spatial structural information more effectively. Therefore, ConvLSTM network can be used in many spatio-temporal sequence forecasting problems. Besides, our experiment also studies the influence of the subregion size in the prediction problem. The result shows that the prediction accuracy may become lower with the smaller subregion size, which means the size of the subregion is an important factor in predicting the taxi demand. For future work, we will take more factors into consideration, such as weather, traffic condition and so on. We also intend to improve our models with better network structures.

# References

1. Kong, X., Xu, Z., Shen, G., Wang, J., Yang, Q., Zhang, B.: Urban traffic congestion estimation and prediction based on floating car trajectory data. Future Gener. Comput. Syst. **61**, 97–107 (2016)
2. Zhang, D., et al.: Understanding taxi service strategies from taxi GPS traces. IEEE Trans. Intell. Transp. Syst. **16**(1), 123–135 (2015)
3. Yang, Q., Gao, Z., Kong, X., Rahim, A., Wang, J., Xia, F.: Taxi operation optimization based on big traffic data. In: Proceedings of 12th IEEE International Conference on Ubiquitous Intelligence and Computing, Beijing, China, pp. 127–134 (2015)
4. Zhao, K., Khryashchev, D., Freire, J., Silva, C., Vo, H.: Predicting taxi demand at high spatial resolution: approaching the limit of predictability. In: Proceedings of IEEE BigData, December 2016, pp. 833–842 (2016)
5. Li, X., et al.: Prediction of urban human mobility using large-scale taxi traces and its applications. Front. Comput. Sci. **6**(1), 111–121 (2012)
6. Kong, X., Xia, F., Wang, J., Rahim, A., Das, S.: Time-location-relationship combined service recommendation based on taxi trajectory data. IEEE Trans. Ind. Inform. **13**(3), 1202–1212 (2017)
7. Hochreiter, S., Schmidhuber, J.: Long short-term memory. Neural Comput. **9**(8), 1735–1780 (1997)
8. Xu, J., Rahmatizadeh, R., Boloni, L.: Real-time prediction of taxi demand using recurrent neural networks. IEEE Trans. Intell. Transp. Syst. **99**(1), 1–10 (2017)
9. Shi, X., Chen, Z., Wang, H., Yeung, D.-Y., Wong, W.-K., Wong, W.-C.: Convolutional LSTM network: a machine learning approach for precipitation nowcasting. In: NIPS (2015)
10. Graves, A.: Supervised Sequence Labelling with Recurrent Neural Networks, vol. 385. Springer, Heidelberg (2012). https://doi.org/10.1007/978-3-642-24797-2
11. Kim, S., Hong, S., Joh, M., Song, S.-K.: DeepRain: ConvLSTM network for precipitation prediction using multichannel radar data. In: IWOCI, September 2017
12. NYC Taxi & Limousine Commission: Taxi and Limousine Commission (TLC) Trip Record Data. http://www.nyc.gov/html/tlc/html/about/trip_record_data.shtml. Accessed Dec 2016
13. Lv, Y., Duan, Y., Kang, W., Li, Z., Wang, F.-Y.: Traffic flow prediction with big data: a deep learning approach. IEEE Trans. Intell. Transp. Syst. **16**(2), 865–873 (2015)
14. Wikipedia. https://en.wikipedia.org/wiki/Symmetric_mean_absolute_percentage_error. Accessed 20 May 2018
15. Vanguard Software Homepage. http://www.vanguardsw.com/business-forecasting-101/symmetric-mean-absolute-percent-error-smape/. Accessed 20 May 2018

# Prediction of Molecular Packing Motifs in Organic Crystals by Neural Graph Fingerprints

Daiki Ito[1](✉), Raku Shirasawa[2], Shinnosuke Hattori[2], Shigetaka Tomiya[2], and Gouhei Tanaka[1]

[1] Department of Electrical Engineering and Information Systems,
The University of Tokyo, Tokyo 113-8656, Japan
`satid110@sat.t.u-tokyo.ac.jp`
[2] Materials Analysis Center, Advanced Technology Research Division,
Sony Corporation, Atsugi, Kanagawa 243-0014, Japan

**Abstract.** Material search is a significant step for discovery of novel materials with desirable characteristics, which normally requires exhaustive experimental and computational efforts. For a more efficient material search, neural networks and other machine learning techniques have recently been applied to materials science in expectation of their potentials in data-driven estimation and prediction. In this study, we aim to predict molecular packing motifs of organic crystals from descriptors of single molecules using machine learning techniques. First, we identify the molecular packing motifs for molecular crystals based on geometric conditions. Then, we represent the information on single molecules using the neural graph fingerprints which are trainable descriptors unlike conventional untrainable ones. In numerical experiments, we show that the molecular packing motifs are better predicted by using the neural graph fingerprints than the other tested untrainable descriptors. Moreover, we demonstrate that the key fragment of molecules in crystal motif formation can be found from the trained neural graph fingerprints. Our approach is promising for crystal structure prediction.

**Keywords:** Crystal structure prediction · Machine learning
Supervised learning · Neural graph fingerprints

## 1 Introduction

Following the great success of deep learning in neural networks [1], the application fields of artificial intelligence and machine learning have been increasingly expanded. These techniques relying on learning algorithms and a plenty of data are expected to provide a powerful information processing platform, which can replace conventional rule-based computation and change manual procedures to automatic processing. Materials science is one of the targets of machine learning. A significant issue in materials science is to efficiently find out novel materials

© Springer Nature Switzerland AG 2018
L. Cheng et al. (Eds.): ICONIP 2018, LNCS 11305, pp. 26–34, 2018.
https://doi.org/10.1007/978-3-030-04221-9_3

that could lead to technological innovations in science and industry. However, material search based on material experiments and/or *ab initio* calculations of material properties often requires substantial time and efforts. To overcome this problem, materials informatics [2] utilizes data science to construct predictive models from existing materials data and thereby accelerate materials discovery [3]. In recent years, much attention has been paid to materials property predictions from atomistic and molecular information using machine learning methods [4–7].

The properties of crystalline materials, such as energies and electric characteristics, are highly sensitive to how molecules are assembled into a crystal structure under intra- and inter-molecular interactions. The design of crystalline materials with targeted structures and properties is the goal of crystal engineering [8]. Therefore, crystal structure prediction (CSP) is a significant step for materials property prediction [9]. For instance, the recent advances in the computational methods for CSP can be seen from the report on the latest CSP blind test hosted by the Cambridge Crystallographic Data Centre (CCDC) [10], where the aim is to accurately predict possible molecular crystal structures from a given molecule. Most contributors to this blind test have tried to predict the crystal structures by calculating the structure that minimizes its energy under empirical assumptions. At present, there are few machine learning-based approaches to CSP.

In this study, we address the prediction problem of structure types in organic crystals as a first step to develop machine learning methods for CSP. It is known that there are several basic molecular packing motifs of polycyclic aromatic hydrocarbons (PAHs) [11]. We first identify the molecular packing motif for organic crystal data from the Cambridge Structural Database (CSD) [12] based on geometric conditions. Then, we aim to predict the molecular packing motif from the information on single molecules in a supervised learning framework. The prediction ability depends on the information representation, or the descriptors, of the single molecules. In this study, we mainly focus on a trainable descriptor called the neural graph fingerprint (NGF) [13] to predict the motif type with a machine learning method. We compare the NGF with two other untrainable descriptors in the prediction accuracy.

In Sect. 2, we describe the method for classification of molecular packing motifs from the crystal data and the descriptors of molecules for motif prediction. In Sect. 3, we show the numerical results for the motif prediction. In Sect. 4, we summarize this work and give a brief discussion.

## 2 Methods

### 2.1 Molecular Packing Motifs

An assembly of molecules often forms a regularly arranged structure in a crystal, which can be classified into molecular packing motifs [11]. Predicting a molecular packing motif for a given molecule is useful for predicting structure and property of crystals, because it is associated with electronic, optoelectronic, and energetic properties of the crystalline materials. In this study, we aim to predict the molecular packing motif from descriptors of single molecules.

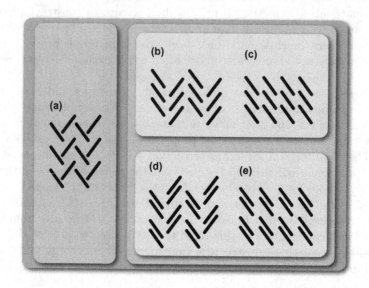

**Fig. 1.** Schematic illustrations of five packing motifs in molecular crystals with layered structures. Each segment represents a single molecule. (a) Herringbone. (b) Pi-stacking flip. (c) Pi-stacking parallel. (d) Dimer herringbone. (e) Dimer pi-stacking.

We classify the molecular crystals with layered structures into five motifs as shown in Fig. 1: (a) *Herringbone* is characterized by tilted edge-to-face interactions. (b) *Pi-stacking flip* (or $\gamma$ [8,11]) is a flattened-out herringbone with stacks of parallel molecules. (c) *Pi-stacking parallel* (or $\beta$ [8,11]) is a layered structure of parallel molecules. (d) *Dimer herringbone* (or sandwich [8,11]) is a herringbone motif made up of dimer molecules. (e) *Dimer pi-stacking* is a sheet-like motif made up of dimer molecules.

In addition to these five classes, we consider the sixth class "the other", into which the crystals with non-layered structures are categorized. As shown in Fig. 2, we formulated an If-Then rule to identify the motif based on the distances and angles between nearest neighboring molecules and those between 2nd nearest neighboring ones. The threshold values used in this rule were empirically determined. According to this flowchart, we identified the motif types for organic crystal structure data extracted from the CSD.

## 2.2 Molecular Representations

In materials informatics approaches, data representation is one of the key components governing the performance of predictive models [3]. To predict the molecular packing motif for each organic crystal, it is required to choose appropriate

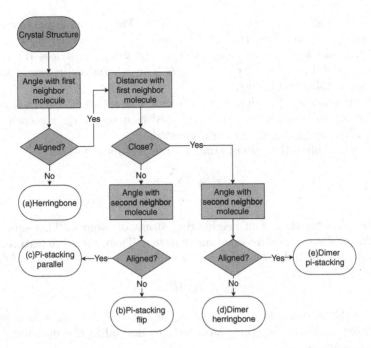

**Fig. 2.** Flowchart showing how to determine the molecular packing motif from the crystal structure.

descriptors of the single molecules by considering what factors are important for the crystal structures. Molecular descriptors represent a set of features of the molecules. There have been proposed many representations of molecular structures [14], including a variety of molecular fingerprints.

Recently, neural network models on graphs have been developed to handle graphic representations of molecules, where the vertices correspond to atoms and the edges correspond to bonds. Duvenaud et al. [13] proposed convolutional networks on graphs for learning molecular fingerprints that reflect a local structural information in a molecule. The proposed NGFs are represented as real-valued vectors, in contrast to the binary vector fingerprints [15,16]. By replacing non-differentiable operations in the fingerprint generation procedure of binary fingerprints with differentiable counterparts, NGFs can be trained for better prediction. This graph-based algorithm is regarded as a special case of the unified framework called message passing neural networks (MPNNs) [17]. We apply this method to the prediction of molecular packing motifs. For comparison, we also test the two other descriptors, the Coulomb matrix and the ellipsoid model. The three descriptors are individually described below.

(i) **Neural Graph Fingerprints (NGFs)** [13]: The input information for generating an NGF is a graph representation of a molecule with atom features. We denote the number of atom features by $F$, the number of layers (called the radius) by $R$, and the size of fingerprints by $L$. The trainable parameters are the hidden weight matrices $H_1^1, \ldots, H_R^K$, where the size of $H_l^k$ is $F \times F$ for $l = 1, \ldots, R$ and $k = 1, \ldots, K$, and the output weight matrices $W_1, \ldots, W_R$ where the size of $W_l$ is $F \times L$ for $l = 1, \ldots, R$. The generation process of an NGF is described as follows (Fig. 3a).

At the initial state, the finger print vector $\mathbf{f}$ is set at $\mathbf{0}$. For each atom $a$ in a molecule, the atom feature is represented as a vector $\mathbf{r}_a$. For each layer $l$ ($l = 1, \ldots, R$), the following steps are repeated.

First, we calculate the sum of atom feature vectors as follows:

$$\mathbf{v} = \mathbf{r}_a + \sum_{i \in n(a)} \mathbf{r}_i, \tag{1}$$

where $n(a)$ denotes the set of neighboring atoms of atom $a$. This summation reflects the local structural information of atom $a$. Then, the atom feature vector is updated with the hidden weights and a nonlinear transformation as follows:

$$\mathbf{r}_a = \sigma(\mathbf{v} H_l^k), \tag{2}$$

where $k$ is the degree (the number of bonds) of atom $a$ and $\sigma$ is a smooth differentiable nonlinear function. Moreover, to normalize the updated feature vector, we compute

$$\mathbf{i} = \mathrm{softmax}(\mathbf{r}_a W_l), \tag{3}$$

and update the fingerprint vector as follows:

$$\mathbf{f} \leftarrow \mathbf{f} + \mathbf{i}. \tag{4}$$

After the loops with respect to $l$, we obtain the NGF $\mathbf{f}$.

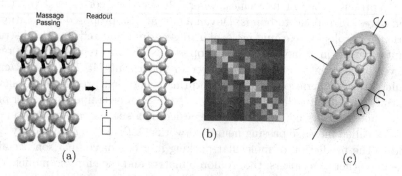

**Fig. 3.** Schematic illustrations of the molecular descriptors. (a) The neural graph fingerprints [13]. (b) The Coulomb matrix. (c) The ellipsoid model.

**(ii) Coulomb Matrix:** This representation of molecules was developed to predict atomization energies and electronic properties of molecules [4,6]. The Coulomb matrix $C = (C_{ij})$ is described as follows:

$$C_{ij} = \begin{cases} 0.5Z_i^{2.4} & \text{for } i = j \\ \frac{Z_i Z_j}{|R_i - R_j|} & \text{for } i \neq j \end{cases}, \tag{5}$$

where $Z_i$ represents the nuclear charge of atom $i$ and $R_i$ represents its Cartesian coordinate in space (Fig. 3b). The off-diagonal elements correspond to the Coulomb repulsion between atoms $i$ and $j$, while the diagonal elements encode a polynomial fit of atomic energies to nuclear charge. The matrices are transformed into vectors of a sufficiently large fixed size.

**(iii) Ellipsoid Model:** The way of packing molecules depends on the shape and the geometric distortion of the molecules [18], which can influence the crystal motif formation. The ellipsoid model represents the approximate shape of a molecule using the moment of inertia given by three-dimensional vectors (Fig. 3c). The degree of anisotropy of a molecule is reflected in this representation.

### 2.3   Machine Learning Experiments

The NGF vectors with size $L$ are fed into a fully connected linear layer and subsequently transformed by a weight matrix $U$ into an output vector representing the motif class. In the two other methods, the descriptor vectors are fed into a one-hidden-layer fully connected neural network with sigmoid activation functions and then transformed by a weight matrix $U$ into an output vector. While the weight matrices in Eqs. (2)–(3) and the weight matrix $U$ are trained in the NGF, only the weight matrix $U$ is trained in the two other methods. All models were trained with the Adam algorithm [19]. The validation set was used to choose the best model to evaluate and avoid models that are overfitting. The Bayesian optimization technique [20] was used to optimize the hyper-parameters. The experiments were performed with the codes for computing neural fingerprints and finding strongly activated motifs, available at https://github.com/HIPS/ neural-fingerprint [13].

## 3   Results

Referring to Devenaud et al. [13], we set the hyper-parameters for the NGF at $L = 2014$ and $R = 3$ and the number of atom features at $F = 20$. The atom features include the atom type as a one-hot vector, its degree, the number of attached hydrogens, the implicit valence, and the aromaticity indicator as a binary number. The size of the weight matrix $U$ in the full connection layer is $L \times 6$ for all the descriptors. The total number of crystal data is 9371. The motif class is identified for each of these data. Some data were used for training and the remaining data were kept for testing. The prediction accuracy was evaluated as the fraction of correct prediction for the testing data.

The results of the experiment are summarized in Table 1. At the best case with 6500 training data, the accuracy is 64% for the NGF, 37% for the Coulomb matrix, and 40% for the ellipsoid model. This result shows that the graph representation of molecules is much better than the two other molecular expressions for motif prediction. This benefit is brought about by the training of the NGFs (descriptors), which can deal with graphs with any size and any topology and takes the local structural property into consideration. We notice an unexpected result that the ellipsoid model represented as a 3-dimensional vector yields better accuracy than the Coulomb matrix represented as an 801-dimensional vector.

**Table 1.** Comparison of molecular representations.

| Representation | Train loss | Train accuracy | Test loss | Test accuracy |
|---|---|---|---|---|
| NGFs | 1.01 | 0.76 | 1.23 | 0.64 |
| Coulomb matrix | 1.40 | 0.45 | 1.50 | 0.37 |
| Ellipsoid model | 1.38 | 0.47 | 1.43 | 0.41 |

**Fig. 4.** A strongly activated fragment in molecules for herringbone crystals.

The other advantage of the NGF is its interpretability [13]. It is possible to extract fragments of molecules that are effective for motif prediction from the trained models. An automatic extraction of essential fragments of molecules for specific motif formation is useful for getting an insight into the key factors for CSP. We first found the elements of the NGFs that are important for each motif from the trained weight matrix and then identified the key fragment containing the atoms that most contribute to those fingerprint elements. Figure 4 demonstrates a strongly activated fragment for the herringbone motif, which is a part of the aromatic ring. This fragment is found in 91 molecules among 364 herringbone crystals.

Compared with the conventional method based on energy minimization, the method based on machine learning is advantageous in terms of the computational time for prediction and the extraction of important fragments. On the other hand, the motif classification is not enough to precisely predict the crystal structure itself.

## 4    Summary and Discussion

We have considered the prediction problem of molecular packing motifs in organic crystals from single molecules using a machine learning framework. First, we have classified the crystal structures in the dataset into several packing motifs by a rule-based method. Next, we have used the three kinds of molecular descriptors to predict the motif class. The results have shown that the prediction performance of the neural graph fingerprint considerably outperforms those of the Coulomb matrix and the ellipsoid model. It suggests that the graph-based molecular representation including local structural information is advantageous for predicting crystal structures. Moreover, we have shown that the neural graph fingerprint is beneficial in revealing the essential fragment correlated with the crystal packing motif.

As demonstrated in this work, the neural networks handling graph structures are promising for CSP in the performance improvement and the specification of key fragments in molecules. For improving the prediction accuracy, it would be effective to use a larger number of training data, select more appropriate features of atoms and edges, and refine the prediction model. The performance comparison between the ellipsoid model and the Coulomb matrix suggests that the global shape of the molecule is also important for the crystal structure formation.

## References

1. Schmidhuber, J.: Deep learning in neural networks: an overview. Neural Netw. **61**, 85–117 (2015)
2. Rajan, K.: Materials informatics. Mater. Today **8**(10), 38–45 (2005)
3. Ward, L., Wolverton, C.: Atomistic calculations and materials informatics: a review. Curr. Opin. Solid State Mater. Sci. **21**(3), 167–176 (2017)
4. Rupp, M., Tkatchenko, A., Müller, K.R., Von Lilienfeld, O.A.: Fast and accurate modeling of molecular atomization energies with machine learning. Phys. Rev. Lett. **108**(5), 058301 (2012)
5. Pilania, G., Wang, C., Jiang, X., Rajasekaran, S., Ramprasad, R.: Accelerating materials property predictions using machine learning. Sci. Rep. **3**, 2810 (2013)
6. Montavon, G., et al.: Machine learning of molecular electronic properties in chemical compound space. New J. Phys. **15**(9), 095003 (2013)
7. Ramakrishnan, R., Dral, P.O., Rupp, M., von Lilienfeld, O.A.: Big data meets quantum chemistry approximations: the δ-machine learning approach. J. Chem. Theor. Comput. **11**(5), 2087–2096 (2015)

8. Campbell, J.E., Yang, J., Day, G.M.: Predicted energy-structure-function maps for the evaluation of small molecule organic semiconductors. J. Mater. Chem. C **5**(30), 7574–7584 (2017)

9. Day, G.M., Gorbitz, C.H.: Introduction to the special issue on crystal structure prediction. Acta Crystallogr. Sect. B: Struct. Sci. Cryst. Eng. Mater. **72**, 435–436 (2016)

10. Reilly, A.M., et al.: Report on the sixth blind test of organic crystal structure prediction methods. Acta Crystallogr. Sect. B: Struct. Sci. Cryst. Eng. Mater. **72**(4), 439–459 (2016)

11. Desiraju, G.R., Gavezzotti, A.: Crystal structures of polynuclear aromatic hydrocarbons. Classification, rationalization and prediction from molecular structure. Acta Crystallogr. Sect. B: Struct. Sci. **45**(5), 473–482 (1989)

12. Groom, C.R., Bruno, I.J., Lightfoot, M.P., Ward, S.C.: The cambridge structural database. Acta Crystallogr. Sect. B: Struct. Sci. Cryst. Eng. Mater. **72**(2), 171–179 (2016)

13. Duvenaud, D.K., et al.: Convolutional networks on graphs for learning molecular fingerprints. In: Advances in Neural Information Processing Systems, pp. 2224–2232 (2015)

14. Bender, A., Glen, R.C.: Molecular similarity: a key technique in molecular informatics. Org. Biomol. Chem. **2**(22), 3204–3218 (2004)

15. Glen, R.C., Bender, A., Arnby, C.H., Carlsson, L., Boyer, S., Smith, J.: Circular fingerprints: flexible molecular descriptors with applications from physical chemistry to adme. IDrugs **9**(3), 199 (2006)

16. Rogers, D., Hahn, M.: Extended-connectivity fingerprints. J. Chem. Inf. Model. **50**(5), 742–754 (2010)

17. Gilmer, J., Schoenholz, S.S., Riley, P.F., Vinyals, O., Dahl, G.E.: Neural message passing for quantum chemistry. arXiv preprint arXiv:1704.01212 (2017)

18. Mingos, D.M.P., Rohl, A.L.: Size and shape characteristics of inorganic molecules and ions and their relevance to molecular packing problems. J. Chem. Soc. Dalton Trans. **12**, 3419–3425 (1991)

19. Kingma, D.P., Ba, J.: Adam: a method for stochastic optimization. arXiv preprint arXiv:1412.6980 (2014)

20. Snoek, J., Larochelle, H., Adams, R.P.: Practical Bayesian optimization of machine learning algorithms. In: Advances in Neural Information Processing Systems, pp. 2951–2959 (2012)

# A Multi-indicator Feature Selection for CNN-Driven Stock Index Prediction

Hui Yang, Yingying Zhu[✉], and Qiang Huang

College of Computer Science and Software Engineering, Shenzhen University,
Shenzhen 518060, China
zhuyy@szu.edu.cn

**Abstract.** Stock index prediction is regarded as a challenging task due to the phenomena of non-linearity and random drift in trends of stock indices. In practical applications, different indicator features have significant impact when predicting stock index. In addition, different technical indicators which contained in the same matrix will interfere with each other when convolutional neural network (CNN) is applied to feature extraction. To solve the above problem, this paper suggests a multi-indicator feature selection for stock index prediction based on a multi-channel CNN structure, named MI-CNN framework. In this method, candidate indicators are selected by maximal information coefficient feature selection (MICFS) approach, to ensure the correlation with stock movements while reduce redundancy between different indicators. Then an effective CNN structure without sub-sampling is designed to extract abstract features of each indicator, avoiding mutual interference between different indicators. Extensive experiments support that our proposed method performs well on different stock indices and achieves higher returns than the benchmark in trading simulations, providing good potential for further research in a wide range of financial time series prediction with deep learning based approaches.

**Keywords:** Stock index prediction · Feature selection
Maximal information coefficient · Convolutional neural networks

## 1 Introduction

Stock index prediction has been an important issue in the fields of finance, engineering and mathematics due to its potential financial gain. The prediction of stock index is regarded as a challenging task of financial time series prediction. There has been so much work done on ways to predict the movements of stock price. In the past years, most research studies focused on the time series models and statistical methods to forecast future trends based on the historical data, such as ARIMA [1], ARCH [2], GARCH [4], etc. With the great development of computer science, many recent works have been proposed based on the machine learning methods, such as neural networks (NN) [8], bayesian approach [12] and support vector machine (SVM) [19], to predict the stock index trends.

© Springer Nature Switzerland AG 2018
L. Cheng et al. (Eds.): ICONIP 2018, LNCS 11305, pp. 35–46, 2018.
https://doi.org/10.1007/978-3-030-04221-9_4

Recently, convolutional neural network (CNN) is gradually applied in the field of stock market, and some methods [7,15,18] based on CNN have shown that CNN can be an effective tool for feature extraction whether in the task of predicting specific price level or predicting the movements of stock. These CNN-driven methods have shown state-of-the-art performance. Gunduz et al. [7] proposed a CNN architecture with a specifically ordered feature set to predict the intraday direction of stocks. In the feature set, each instance was transformed into 2D-matrix by taking into account different indicators, price and temporal information. Sezer et al. [15] proposed a CNN-TA stock trading model, and $15 \times 15$ sized 2-D images were constructed using 15 different technical indicators. However, it is worth noting that there are still some common disadvantages hindering the current CNN-driven stock index prediction methods. First, they used fixed technical indicators as the input of CNN for stock forecasting. But different stock indices represent different industries, which present different characteristics and market cycles. Therefore, the adoption of fixed technical indicators is not adaptable to the prediction of different stock indices. In addition, when convolving the indicator matrix, different indicators will interfere with each other and cause confusion information in the feature maps. Because different technical indicators are fused in the same matrix.

To solve these problems, a CNN-driven multi-indicator stock index prediction framework, named MI-CNN, is presented in this paper, which applied CNN to extract abstract features in different indicators independently. In the MI-CNN framework, we utilize maximal information coefficient feature selection (MICFS) to filter more effective technical indicators for different stock indices intelligently, instead of using fixed indicators to predict all kinds of stock indices. Then a multi-channel CNN structure is proposed to extract features from each independent technical indicator, rather than extracting all indicator features in a single matrix confusedly. Our MI-CNN framework is proved to be effective on various stock indices and numerous experiments are illustrated in this paper. The average prediction accuracy and returns achieve 60.02% and 31.07% in the experiments.

The remainder of this paper is structured as follows. In Sect. 2, we briefly review the related work in stock index prediction tasks. In Sect. 3, we describe the architecture and detailed design of the framework. Then the experiments and the corresponding analysis are shown in Sect. 4. Finally, some concluding remarks are drawn and future research directions are discussed from Sect. 5.

## 2   Related Work

Financial time series modeling is regarded as one of the most challenging forecasting problems. In [17], Kevin indicated that the change in the stock price was better forecasted by the non-linear methods when compared with linear regression models. In [5], it has shown that forecasting price movements can often result in more trading results. Oriani et al. [13] evaluated the impact of technical indicators on stock forecasting and concluded that lagging technical indicators can improve the accuracy of the stock forecasts compared to that made with the original series of closing price.

The purpose of feature selection method is to reduce data complexity and improve prediction accuracy. Feature subset selection methods can be classified into two categories: the filter approach and the wrapper approach [9]. Lee proposed a F-score and supported sequential forward search feature selection method, which combined the advantages of filter methods and wrapper methods to select the optimal feature subset from the original feature set [11]. Su et al. proposed an integrated nonlinear feature selection method to select the important technical indicators objectively in forecasting stock price [16]. The results showed that the proposed method outperforms the other models in accuracy, profit evaluation and statistical test.

Recently, more and more practice shows promising performance in different ways of combining CNN and stock prediction tasks together. Tsantekidis et al. [18] proposed a deep learning methodology, based on CNN, that predicted the price movements of stocks, using as input large-scale, high-frequency time-series derived from the order book of financial exchanges. Results showed that CNN is better suited for this kind of task in finance. In [6], researchers extracted commonly used indicators from financial time series data and used them as their features of artificial neural network (ANN) predictor. They generated $28 \times 28$ images by taking snapshots that were bounded by the moving window over a daily period.

## 3   Proposed Framework

The architecture of the MI-CNN framework is first briefly described in Fig. 1. More specifically, the system selects several effective indicators through MICFS from given stock index data. Then potential features of each indicator are extracted using a special CNN structure. After that, extracted features are input into the ANN model to provide prediction results. Finally, a straight trading strategy [3] is applied according to the final prediction results. We will introduce the details of the trading strategy in the experimental section.

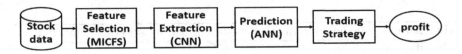

**Fig. 1.** Block diagram of complete framework

### 3.1   Stock Feature Selection

Stock market data and several common-use technical indicators are employed as input features in this study. Because most of the technical indicators are calculated from basic stock market data, the redundancy of information is unavoidable between different technical indicators. Therefore, we apply the MICFS approach to filter out indicators which are most relevant to the movements of given stock index, while the correlation between the selected indicators is minimal.

Maximal information coefficient (MIC) is a measure of dependence for two-variable relationships that captures a wide range of associations both functional and not [14]. MIC belongs to a larger class of maximal information-based non-parametric exploration (MINE) statistics for identifying and classifying relationships. Let $D$ be a set of ordered pairs. For a grid $G$, let $D|_G$ denote the probability distribution induced by the data $D$ on the cells of $G$. And $I$ denote mutual information (MI):

$$I(x;y) = \iint p(x,y) log \frac{p(x,y)}{p(x)p(y)} dxdy \tag{1}$$

Let $I^*(D, x, y) = \max_G I(D|G)$, where the maximum is taken over all $x$-by-$y$ grids $G$ (possibly with empty rows/columns). MIC is defined as

$$MIC(D) = \max_{xy < B(|D|)} \frac{I^*(D, x, y)}{log_2 min\{x, y\}} \tag{2}$$

where $B$ is a growing function satisfying $B(n) = o(n)$, and the authors heuristically suggest $B(n) = n^{0.6}$ as a default setting.

In this study, we don't limit the number of selected indicators until the best feature subset is constructed. For each indicator $f$, the MICFS is described by:

$$J(f) = \frac{1}{n} \sum_{i=1}^{n} MIC(C; f_i) - \frac{1}{n} \beta \sum_{s \in S} \sum_{i=1}^{n} MIC(s_i : f_i) \tag{3}$$

where $MIC(C; f_i)$ is the MIC of labels $C$ and feature $f$, and $f_i$ is the $i$th series of feature $f$, $MIC(s_i; f_i)$ is the MIC of candidate feature $f$ and the selected feature $s$ in feature subset $S$, and coefficient $\beta \in [0, 1]$ indicates the effect of selected feature redundancy to the result. $n$ is the number of series contained in each indicator, and $n = 3$ in our study. The basic steps of feature selection are summarized by the pseudo code listed below.

---

**Algorithm 1.** Maximal information coefficient feature selection algorithm

---
**Input:**
     Training dataset $D = \{F, C\}, F = (f_1, f_2, ..., f_n)$
**Output:**
     Selected feature subset $S, S \subseteq F$
**Begin** Set $S = \emptyset$, $\beta = 1$, $B(n) = n^{0.6}$
**for** $i = 1 : n$ **do**
   Calculate $J$ for each feature $f$ and $f \notin S$
   Select the maximum $J$ and put corresponding feature $f$ into subset $S_i$
   Train the model using $S_i$, and obtain the accuracy $P(S_i)$
**end for**
Select the best $P(S)$ and output the feature subset $S$
**End**

---

Eight popular indicators are chosen as candidate features for MICFS and each indicator contains three common-use series. An open-source library Technical Analysis Library (TA-Lib) is utilized to calculate technical indicators above. The details of all the eight technical indicators are depicted in Table 1.

**Table 1.** Candidate technical indicators

| Technical indicator | Concrete series | Description |
|---|---|---|
| *Price* | *high, low, close* | Highest price, lowest price and close price |
| *Vol* | $Vol, Vol_5, Vol_{10}$ | Trading volume of stock index |
| *MACD* | $MACD, MACD_{hist}, MACD_{signal}$ | Moving Average Convergence and Divergence |
| *RSI* | $RSI_5, RSI_{10}, EMA_n$ | Relative Strength Index |
| *KD* | $slow_K, slow_D, fast_k$ | Stochastic Index |
| *WR* | $WR_5, WR_{10}, WR_{20}$ | Larry Williams R |
| *ROC* | $ROC_5, ROC_{10}, ROC_{20}$ | Rate of Change |
| *CCI* | $CCI_5, CCI_{10}, CCI_{20}$ | Commodity Channel Index |

## 3.2 CNN-Driven Stock Feature Extraction

In normal conditions, CNN is principally utilized to deal with image-related problem because of the ability to discern the spatial correlation of neighboring pixels. CNN can automatically extract the characteristic relationship between adjacent data elements and reconstruct the feature vectors [10]. The financial time series forecasting problem can be implicitly converted into an image classification problem when technical analysis data is shaped into two-dimension matrices [15]. As for the task of stock prediction in this study, market data and technical indicators are continuous-discrete time. As a consequence, vectors in each indicator series are relevant while different indicators are independent of each other. What's more, there are potential characteristics between the series of the same technical indicator with different computing periods. It raises the difficulty of extracting features from the stock data for predicting stock movements. Considering the above issues and the characters of CNN, a more applicable multi-channel CNN-driven framework is proposed to extract features from technical indicators in stock prediction task.

Figure 2 illustrates the sketch of the CNN architecture when three indicators are selected. Continuous time series of each indicator are formed as the shape of two-dimension matrices:

$$Indicator = \begin{bmatrix} f_{11} \cdots f_{1n} \\ f_{21} \cdots f_{2n} \\ f_{31} \cdots f_{3n} \end{bmatrix}_{3 \times n} \quad (4)$$

where $n$ represents a continuous date span of indicator and we fix $n = 10$. For example, as for the indicator $RSI$, three series $RSI_5$, $RSI_{10}$ and $RSI_{20}$ are formed as a $3 \times n$ matrix. For each indicator, we use the same convolution architecture, two convolutional layers without subsampling, to extract features, which can reflect the extracted abstract features in a high level. Finally, the extracted features are integrated and put into the final prediction model, fully connected ANN. The purpose of our design is to make CNN automatically extract underlying features of each indicator. At the same time, it will not be subject to any other indicator when it extracts features of a certain indicator, owing to the independent CNN structure. The number of CNN channels depends on how many indicators are selected in feature subset $S$.

In the field of image, subsampling plays the role of dimensionality reduction and invariance. But in this study, the series of each indicator takes on specific meanings, and the different series, such as $RSI_5$ and $RSI_{10}$, are independent of each other. If subsampling, such as the maxpool layer, is adopted, the information in the indicator series would be lost and cause information loss probably. To avoid the loss of extracted features, there are two convolutional layers without subsampling applied in this study.

In the two convolutional layers, the sizes of the convolution kernels are $3 \times 3$ and $3 \times 2$, and the stride length is 1. Zero-padding strategy is applied in the first convolutional layer. In this way, when extracting features, it can not only extract momentum features between horizontally adjacent data points in an identical series, but also extract other underlying features between every two series and between total three series that are vertically adjacent. The number of feature maps in the two convolutional layers are 16 and 32, respectively.

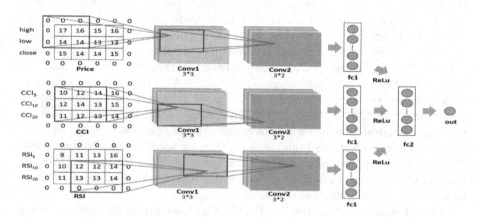

**Fig. 2.** Architecture of CNN when 3 indicators are selected

Fully connected ANNs are deployed after the second convolutional layers in each feature extraction channel. Then the ANNs are merged in the next hidden layer. The outputs are the movements of a given stock index. Given 1 shows the stock index price will rise next day. By contrast, given 0 indicates the opposite way. Specifically, the rectified linear units (ReLu) activation function is applied in ANN and the learning rate $\alpha$ is set to 0.01. The back-propagation algorithm is used to train the model. The parameters of the model are learned by minimizing the categorical cross entropy loss:

$$C(W) = -\sum_{i=1}^{L} y_i \cdot logy_i^{'} \tag{5}$$

where $L$ is the number of different labels and the notation $W$ refers to the parameters of weights. The ground truth vector is denoted by $y$, while $y^{'}$ is the predicted label distribution. In order to overcome the over fitting phenomenon, we punish the weights $W$ with L2 regularization, and the L2 regularization is calculated as:

$$C^{'} = C + \frac{\lambda}{2n} \sum_{w} w^2 \tag{6}$$

where $C$ denotes the original cost function, $\lambda$ is regularization coefficient and $n$ is the scale of training set. $w$ is the weights need to be punished.

## 4   Experiment

### 4.1   Experiments on SPY

The basic experiments are developed on the S&P 500 Index ETF (SPY), from January 2008 to December 2017. The stock market data is published on Yahoo Finance. Afterwards, the dataset is divided into two sets. Data with eight years is served as training dataset while the remaining data with two years is used as the test dataset. The input data is scaled to [0, 1], using the min-max normalization approach. We employ accuracy and returns as our evaluation metrics:

$$Accuracy = \frac{TP + TN}{TP + FP + TN + FN} \tag{7}$$

$$Returns = \frac{C_{final} - C_{initial}}{C_{initial}} \times 100\% \tag{8}$$

$TP, FP, TN$ and $FN$ represent respectively the true positive, false positive, true negative and false negative [7]. $C_{initial}$ and $C_{final}$ present the capital at the beginning and at the end of transaction respectively.

The process of feature selection is analyzed in accordance with the execution phase of the MICFS approach. Table 2 illustrates the $J$ values of candidate indicators in each iteration of MICFS approach, where $J(f)$ is calculated by Eq. 3. In each iteration, the indicator with the largest $J$ value will be selected into the feature subset $S$. In Table 3, each row illustrates the selected feature subset $S$ and the $J$ value of the selected feature at this iteration. It also illustrates the prediction accuracy $P(S)$ on the training set when the current feature subset $S$ is used as input data for the MI-CNN model. As shown in the table, when the feature subset contains Price, CCI and MACD, the prediction accuracy reaches the highest value.

**Table 2.** Execution detail of MICFS approach on SPY

| Iteration | 1 | 2 | 3 | 4 | 5 | 6 | 7 |
|---|---|---|---|---|---|---|---|
| J(Price) | **0.1038** | - | - | - | - | - | - |
| J(Vol) | 0.0950 | −0.4485 | −0.5219 | −0.6817 | −0.3031 | **−0.4245** | - |
| J(MACD) | 0.0943 | −0.0184 | **−0.1491** | - | - | - | - |
| J(KDJ) | 0.0996 | −0.0283 | −0.1791 | −0.6812 | **−0.2811** | - | - |
| J(RSI) | 0.0958 | −0.0577 | −0.2730 | −0.5399 | −0.5112 | −0.7858 | **−0.9583** |
| J(WR) | 0.0987 | −0.0157 | −0.3337 | −0.5030 | −0.4199 | −0.6485 | **−0.7802** |
| J(ROC) | 0.0959 | −0.0462 | −0.2261 | **−0.3928** | - | - | - |
| J(CCI) | 0.1010 | **0.0212** | - | - | - | - | - |

**Table 3.** Results of MICFS approach on SPY

| Selected feature subset S | Maximum J(f) | P(S) |
|---|---|---|
| Price | 0.1038 | 59.75% |
| Price, CCI | 0.0212 | 63.32% |
| **Price, CCI, MACD** | −0.1491 | **66.91%** |
| Price, CCI, MACD, ROC | −0.3928 | 64.51% |
| Price, CCI, MACD, ROC, KD | −0.2811 | 63.34% |
| Price, CCI, MACD, ROC, KD, Vol | −0.4245 | 60.30% |
| Price, CCI, MACD, ROC, KD, Vol, WR | −0.7802 | 59.65% |
| Price, CCI, MACD, ROC, KD, Vol, WR, RSI | −1.5551 | 57.65% |

We apply the MICFS method to conduct experiments on several stock indices introduced in [3]. Figure 3 presents the experimental results of MICFS on different stock indices. Since different stock indices represent different industries and have their own characteristics, the number of features in the optimal feature subset is different when forecasting stock trends. It shows that the MICFS method can select effective features for different stock indices.

The signals produced from the final prediction model are applied to implement trading simulations. A simple trading strategy is conducted with the given signals.

$$Action = \begin{cases} buy & signal_c = 1 \text{ and } signal_{c-1} = 0 \\ sell & signal_c = 0 \text{ and } signal_{c-1} = 1 \\ sit & others \end{cases} \qquad (9)$$

where $signal_c$ represents the signal of current date and $signal_{c-1}$ represents the signal of the last date. We adopt the T + 1 trading system in the trading simulations. The capital curves of trading results are shown in Fig. 4; it displays the accumulated capitals trading on SPY index during the year of 2016 and 2017. In order to simplify the transaction process, the transaction fees are not took into account. As shown in this picture, the red line presents the capital trading with our MI-CNN framework, while the blue line indicates the capital trading with Buy&Hold strategy, where we simply buy the index at the beginning of transaction and sell it at the end. The Buy&Hold strategy is usually considered as a benchmark to compare with different trading strategies. Comparing with

Buy&Hold strategy, our method obtains higher returns, exactly 59.16%. Overall, our system performs well on the SPY stock index.

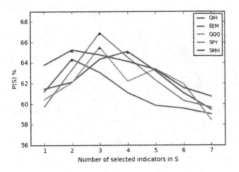

**Fig. 3.** MICFS results with different stock indices

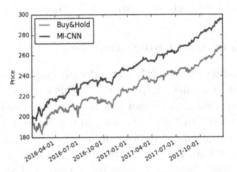

**Fig. 4.** Trading simulation on SPY

## 4.2  Comparison with Other Models

In this part, several popular prediction models are compared with our MI-CNN framework. The classifier that often involved in other decision support systems contains SVM, ANN and so on. These methods are used for contrasting with MI-CNN and the simulations are conducted on SPY with confirmed indicators. The structure of ANN is designed as a three-layer network, and the number of neurons in the hidden layer is determined at $2N + 1$ [3], where $N$ is the dimension of input vectors. Finally, a single CNN (SCNN) structure is used to extract features in the indicator matrix, where the selected indicators are converted into a 2-D matrix together [15]. The contrast results are illustrated in Fig. 5(a). The accumulated capital curves are displayed in Fig. 5(b).

The histogram shows the proposed MI-CNN presents the best performance with both the test accuracy and returns. Then the SCNN model ranks in the

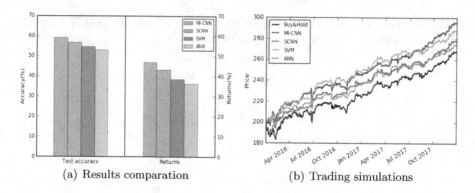

(a) Results comparation                    (b) Trading simulations

**Fig. 5.** Comparing results of different models

second and SVM ranks in the following. ANN performs worst that may caused by the vector-ordered representation of input data. Moreover, the high prediction accuracy, in general, conducts more fruitful returns in the real stock market. So the MI-CNN conducts the highest returns in the trading simulation.

### 4.3   Experiments on Other Stock Indices

To prove the adaptability of MI-CNN method with various stock indices, several stock indices in different industries are tested in this section. These stock indices are a good mix of large, mid, and small stocks, which has been introduced in [3]. In order to compare with the existing work in the latest research, the trading simulations are conducted on the dataset in [3]. The testing result with different stock indices are shown in Table 4. In terms of prediction accuracy, our MI-CNN is generally higher than their experiment results. The returns of trading simulation are better than theirs partly. We can see the accumulated returns

**Table 4.** Experiment results on different indices

| Stock index | Accuracy in [3] | Accuracy of MI-CNN | Returns of Buy&Hold | Returns in [3] | Returns of MI-CNN |
| --- | --- | --- | --- | --- | --- |
| SPY | 61.11% | **61.94%** | 13.69% | 24.75% | **27.37%** |
| EEM | 55.56% | **60.31%** | 17.07% | **36.48%** | 33.04% |
| EFA | 54.76% | **61.50%** | 8.13% | 30.09% | **32.37%** |
| FXI | 52.78% | **58.71%** | 4.54% | 28.07% | **35.75%** |
| IWM | 53.79% | **59.11%** | 25.60% | **31.48%** | 29.01% |
| OIH | 53.57% | **57.90%** | 21.71% | 24.17% | **25.68%** |
| QQQ | 58.33% | **60.69%** | 19.29% | 29.09% | **32.31%** |
| SMH | 52.78% | **60.07%** | 18.74% | 29.66% | **33.09%** |
| Average | 55.33% | **60.02%** | 16.09% | 29.22% | **31.07%** |

conducted with our MI-CNN framework are universally higher than that with Buy&Hold strategy. The average prediction accuracy and returns of MI-CNN are 60.02% and 31.07% respectively.

## 5    Conclusion

In this paper, we propose a CNN-driven multi-indicator framework for stock index movement prediction. We employ MICFS to select effective indicators for different stock indices by considering the correlation between indicators and labels, and considering the redundancy between different indicators simultaneously. Feature extraction is carried out for several selected indicators respectively, which avoids interference between different indicators. For multi-indicator, we design a multi-channel CNN structure to extract underlying features from different indicators. In the experiments, the average prediction accuracy and average returns achieve 60.02% and 31.07% respectively. Experimental results demonstrate that our MI-CNN framework can successfully select features and effectively improve the prediction and transaction results.

There is much room for improvement in the future works. On the one hand, the source and representation of input data are potentially diverse. Finance data and news text could be taken into account. On the other hand, the features extracted by CNN can be considered as an important reference for different trading strategies.

**Acknowledgements.** This work was supported by: (i) National Natural Science Foundation of China (Grant No. 61602314); (ii) the Natural Science Foundation of Guangdong Province of China (Grant No. 2016A030313043); (iii) Fundamental Research Project in the Science and Technology Plan of Shenzhen (Grant No. JCYJ20160331114551175).

## References

1. Bollerslev, T.: Generalized autoregressive conditional heteroskedasticity. J. Econom. **31**(3), 307–327 (1986)
2. Box, G.E., Jenkins, G.M., Reinsel, G.C., Ljung, G.M.: Time Series Analysis: Forecasting and Control. Wiley, Hoboken (2015)
3. Chiang, W.C., Enke, D., Wu, T., Wang, R.: An adaptive stock index trading decision support system. Expert Syst. Appl. **59**, 195–207 (2016)
4. Donaldson, R.G., Kamstra, M.: An artificial neural network-garch model for international stock return volatility. J. Empirical Finance **4**(1), 17–46 (1997)
5. Enke, D., Thawornwong, S.: The use of data mining and neural networks for forecasting stock market returns. Expert Syst. Appl. **29**(4), 927–940 (2005)
6. Gudelek, M.U., Boluk, S.A., Ozbayoglu, A.M.: A deep learning based stock trading model with 2-D CNN trend detection. In: 2017 IEEE Symposium Series on Computational Intelligence (SSCI), pp. 1–8, November 2017. https://doi.org/10.1109/SSCI.2017.8285188

7. Gunduz, H., Yaslan, Y., Cataltepe, Z., Gunduz, H., Yaslan, Y., Cataltepe, Z.: Intraday prediction of Borsa Istanbul using convolutional neural networks and feature correlations. Knowl.-Based Syst. **137**, 138–148 (2017)
8. Guresen, E., Kayakutlu, G., Daim, T.U.: Using artificial neural network models in stock market index prediction. Expert Syst. Appl. **38**(8), 10389–10397 (2011)
9. Huang, C.L., Tsai, C.Y.: A hybrid SOFM-SVR with a filter-based feature selection for stock market forecasting. Expert Syst. Appl. **36**(2), 1529–1539 (2009)
10. Krizhevsky, A., Sutskever, I., Hinton, G.E.: Imagenet classification with deep convolutional neural networks. In: Advances in Neural Information Processing Systems, pp. 1097–1105 (2012)
11. Lee, M.C.: Using support vector machine with a hybrid feature selection method to the stock trend prediction. Expert Syst. Appl. **36**(8), 10896–10904 (2009)
12. Malagrino, L.S., Roman, N.T., Monteiro, A.M.: Forecasting stock market index daily direction: a Bayesian network approach. Expert Syst. Appl. **105**, 11–22 (2018)
13. Oriani, F.B., Coelho, G.P.: Evaluating the impact of technical indicators on stock forecasting. In: 2016 IEEE Symposium Series on Computational Intelligence (SSCI), pp. 1–8. IEEE (2016)
14. Reshef, D.N., Reshef, Y.A., Finucane, H.K., Grossman, S.R., McVean, G., Turnbaugh, P.J., Lander, E.S., Mitzenmacher, M., Sabeti, P.C.: Detecting novel associations in large datasets. Science **334**(6062), 1518 (2011)
15. Sezer, O.B., Ozbayoglu, A.M.: Algorithmic financial trading with deep convolutional neural networks: time series to image conversion approach. Appl. Soft Comput. **70**, 525–538 (2018)
16. Su, C.H., Cheng, C.H.: A Hybrid Fuzzy Time Series Model Based on ANFIS and Integrated Nonlinear Feature Selection Method for Forecasting stock. Elsevier Science Publishers B. V., Amsterdam (2016)
17. Swingler, K.: Applying Neural Networks: A Practical Guide. Morgan Kaufmann, Burlington (1996)
18. Tsantekidis, A., Passalis, N., Tefas, A., Kanniainen, J., Gabbouj, M., Iosifidis, A.: Forecasting stock prices from the limit order book using convolutional neural networks. In: Business Informatics, pp. 7–12 (2017)
19. Wen, Q., Yang, Z., Song, Y., Jia, P.: Automatic stock decision support system based on box theory and SVM algorithm. Expert Syst. Appl. Int. J. **37**(2), 1015–1022 (2010)

# Uplift Prediction with Dependent Feature Representation in Imbalanced Treatment and Control Conditions

Artem Betlei[1,2]([⊠]), Eustache Diemert[1], and Massih-Reza Amini[2]

[1] Criteo Research, Grenoble, France
{a.betlei,e.diemert}@criteo.com
[2] UGA/CNRS LIG, Grenoble, France
Massih-Reza.Amini@univ-grenoble-alpes.fr

**Abstract.** Uplift prediction concerns the causal impact of a treatment over individuals and it has attracted a lot of attention in the machine learning community these past years. In this paper, we consider a typical situation where the learner has access to an imbalanced treatment and control data collection affecting the performance of the existing approaches. Inspired from transfer and multi-task learning paradigms, our approach overcomes this problem by sharing the feature representation of observations. Furthermore, we provide a unified framework for the existing evaluation metrics and discuss their merits. Our experimental results, over a large-scale collection show the benefits of the proposed approaches.

**Keywords:** Uplift prediction · Causal inference
Digital advertising · Supervised learning

## 1 Introduction

Uplift prediction is mostly studied in digital advertising and personalized medicine. In the former, the treatment is exposure to different ads [7] while in the latter, the treatment is usually a medication [3]. In both cases the aim is to predict if the treatment over an individual *would be* more preferable or not.

The ultimate goal of uplift models is to lead a policy that makes decisions over future instances. Such a decision could, for example, be to focus the advertising budget on users on whom it will be the most profitable (and possibly less annoying). One can also imagine to use such a model as a building block for a reinforcement learning algorithm that would take advantage of the predict uplift to choose relevant actions given a state. In any of these applications it is essential to enforce a causal interpretation of the uplift model in order to inform further actions.

In this paper we focus on uplift approaches that would scale to industrial applications, especially in terms of learning and inference time.
Our main contributions are threefold.

© Springer Nature Switzerland AG 2018
L. Cheng et al. (Eds.): ICONIP 2018, LNCS 11305, pp. 47–57, 2018.
https://doi.org/10.1007/978-3-030-04221-9_5

1. First, we introduce two novel approaches that tackle the case of imbalanced treatment and control datasets and discuss their merits (Sect. 3)
2. we then provide a unified view of existing evaluation metrics and indicate which one should be preferred for a given application (Sect. 4.2)
3. Finally, we evaluate the proposed approaches on a real-life collection and produce palpable evidence of their practical usefulness (Sect. 5).

## 2   Problem Formulation

The causal uplift $U(x)$ is the expected difference indicating if an individual *should* take a treatment or not. We formalize it using Pearl's causal inference framework [5] as

$$U(x) = \mathbb{E}[Y|X = x, do(T = 1)] - \mathbb{E}[Y|X = x, do(T = 0)]. \tag{1}$$

Conversely, the conditional uplift $u(x)$ in Eq. 2 is the expected difference in outcome *when* the individual has taken the treatment or not: that is when we observe it after the fact:

$$u(x) = \mathbb{E}[Y|X = x, T = 1] - \mathbb{E}[Y|X = x, T = 0]. \tag{2}$$

Causal and conditional uplifts are equivalent if treatment was administered at random:

$$T \perp\!\!\!\perp X \Rightarrow U(x) \equiv u(x)$$

Note that it is always possible to learn a predictor of $u(x)$ using traditional approaches in supervised learning, even though we only observe treatment and outcome coming from a natural experiment. But in order to interpret the uplift predictions as causal (especially for taking actions like exposing users to ads or taking medicine) the model must be learned on data for which $U(x) \equiv u(x)$. Therefore we assume a dataset composed of i.i.d. samples of the joint covariates $X$, label $Y$ and treatment $T$ variables:

$$\mathcal{D} = \{X_i, Y_i, T_i\}_{i=1...n} \; ; \; T_i \perp\!\!\!\perp X_i, \forall i$$

Learning algorithms have access to $\mathcal{D}$ and can learn any distributions (we will see that there are multiple possible choices). We consider a binary outcome: at inference time the model performs a prediction of the form $\hat{u}(x) = \hat{P}_T(Y = 1|X = x) - \hat{P}_C(Y = 1|X = x)$, that is with the same $x$ the model predicts the difference between two potential outcomes, if the subject is treated or not, respectively.

## 3   Proposed Approaches

In this section, we present two uplift prediction methods for large-scale data case which attempt to take advantage of the relatedness of response in treatment and control groups during the learning process. We posit that this general idea is useful when the amount of data varies drastically between groups.

**Dependent Data Representation (DDR)** approach is based on a Classifier Chains method [9] originally developed for multi-label classification problems. The idea is that if there are $L$ different labels, one can build $L$ different classifiers, each of which solves the problem of binary classification and at the training process each next classifier uses predictions of the previous ones as extra features.

We use the same idea for our problem in two steps. At the beginning we train a first classifier on control data:

$$P_C = P(Y = 1 | X = x, T = 0),$$

then we use predictions $P_C$ as an extra feature for the classifier learning on the treatment data, effectively injecting a dependency between the two datasets:

$$P_T = P(Y = 1 | X = x, \hat{P}_C(x) = p, T = 1).$$

To obtain uplift for each individual we compute the difference:

$$\hat{u}^{DDR}(x) = \hat{P}_T(x, \hat{P}_C(x)) - \hat{P}_C(x)$$

Intuitively, the second classifier is learning the difference between the expected outcome in treatment and control, that is the uplift itself. Examination of the weights of this uplift classifier could also lead to interesting information on the role of different features in explaining the treatment outcome.

**Shared Data Representation (SDR)** approach for uplift prediction is based on a popular implementation of the multi-task framework [1]. A predictor is learned on a modified features representation that allows to learn related tasks jointly and with a single loss. We specialize this approach considering predicting outcomes in control and treatment groups as the related tasks.

The general form of the model is given by

$$P(Y | T = t, X = x) = f(\langle w_0, x \rangle + \mathbb{1}_{[t=1]} \langle w_t, x \rangle + \mathbb{1}_{[c=1]} \langle w_c, x \rangle) \qquad (3)$$

with $f$ an arbitrary link function. Practically speaking we augment the dataset by stacking the original features with a conjunction of the treatment group indicator and the same features. Letting $\mathbf{D}_T$ and $\mathbf{D}_C$ be the covariates from treatment and control groups respectively such that $\mathbf{D}_T \cup \mathbf{D}_C = \mathbf{D}$ we obtain the following shared learning representation:

$$\mathbf{D}_{train}^{SDR} = \begin{bmatrix} \mathbf{D}_T & \mathbf{D}_T & 0 \\ \mathbf{D}_C & 0 & \mathbf{D}_C \end{bmatrix}$$

So a single vector of weights $\mathbf{w}$ is learned jointly as $\mathbf{w} = [\mathbf{w}_0 \ \mathbf{w}_T \ \mathbf{w}_C]$ where $\mathbf{w}_0$ is a vector of weights that relate to the original features and $\mathbf{w}_T$ and $\mathbf{w}_C$ are corresponding to treatment/control conjunction features.

At inference time we compute the difference between predicted probabilities using two representations of the individual features, corresponding to the counterfactual outcomes:

$$\hat{u}^{SDR}(x) = \hat{P}(Y = 1 | \begin{bmatrix} x & x & 0 \end{bmatrix}) - \hat{P}(Y = 1 | \begin{bmatrix} x & 0 & x \end{bmatrix})$$

An advantage of this method is the possibility to assign different regularization penalties for $\mathbf{w}_0$ ($\lambda_0$) and $\mathbf{w}_T$ / $\mathbf{w}_C$ ($\lambda_1$). In this way it is possible to control the strength of the connection between the tasks. As reported by Chapelle, it is equivalent to rescaling the conjunction features by $\sqrt{\lambda_0 / \lambda_1}$. Intuitively this model allows to learn a common set of weights for predicting the global, average outcome whilst keeping enough capacity to express the peculiar influence of features in the treatment or control conditions.

## 4    Related Work

We review existing uplift prediction methods first to highlight links and differences with the proposed methods and will then proceed to the evaluation metrics.

### 4.1    Learning Approches

Here we describe current methods for uplift prediction and explain the advantages and drawbacks of them.

The most basic method to predict uplift is **Two-Model** method, which uses two separate probabilistic models - first one fits on treatment group and predicts probability $P_T(Y = 1|X)$ while second one uses control group and predicts $P_C(Y = 1|X)$. Uplift then can be computed as $\hat{u}^{2m}(x) = \hat{P}_T(Y = 1|X = x) - \hat{P}_C(Y = 1|X = x)$. For this method any classification model can be used and if both of classifiers perform well, uplift model will also perform highly. At the same time the main goal of the models is to predict outcomes separately but not exactly the uplift. In cases where the average response is low and/or noisy there is a risk that the difference of predictions would be very noisy too.

DDR can be seen as an extension of the two-model approach, the difference in interpretation is that adding an extra feature to the classifier learned on treatment group we add a knowledge about control group, thus we learn directly to transfer uplift to the unobserved, counter-factual case.

Jaskowski and Jaroszewicz [3] propose a **Class variable transformation or Revert Label** method for adapting standard classification models to the uplift case. Authors create a new label $Z$ as follows: $Z = YT + (1 - Y)(1 - T)$, and for uplift prediction in case of balanced treatment-control subgroups they obtain: $P_T(Y = 1|X) - P_C(Y = 1|X) = 2P(Z = 1|X) - 1$.

As in the two-model method, any classifier can be used to predict $P(Z = 1|X)$. New label $Z$ unites two subgroups with different outcomes, so it is not clear how difficult it would be for a model to find optimal weights, especially on the imbalanced outcomes case. Another drawback is that all capacity of a model is spent for direct uplift prediction, without any assumptions connected with treatment and control subgroups.

**Other methods** include some transformed variants of SVM [4,13] and tree-based algorithms [8,10,11]. SVM algorithms designed for uplift prediction have specific tasks such as construction of two separating hyperplanes instead of one or optimizing of ranking measure between pairs of examples. Tree-based methods incur finding splits in the data that optimize local variants of uplift. Both families of methods have in common that they are generally not trivial to scale in terms of either learning or inference time.

## 4.2 Metrics

We now describe the two major uplift metrics. As both are based on an ordering of samples according to their predicted uplift scores we assign following notations: for a given model let $\pi$ be the ordering of the dataset satisfying:

$$\hat{u}^{\pi}(x_i) \geq \hat{u}^{\pi}(x_j), \ \forall i < j$$

we note $\pi(k)$ the first $k$ samples sorted according to the descending predicted uplift $\hat{u}^{\pi}(x)$:

$$\pi(k) = \{d_i \in \mathcal{D}\}_{i=1,\ldots,k}$$

thus satisfying $\hat{u}(x_i) \geq \hat{u}(x_j), \forall i < j$ and $\hat{u}(x_l) \leq \hat{u}(x_i) \forall l > k, i \leq k$.

To define the uplift prediction performance let $R_{\pi}(k)$ be an amount of positive outcomes among the first $k$ data points:

$$R_{\pi}(k) = \sum_{d_i \in \pi(k)} \mathbb{1}[y_i = 1],$$

and we define $R_{\pi}^{T}(k)$ and $R_{\pi}^{C}(k)$ as the numbers of positive outcomes in the treatment and control groups respectively among the first $k$ data points:

$$R_{\pi}^{T}(k) = R_{\pi}(k)|T = 1, R_{\pi}^{C}(k) = R_{\pi}(k)|T = 0$$

To define a baseline performance let also $\bar{R}^{T}(k)$ and $\bar{R}^{C}(k)$ be the numbers of positive outcomes assuming a uniform distribution of positives:

$$\bar{R}^{T}(k) = k \cdot \mathbb{E}[Y|T = 1], \bar{R}^{C}(k) = k \cdot \mathbb{E}[Y|T = 0].$$

Finally, let $N_{\pi}^{T}(k)$ and $N_{\pi}^{C}(k)$ be the numbers of data points from treatment and control groups respectively among the first $k$.

**Area Under Uplift Curve** (*AUUC*) [4] is based on the *lift curves* [12] which represent the proportion of positive outcomes (the sensitivity) as a function of the percentage of the individuals selected. Uplift curve is defined as the difference in lift produced by a classifier between treatment and control groups, at a particular threshold percentage $k/n$ of all examples.

$AUUC$ is obtained by subtracting the respective Area Under Lift ($AUL$) curves:

$$AUUC_\pi(k) = AUL_\pi^T(k) - AUL_\pi^C(k) = \underbrace{\sum_{i=1}^{k}\left(R_\pi^T(i) - R_\pi^C(i)\right)}_{uplift} - \underbrace{\frac{k}{2}\left(\bar{R}^T(k) - \bar{R}^C(k)\right)}_{baseline} \quad (4)$$

The total $AUUC$ is then obtained by cumulative summation:

$$AUUC = \int_0^1 AUUC_\pi(\rho)d\rho \approx \frac{1}{n}\sum_{k=1}^{n} AUUC_\pi(k)dk \quad (5)$$

Uplift curves always start at zero and end at the difference in the total number of positive outcomes between subgroups. Higher $AUUC$ indicates an overall stronger differentiation of treatment and control groups.

**Qini coefficient** [7] or $Q$ is a generalization of the Gini coefficient for the uplift prediction problem. Similarly to $AUUC$ it is based on Qini curve, which shows the cumulative number of the incremental positive outcomes or uplift (vertical axis) as a function of the number of customers treated (horizontal axis). The formluation is as follows:

$$Q_\pi(k) = \underbrace{\sum_{i=1}^{k}\left(R_\pi^T(i) - R_\pi^C(i)\frac{N_\pi^T(k)}{N_\pi^C(k)}\right)}_{uplift} - \underbrace{\frac{k}{2}\left(\bar{R}^T(k) - \bar{R}^C(k)\right)}_{baseline} \quad (6)$$

A perfect model assigns higher scores to all treated individuals with positive outcomes than any individuals with negative outcomes. Thus at the beginning perfect model climbs at 45°, reflecting positive outcomes which are assumed to be caused by treatment. After that the graph proceeds horizontally and then climbs at 45° down due to the negative effect. In contrast, random targeting results in a diagonal line from $(0,0)$ to $(N, n)$ where $N$ is the population size and $n$ is the number of positive outcomes achieved if everyone is targeted. Real models usually fall somewhere between these two curves, forming a broadly convex curve above the diagonal. Given these curves we can now define the Qini coefficient $Q$ for binary outcomes as the ratio of the actual uplift gains curve above the diagonal to that of the optimum Qini curve:

$$Q_\pi = \frac{\sum_{k=1}^{n} Q_\pi(k)dk}{\sum_{k=1}^{n} Q_{\pi^*}(k)dk} \quad (7)$$

where $\pi^*$ relates for the optimal ordering. Therefore $Q$ theoretically lies in the range $[-1, 1]$.

**Choice of metric** for this task can seem unclear at first since both Eqs. 4 and 6 share the same high level form: a cumulative sum of uplifts in increasing share of the population penalized by subtracting a baseline corresponding to a random model.

A first difference is that Qini corrects uplifts of selected individuals with respect to the number of individuals in treatment/control using the $N_\pi^T(k)/N_\pi^C(k)$ factor. Imagine a model selecting majorly treated individuals at a given $k$. The uplift part of $AUUC(k)$ can be maximized by accurately selecting positive among treated, even if there is a large proportion of positives in selected control individuals. Contrarily, $Q(k)$ would penalize such a situation. We observe in practice that Qini tend to be harder to maximize but should be preferred for model selection as it is robust to this group selection effect. Also, given that at inference time uplift models are used to predict both counter-factual outcomes we should prefer a metric that evaluates accordingly.

A second advantage of Qini is that it is normalized (7) and thus more comparable when datasets are updated over time, a typical case in some applications. We report Qini metrics in the rest of this paper.

## 5    Experiments

In this section we define a benchmark for the experiments, present a comparison between proposed and other uplift prediction methods.

### 5.1    Benchmark

It is difficult to obtain data for learning an unbiased uplift prediction model (i.e. data from random treatment assignment). We only know of two unbiased, large scale datasets. **Hillstrom dataset** [2] contains results of an e-mail campaign for an Internet based retailer. The dataset contains information about 64,000 customers involved in an e-mail test who were randomly chosen to receive men's, women's merchandise e-mail campaigns or not receive an e-mail. We use the no-email vs women e-mail split with "visit" as outcome as in [8]. Our second dataset is **CRITEO-UPLIFT1**[1] which is constructed by assembling data resulting from incrementality tests, a particular randomized trial procedure where a random part of the population is prevented from being targeted by advertising. It consists of 25M rows, each one representing a user with 12 features, a treatment indicator and 2 labels (visits and conversions).

For the experiments we firstly preprocess datasets, specifically we binarize categorical variables and normalize the features, for the classification we use Logistic Regression model from `Scikit-Learn` [6] Python library as it has fast learning and inference processes. Then we do each experiment in the following way: we do 50 stratified random train/test splits both for treatment and control groups with a ratio 70/30, during learning process we tune parameters of each model on a grid search. For DDR and SDR we use the regularization trick that we explained earlier, we tune additional regularization terms on a grid search as well. To check statistical significance we use two-sample paired t-test at 5% confidence level (marked in bold in the tables when positive).

---

[1] this dataset will be released shortly at http://research.criteo.com/outreach.

## 5.2   Performance of Dependent Data Representation

We compare DDR with a Two-Model as first is an extention of the second, results are shown on Table 1. We use Hillstrom dataset with a "visit" outcome and cover three cases: firstly we compare approaches on a full dataset, then reduce control group randomly choosing 50% of it and for the last experiment we randomly choose 10% of control group to check how methods will perform with imbalanced data case. Indeed it is usually the case that the control group is kept to a minimum share so as not to hurt global treatment efficiency (e.g. ad revenue). As we can see, DDR significantly outperform Two-Model on imbalanced cases.

**Table 1.** Performances of Two-Model and DDR approaches measured as mean $Q$.

|            | Balanced T/C | Imbalanced T/C (50% of C group) | Highly imbalanced T/C (10% of C group) |
|------------|--------------|---------------------------------|----------------------------------------|
| Two model  | 0.06856      | 0.06292                         | 0.03979                                |
| DDR        | 0.06866      | **0.06444**                     | **0.04557**                            |

**Different directions of DDR.** As DDR approach is based on a consecutive learning of two classifiers, there are two ways of learning - to fit first model on treatment group and then use output as a feature for the second one and fit it on a control part (we denote it as $T \to C$), or vice versa ($C \to T$). Table 2 indicates that both approaches are comparable in the balanced case but $C \to T$ direction is preferable in other cases (at least with this dataset). Since the test set has more treated examples it makes sense that the stronger predictor obtained on this group by using information from predicted uplift on control performs best.

**Table 2.** Comparison of directions of learning in DDR approach ($Q$).

|                    | Balanced T/C | Imbalanced T/C (50% of C group) | Highly imbalanced T/C (10% of C group) |
|--------------------|--------------|---------------------------------|----------------------------------------|
| DDR ($T \to C$)    | 0.06895      | 0.06394                         | 0.03979                                |
| DDR ($C \to T$)    | 0.06866      | 0.06444                         | 0.04557                                |

**Complexity of Treatment Effect with DDR**

To investigate complexity of the link between treatment and control group we use a dummy classifier (predicting the average within-group response) successively for one of treatment or control group while still using the regular model for the remaining group. Intuitively if the treatment effect is a constant, additive uplift then a simple re-calibration using a dummy model should be good enough. Conversely if there is a rich interaction between feature and treatment to explain outcome a second, a dummy classifier would perform poorly. Table 3 indicates that the rich interaction hypothesis seems more plausible in this case, with maybe an even richer one in treated case.

**Table 3.** Comparison between different variants of DDR approach.

|                        | Balanced T/C |
|------------------------|--------------|
| DDR                    | 0.06866      |
| DDR (dummy for C group)| 0.04246      |
| DDR (dummy for T group)| 0.01712      |

## 5.3 Performance of Shared Data Representation

Here we compare SDR approach with Revert Label because of a similar nature of the uplift prediction. Revert Label model is learned with samples reweighting as in the original paper. Table 4 indicates that SDR significantly outperforms Revert Label on imbalanced cases. Note that due to heavy down-sampling in the imbalanced cases it is not trivial to compare $Q$ values between columns.

**Table 4.** Performances of Revert Label and SDR approaches measured as mean $Q$.

|              | Balanced T/C | Imbalanced T/C (50% of C group) | Highly imbalanced T/C (10% of C group) |
|--------------|--------------|----------------------------------|------------------------------------------|
| Revert label | 0.06879      | 0.06450                          | 0.05518                                  |
| SDR          | 0.06967      | **0.06945**                      | **0.08842**                              |

### Usefullness of Conjunction Features

In order to check usefulness of conjunctions features with SDR we compare it with a trivial variant in which we simply add an indicator variable for treatment instead of the whole feature set. This allows the model to learn only a simple re-calibration of the prediction for treated/control. Table 5 indicates that it strongly degrades model performance.

**Table 5.** Comparison between variants of SDR in balanced treatment/control conditions.

|   | SDR (standard) | SDR (T/C indicator) |
|---|----------------|---------------------|
| Q | **0.06967**    | 0.02706             |

**Performance in Imbalanced Outcome Condition**
We also compare SDR approach with Revert Label on CRITEO-UPLIFT1 dataset with conversion as outcome on a random sample of 50,000. Ratio between C and T group is 0.18 so it is highly imbalanced case as well but the outcome is also imbalanced with average level at only .00229. Table 6 indicates that SDR again significantly outperforms Revert Label in this setting.

**Table 6.** Performances of Revert Label and SDR in highly imbalanced conditions for both treatment and outcome.

|   | Revert Label | SDR |
|---|---|---|
| Q | 0.25680 | **0.54228** |

# 6    Conclusion

We proposed two new approaches for the Uplift Prediction problem based on dependent and shared data representations. Experiments show that they outperform current methods in imbalanced treatment conditions. In particular they allow to learn rich interaction between the features and treatment to explain response. Future research would include learning more complex (highly non-linear) data representations permitting even richer interactions between features and treatment.

# References

1. Chapelle, O., Manavoglu, E., Rosales, R.: Simple and scalable response prediction for display advertising. ACM Trans. Intell. Syst. Technol. **5**(4), 61:1–61:34 (2014)
2. Hillstrom, K.: The MineThatData e-mail analytics and data mining challenge (2008)
3. Jaskowski, M., Jaroszewicz, S.: Uplift modeling for clinical trial data. In: ICML Workshop on Clinical Data Analysis (2012)
4. Kuusisto, F., Costa, V.S., Nassif, H., Burnside, E., Page, D., Shavlik, J.: Support vector machines for differential prediction. In: Calders, T., Esposito, F., Hüllermeier, E., Meo, R. (eds.) ECML PKDD 2014. LNCS (LNAI), vol. 8725, pp. 50–65. Springer, Heidelberg (2014). https://doi.org/10.1007/978-3-662-44851-9_4
5. Pearl, J.: Causality: Models, Reasoning, and Inference. Cambridge University Press, New York (2000)
6. Pedregosa, F., et al.: Scikit-learn: machine learning in Python. J. Mach. Learn. Res. **12**, 2825–2830 (2011)
7. Radcliffe, N.J.: Using control groups to target on predicted lift: building and assessing uplift model. Direct Mark. Anal. J. **3**, 14–21 (2007)
8. Radcliffe, N.J., Surry, P.D.: Real-world uplift modelling with significance-based uplift trees (2011)

9. Read, J., Pfahringer, B., Holmes, G., Frank, E.: Classifier chains for multi-label classification. Mach. Learn. **85**(3), 333 (2011)
10. Rzepakowski, P., Jaroszewicz, S.: Decision trees for uplift modeling with single and multiple treatments. Knowl. Inf. Syst. **32**, 303–327 (2012)
11. Jaroszewicz, S.S.M., Rzepakowski, P.: Ensemble methods for uplift modeling. Data Min. Knowl. Discov. **29**(6), 1531–1559 (2015)
12. Tufféry, S.: Data Mining and Statistics for Decision Making (2011)
13. Zaniewicz, L., Jaroszewicz, S.: Support vector machines for uplift modeling. In: Proceedings of the 2013 IEEE 13th International Conference on Data Mining Workshops, ICDMW 2013, Washington, DC, USA, pp. 131–138 (2013)

# Applying Macroclimatic Variables to Improve Flow Rate Forecasting Using Neural Networks Techniques

Breno Santos[✉], Bruna Aguiar, and Mêuser Valença

UPE, ECOMP, Benfica St. 455, Recife, PE 50720-001, Brazil
{bss2,bcga,meuser}@ecomp.poli.br
http://w2.portais.atrio.scire.net.br/upe-ppgec/index.php/en/

**Abstract.** Since 2013, the São Francisco River has being through a low hydraulicity period. In other words, the rain intensity is below average. Consequently, it has being necessary to operate at a minimal flow rate. It is far below the ones established at the operation licence, which is 1300 $m^3/s$. Due to this hydraulic crisis, the actual operational flow rate is 700 $m^3/s$ at São Francisco River, characterizing this situation as critical. In this work, it was proposed to use Reservoir Computing (RC), Long Short Term Memory (LSTM) and Deep Learning to predict Sobradinho's flow rate for 1, 2 and 3 months ahead using macroclimatic variables. After having the results for each one of them, a comparison was made and statistical tests where executed for evaluation.

**Keywords:** Artificial Neural Network · ANN · Reservoir Computing RC · Long Short Term Memory · LSTM · Deep learning · Flow rate Prediction · Forecast · São Francisco River · Macrolimatic variables

## 1 Introduction

A water reservoir is used to store water from wet periods, when the natural affluence is greater than the demand, to be used in the dry periods, when it is low compared to the demand. Thus, the decision process in operation of reservoirs seeks to establish the optimal value of the volume of water to be withdrawn from the reservoir at each time of operation [1]. In order for the operation of these reservoirs to be efficient, it is essential to know in advance the volume of water expected in that period so that it is possible to calculate the flow required in relation to the demand.

Flow forecasting is a key tool in water resource studies, since the models used to perform this task provide estimates of the future flow of water from the reservoir. A flow forecast is an important component for sustainable river basin planning and management [2]. With information of this type, it is possible to minimize the damage caused by extreme changes, such as floods or droughts, and to optimize the generation of energy in a hydroelectric plant, for example.

© Springer Nature Switzerland AG 2018
L. Cheng et al. (Eds.): ICONIP 2018, LNCS 11305, pp. 58–69, 2018.
https://doi.org/10.1007/978-3-030-04221-9_6

The control of water storage for electric power generation in the Brazilian Northeast is concentrated in two accumulation dams situated on the São Francisco River, which are the Três Marias and Sobradinho Power Plants [3]. Since 2013, the São Francisco River has been undergoing a period of low hydraulicity, that is, the rainfall intensity is below average. This has made it necessary to practice minimum flows much lower than the planning established in the operating license, which is 1300 m$^3$/s. Due to this hydraulic crisis, it is practicing in the lower São Francisco flows of the order of 700 m$^3$/s, characterizing this situation as critical [4]. In this way, it is of fundamental importance for the managers of the basin to have a forecast of flow more accurate and as advanced as possible.

In addition, since macroclimatic information has been used as an indicator element of critical situations for several river basins, including for the São Francisco River basin, they will also be used in this work. These variables present relevant climate information such as precipitation rate and air temperature, for example.

Thus, in order to contribute to this scenario, this work proposes to investigate and analyze the performance in the prediction of flow rate using Artificial Neural Network (RNA) techniques together with macroclimatic variables. Subsequently, the results will be compared to each other and it will be possible to know what kind of architecture might be suitable to solve the problem.

## 2  Macrolimatic Variables

In this section, each selected macroclimatic variable will be presented, highlighting the definitions, importance and weather impact.

### 2.1  Air Temperature

The air temperature is a measure to define if the air is hot or cold. It is the most common measure in climatology. It describes the kinetic energy of the air gases. As the air molecules move faster, the air temperature increases. It is commonly measured in Fahrenheit (°F) or Celsius (°C) [5].

The air temperature of a given location is the result of a momentary combination of certain factors. In the urban environment, the spatial and temporal scale of the factors involved in the climatic configuration presents particularities derived from both the greater heterogeneity related to the use and occupation of the soil, and the greater speed and diversity of human activities in relation to the agricultural and rural environment [6].

The knowledge of air temperature is fundamental in several areas of research, especially in meteorology, oceanography, climatology and hydrology [7]. The air temperature acts on the evapotranspiration process, due to the fact that the solar radiation absorbed by the atmosphere and the heat emitted by the cultivated surface raise the air temperature [8]. It affects the rate of evaporation, relative humidity, precipitation, and wind speed/direction, and thus impacts the amount of water that will arrive at the [5] power plants.

## 2.2   Outgoing Longwave Radiation (OLR)

Virtually all energy in the Earth's system comes from solar radiation (Short-wave Radiation), there being a near-perfect balance between incident solar radiation and Earth-to-space radiation (Long-wave Radiation) [9]. Thus, Outgoing Longwave Radiation (OLR) is an electromagnetic radiation in the form of heat that is emitted into space by the planet Earth and its atmosphere. Its energy is measured in $W/m^2$ [10].

OLR at a given site is affected primarily by surface temperature, surface spectral emissivity, atmospheric vertical temperature and water vapor. In the heights, they would be the spectral emissivity of several layers of clouds [10].

It is believed that precipitation is directly related to OLR, since for the tropical region, where the Sea Surface Temperatures (SST) varies modestly over the annual cycle, the greatest OLR variations result from changes in the amount and height of the clouds. This direct connection with clouds caused OLR to be used to quantitatively estimate precipitation [9]. This was the main reason for adding this variable in this work.

## 2.3   Precipitation Rate

Precipitation is the phase of this cycle responsible for transporting the waters from the atmosphere to the earth's surface. The quantitative knowledge of its spatial variability over the regions, or watersheds, must be understood as essential to the efficient planning and management of water resources [11].

The availability of precipitation in a basin during the year is a determining factor to quantify, among others, the need for crop irrigation, domestic/industrial water supply and hydroelectric power generation. Determination of precipitation intensity is important for flood control and soil erosion. Due to its capacity to produce runoff, rainfall is the most important type of precipitation for hydrology [12].

## 2.4   Sea Surface Temperature (SST)

Sea surface temperature is a widely used indicator in the atmospheric sciences for analyzing patterns of climate variability. Its measurement is carried out through readings of photographs captured by satellites, boats and meteorological buoys (both on the surface and submerged), infrared or photographs made by aircraft, among others. These measures are interpreted, analyzed, interpolated, re-analyzed and published in the public domain for the use of atmospheric science institutions found throughout the world [13].

The ocean interferes with the climate system because of its large capacity to store heat, and, like the atmosphere, it distributes energy from the solar radiation that reaches the equator towards the poles. The variability patterns of sea surface temperature at interannual and interdecadal time scales are the result of the combination of oceanic-atmospheric coupled processes [13].

# 3   Neural Network Techniques

In general, it is possible to define an Artificial Neural Network (ANN) as a system built by interconnected processing elements, called neurons, which are arranged in layers (one input layer, one or many intermediate layers and one output layer) and are responsible for the network non-linearity and memory [14].

ANN is a popular method among the machine learning based load forecasting methods. There are several ANN based forecasting methods reported in the literature [15].

For this research, three neural network techniques were selected: Reservoir Computing (RC), Long Short Term Memory (LSTM) and Deep Learning. Despite being all neural networks, each one of them has its own peculiarity and will be explained in this section.

## 3.1   Reservoir Computing (RC)

Additionally to feedforward models, for instance the Multi-Layer Perceptron, largely applied in time-series prediction, Recurrent Neural Networks (RNR) began to appear. In this network architecture, recurring connections were implemented to the existing feedforward topology. As a result of the presence of these connections, the system becomes more complex, dynamic and even more adequate for solving temporal problems [16].

In 2001, suggested by Wolfgang Maass under the name Liquid State Machine (LSM) and Jaeger [17] with the name Echo State Networks (ESN), a new proposal for RNR design and training [18] was divulged. Verstraeten then proposed the combination of these two approaches in a single term called Reservoir Computing (RC) [19].

The reservoir and the linear output layer are the main parts of a RC system.

With a recurring topology of processing nodes, reservoir is a non-linear dynamic system. The connections are randomly generated and are globally rescaled aiming to achieve a suitable dynamic state. Due to the fixed weights of the reservoir, the training is not necessary for the reservoir layer. That is an important property of this architecture. The training occurs only at the output layer and for this reason it has an output function which might be a linear classifier or a regression algorithm, for example [19].

Capable of internally creating the memory needed to store the history of the input patterns through their recurring connections, RNRs are a powerful computational model [16]. This particularity is interesting for practical problems involving time series, since in this type of scenario the historical values are essential for the predictions that one wishes to make. Thus, if RNRs are capable of storing these values, they theoretically present better results than if another technique were used without this feature. For this reason, this technique was chosen in this work to be used in the comparison of the results.

## 3.2   Long Short-Term Memory (LSTM)

The Long Short-Term Memory, or LSTM, is a recurring neural network trained using backpropagation through time (BPTT), which is the application of the backpropagation algorithm for the RNRs applied in time series [20], and overcomes the problem of gradient disappearance. This problem occurs when each weight of the neural network receives an update proportional to the error and this update is very small, causing that the weight does not change. In the worst case, this may even impede the learning of the neural network. Thus, by overcoming this problem, LSTM can be used to create large recurring networks which, in turn, can be used to solve difficult sequence problems in machine learning [21].

Unlike other networks that have neurons, LSTMs have memory blocks connected by their layers. Each block has components that make it smarter than a classic neuron plus a memory for recent sequences. This block contains gates that administer its state and its output. It operates on an input sequence and each port within a block uses sigmoid activation units to control whether they are triggered or not, causing the state change and addition of information to flow through the conditional block [21].

Gates within each unit:

- Memory Gate: decides what information from the block should be disregarded. If it is closed, no old memory will be kept. If it is completely open, all the old memory will pass on;
- Input Gate: decides which input values should update the memory, i.e. how much of the new memory should influence the old memory;
- Output Gate: decides what should be generated from the input and the memory block.

Each unit is like a mini state machine where the unit doors have weights that are calculated during the training procedure. In this way, sophisticated learning and memory can be obtained from only one layer of LSTMs [21].

The LSTM can handle noise, distributed representations and continuous values. In contrast to finite-state automata or hidden Markov models, the LSTM does not require an a priori choice of a finite number of states. In principle, it can handle unlimited state numbers. For this network, there seems to be no need for fine-tuning parameters. It works well on a wide range of parameters such as learning rate, input gate bias, and bias of the output gate. In addition, constant error propagation within the memory cells results in the LSTM's ability to fix very long delays [22].

Taking into account the facts mentioned above, it became relevant to choose the LSTM for the possible solution of the problem presented in this work.

## 3.3   Deep Learning

Since 2006, deep structured learning, more commonly called deep learning or hierarchical learning, has emerged as a new area of research in machine learning. Over the last few years, techniques developed from deep learning research are already impacting a large amount of information and signal processing work, increasing scopes that include key aspects of machine learning and artificial intelligence [23].

Deep learning can be defined as a class of machine learning techniques that exploits multiple layers of nonlinear information processing for supervised or unsupervised extraction and transformation of attributes, pattern analysis, and classification [23].

Modern deep learning provides a powerful framework for supervised learning. By adding more layers and more units to a layer, a deep network can represent functions of increasing complexity [24].

However, they are often used with inadequate approximations in inference, learning, prediction, and topology design, all stemming from the intractability inherent of these tasks for most real-world applications [23].

Deep learning is based on the philosophy of connectionism: while an individual biological neuron or an individual characteristic in a machine learning model is not intelligent, a large population of these neurons or characteristics acting together can expose intelligent behavior. It is important to emphasize the fact that the number of neurons must be large. One of the major factors responsible for improving the accuracy of the neural network and for improving the complexity of the tasks that can solve between the 1980s and today is the dramatic increase in the size of the networks we use. Because the size of neural networks is critical, deep learning requires high-performance hardware and software infrastructure [24].

Considering the properties of the deep networks, it was decided to integrate this model in the comparative of this work in order to verify if this type of network is able to extract the necessary characteristics to obtain an accurate forecast with the use of macroclimatic variables.

## 4 Methodology

### 4.1 Database

The data bases used in the experiments were created from the monthly average Sobradinho flow, provided by the Brazilian Companhia Hidroelétrica do São Francisco (Chesf) or Hydroelectric Company of San Francisco, and the macroclimatic variables extracted from the National Oceanic and Atmospheric Administration (NOAA) [25]. Chesf is a subsidiary company of Eletrobras [26] and its main activity is the generation, transmission and sale of electric power [27]. NOAA is an American scientific agency within the US Department of Commerce that focuses on the conditions of the oceans, the main waterways and the atmosphere [25].

Since data sets might contain redundant or irrelevant information and inconsistent formats of data may disturb the extraction process [28], variable selection techniques were executed. To select the macroclimatic variables and their geographical locations that are most related to our problem, a search was performed based on the Linear Correlation [29] and Entropy (Random Forest algorithm

[30]). The variables selected and the number of filtered locations are described in Table 1. Both Chesf and NOAA data are monthly from January 1948 through December 2016.

**Table 1.** Macroclimatic variables

| Variable name | Locations |
|---|---|
| Air temperature | 2592 |
| Outgoing Longwave Radiation (OLR) | 2592 |
| Precipitation rate | 2295 |
| Sea Surface Temperatures (SST) | 2281 |

Several databases were built for the experiments, each with Sobradinho flow and different combinations of the 4 selected variables, resulting in 15 different bases. For example, one base showed the flow with only the air temperature, another contained the flow, air temperature and OLR, and so on.

In addition to the macroclimatic variables, the last 4 Sobradinho outflows were also used to predict the next 3 months.

## 4.2  Pre-processing of Data

Normalization of values is the first step in the stage of pre-processing data. It avoids the high values from overly affecting the ANN's calculations and at the same time it prevents low values going unseen. It is important to assure that the variables at distinct intervals are presented with the same consideration for training. Furthermore, the variables values should be proportionally adapted to the borderline of the activation function applied in the output layer. If it is the logistic sigmoid, the values are defined between [0 and 1], later the data are commonly normalized between [0.10 and 0.90] and [0.15 to 0.85] [14].

To calculate the normalized value, the following equation is used:

$$y = \frac{(b-a)(x_i - x_{min})}{(x_{max} - x_{min})} + a \tag{1}$$

where y = output normalized value; a e b = limits chosen; $x_i$ = original value; $x_{min}$ = minimum value of x and $x_{max}$ = maximum value of x. The chosen limits were a = 0.15 e b = 0.85.

## 4.3  Measure Network Performance

The Mean Absolute Percentage Error (MAPE) was defined to measure the network performance in this work. MAPE expresses accuracy as a percentage:

$$MAPE = \frac{100\%}{n} \sum_{t=1}^{n} \left| \frac{A_t - F_t}{A_t} \right| \tag{2}$$

where $A_t$ = actual value; $F_t$ = forecast value and n = number of fitted points.

## 4.4   Neural Networks Parameters

Each neural network used in this work has several parameters that require configuration. These settings may not be considered ideal due to the recentness of this field of research and is often performed empirically. Proper weights initialization will place the weights close to a good solution with reduced training time and increase the possibility of reaching a good solution [31].

The number of entries is the same for all topologies in order to run statistical tests in the future. This number is the sum of all the geographical positions described in Table 1 that are related to the variables used in the chosen base and the flows of the previous 4 months. For example, if the base in question is the one that uses precipitation and air temperature, there will be: 2592 + 2295 + 4 = 4891 entries. The number of outputs is related to the number of months to be predicted, which in this case are 3. All the ANN techniques used in this work will present this same number of neurons in the output layer.

**RC.** During this work, the RC algorithm used was developed by students of the University of Pernambuco (UPE).

RC network parameters:

- Number of neurons in the reservoir: 100;
- Activation function of the reservoir: logistic sigmoid;
- Activation function of the output layer: linear;
- Initialization of weights: warm up;
- Connection rate of the reservoir: 20%;
- Number of warm up cycles: 10 cycles;
- Stopping criterion: cross-validation (50% training, 25% cross-validation and 25% testing).

**LSTM.** The LSTM implemented in this work is based on the Brownlee [21] model created with Keras [32] in Python [33]. It has been adapted to support intermediate layers and increase the number of neurons in the output layer.

LSTM network parameters:

- Number of neurons in each intermediate layer: for the first intermediate layer is 512 and for each additional layer in the model, this number is reduced by half the previous layer;
- Initialization of weights: random;
- Batch size: 100;
- Activation function: Adam optimizer [34];
- Stopping criterion: 50 epochs.

**Deep Learning.** The deep network implemented in this work is based on the Heinz [35] model which is a deep MLP created with Tensorflow [36] in Python [33]. The model was adapted to support multiple intermediate layers and increase the number of neurons in the output layer.

Deep learning network parameters:

- Number of neurons in each intermediate layer: for the first intermediate layer is 1024 and for each additional intermediate layer in the model, this number is reduced by half of the previous one;
- Initialization of weights: TensorFlow's initializer (*tf.variance_scaling_initializer()*) which is the default bootstrap strategy since the distribution is uniform [36];
- Activation function: Adam optimizer [34];
- Stopping criterion: 1000 epochs.

### 4.5   Statistical Tests

To evaluate which one of the techniques has the lowest MAPE when predicting flow rate or even if they are statistically equivalent, statistical tests were performed on each neural network result after 30 training cycle [37].

The Wilcoxon Rank-Sum test was chosen, among various tests in the literature, as a result of being a non-parametric statistical hypothesis test that can be used to determine whether two dependent samples were selected from populations having the same distribution.

## 5   Results

For each neural network architecture chosen in this work, a ranking of the medians of the 30 calculated MAPE was created. The best results for all the databases were those that presented only one macroclimatic variable at a time. Whenever a second, third or fourth variable was added to the base, MAPE increased.

Among those results, the ones that presented lower MAPE were those that only used the precipitation rate as macroclimatic entry. The best results for each network can be observed in Table 2.

**Table 2.** MAPE results

| Network | Variable combination | MAPE median |
|---------|---------------------|-------------|
| RC      | Precipitation rate  | 8.67%       |
| LSTM    | Precipitation rate  | 10.43%      |
| DeepMLP | Precipitation rate  | 6.59%       |

The Wilcoxon test was run using the software R [38], which uses a significance level of 0.05 and the result pointed out that p-value is much smaller than the level of significance, as shown in Table 3.

Thus, the hypothesis which supports that all the architectures are considered statistically equivalent is rejected.

**Table 3.** P-value results

| Networks | P-value |
|---|---|
| RC e LSTM | $5.55^{-10}$ |
| RC e DeepMLP | $1.10^{-6}$ |
| LSTM e DeepMLP | $4.98^{-9}$ |

As the results are not equivalent, the best will be the one with the lowest MAPE. In this case, it is also the one that presents the lowest median of MAPE: the DeepMLP. Since the output is a forecast to predict 1, 2 and 3 months ahead and they are very similar, we will only display the comparison graphs for the 3 months ahead, as it is shown in Fig. 1.

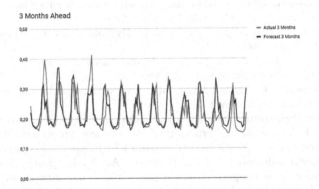

**Fig. 1.** Sobradinho's 3 months ahead forecast using precipitation rate and DeepMLP architecture.

## 6   Conclusion and Future Work

The objective of this work was to predict Sobradinho flow for one, two and three months ahead using three different neural network techniques combined with macroclimatic variables. In order to reach this objective, a Reservoir Computing previously implemented by the students of the University of Pernambuco was used as well as an LSTM (Keras) and a deep learning model (Tensorflow) that were implemented during this research.

In addition, a database provided by Chesf was used in combination with the macroclimatic variables collected by the NOAA. After a considerable number of simulations were performed for each neural network model, the results were statistically compared. The results proved that the lowest MAPE was indeed achieved by the DeepMLP.

As future work, an improvement on the parameters' selection of all networks can be performed in order to find greater settings. More accurate values, as well as other macroclimatic variables, can positively impact the network performance (specially for LSTM). To achieve lower MAPE values it is possible to consider changing the weight initialization technique, trying other optimizers for an improved convergence and adjusting both depth and wideness of the network.

It is also important to investigate the phenomenon of the accuracy decreasing when more variables are simultaneously added to the database. Finally, it is relevant to apply this methodology to other databases aiming to determinate if the deep model can still achieve the same results.

# References

1. Valença, M.J.S., Ludermir, T.B., Valença, A., Vasconcelos, I.: Sistema de Apoio a Decisão para a Operação Hidráulica de Sobradinho Incorporando Tendências Macro-Climáticas Utilizando Redes Neurais (2009)
2. Brooks, K.N., Ffolliott, P.F., Magner, J.A.: Hydrology and the Management of Watersheds, 3 edn. (2003)
3. Valença, M.J.S.: Proposta de Alerta Ambiental baseada na Previsão de Volumes Máximos Afluentes ao reservatório de Sobradinho
4. Agência Nacional de Águas (ANA): http://www2.ana.gov.br/Paginas/servicos/saladesituacao/v2/saofrancisco.aspx/
5. What is Air Temperature? https://www.fondriest.com/news/airtemperature.htm
6. Mendonça, F., Dubreuil, V.: Termografiada superfície e temperatura do ar na rmc (regiãometropolitana de curitiba/pr). Editora UFPR, pp. 25–35 (2005)
7. Cavalcanti, E.P., de Silva, V.P.R., de Sousa, A.S.F.: Programa computacional para a estimativa da temperatura do ar para a região nordeste do Brazil. Revista Brasileira de Engenharia Agrícola e Ambiental, pp. 140–147 (2006)
8. Embrapa:     http://www.agencia.cnptia.embrapa.br/Agencia22/AG01/arvore/AG01_79_24112005115223.html
9. Bomventi, T.N., Wainer, I.E.K.C., Taschetto, A.S.: Relação entre a radiação de onda longa, precipitação e temperatura da superfície do mar no oceano atlântico tropical. Revista Brasileira de Geofísica **24**, 513–524, 12 2006. http://www.scielo.br/scielo.php?script=sci_arttext&pid=S0102-261X2006000400005&nrm=iso
10. Susskind, J., Molnar, G., Iredell, L.: Contributions to Climate Research Using the AIRS Science Team Version-5 Products (2011)
11. de Stefano Ermenegildo, L.F., Pereira, S.B., Arai, F.K., Rosa, D.B.C.J.: Vazão específica e precipitação média na bacia do ivinhema. Dourados, pp. 428–432 (2012)
12. Tucci, C.E.M.: Hidrologia: ciência e aplicação. ABRH, 3 edn. (2002)
13. Rodrigues, A.L.: Informações macroclimáticas aplicadas na previsão de vazões. São Paulo (2016)
14. Valença, M.J.S.: Fundamentos das Redes Neurais. Livro Rápido, Brazil (2016)
15. Ye Rena, P.N., Suganthana, N.S., Amaratungac, G.: Random vector functional link network for short-term electricity load demand forecasting. Inf. Sci. **367–368**, 1078–1093 (2016)
16. Ferreira, A.A., Ludermir, T.B.: Um Método para Design e Treinamento de Reservoir Computing Aplicado à Previsão de Séries Temporais. Pernambuco (2011)
17. Jaeger, H.: The "echo state" approach to analyzing and training recurrent neural networks (2011)

18. Mass, W., Natschläger, T., Markram, H.: Real-time computing without stable states: a new framework for neural computation based on perturbations. Neural Comput. **14**, 2531–2560 (2002)
19. Verstraeten, D.: Reservoir computing: computation with dynamical systems. Belgium (2009)
20. A Gentle Introduction to Backpropagation Through Time. https://machinelearningmastery.com/gentle-introduction-backpropagation-time/
21. Time Series Prediction with LSTM Recurrent Neural Networks in Python with Keras. https://machinelearningmastery.com/time-series-prediction-lstm-recurrent-neural-networks-python-keras/
22. Hochreiter, S., Schmidhuber, J.: Long short-term memory. Neural Comput. 1735–1780 (1997)
23. Deng, L., Yu, D.: Deep Learning Methods and Applications, Foundations and Trends in Signal Processing, 7th edn. Now Publishers Inc., Boston (2014)
24. Goodfellow, I., Bengio, Y., Courville, A.: Deep Learning. MIT Press, Cambridge (2016). http://www.deeplearningbook.org
25. National Oceanic and Atmospheric Administration (NOAA). http://www.noaa.gov/
26. Eletrobras. http://eletrobras.com/pt/Paginas/home.aspx
27. Companhia Hidroelétrica do São Francisco (Chesf). http://www.chesf.gov.br
28. Wang, L., Fu, X.: Data Mining with Computational Intelligence. Springer, Heidelberg (2005). https://doi.org/10.1007/3-540-28803-1
29. Hartshorn, S.: Linear Regression And Correlation: A Beginner's Guide, 1st edn. Amazon Digital Services LLC, Seattle (2017)
30. Hartshorn, S.: Machine Learning With Random Forests And Decision Trees: A Visual Guide For Beginners, 1st edn. Amazon Digital Services LLC, Seattle (2016)
31. Kok Keong Teo, L.W., Lin, Z.: Wavelet packet multi-layer perceptron for chaotic time series prediction: effects of weight initialization. In: International Conference on Computational Science, vol. 2074, pp. 310–317, 7 2001
32. Keras. https://keras.io/
33. Python. https://www.python.org/
34. Kingma, D.P., Ba, J.L.: Adam: a method for stochastic optimization. In: International Conference for Learning Representations (2014)
35. A simple deep learning model for stock price prediction using TensorFlow. https://medium.com/mlreview/a-simple-deep-learning-model-for-stock-price-prediction-using-tensorflow-30505541d877
36. Tensorflow. https://www.tensorflow.org/
37. Moreno, A.M., Juristo, N.: Basics of Software Engineering Experimentation. Kluwer Academic Publisher, Dordrecht (2001)
38. Venables, W.N., Smith, D.M., the R Core Team: An Introduction to R. Network Theory Ltd, Hershey (2009)

# Predicting Functional Interactions Among DNA-Binding Proteins

Matloob Khushi[1,2,3,4](✉), Nazim Choudhury[3], Jonathan W. Arthur[2], Christine L. Clarke[4], and J. Dinny Graham[4,5]

[1] School of Information Technology, The University of Sydney, Sydney, NSW 2006, Australia
mkhushi@uni.sydney.edu.au
[2] Children's Medical Research Institute, The University of Sydney, Westmead, NSW 2145, Australia
[3] Faculty of Engineering and IT, The University of Sydney, Sydney, NSW 2006, Australia
[4] Centre for Cancer Research, The Westmead Institute for Medical Research, The University of Sydney, Westmead, NSW 2145, Australia
[5] Westmead Breast Cancer Institute, Westmead Hospital, Westmead, NSW 2145, Australia

**Abstract.** Perturbation of the binding pattern of one or more DNA-binding proteins, called transcription factors, plays a role in many diseases including, but not limited to, cancer. This has prompted efforts to characterise transcription cofactors i.e., transcription factors that work together to regulate gene expression. The Overlap Correlation Value (OCV), ranging from 0 (no correlation) to 1 (highly correlated), has been previously reported as a measure of the statistical significance in the overlap of binding sites of two transcription factors and thus a measure of the extent to which they may act as cofactors. In this study, we examined the variation in the OCV due to the peak caller employed to identify transcription factor binding sites. We identified that the significance of correlation between two transcription factors was unaffected by the peak-caller employed to identify transcription factor binding sites (Spearman R = 0.98). Furthermore, we used OCV measurements to develop a novel network map to study the correlation between twelve breast cancer cell-line datasets. Our proposed novel map revealed that transcription factor FOXA1 influenced the binding of six other transcription factors: JUND, P300, estrogen receptor alpha (ERα), GATA3, progesterone receptor (PR), and XBP1. Our model identified that binding sites that were targeted by PR were different under progesterone agonist (R5020 or ORG2058) or antagonist (RU486) treatment. Interestingly ERα had a significant OCV with PR when stimulated by anti-progestin, while it showed no significant overlap with PR when simulated with progestin. Our proposed network map drawn using OCV measurements is feature rich, more meaningful, and is better interpretable then using Venn diagram. The network map can be used in all scientific domains.

**Keywords:** Bioinformatics · DNA-binding proteins · Transcription factors

© Springer Nature Switzerland AG 2018
L. Cheng et al. (Eds.): ICONIP 2018, LNCS 11305, pp. 70–80, 2018.
https://doi.org/10.1007/978-3-030-04221-9_7

# 1   Introduction

Transcription factors (TFs) regulate gene expression by either binding directly to specific DNA sequences or through the recruitment of other DNA binding co-factors by a tethering mechanism [1–3]. Therefore, if two TFs that are endogenously co-expressed, co-locate to the same genomic locations, or if the genomic locations of the TFs are very close to each other, it is reasonable to hypothesize that the two transcription factors may act together to regulate gene expression, by forming a complex or otherwise interacting, rather than acting independently. Targeting transcription factors via protein-protein interactions can also offer a novel strategy for cancer therapy. For example, in many human cancers, MDM2 binds to tumour suppressor transcription factor p53 and impairs p53 function. This led to the discovery of Nutil, a small molecule inhibitor that perturbs the interaction between MDM2 and p53 and thus restores p53 function [4]. Thus, by identifying interacting or partner TFs, new targets for therapy can also be identified.

Another example of TF cooperation can be seen in human liver hepatocellular carcinoma cells (HepG2) where the analysis of binding sites for FOXA1 and FOXA3 identified co-location of FOXA1, FOXA2, and FOXA3, suggesting that these FOXA family factors formed a complex. Further analysis using co-immunoprecipitation identified that FOXA2 interacted with FOXA1 and FOXA3. However, FOXA1 and FOXA3 did not interact [5].

Previously various computational approaches such as the definition of degenerate consensus binding sites using positional weight matrices (PWMs) were proposed to identify overlapping motifs [2, 3]. Transcription factors usually bind to thousands of genomic locations and their binding is influenced by various factors in addition to their consensus motifs. Therefore, in recent years chromatin immunoprecipitation (ChIP) with sequencing (ChIP-seq) has been a commonly used method for identifying transcription factors binding sites (TFBS) genome-wide and predicting their cofactor associations [6]. Briefly, raw sequencing reads are aligned to the reference genome assembly. In theory, as the origins of the sequence reads are the binding sites of the transcription factor isolated by ChIP, the mapped reads will aggregate to the positions within the genome where the transcription factor binds. Thus, the transcription factor binding sites may be identified by software tools, known as "peak callers" that identify these regions of aggregation that can be visualized as "peaks" of mapped reads. We refer to sets of TFBS as genomic region datasets, having information about chromosome, start and end positions. In statistical colocalisation analysis (overlap or spatial proximity), if the genomic regions of two TFs significantly overlap then it can be inferred that the two TFs interact either directly or via a tethering mechanism [7, 8].

It is widely acknowledged that different peak-calling tools employ different computational algorithms to call peaks and therefore their results can vary in the number and locations of the called peaks. This, in turn, can affect any downstream co-occurrence analysis [9]. Therefore, firstly we investigated the variation in significance of overlap among datasets (genomic regions) when the datasets were generated with two widely used peak-callers, HOMER and MACS [10, 11]. In addition, we propose a novel network map drawn using OCV by which we modelled transcription factor co-localisation networks in the T-47D breast cancer cell line by calculating the statistical significance of overlap of various transcription factor binding sites (TFBS).

## 2  Methods

Our previously published BiSA tool reports the statistical significance of overlap in the genomic binding sites of two transcription factors through a summary statistic called the Overlap Correlation Value (OCV) [12, 13]. Briefly the BiSA tool analysis outputs a text file with all the query regions in the first column and an overlapping (or closest) reference region in the second column, the significance (p-value) [14] of each overlap is saved in the last column. The OCV is defined as:

$$OCV = \frac{n}{q} \tag{1}$$

Where n is the number of regions in query dataset having p-values less than a defined significance (say 0.05) and q is the total number of regions in query dataset. Therefore, the OCV is changed if the total number of query regions is changed or if n is changed due to change in reference regions. The OCV ranges from 0 to 1. The significance of the overlap becomes stronger as the value of the OCV moves closer to one.

### 2.1  Investigating Variation in OCV Due to Peak-Callers

To investigate the influence of peak-caller tool on the variation in OCV, ChIP-Seq datasets for the HepG2 liver cancer cell line were downloaded from the ENCODE (www.encodeproject.org) or Gene Expression Omnibus (GEO) websites, and, if necessary, converted to FASTQ format using the SRA Toolkit. FastQC (v0.11.3) was used to check the quality of the sequencing reads. In accordance with the ENCODE ChIP-Seq guidelines and practices [15], datasets with less than 10,000,000 mapped reads with a minimum average Phred quality score of 20 were removed. Moreover, the remaining datasets were checked for quality control by using Cutadapt (v1.9.dev4) [16]. Reads were trimmed using a quality threshold of 20 and reads which became shorter than 25 bp after trimming were removed. The processed sequencing reads were then mapped to the human genome (hg38) with Bowtie (v1.1.2) [17]. Subsequently, duplicated reads were removed using the MarkDuplicates tool available in the PicardTools package. Peak-calling was performed using both MACS 2 [11] and HOMER [10], employing their standard parameters. For this analysis, we selected transcription factors for which we called a minimum of 10,000 genomic transcription factor binding sites (TFBS). Finally, sixteen transcription factors were selected (Table 1), namely ARID3A, CEBPB, CTCF, ELF1, FOXA2, HDAC2, HNF4G, JUND, MAZ, SMC3, TBP, USF1, YY1, ZBTB7A, ZNF143, and ZNF384. We calculated the OCV pair-wise, selecting one dataset of TFBS as the query factor while the other acted as the reference.

# 3   Results

## 3.1   Validation of the Overlap Correlation Value (OCV)

We previously proposed the use of OCV analysis to study the correlation among TFBS. However, the OCV is subject to change due to changes in the location and number of binding sites. Therefore, we investigated the impact of the peak caller employed, on the resulting OCV.

The most common peak caller used to generate the datasets in the BiSA database was MACS. Approximately 82% (825/1005) of the ChIP-seq datasets within the BiSA database were generated using MACS. Furthermore, many of the remaining studies validated their peak-calling with MACS. The second most commonly used peak-caller was HOMER, with 4.2% (42/1005) datasets in the database generated with this peak-caller. As these were the two most widely used peak-calling applications, we investigated the change in OCV due to variation in calling peaks with these two widely used tools.

**Table 1.**  Transcription factor bindings sites (TFBS) and sample ID for sixteen selected HepG2 datasets.

| Transcription factor | Number of TFBS | | Difference $b - a$ | ENCODE sample ID |
| --- | --- | --- | --- | --- |
| | HOMER ($a$) | MACS ($b$) | | |
| ARID3A | 32378 | 24761 | 7617 | ENCFF000XOS |
| CEBPB | 56466 | 50112 | 6354 | ENCFF000XQN |
| CTCF | 54064 | 46853 | 7211 | ENCFF000PHE |
| ELF1 | 36277 | 31906 | 4371 | ENCFF000PHM |
| FOXA2 | 25543 | 16435 | 9108 | ENCFF000PIT |
| HDAC2 | 42317 | 22265 | 20052 | ENCFF000PJD |
| HNF4G | 41361 | 34475 | 6886 | ENCFF000PKH |
| JUND | 44180 | 42006 | 2174 | ENCFF000PKR |
| MAZ | 26556 | 26044 | 512 | ENCFF000XUN |
| SMC3 | 34383 | 27727 | 6656 | ENCFF000XXY |
| TBP | 21801 | 23413 | −1612 | ENCFF000XZI |
| USF1 | 22817 | 15203 | 7614 | ENCFF000PSA |
| YY1 | 27589 | 20516 | 7073 | ENCFF000PSD |
| ZBTB7A | 35203 | 22865 | 12338 | ENCFF000PTF |
| ZNF143 | 31418 | 26687 | 4731 | ENCFF001YXH |
| ZNF384 | 37907 | 31755 | 6152 | ENCFF001YXE |

Generally, HOMER called a greater number of peaks (TFBS) than MACS, with the exception of the TBP dataset (Table 1). For example, HOMER called 56,466 peaks for CEBPB while for the same dataset MACS called 50,112 peaks. This confirmed that the number of TFBS generated using the two peak-callers varied. We investigated that how much this variation of total number of regions affected the OCV. Using BiSA, the pairwise OCVs for all transcription factors were calculated for four combinations. (i) Both (query and reference) datasets were generated using MACS (ii) MACS generated datasets as query and HOMER generated datasets as reference, (iii) HOMER generated datasets as query and MACS generated datasets as reference and (iv) both datasets generated using HOMER.

We identified minor variations in calculated OCVs among the four groups. The change was negligible and was not noticeable on the heat plots (not shown). Since MACS was the most widely used peak-caller, we studied the correlation between OCVs obtained from MACS generated datasets against the other three groups. Using the D'Agostino-Pearson omnibus test we identified that data were not normally distributed, therefore, a non-parametric Spearman correlation was used to identify correlations among the four groups. We identified a high correlation among the groups ($R \geq 0.98$). This high value of correlation showed that the correlation between the locations of binding sites of the two factors was independent of the two peak-callers.

## 3.2    Transcription Factor Networks in T-47D Breast Cancer Cell Line

We further studied the application of the OCV to disease-specific datasets to evaluate how well the method described the existing knowledge and if we can hypothesize novel interactions based on the results. We studied publicly available T-47D breast cancer cell line datasets, as breast cancer is one of the leading causes of cancer related deaths in the world [18]. T-47D is an ERα and PR positive breast cancer cell line. These steroid hormone receptors play a critical role in the development and progression of breast cancer, therefore, using the OCV, we studied the correlation between the cistromes of ERα, PR and nine other transcription factors in T-47D breast cancer cells from published studies in BiSA database. Briefly, progesterone receptor (PR) binding sites were collected from three studies, Yin et al., Ballare et al. and Clarke and Graham [19–21]. Yin et al. generated PR binding sites by treatment with anti-progestin RU486 (mifepristone), labelled as PR-RU486 in our analysis.

**Table 2.** Datasets in the first column were selected as query and the datasets from other columns were selected as reference in the calculations of OCV. Treatment is shown after a hyphen against the transcription factor name.

| Query datasets | Reference datasets | | | | | | | | | | | |
|---|---|---|---|---|---|---|---|---|---|---|---|---|
| | FOXA1-DMSO | FOXA1-E2 | PR-R5020 | PR-ORG2058 | PR-RU486 | JUND | CTCF | P300 | ERα-E2 | GATA3 | JARID1B | XBP1 |
| FOXA1-DMSO | 1 | 0.37 | 0.25 | 0.23 | 0.47 | 0.16 | 0.13 | 0.27 | 0.2 | 0.25 | 0.31 | 0.16 |
| FOXA1-E2 | 0.57 | 1 | 0.23 | 0.22 | 0.39 | 0.14 | 0.14 | 0.27 | 0.23 | 0.24 | 0.36 | 0.14 |
| PR-R5020 | 0.42 | 0.26 | 1 | 0.7 | 0.53 | 0.15 | 0.1 | 0.25 | 0.22 | 0.25 | 0.21 | 0.19 |
| PR-ORG2058 | 0.34 | 0.22 | 0.64 | 1 | 0.44 | 0.14 | 0.09 | 0.21 | 0.21 | 0.22 | 0.17 | 0.17 |
| PR-RU486 | 0.65 | 0.32 | 0.37 | 0.33 | 1 | 0.15 | 0.11 | 0.25 | 0.2 | 0.25 | 0.28 | 0.16 |
| JUND | 0.96 | 0.67 | 0.43 | 0.45 | 0.86 | 1 | 0.05 | 0.78 | 0.56 | 0.74 | 0.28 | 0.37 |
| CTCF | 0.26 | 0.17 | 0.11 | 0.1 | 0.18 | 0.06 | 1 | 0.1 | 0.08 | 0.11 | 0.75 | 0.1 |
| P300 | 0.94 | 0.61 | 0.43 | 0.4 | 0.8 | 0.37 | 0.11 | 1 | 0.45 | 0.52 | 0.5 | 0.32 |
| ERα-E2 | 0.65 | 0.57 | 0.39 | 0.41 | 0.58 | 0.29 | 0.08 | 0.47 | 1 | 0.5 | 0.29 | 0.28 |
| GATA3 | 0.63 | 0.39 | 0.32 | 0.32 | 0.55 | 0.25 | 0.11 | 0.36 | 0.33 | 1 | 0.28 | 0.23 |
| JARID1B | 0.44 | 0.32 | 0.18 | 0.16 | 0.3 | 0.09 | 0.35 | 0.21 | 0.14 | 0.17 | 1 | 0.12 |
| XBP1 | 0.6 | 0.35 | 0.36 | 0.35 | 0.55 | 0.22 | 0.12 | 0.37 | 0.28 | 0.34 | 0.36 | 1 |

Ballare et al. and Clarke and Graham treated with agonist (10 nM R5020 (60 min) and 10 nM ORG2058 (45 min), labelled as PR- R5020 and PR-ORG2058, respectively in our analysis. Estrogen receptor alpha (ERα), JUND, CTCF, and P300 were taken from Joseph et al. and Gertz et al. [22–24]. FOXA1 datasets were taken from a study by Joseph et al. [22]. GATA, JARID1B, and XBP1 TFBS were collected from Adomas et al., Yamamoto et al., and Chen et al. respectively [25–27].

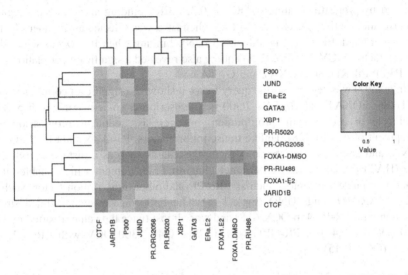

**Fig. 1.** Hierarchical clustering heat map showing correlation of 12 datasets in T47D cells using OCV calculated in Table 2.

A total of 12 datasets for the T-47D cell line were selected to study the statistical significance of their co-localisation with each other (Table 2) and a hierarchical clustering heat map was drawn using the R package gplots (Fig. 1). The OCV is changed when the query and reference datasets are swapped (Eq. 1), and a high correlation (>0.5) can become weaker (<0.5). We showed this relationship graphically by an arrow going from one factor to another (Fig. 2). When the OCV of a query factor is equal to or greater than 0.5 we consider the binding of a query factor to be highly correlated and its binding might be enhanced/facilitated by the binding of a reference factor by either direct interaction, tethering mechanism or by remodelling chromatin. This is illustrated graphically by drawing an arrow from the reference factor towards the query factor. For example, when JUND was selected as the query transcription factor against P300 as the reference transcription factor, the OCV was significant (0.78), however, when P300 was selected as the query against JUND as the reference, then the OCV did not meet the threshold (0.37). Thus, the relationship between the two transcription factors was drawn as a one-way arrow pointing from P300 to JUND (Fig. 2-i). This notation provides rapid visualisation of the relationship between the query and the reference factor depicting that the majority of binding sites for the query factor overlap with the reference factor while the reference may bind uniquely to a large set of additional sites.

A two-way arrow was used to represent two-way co-localisation. For example, there was significant two-way co-localisation of PR binding sites when treated with R5020 or ORG2058 (Fig. 2-ii). Using the above concept a transcription factor directionality network map was drawn (Fig. 2). This network map visualisation facilitates easy identification of potentially interesting interactions between transcription factors.

ERα binding stimulated by E2 revealed a significant correlation (OCV = 0.58) with PR binding stimulated with RU486 (anti-progestin) in comparison to PR binding stimulated by progestin treatment (OCV = 0.33). ERα binding also showed a significant correlation with FOXA1 and GATA3 when ERα was selected as the query factor. On the other hand, ERα binding influenced JUND binding when JUND was selected as the query factor (OCV = 0.56). JUND query also revealed a significant correlation with ERα, P300, PR-RU486, FOXA1 and GATA3.

When P300 was selected as the query factor it showed a significant correlation with PR-RU486, FOXA1 and GATA3. P300 is a transcription co-activator which plays a critical role in cell growth, proliferation, oncogenesis, apoptosis progression and development of disease [28]. Unlike transcription factors, P300 does not bind directly to DNA, and contains a number of structural domains such as a histone acetyltransferase (HAT) domain, and a C-terminal glutamine-rich domain which enables it to interact with nuclear receptors such as ERα/PR and other transcription factors such as GATA3, JARID1B and JUND (Figs. 1, 3). The P300 dataset demonstrated strong correlation with PR-RU486 OCV (0.8) and was almost twice the value reported by PR-R5020 (OCV = 0.43) or PR-ORG2058 (OCV = 0.4). P300 OCV with ERa-E2 was also low (OCV 0.45).

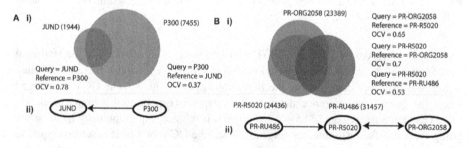

**Fig. 2.** Representation of the degree of transcription factor overlap as a network diagram. Treatment to transcription factor is shown with a hyphen, TFBS are shown within brackets. (A) When comparing the binding sites of JUND and P300, the OCV meets the 0.5 threshold when JUND is selected as the query (0.78) but not when P300 was selected as the query. The JUND-P300 relationship is thus shown as a one-way arrow pointing to JUND. (B) When considering the effect of three treatments on the transcription factor binding sites for PR, there are six possible combinations. The three relationships having an OCV over 0.5 are shown. The relationship between PR-R5020 and PR-ORG2058 met the threshold when either factor was selected as a query factor. Thus, this is shown by a two-way arrow.

FOXA1 is a pioneer factor that facilitates the binding of ERα and other factors [29]. When the FOXA1 dataset was selected as the query against all other factors the OCV

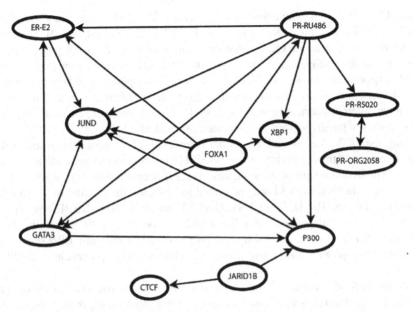

**Fig. 3.** Transcription factor network in T47D breast cancer cell line. Arrow head points towards a query factor whose OCV was greater than 0.5.

was not significant. This statistical analysis confirmed previous findings that FOXA1 was a pioneer factor and its binding was independent of binding of other factors. On the other hand, co-location of JUND, P300 and ERα with FOXA1 datasets (DMSO and E2 treatment) revealed a significant OCV. Co-location of XBP1, GATA3 and PR-RU488 only revealed a significant co-location with the FOXA1-DMSO dataset. Therefore in summarising these relationships in Fig. 2, we used the FOXA1- DMSO dataset to show the relationship with other factors that significantly facilitate binding of JUND, P300, ERα, GATA3, PR-RU486, and XBP1.

CTCF is a silencing factor and its co-location with cohesin components SA1 and RAD21 is known [30]. CTCF binding revealed no significant overlap with other factors except JARID1B. Therefore, our data are consistent with other studies that demonstrate that CTCF forms homodimers by binding to itself, and it's binding to DNA results in tightly bound chromatin where these regions become unavailable for binding of other factors [31, 32]. JARID1B also known as PLU-1 was found highly expressed in some cancers including breast cancer [33], therefore, the significant co-location (OCV = 0.75) of CTCF with JARID1B identified an interesting biological correlation which should be analysed further.

## 4   Discussion

In this study we showed that the calculation of OCV is insensitive to employed peak-callers even though they report slightly different DNA genomic regions. A high value of OCV identifies globally interacting partner factors, as we previously reported a high value (OCV = 0.79) for interacting factors such as SA1 and CTCF and a low value (OCV = 0.21) for non-interacting factors such as ZNF263 and c-Fos [12]. However, we acknowledge that a low OCV value between two factors does not prove that the factors do not have any interaction in regulating a specific subset of genes. Other possible explanations for low OCV could be that the cooperation of factors could be limited to certain genomic locations such as promoter or enhancer regions, the factors might have different partners or work independently and happen to co-locate on HOT (high-occupancy target) regions as suggested by Li et al. [34].

Using the OCV, we studied the global overlap of 12 datasets of various factors under different treatment conditions and used this information to generate a network diagram. The network map of these factors revealed interesting biological interactions among various factors. FOXA1 was predicted to influence the binding of seven other transcription factors: JUND, P300, ERa, GATA3 and PR-RU486, XBP1, however, its own binding was independent of other factors. Interestingly, ERα met the OCV threshold with PR when stimulated by anti-progestin (RU-486) while ERα showed no significant overlap with PR when simulated with synthetic progestins (R5020 or ORG2058).

RU486 (Mifepristone) also abbreviated as MFP, is a selective PR modulator. This synthetic compound binds to PR and exhibits phenotypes ranging from agonism and antagonism [35]. Our analysis showed that the binding sites targeted by PR were different according to whether the cells were treated with progestin or mifepristone. However, there were 8801 common regions between PR-RU486 and PR-R5020. This supports previous results that mifepristone acts as a partial agonist in the absence of progesterone.

The ERα treated with E2 significant overlap with PR treated with RU486 (OCV = 0.65) may indicate an important biological cooperation among the two factors. We further explored the cis-regulatory interaction between agonist treated ERα and PR in detail and identified that their colocation and convergence on a subset of genomic locations are statistically significant [36]. In addition, statistically significant binding of CTCF with JARID1B identified an important biological correlation which should be explored further in laboratory.

Conventionally, researchers use Venn diagrams to represent shared properties visually among few factors, however, the diagram does not work well for large number of DNA-binding proteins (transcription factors). We proposed novel OCV-based arrow directionality network diagram (Fig. 3) to show interaction among TFs. An arrow pointing to a factor shows that the TFBS of the pointed factor highly correlate to other factor. Our proposed network map is easy to expand for large number of TFs and is more feature rich, meaningful, and is better interpretable then using Venn diagram. We believe that he network map is useful in all scientific domains.

# References

1. Wang, J., et al.: Sequence features and chromatin structure around the genomic regions bound by 119 human transcription factors. Genome Res. 22(9), 1798–1812 (2012)
2. Hu, Z., Hu, B., Collins, J.F.: Prediction of synergistic transcription factors by function conservation. Genome Biol. 8(12), R257 (2007)

3. Hannenhalli, S., Levy, S.: Predicting transcription factor synergism. Nucleic Acids Res. **30** (19), 4278–4284 (2002)
4. Vassilev, L.T., et al.: In Vivo activation of the p53 pathway by small-molecule antagonists of MDM2. Science **303**(5659), 844–848 (2004)
5. Motallebipour, M., et al.: Differential binding and co-binding pattern of FOXA1 and FOXA3 and their relation to H3K4me3 in HepG2 cells revealed by ChIP-seq. Genome Biol. **10**(11), R129 (2009)
6. Park, P.J.: ChIP-seq: advantages and challenges of a maturing technology. Nat. Rev. Genet. **10**(10), 669–680 (2009)
7. Simovski, B., et al.: Coloc-stats: a unified web interface to perform colocalization analysis of genomic features. Nucleic Acids Res. **46**(W1), W186–W193 (2018)
8. Stavrovskaya, E.D., et al.: StereoGene: rapid estimation of genome-wide correlation of continuous or interval feature data. Bioinformatics **33**(20), 3158–3165 (2017)
9. Thomas, R., et al.: Features that define the best ChIP-seq peak calling algorithms. Brief. Bioinform. **18**, 441–450 (2016)
10. Heinz, S., et al.: Simple combinations of lineage-determining transcription factors prime cis-regulatory elements required for macrophage and B cell identities. Mol. Cell **38**(4), 576–589 (2010)
11. Zhang, Y., et al.: Model-based analysis of ChIP-Seq (MACS). Genome Biol. **9**(9), R137 (2008)
12. Khushi, M., et al.: Binding sites analyser (BiSA): software for genomic binding sites archiving and overlap analysis. PLoS One **9**(2), e87301 (2014)
13. Khushi, M.: Benchmarking database performance for genomic data. J. Cell. Biochem. **116** (6), 877–883 (2015)
14. Chikina, M.D., Troyanskaya, O.G.: An effective statistical evaluation of ChIPseq dataset similarity. Bioinformatics **28**(5), 607–613 (2012)
15. Landt, S.G., et al.: ChIP-seq guidelines and practices of the ENCODE and modENCODE consortia. Genome Res. **22**(9), 1813–1831 (2012)
16. Martin, M.: Cutadapt removes adapter sequences from high-throughput sequencing reads. EMBnet. J. **17**(1), 10–12 (2011)
17. Langmead, B., et al.: Ultrafast and memory-efficient alignment of short DNA sequences to the human genome. Genome Biol. **10**(3), R25 (2009)
18. Jemal, A., et al.: Global cancer statistics. CA Cancer J. Clin. **61**(2), 69–90 (2011)
19. Yin, P., et al.: Genome-wide progesterone receptor binding: cell type-specific and shared mechanisms in T47D breast cancer cells and primary leiomyoma cells. PLoS One **7**(1), e29021 (2012)
20. Ballare, C., et al.: Nucleosome-driven transcription factor binding and gene regulation. Mol. Cell **49**(1), 67–79 (2013)
21. Clarke, C.L., Graham, J.D.: Non-overlapping progesterone receptor cistromes contribute to cell-specific transcriptional outcomes. PLoS One **7**(4), e35859 (2012)
22. Joseph, R., et al.: Integrative model of genomic factors for determining binding site selection by estrogen receptor-alpha. Mol. Syst. Biol. **6**, 456 (2010)
23. Gertz, J., et al.: Genistein and bisphenol A exposure cause estrogen receptor 1 to bind thousands of sites in a cell type-specific manner. Genome Res. **22**(11), 2153–2162 (2012)
24. Gertz, J., et al.: Distinct properties of cell-type-specific and shared transcription factor binding sites. Mol. Cell **52**(1), 25–36 (2013)
25. Adomas, A.B., et al.: Breast tumor specific mutation in GATA3 affects physiological mechanisms regulating transcription factor turnover. BMC Cancer **14**, 278 (2014)
26. Yamamoto, S., et al.: JARID1B is a luminal lineage-driving oncogene in breast cancer. Cancer Cell **25**(6), 762–777 (2014)

27. Chen, X., et al.: XBP1 promotes triple-negative breast cancer by controlling the HIF1 alpha pathway. Nature **508**(7494), 103–107 (2014)
28. Ghosh, A.K., Varga, J.: The transcriptional coactivator and acetyltransferase p300 in fibroblast biology and fibrosis. J. Cell. Physiol. **213**(3), 663–671 (2007)
29. Jin, H.J., et al.: Cooperativity and equilibrium with FOXA1 define the androgen receptor transcriptional program. Nat. Commun. **5**, 3972 (2014)
30. Lee, B.K., Iyer, V.R.: Genome-wide studies of CCCTC-binding factor (CTCF) and cohesin provide insight into chromatin structure and regulation. J. Biol. Chem. **287**(37), 30906–30913 (2012)
31. Yusufzai, T.M., et al.: CTCF tethers an insulator to subnuclear sites, suggesting shared insulator mechanisms across species. Mol. Cell. **13**(2), 291–298 (2004)
32. Holwerda, S.J., de Laat, W.: CTCF: the protein, the binding partners, the binding sites and their chromatin loops. Philos. Trans. R. Soc. Lond. B Biol. Sci. **368**(1620), 20120369 (2013)
33. Yamane, K., et al.: PLU-1 is an H3K4 demethylase involved in transcriptional repression and breast cancer cell proliferation. Mol. Cell **25**(6), 801–812 (2007)
34. Li, H., et al.: Functional annotation of HOT regions in the human genome: implications for human disease and cancer. Sci. Rep. **5**, 11633 (2015)
35. Benagiano, G., Bastianelli, C., Farris, M.: Selective progesterone receptor modulators 2: use in reproductive medicine. Expert Opin. Pharmacother. **9**(14), 2473–2485 (2008)
36. Khushi, M., Clarke, C.L., Graham, J.D.: Bioinformatic analysis of cis-regulatory interactions between progesterone and estrogen receptors in breast cancer. PeerJ **2**, e654 (2014)

# BayesGrad: Explaining Predictions of Graph Convolutional Networks

Hirotaka Akita[1]($\boxtimes$), Kosuke Nakago[2], Tomoki Komatsu[2], Yohei Sugawara[2], Shin-ichi Maeda[2], Yukino Baba[3], and Hisashi Kashima[1]

[1] Kyoto University, Kyoto, Japan
h_akita@ml.ist.i.kyoto-u.ac.jp, kashima@i.kyoto-u.ac.jp
[2] Preferred Networks, Inc., Tokyo, Japan
{nakago,komatsu,suga,ichi}@preferred.jp
[3] University of Tsukuba, Tsukuba, Japan
baba@cs.tsukuba.ac.jp

**Abstract.** Recent advances in graph convolutional networks have significantly improved the performance of chemical predictions, raising a new research question: "how do we explain the predictions of graph convolutional networks?" A possible approach to answer this question is to visualize evidence substructures responsible for the predictions. For chemical property prediction tasks, the sample size of the training data is often small and/or a label imbalance problem occurs, where a few samples belong to a single class and the majority of samples belong to the other classes. This can lead to uncertainty related to the learned parameters of the machine learning model. To address this uncertainty, we propose BayesGrad, utilizing the Bayesian predictive distribution, to define the importance of each node in an input graph, which is computed efficiently using the dropout technique. We demonstrate that BayesGrad successfully visualizes the substructures responsible for the label prediction in the artificial experiment, even when the sample size is small. Furthermore, we use a real dataset to evaluate the effectiveness of the visualization. The basic idea of BayesGrad is not limited to graph-structured data and can be applied to other data types.

**Keywords:** Machine learning · Deep learning · Interpretability
Cheminformatics · Graph convolution

## 1 Introduction

The applications of deep neural networks are expanding rapidly in various fields, including chemistry and biology. Graph convolutional neural networks, which can handle graph-structured data (e.g., chemical compounds) as inputs, have opened the door to end-to-end learning for chemical prediction. Many variants of graph

---

H. Akita—The work was done while the author was an intern at Preferred Networks, Inc.

L. Cheng et al. (Eds.): ICONIP 2018, LNCS 11305, pp. 81–92, 2018.
https://doi.org/10.1007/978-3-030-04221-9_8

convolutional neural networks have been proposed, which are now improving the performance of various chemical prediction tasks, including physical property prediction [4], toxicity prediction [7], solubility and drug efficiency prediction [2], and total energy prediction [11].

Deep neural networks automatically learn useful *features* for prediction, which sometimes outperform hand-engineered features carefully designed by domain experts, enabling these neural networks to find new knowledge about molecular properties. However, the complex non-linear operations in deep neural networks make it prohibitively difficult to understand their behaviors.

*Sensitivity map*, also known as *saliency map* or *pixel attribution map*, is a common approach used to explain the reasons for the predictions of neural networks. The map assigns an *importance score* to each substructure of an instance, which reflects the influence of the substructure on the final prediction, and visualizes high-scored substructures. The gradients are commonly used to measure the importance. A naive way entails using the size of the norm of the gradient [13] (we call this approach *VanillaGrad*). Sensitivity maps generated by this approach are typically noisy. As a result, SmoothGrad has been proposed to address this issue by adding noise to input samples and taking the mean values of the gradients [14].

The existing approaches do not take into account the uncertainty in the prediction of the model. The uncertainty becomes particularly apparent in the chemical domain, because the sample size of the chemical dataset is often small and/or a imbalance problem occurs, where only a few samples belong to a single class and the majority of the samples belong to the other classes. In such cases, it is difficult to estimate which substructures are responsible for a prediction.

In this paper, we propose *BayesGrad*, a novel sensitivity map algorithm that can deal with model uncertainty. Our key idea is to quantify the uncertainty of a prediction utilizing its Bayesian predictive distribution. We implement the idea using the dropout, a common regularization technique for deep neural networks, because the outputs obtained using this technique approximate the expected value with respect to the Bayesian predictive distribution [3,9].

We conducted experiments using a synthetic compound dataset labeled with a particular substructure, and quantitatively evaluated the validity of our importance score. BayesGrad achieved superior performance, especially when the number of training data is small. We also use real datasets to visualize the bases of the predictions, and found that the visualized substructure is consistent with the known results. Although we present the formalization of BayesGrad in the context of graph-structured data and demonstrate its efficiency in the chemical domain, BayesGrad is a general framework and, thus, can be applied to other data types such as images.

Our contributions to the literature are summarized as follows:

1. **Bayesian approximation for sensitivity map visualization:** We propose a novel method that uses the dropout technique to quantify model uncertainty.

2. **Application of gradient-based sensitivity map visualization for graphs:** Most of the existing gradient-based sensitivity map algorithms are evaluated on image classification tasks.
3. **Quantitative evaluation in the chemical domain:** We quantitatively evaluated the performance of the gradient-based method to visualize the basis of a prediction.

## 2   Preliminaries

We begin with the problem setting for sensitivity map visualization in graph prediction, followed by a brief review of several existing sensitivity map generation methods.

### 2.1   Problem Definition

We assume that we have an (already trained) regression or classification model $f : \mathcal{G} \to \mathbb{R}$, where $G = (V, E) \in \mathcal{G}$ is a graph consisting of a set of nodes $V$ and a set of edges $E$, and the output of the model indicates the regression result or the classification score. Note that in the case of the binary classification model, the output of $f$ is the raw score in $\mathbb{R}$, not a value transformed by the sigmoid function. In the graph neural network $f$, a node $v_i$ is associated with a feature vector $\phi_i$.

Given the model $f$ and a target input graph $G$, our goal is to assign an importance score $s_i$ to each node $v_i \in V$.

### 2.2   VanillaGrad

There have been several recent attempts to interpret the predictions made by complex neural network models. Although these methods focus on images, they are easily applied to graphs. The gradient of $f$ with respect to feature $\phi_i$ is often used as the importance score of an input (i.e., a node) [13]:

$$s_i = \left\| \frac{\partial f(\phi; W)}{\partial \phi_i} \right\|. \tag{1}$$

To simplify the calculation, we often use the 2-norm, but we can also use another norm such as the 1-norm. We call the importance score defined in Eq. (1) *VanillaGrad*.

### 2.3   SmoothGrad

It is known that sensitivity maps generated by VanillaGrad are likely to be noisy. To address the problem, SmoothGrad [14] calculates the expected value of the gradient (1) over the Gaussian noise added to the input:

$$s_i = E_{\epsilon \sim \mathcal{N}(0, \sigma^2)} \left[ \left\| \frac{\partial f(\phi + \epsilon; W)}{\partial \phi_i} \right\| \right]. \tag{2}$$

We approximate the value of Eq. (2) using sampling. SmoothGrad first generates $M$ noisy inputs $\{\check{\phi}_i^m\}_{m=1}^M$ by adding noise to the original input $\phi_i$; that is, $\check{\phi}_i^m$ is given as:

$$\check{\phi}_i^m = \phi_i + \epsilon^m, \tag{3}$$

where $\epsilon^m$ is a sample from a Gaussian distribution $\mathcal{N}\left(0, \sigma^2\right)$ with a mean of zero a variance of $\sigma^2$. The importance score of a noisy input $\check{\phi}_i^m$ is then calculated as

$$\check{s}_i^m = \left\| \frac{\partial f(\check{\phi}^m; W)}{\partial \phi_i} \right\|. \tag{4}$$

Finally, the importance score of the original input $s_i$ is estimated as the average of $\{\check{s}_i^m\}_{m=1}^M$:

$$s_i \approx \frac{1}{M} \sum_{m=1}^M \check{s}_i^m, \tag{5}$$

which is called *SmoothGrad*. Note that both of the variance of the Gaussian noise $\sigma^2$ and the sample size $M$ are hyperparameters to be tuned. In the original paper, $\sigma$ is tuned as a relative scale from the range of the input value $\phi$. However we used a fixed value of $\sigma$ for each input, because this was more stable in our experiment.

## 2.4   Importance Score Calculation Using Signed Values

In the previous discussion, the sensitivity map only gives how much each atom impacts on the prediction, but does not give whether the atoms have positive or negative effects on the prediction. To address this, Shrikumar *et al.* [12] used the product of the input and the gradient instead of the norm to evaluate how the atoms affect the output:

$$(\phi_i - b_i)^\top \frac{\partial f(\phi; W)}{\partial \phi_i}, \tag{6}$$

where $b$ denotes the baseline vector. The above formula represents the effect on function $f$ when we change the $i$-th input from $\phi_i$ to $b_i$. It can be understood as (the negative of) the first-order term of the Taylor expansion of $f$ at $\phi$, which is evaluated at $b$. Note that there is a freedom of choice of the baseline $b$; it is often set to 0, which corresponds to a black image in the image domain. As we discuss in the experimental section, this technique gives us richer information in certain cases.

## 3   Proposed Method

Existing approaches do not consider the uncertainty of the prediction by the model. To address this issue, we propose BayesGrad, which quantifies the uncertainty of the sensitivity map using Bayesian inference. We first describe the formulation of BayesGrad, and then explain the practical implementation using the dropout technique.

## 3.1 BayesGrad

The existing methods are formulated in the framework of the maximum likelihood estimation of the neural network parameter $W$. However, the learned $W$ is not necessarily stable and can vary with addition or deletion of a small portion of the training data sample.

In our formulation, we consider the uncertainty of the parameter $W$ by using the posterior of the neural network parameter $p(W|\mathcal{D})$ given the training data $\mathcal{D}$. We consider the expected value of the importance score with respect to $p(W|\mathcal{D})$:

$$\bar{s}_i = E_{W \sim p(W|\mathcal{D})} \left[ \left\| \frac{\partial f(\phi; W)}{\partial \phi_i} \right\| \right]. \tag{7}$$

We approximate this using sampling as

$$\bar{s}_i \approx \frac{1}{M} \sum_{j=1}^{M} s_i^{(j)}, \tag{8}$$

where $s_i^{(j)}$ is the $j$-th importance score computed from the $j$-th sample $W^{(j)} \sim p(W|\mathcal{D})$ $(j = 1, \cdots, M)$ as

$$s_i^{(j)} = \left\| \frac{\partial f(\phi; W^{(j)})}{\partial \phi_i} \right\|. \tag{9}$$

We call the importance score computed by Eq. (8) *BayesGrad*. BayesGrad has a sample size $M$ as a hyperparameter.

## 3.2 Dropout as a Bayesian Approximation

In order to implement BayesGrad, we need to take samples from the posterior distribution $p(W|\mathcal{D})$. In general, the exact computation of the posterior is intractable, in which case we resort to approximation methods such as Markov chain Monte Carlo methods or variational Bayesian approximations. In particular, we utilize the dropout technique which can be interpreted as a variational Bayesian method because of its relatively small computational cost. Dropout is originally introduced as a regularization technique to prevent overfitting [5,15], but recent studies show that dropout can be viewed as a kind of variational Bayesian inference [3,9]. We use the "Dropout as a Bayesian Approximation (DBA)" technique to calculate the uncertainty of the model using dropout. It approximates the posterior distribution $p(W|\mathcal{D})$ using variational distribution $q(W; \eta)$, where $\eta$ is a parameter to best approximate $p(W|\mathcal{D})$. In DBA, $q(W; \eta)$ has a special form. $W \sim q(W; \eta)$ is given as an Hadamard product of an adjustable constant matrix $\tilde{W}$ and the random mask matrix, and the stochasticity lies in each element of the random mask matrix that take the values either zero or one, typically with equal probability.

### 3.3 Comparison Between SmoothGrad and BayesGrad

In contrast with SmoothGrad that takes the expectation of the gradient over possible fluctuations in the input variable, BayesGrad smooths gradients over fluctuations in the model parameter $W$ that follows the (approximate) posterior distribution $p(W|\mathcal{D})$. Validity of adding the Gaussian noise to the input depends on the task. In the image domain, even if some noise is added to the original image, the noisy image still looks similar to the original one, and is still considered natural. However, in the chemical domain, the input is the feature vector of each atom which is originally a discrete object; hence, the noisy input does not correspond to a real atom anymore. The similar discussion also applies in other domain as well, e.g., word embedding in natural language processing.

Another benefit of BayesGrad is emphasized when the training data is few. Model training tends to be unstable in such cases, and the model predictions tend to be stochastic. BayesGrad can treat this type of uncertainty by exploiting the Bayesian inference.

Note that the idea behind SmoothGrad taking the expectation in the input space and that of BayesGrad taking the expectation in the model space are not mutually exclusive, and we can combine both techniques to calculate a sensitivity map (as *BayesSmoothGrad*).

## 4  Experiments

We demonstrate the effectiveness of our approach in the chemical domain, where the sample size could be small and there is high demand for substructure visualization. We first validate the methods using a synthetic dataset where the ground-truth substructures correlated with the target label are known. In addition, we demonstrate the method using the real datasets and discuss its effectiveness.

We used Chainer Chemistry which is an open-source deep learning framework providing major graph convolutional network algorithms [10]. We slightly modified the neural fingerprint method [2] and the gated-graph neural network (GGNN) [8] in the library by including the dropout function to perform BayesGrad. Our code used in this experiment is available at https://github.com/pfnet-research/bayesgrad. Please refer the code for how to reproduce our result, including the hyperparameter configuration.

### 4.1  Quantitative Evaluation on Tox21 Synthetic Data

Tox21 [6] is a collection of chemical compounds including 11,757 training, 295 validation, and 645 test data samples. Each compound is associated with some of 12 toxicity type labels; we used only the training and validation data in our experiment since the test dataset has no label information.

Since the original tox21 dataset does not have the information of what substructure actually contribute to their toxicity labels, We first used synthetic labels to quantitatively evaluate the different evidence visualization methods. We assigned the label 1 to compounds that contain pyridine ($C_5H_5N$) and 0 to the remainder, which resulted in 760 label-1 compounds and 10,997 label-0 compounds in the training dataset.

We trained the GGNN with the dropout function to predict whether the input compound contains pyridine. The GGNN has a gating architecture that enables the model to set the weights for important information. After training the model, the ROC-AUC scores for both the training and the validation data were as high as 0.99, which suggests that the model was successfully trained.

The validation dataset was used for testing. There are 28 molecules that contain a pyridine substructure in the validation dataset. We expect atoms that belong to pyridine rings to have a higher importance score than the others; hence, we selected the atoms in descending order of the importance scores after calculating the importance scores by each method. We calculated the gradient of the output of the pre-final layer just before applying the sigmoid function that gives probability values, because the sigmoid function squashes the gradient and therefore the performance became worse if we took the gradient after the sigmoid function.

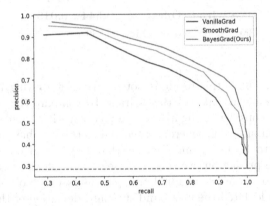

**Fig. 1.** The precision-recall curve for each algorithm. All methods record high precision-recall curve.

Figure 1 shows the precision-recall curve, where the precision indicates the proportion of the atoms consisting of pyridine rings in the extracted substructure, and the recall indicates the proportion of the extracted atoms in all the atoms in the pyridine rings. We used $M = 100$ for the SmoothGrad and BayesGrad calculations.

Figure 2 shows the sensitivity map visualization for each method. All of the methods successfully extracted the substructure containing the pyridine ring. This result implies that the gradient-based sensitivity map calculation is effective in extracting the substructure responsible for the target label in chemical prediction tasks. Note that even though BayesGrad seems to outperform SmoothGrad

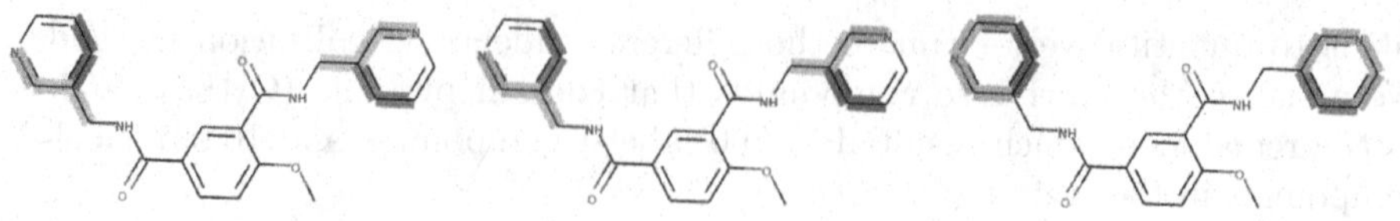

(a) VanillaGrad            (b) SmoothGrad            (c) BayesGrad

**Fig. 2.** Examples of extracted substructure. The important atoms are highlighted. All methods successfully focuses on pyridine ($C_5H_5N$) substructure at the top-left and the top-right.

**Table 1.** PRC-AUC score between algorithms. The value before and after $\pm$ represent the mean and the standard deviation of PRC-AUC score calculated by 30 different models. Fixed value of $\sigma = 0.15$ is used for SmoothGrad, $M = 100$ is used for both SmoothGrad and BayesGrad.

| Algorithm | PRC-AUC score |
|---|---|
| VanillaGrad | $0.506 \pm 0.044$ |
| SmoothGrad | $0.514 \pm 0.042$ |
| **BayesGrad (Ours)** | $\mathbf{0.544 \pm 0.019}$ |
| BayesSmoothGrad (Ours) | $0.536 \pm 0.028$ |

or VanillaGrad in Fig. 1, this result is not deterministic owing to the stochastic behavior of SmoothGrad and BayesGrad. To compare the performance of the methods, we consider a slightly difficult case with a small dataset. This reflects a practical situation where limited data are available and the model's prediction tends to be uncertain. To test that BayesGrad can deal with the uncertainty of the prediction, we randomly select 30 different subset consisting of 1000 compounds from the original training dataset and obtained 30 different models. We calculated the mean and standard deviation of their PRC-AUC scores. The results are summarized in Table 1. BayesGrad records statistically higher PRC-AUC scores than both VanillaGrad and SmoothGrad. We also tested BayesSmoothGrad method, which uses both dropout and noise; however, its performance did not improve in this experiment.

## 4.2   Visualization on Tox21 Actual Data

We also performed a toxicity prediction task experiment using the Tox21 dataset where each compound has some of 12 toxicity labels We trained the prediction model for each of the labels, and visualized the grounds for prediction of the label *SR-MMP* with the highest prediction accuracy (0.889 ROC-AUC in test data).

Figure 3 shows some interesting results; Tyrphostin 9 (Fig. 3(a)) is a tyrosine kinase inhibitor and is known to be a potent uncoupler of oxidative phosphorylation, which has a strong influence on the mitochondrial membrane potential (*SR-MMP*). Terada *et al.* [16] examined the effect of the mitochondrial function of the acid-dissociable group using Tyrphostin 9 and a derivative, modified by methylation of its phenolic OH group. They confirmed that the acid-dissociable group is essential for uncoupling. We computed sensitivity maps for these compounds, as shown in Fig. 3(a) and (b). Our visualization results are consistent with their experimental results. We also found similar compounds in the Tox21 dataset, with an acid-dissociable group, as shown in Fig. 3(c) and (d). We confirmed that our visualization method has the potential to detect these essential substructures accurately.

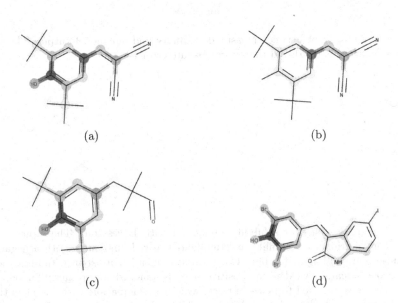

(a)                                                  (b)

(c)                                                  (d)

**Fig. 3.** Visualizing the sensitivity map for SR-MMP toxicity prediction with BayesGrad. (a) Chemical structure of Tyrphostin 9 and our model highlighted the phenolic OH (acid-dissociable) group, which is confirmed to be essential for uncoupling [16]. (b) O-methylated derivative of Tyrphostin 9 does not induce uncoupling. (c), (d) We found some compounds with similar sub-structure. Our model detected the same phenolic OH group.

## 4.3    Evaluation on Solubility Dataset

Solubility is an important property in drug design because sufficient water-solubility is necessary for drug absorption. It is well known that some functional groups such as the hydroxyl group and primary amine group contribute to the hydrophilic nature, whereas other groups such as the phenyl group and ethyl group contribute to the hydrophobic nature. In addition, molecular weight has a strong correlation with solubility. Medicinal chemists need to modify the chemical structure by adding charged substituents, reducing the hydrophobic groups

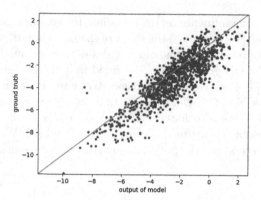

**Fig. 4.** The scatter plot between measured solubility and our model output. The correlation coefficient is 0.851 and the mean absolute error is 1.024.

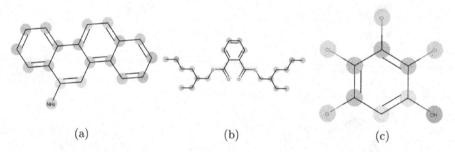

(a)                                (b)                                (c)

**Fig. 5.** Sensitivity maps for solubility prediction with BayesGrad. The atoms with positive contributions to solubility are highlighted in red, and those with a negative contributions are highlighted in blue. Our results are mostly consistent with fundamental physicochemical knowledge. (a) Positive score is assigned to a primary amine and negative scores is assigned to benzene rings, which are compatible with the facts that polycyclic aromatic hydrocarbon (PAH) has a hydrophobic nature and primary-amine is negatively charged, respectively. (b) Ester is detected as the important substructure for hydrophilicity, which is known to have low polarity. (c) Halogen substituents make a compound more lipophilic and less water-soluble. On the other hand, the hydroxyl group is negatively charged and expected to contribute to hydrophilicity. (Color figure online)

and the molecular weight, to improve solubility. However, if the molecule has a complicated structure, it becomes difficult to identify which part of structure is significant for the chemical property. Thus, our motivation is to provide a way to visualize which parts of a chemical structure are significant for the solubility. We demonstrated the effectiveness of our approach using a publicly available dataset.

We used the ESOL dataset [1] to evaluate our approach. This dataset contains 1,127 compounds with measured log solubility. We used the signed importance score explained in Sect. 2.4 to discriminate the positive/negative contributions to solubility. The choice of the baseline $b$ is not trivial in chemical prediction tasks, where we consider the embedded feature vector space of atom representation as input. In our experiment, we used the baseline $b = 0$, which corresponds to the mean of the prior distribution of the embedded feature vector. We used the neural fingerprint model [2] to evaluate this task.

Figure 4 shows the prediction result, where the model achieved good performance for the solubility prediction. Figure 5 shows examples of the visualization for solubility prediction. Our approach accurately assigns positive importance scores to the hydrophilic atoms, and negative scores to the hydrophobic atoms, even for such compounds with complicated structures.

## 5   Conclusion

We proposed a method to visualize a sensitivity map of chemical prediction tasks. While existing methods focus on the visualization on image domain, our quantitative evaluation with the tox21 dataset showed that BayesGrad outperforms the existing methods. BayesGrad exploits the Bayesian inference technique to handle the uncertainty in predictions, which contributes to a robust sensitivity map, especially for small datasets. Furthermore, we obtained the promising experimental results on the real datasets, which accord with the well-known chemical properties.

Elucidating the chemical mechanism is challenging research. We believe the proposed algorithm will lead to a better understanding of the chemical mechanism. Our idea is easily applicable to other deep neural networks in other domains, which we leave to future research.

**Acknowledgement.** This research was supported by JSPS KAKENHI Grant Number 15H01704, Japan.

## References

1. Delaney, J.S.: ESOL: estimating aqueous solubility directly from molecular structure. J. Chem. Inf. Comput. Sci. **44**(3), 1000–1005 (2004). pMID: 15154768
2. Duvenaud, D.K., et al.: Convolutional networks on graphs for learning molecular fingerprints. In: Advances in Neural Information Processing Systems, vol. 28, pp. 2224–2232 (2015)
3. Gal, Y., Ghahramani, Z.: Dropout as a Bayesian approximation: representing model uncertainty in deep learning. In: Proceedings of the 33rd International Conference on International Conference on Machine Learning, ICML, pp. 1050–1059 (2016)
4. Gilmer, J., Schoenholz, S.S., Riley, P.F., Vinyals, O., Dahl, G.E.: Neural message passing for quantum chemistry. In: Proceedings of the 34th International Conference on Machine Learning, ICML, pp. 1263–1272 (2017)
5. Hinton, G.E., Srivastava, N., Krizhevsky, A., Sutskever, I., Salakhutdinov, R.: Improving neural networks by preventing co-adaptation of feature detectors. arXiv preprint arXiv:1207.0580 (2012)

6. Huang, R., et al.: Tox21Challenge to build predictive models of nuclear receptor and stress response pathways as mediated by exposure to environmental chemicals and drugs. Front. Environ. Sci. **3**, 85 (2016)
7. Kearnes, S., McCloskey, K., Berndl, M., Pande, V., Riley, P.: Molecular graph convolutions: moving beyond fingerprints. J. Comput.-Aided Mol. Des. **30**(8), 595–608 (2016)
8. Li, Y., Tarlow, D., Brockschmidt, M., Zemel, R.: Gated graph sequence neural networks. In: Proceedings of the International Conference on Learning Representations, ICLR (2016)
9. Maeda, S.: A Bayesian encourages dropout. arXiv preprint arXiv:1412.7003 (2014)
10. pfnet research: chainer-chemistry. https://github.com/pfnet-research/chainer-chemistry
11. Schütt, K., Kindermans, P.J., Felix, H.E.S., Chmiela, S., Tkatchenko, A., Müller, K.R.: SchNet: a continuous-filter convolutional neural network for modeling quantum interactions. In: Advances in Neural Information Processing Systems, vol. 30, pp. 992–1002. Curran Associates, Inc. (2017)
12. Shrikumar, A., Greenside, P., Kundaje, A.: Learning important features through propagating activation differences. arXiv preprint arXiv:1704.02685 (2017)
13. Simonyan, K., Vedaldi, A., Zisserman, A.: Deep inside convolutional networks: visualising image classification models and saliency maps. arXiv preprint arXiv:1312.6034 (2013)
14. Smilkov, D., Thorat, N., Kim, B., Viégas, F., Wattenberg, M.: SmoothGrad: removing noise by adding noise. arXiv preprint arXiv:1706.03825 (2017)
15. Srivastava, N., Hinton, G., Krizhevsky, A., Sutskever, I., Salakhutdinov, R.: Dropout: a simple way to prevent neural networks from overfitting. J. Mach. Learn. Res. **15**, 1929–1958 (2014)
16. Terada, H., Fukui, Y., Shinohara, Y., Ju-ichi, M.: Unique action of a modified weakly acidic uncoupler without an acidic group, methylated SF 6847, as an inhibitor of oxidative phosphorylation with no uncoupling activity: possible identity of uncoupler binding protein. Biochimica et Biophysica Acta **933**, 193–199 (1988)

# Deep Multi-task Learning for Air Quality Prediction

Bin Wang[1,2(✉)], Zheng Yan[2], Jie Lu[2], Guangquan Zhang[2], and Tianrui Li[1]

[1] School of Information Science and Technology, Southwest Jiaotong University,
Chengdu, China
wangbin@my.swjtu.edu.cn, trli@swjtu.edu.cn
[2] Faculty of Engineering and Information Technology,
University of Technology, Sydney, Sydney, Australia
{yan.zheng,jie.lu,guangquan.zhang}@uts.edu.au

**Abstract.** Predicting the concentration of air pollution particles has been an important task of urban computing. Accurately measuring and estimating makes the citizen and governments can behave with suitable decisions. In order to predict the concentration of several air pollutants at multiple monitoring stations throughout the city region, we proposed a novel deep multi-task learning framework based on residual Gated Recurrent Unit (GRU). The experimental results on the real world data from London region substantiate that the proposed deep model has manifest superiority than shallow models and outperforms 9 baselines.

**Keywords:** Deep learning · Recurrent neural networks
Neural networks · Air quality prediction · Urban computing

## 1 Introduction

In recent years, along with economic development, air pollution in developing countries has become a serious issue [1]. Air pollutants comprise molecule (e.g., $PM_{2.5}$ and $PM_{10}$) and harmful gas (e.g. $NO_2$) are threatening the public health [4]. For monitoring real-time air pollution, Chinese governments have built the amount of air quality monitoring stations and collect air quality data every hour in recent years [19]. Besides monitoring, there is a rising demand for forecasting future air quality index (AQI). Accurately measuring and estimating the concentration of air pollution particles makes the citizen and governments can behave with suitable decisions, such as reducing outdoor activities, to remarkably reduce the adverse results of air pollution.

Air quality prediction methods mainly fall into two categories: classical dispersion models and data-driven models [17,18]. Classical dispersion models identify the root cause of air pollution from chemical, emission, climatological and combinations of these factors. These models are most a numerical function of emissions from industry and vehicular, meteorology, and other factors. However, it is very difficult to get all these factors completely and accurately. Thus, the

© Springer Nature Switzerland AG 2018
L. Cheng et al. (Eds.): ICONIP 2018, LNCS 11305, pp. 93–103, 2018.
https://doi.org/10.1007/978-3-030-04221-9_9

prediction accuracy is hard to be guaranteed. Also, the computation complexity is very high.

Data-driven approaches, e.g. artificial neural networks, forecast air pollutions based on massive observed data. Deep learning, as a cutting-edge technique of machine learning, has made great success in computer vision tasks and is very suitable and robust when modeling complicated spatio-temporal data e.g., videos. Inspired by this, more and more researchers begin to improve and apply it to solve urban computing problems and achieve considerable results. In this paper, we propose to use a novel deep learning model to predict the concentration of air pollutants. Following are the major contributions of this paper.

1. We introduce a novel Residual-GRU based on vanilla GRU for short-term air pollutants. Experimental results demonstrate this model can speed up convergence and perform better during the test.
2. We formalize the air pollutants prediction into a multi-task end-to-end framework. Previous related studies are single-task which predict only for one station or only one pollutant but our method can forecast for all stations and all pollutants at one time.

The rest of the paper is organized as follows: In Sect. 2, we first explain some basic variants of recurrent neural networks (RNN) and then introduce the proposed model in more details. Then, we experimentally evaluate our proposed prediction models and compare them with other baselines in Sect. 3. In Sect. 4, we present related works. Finally, conclusions and future works are given in Sects. 5 and 6, respectively.

## 2    Recurrent Neural Networks

In this section, we introduce the different versions of RNN and our proposal, the improved model Residual-GRU. For simplicity, all bias terms are omitted.

**Vanilla RNN.** This model is specially designed to incorporate sequential information and has achieved great success particularly in NLP tasks. The formulas of vanilla RNN are shown as below:

$$h_t = f(W_{hh}h_{t-1} + W_{xh}x_t)$$
$$y_t = f(W_{ht}h_t)$$

where $x_t$ is the input at time $t$, $W$ is the transformation weights, $f$ is the element-wise activation function such as $tanh$, and $h_t$ and $y_t$ is the hidden state and output at time $t$ respectively. A serious drawback of vanilla RNN is it can hardly capture long-term sequential dependency due to vanishing gradients.

**LSTM.** To overcome the drawback of vanilla RNN, LSTM is designed as below:

$$i_t = \sigma(W_{hh}^i h_{t-1} + W_{xh}^i x_t)$$
$$f_t = \sigma(W_{hh}^f h_{t-1} + W_{xh}^f x_t)$$
$$o_t = \sigma(W_{hh}^o h_{t-1} + W_{xh}^o x_t)$$
$$\tilde{C}_t = tanh(W_{hh}^g h_{t-1} + W_{xh}^g x_t)$$
$$C_t = \sigma(f_t \odot C_{t-1} + i_t \odot \tilde{C}_t)$$
$$h_t = tanh(C_t) \odot o_t$$

where $i, f, o$ are input, forget and output gates, respectively. $\odot$ is the Hadamard product. Such a three-gates mechanism can effectively alleviate vanishing gradients problems.

**GRU.** GRU could be regarded as the light LSTM. It only utilizes two gates to control information flow, which is shown as below:

$$z_t = \sigma(W_{hh}^z h_{t-1} + W_{xh}^z x_t)$$
$$r_t = \sigma(W_{hh}^r h_{t-1} + W_{xh}^r x_t)$$
$$\tilde{h}_t = tanh(W_{hh}^h (r_t \odot h_{t-1}) + W_{xh}^h x_t)$$
$$h_t = z_t \odot \tilde{h}_t + (1 - z_t) \odot h_{t-1}$$

where $z_t$ is called update gate and $r_t$ is called reset gate.

**Proposed Residual-GRU.** Based on GRU, inspired from residual convolutional networks [3], the proposed Residual-GRU combines residual learning with GRU. The unique difference between Residual-GRU and GRU resides in $h_t = z_t \odot \tilde{h}_t + (1-z_t) \odot h_{t-1} + W_{xh}^{res} x_t$. By providing a shortcut path (i.e., $W_{xh}^{res} x_t$) between adjacent layer outputs, it could release gradient flow more smoothly and accelerate the training process. Please note that $W_{xh}^{res}$ can be omitted if the dimension of $x_t$ is equal to $h_{t-1}$ and $\tilde{h}_t$, that is, $h_t = z_t \odot \tilde{h}_t + (1-z_t) \odot h_{t-1} + x_t$. In our experiments, we set the their dimension equal, and hence omit the $W_{xh}^{res}$. The graphic framework is shown in Fig. 1. Particularly, the involved hyper-parameters include

1. *Input length* is set as 5, i.e., we use the previous 5-hours vector sequences as inputs to predict next time vector. At each timestamp, the dimension of the input vector is $19 * 3 = 57$, which means 19 monitoring stations times by 3 pollutants concentration ($PM_{2.5}$, $PM_{10}$, $NO_2$). In this way, we can implement a multi-task learning in a unified end-to-end framework.
2. *Number of layers* and *hidden states* are set as 2 and 53 respectively. The last timestamp is furthermore followed by fully connection with 53 nodes for deep representation learning.
3. *Epochs* and *batch size* are set as 100 and 32 respectively.
4. *Activation functions* are all set as *tanh* except the output layer with *sigmoid*.

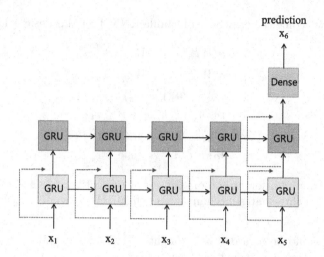

**Fig. 1.** The framework of Residual-GRU, where the dash line means the residual connection and $x_i$ is the input vector. *Dense* means the fully connected layer.

## 3  Experiments

### 3.1  Dataset

The hourly concentration data of three air pollutants, i.e., $PM_{2.5}$, $PM_{10}$, and $NO_2$ at London region from 1st/Mar/2017 to 27th/Mar/2018 were downloaded from KDD CUP of Fresh Air. This dataset includes 9385 timestamps for 19 stations. We first impute the missing value with historical mean and then use maxmin to normalize features into $[0, 1]$. In our experiments, we split data as 0.8/0.2 for training and test.

### 3.2  Baselines

Our baselines are generally separated into 2 categories, i.e., classic machine learning methods and deep learning models. The classic machine learning methods mainly include:

- **SVR.** SVR is the support vector machine designed for regression. The kernel function usually includes 'rbf', 'linear', 'poly' and so on.
- **LASSO.** Lasso is a regression analysis method that performs both variable selection and regularization.
- **RandomForest.** Random forest is a popular ensemble method for classification and regression. The main drawback of it is time-consuming for prediction.

  The deep learning baselines include:

- **Residual-LSTM.** This baseline just replaces the GRU unit with LSTM to demonstrate the superiority of GRU.

- **LSTMs-100-100.** This baseline removes residual connections and consists of 2-layers-depth LSTM and 100 hidden states for each layer by which we can observe the effectiveness of residual learning.
- **LSTM-100.** This 1-layer-depth LSTM is devised to check the effect of depth.
- **Dense-100_LSTM-100.** Different from LSTM-100, we add a time distributed fully connected layer for each time step. Hence this model introduces more powerful learning ability and might be with weaker generalization.

## 3.3 Evaluation Metric

We measure our method by Root Mean Square Error (RMSE) and Mean Absolute Error (MAE).

$$RMSE = \sqrt{\frac{1}{z} \sum_i (x_i - \hat{x}_i)^2} \tag{1}$$

$$MAE = \frac{1}{z} \sum_i |x_i - \hat{x}_i| \tag{2}$$

where $\hat{x}$ and $x$ are the predicted value and ground truth, respectively; $z$ is the number of all predicted values.

## 3.4 Results

We train all deep models on a server with Quadro P5000 GPU and the programming environment is Keras with TensorFlow backend. Table 1 shows the experimental results which illustrate our proposed model is state-of-the-art. Figure 2 reveals the loss variation during training and test. Please note that we do not fine-tune the hyper-parameters exhaustly. There are some important conclusions we can summarise:

1. Above all, we can see that classic machine learning methods are not very suitable than deep learning models. This demonstrates the effectiveness and importance of deep learning for complicate spatio-temporal feature extract.
2. Residual-GRU performs better than Residual-LSTM. This may be GRU module has fewer parameters and hence make it easier for learning in this spatio-temporal task.
3. By comparing LSTMs-100-100 with LSTM-100, we conclude more layers do not always mean better performance, the learning process can be difficult due to the deeper layers.
4. By analyzing LSTM-100-100 and Dense100-LSTM100, we find that even though time distributed fully connected layers induce more powerful learning ability, it degrades with poor generalization.
5. Figure 2a reflects the residual connect can boost convergence, especially in the first few epochs. From Fig. 2b, we can see that Residual-GRU has better learning ability i.e., with less loss at the end of training. Figure 2c demonstrates the proposed has better generalization capability during the test.

6. By observing yellow, blue, and red loss curves from Fig. 2b and c, we know that less training loss does not mean better generalization. However, the proposed model has both the lowest loss in the trade-off. This indicates the effectiveness of Residual-GRU for air pollutants prediction.

**Table 1.** RMSE and MAE among different models

| Model | RMSE | | MAE | |
|---|---|---|---|---|
| | Train | Test | Train | Test |
| **Residual-GRU** | **1.59** | **1.85** | **0.96** | **1.15** |
| Residual-LSTM | 1.62 | 1.87 | 0.98 | 1.16 |
| LSTM-100 | 1.65 | 1.90 | 1.00 | 1.18 |
| LSTMs-100-100 | 1.66 | 1.97 | 1.02 | 1.23 |
| Dense100-LSTM100 | 1.59 | 1.91 | 0.97 | 1.18 |
| SVR-rbf | 10.69 | 10.43 | 8.91 | 8.55 |
| SVR-poly | 10.61 | 10.35 | 8.83 | 8.46 |
| SVR-linear | 10.36 | 10.09 | 8.67 | 8.30 |
| LASSO | 3.88 | 4.64 | 2.33 | 2.76 |
| RandomForest | 2.11 | 2.69 | 1.29 | 1.64 |

## 4    Related Works

### 4.1    Deep Spatio-Temporal Learning

Inspired by deep learning on computer vision, [16] firstly applied deep learning to crowds flow prediction and proposed DeepST. This model adopted 3 deep CNNs to capture closeness, period and seasonal trend respectively. Besides spatio-temporal data, it also fused different source data such as weather for a better generalization. Furthermore, [15] proposed ST-ResNet which is an improved version of DeepST. To add deeper layers and get better precision, it introduced residual convolutional neural networks (CNN). With the similar philosophy as predicting citywide crowds flows, [9] adopted deep CNN on grid-based spatio-temporal data to make transportation network speed prediction. [10] applied ST-ResNet to real-time crime forecasting. [13] proposed a deep multi-view network (DMVST-Net) to predict taxi demand based on the hybrid of CNN, LSTM, and fully connected networks. By fusing different models, [5] proposed fusion convolutional LSTM (FCL-Net) to forecast passenger demand under on-demand ride services. They also suggested a random forest employed for feature selection can

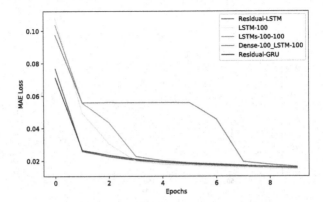

(a) MAE loss of the first 10 epochs during training

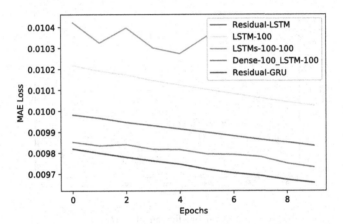

(b) MAE loss of the last 10 epochs during training

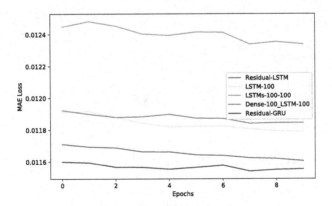

(c) MAE loss of the last 10 epochs during test

**Fig. 2.** Training MAE loss (Color figure online)

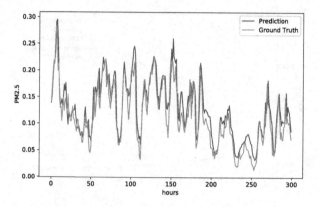

(a) Concentration prediction of $PM_{2.5}$

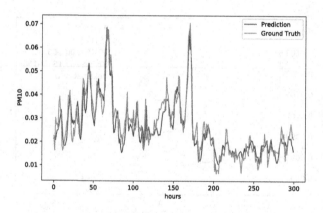

(b) Concentration prediction of $PM_{10}$

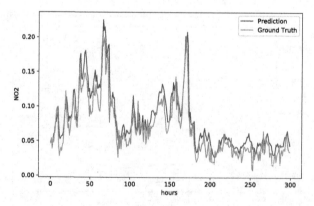

(c) Concentration prediction of $NO_2$

**Fig. 3.** Concentration prediction at one certain station

save training time without losing much accuracy. [12] creatively integrated convolution and LSTM together and introduced ConvLSTM for precipitation nowcasting. [2] proposed LSTM-based spatio-temporal learning for wind speed forecasting. [11] proposed short-term traffic flow forecasting with spatial-temporal correlation in a hybrid deep learning framework (CLTFP).

## 4.2   Air Quality Forecasting

[20] proposed the concept of spatial partition and aggregation and a hybrid shallow model based on multi-view learning to implement real-time air quality prediction over future 48 h. [7] developed a stacked auto-encoder network to extract deep representations of the concentration of PM2.5 and then use a logistic regression to predict. Such a two-stage training and fine-tuning framework is tedious and cannot learn from end-to-end. [6] used two vanilla LSTMs to predict the concentration of $O_3$ and $NO_2$ respectively. They furthermore transformed the predictive results into category label according to the AQI thresholds and measure results from accuracy. [14] introduced a new distributed fusion framework to fuse heterogeneous multi-source data, which can simultaneously capture the individual and overall effects from all influential factors for AQI prediction. [8] proposed GeoMAN using a multi-level attention-based RNN that considers multiple sensors and information fusion. The main contribution is the multi-level attention mechanism, i.e. local attention and global attention. A clear difference between previous works and ours is that the previous study is single-task but Residual-GRU can implement multi-task at one time. To the best of our knowledge, we are the first to utilize residual GRU to implement multi-task learning for air quality prediction.

## 5   Conclusion

In this paper, we propose a deep multi-task learning model to predict air quality. This model integrates residual connections into GRU and can predict multiple concentrations of pollutants at all sites simultaneously. We evaluate our method based on a real-world dataset and extensive experiments show the superiority of our method against 9 baselines.

## 6   Future Works

In the future, we will extend our method to solve the problem of long-term prediction and design more metrics for different considerations. Moreover, we will incorporate more external factors (e.g., weather) and explore new processing techniques such as fuzzy granulation.

**Acknowledgment.** This work was supported by the Natural Science Foundation of China (No. 61773324), the Fundamental Research Funds for the Central Universities (No. 2682015QM02) and the Australian Research Council (No. DP150101645).

# References

1. Akimoto, H.: Global air quality and pollution. Science **302**(5651), 1716–1719 (2003)
2. Ghaderi, A., Sanandaji, B.M., Ghaderi, F.: Deep forecast: deep learning-based spatio-temporal forecasting. arXiv preprint arXiv:1707.08110 (2017)
3. He, K., Zhang, X., Ren, S., Sun, J.: Identity mappings in deep residual networks. In: European Conference on Computer Vision, pp. 630–645 (2016)
4. Kampa, M., Castanas, E.: Human health effects of air pollution. Environ. Pollut. **151**(2), 362–367 (2008)
5. Ke, J., Zheng, H., Yang, H., Chen, X.M.: Short-term forecasting of passenger demand under on-demand ride services: a spatio-temporal deep learning approach. Transp. Res. Part C: Emerg. Technol. **85**, 591–608 (2017)
6. Kök, I., Şimşek, M.U., Özdemir, S.: A deep learning model for air quality prediction in smart cities. In: 2017 IEEE International Conference on Big Data (Big Data), pp. 1983–1990, December 2017
7. Li, X., Peng, L., Hu, Y., Shao, J., Chi, T.: Deep learning architecture for air quality predictions. Environ. Sci. Pollut. Res. **23**(22), 22408–22417 (2016)
8. Liang, Y., Ke, S., Zhang, J., Yi, X., Zheng, Y.: GeoMAN. In: International Joint Conference on Artificial Intelligence (IJCAI-18) (2018)
9. Ma, X., Dai, Z., He, Z., Ma, J., Wang, Y., Wang, Y.: Learning traffic as images: a deep convolutional neural network for large-scale transportation network speed prediction. Sensors **17**(4), 818 (2017)
10. Wang, B., Yin, P., Bertozzi, A.L., Brantingham, P.J., Osher, S.J., Xin, J.: Deep learning for real-time crime forecasting and its ternarization. arXiv preprint arXiv:1711.08833 (2017)
11. Wu, Y., Tan, H.: Short-term traffic flow forecasting with spatial-temporal correlation in a hybrid deep learning framework. arXiv preprint arXiv:1612.01022 (2016)
12. Xingjian, S., Chen, Z., Wang, H., Yeung, D.Y., Wong, W.K., Woo, W.C.: Convolutional LSTM network: a machine learning approach for precipitation nowcasting. In: Advances in Neural Information Processing Systems, pp. 802–810 (2015)
13. Yao, H., et al.: Deep multi-view spatial-temporal network for taxi demand prediction. arXiv preprint arXiv:1802.08714 (2018)
14. Yi, X., Zhang, J., Wang, Z., Li, T., Zheng, Y.: Deep distributed fusion network for air quality prediction. In: Proceedings of the 24th ACM SIGKDD International Conference on Knowledge Discovery and Data Mining. ACM (2018)
15. Zhang, J., Zheng, Y., Qi, D.: Deep spatio-temporal residual networks for city-wide crowd flows prediction. In: Proceedings of the AAAI Conference on Artificial Intelligence, pp. 1655–1661 (2017)
16. Zhang, J., Zheng, Y., Qi, D., Li, R., Yi, X.: DNN-based prediction model for spatio-temporal data. In: Proceedings of the International Conference on Advances in Geographic Information Systems, p. 92 (2016)
17. Zhang, Y., Bocquet, M., Mallet, V., Seigneur, C., Baklanov, A.: Real-time air quality forecasting, Part I: history, techniques, and current status. Atmos. Environ. **60**, 632–655 (2012)
18. Zhang, Y., Bocquet, M., Mallet, V., Seigneur, C., Baklanov, A.: Real-time air quality forecasting, Part II: state of the science, current research needs, and future prospects. Atmos. Environ. **60**, 656–676 (2012)

19. Zheng, Y., Liu, F., Hsieh, H.P.: U-Air: when urban air quality inference meets big data. In: Proceedings of the 19th ACM SIGKDD International Conference on Knowledge Discovery and Data Mining, KDD 2013, pp. 1436–1444. ACM (2013)
20. Zheng, Y., et al.: Forecasting fine-grained air quality based on big data. In: Proceedings of the 21th ACM SIGKDD International Conference on Knowledge Discovery and Data Mining, KDD 2015, pp. 2267–2276. ACM (2015)

# Research on the Prediction Technology of Ice Hockey Based on Support Vector Machine

Mengying Li[(✉)], Shanliang Xue[(✉)], and Sijia Cheng

College of Computer Science and Technology,
Nanjing University of Aeronautics and Astronautics, Nanjing, China
1136642649@qq.com, xuesl@nuaa.edu.cn, 277815008@qq.com

**Abstract.** The influencing factors of the ice hockey match result are complex, and there is a nonlinear relationship with the relevant predictive indicators. The input characteristics and parameter selection of the model have important influence on the prediction performance. Based on this, this paper proposes a support vector machine ice hockey situation prediction model based on principal component analysis, hybrid genetic algorithm and particle swarm optimization. This model uses principal component analysis to perform principal component analysis on the original features, this can reduce the dimensions of the original features effectively. Using hybrid genetic algorithm and particle swarm algorithm to optimize the parameters of support vector machine to establish a predictive model. The simulation results show that when principal component analysis is used to reduce the input features, the running time of the SVM prediction model based on principal component analysis, hybrid genetic algorithm and particle swarm optimization is reduced. Compared to a single genetic algorithm optimization parameter or a particle swarm optimization parameter support vector machine prediction model, the prediction accuracy and stability are significantly improved.

**Keywords:** SVM · Hybrid GAPSO algorithm · Prediction
Principal component analysis · Parameter optimization

## 1 Introduction

Ice hockey is a combination of changeable skating skills and skillful hockey skills, one of the more antagonistic collective ice sports, and a formal event for the Winter Olympic Games [1]. With the approaching of the Beijing Winter Olympics in 2022, the attention of ice hockey competition in China has increased significantly. The success of the competition is influenced by many factors [2], such as physical ability, psychology, technology, tactics and competition environment, in which the technical and tactical levels directly affect the result of the competition. So they are listed as the core factor of winning the competition in the event group theory of sports training. However, there is a complex nonlinear

© Springer Nature Switzerland AG 2018
L. Cheng et al. (Eds.): ICONIP 2018, LNCS 11305, pp. 104–115, 2018.
https://doi.org/10.1007/978-3-030-04221-9_10

relationship between the competition results and these core factors. Therefore, by analyzing these core factors, we can predict the situation of ice hockey match.

Machine learning algorithm has been widely applied to solve various complex forecasting problems. Kong and Xia [3,4] put forward the use of BP neural network to predict the situation of the match, which achievs good prediction results. However, the BP neural network algorithm theory itself has shortcomings, such as the slow learning speed, the selection of network structure without standard, the number of learning samples has great influence on the prediction accuracy, algorithm complexity and generalization ability [5]. Zeng, Gao and et al. [6,7] put forward the use of support vector machine (SVM) to predict sports events and solve the problem of small sample input of raw data, but simply using SVM does not achieve good prediction results. Ardjani, Yu and et al. [8,9] think that the input parameters of the model have great influence on the prediction precision when the support vector machines predict the complex nonlinear relation, so we propose to use the evolutionary algorithm to optimize the parameters of the SVM prediction model. The prediction accuracy is obviously improved, but the stability and time efficiency of the prediction results are significantly reduced.

SVM is a machine learning tool, which can better handle small sample, nonlinear and high dimensional pattern recognition problems. The model parameters are few and perform good in both learning accuracy and generalization learning ability. The penalty factor C and kernel parameter g of SVM have a great influence on the prediction effect, but the theory itself does not give the best method to obtain the value. Genetic algorithm (GA) can search multiple local best optimum and have good performance of global optimization, so it can effectively use historical information to speculate on the advantage set which the performance is expectedly improving of the next generation [10]. Particle swarm optimization (PSO) is a global intelligent optimization method. It has the advantages of fast searching speed, high efficiency, simple algorithm, and is suitable for real value processing [11].

Because of the complex influence factors of the competition results and the nonlinear relationship with the related prediction indexes and the less actual samples, the support vector machine model is used to predict the competition in this paper, beside the theory of principal component analysis is considered [12–14]. The main components are extracted from a large number of original indexes, and the model input is simplified to reduce the operational complexity of SVM in high dimensional space and the reduce the running time of the prediction model. And we also use hybrid GAPSO to optimize the penalty factor C of SVM and the kernel function parameter g. Based on principal component analysis and hybrid GAPSO, a support vector machine hockey racing condition prediction model (PCA-GAPSO-SVM) is proposed and applied to ice hockey prediction.

## 2    Establishment of Ice Hockey Condition Prediction Model Based on Support Vector Machine

### 2.1    Principal Component Analysis

Suppose that $X_1, X_2, ..., X_p$ is the P original index to represent the ice hockey racing prediction model based on support vector machine, denote as $X = (X_1, X_2, ..., X_p)^T$. $\Sigma$ is the covariance matrix of the input vector X of ice hockey prediction model based on support vector machine. The eigenvalues of and corresponding orthogonal unit eigenvectors are $\lambda_1 \geq \lambda_2 \geq ... \geq \lambda_p \geq 0$ and $e_1, e_2, ..., e_p$, The i principal component of the input eigenvector X is $Y_i = e_i^T X = e_{i1} X_1 + e_{i2} X_2 + ... + e_{ip} X_p$, of which $i = 1, 2, ..., p$. The ratio of information extracted from a principal component to total information is called contribution rate, and the contribution rate of Y1 is the largest, which indicates that Y1 has the strongest ability to synthesize original variable information. The sum of the contributions of the former m principal components $Y_1, Y_2, ..., Y_m$ is the cumulative contribution rate.

In practical applications, we usually set m < p, so that the cumulative contribution rate of the former m principal components reaches a higher proportion (for example, 85% to 90%). The former m principal component $Y_1, Y_2, ..., Y_m$ is used to replace the original index $X_1, X_2, ...., X_p$ of the ice hockey racing condition prediction model based on support vector machine, it does not lose much information because the dimension of the original index. The PCA algorithm description is shown in Table 1:

**Table 1.** The frame diagram of prediction algorithm

---

**Input:** sample set $D = \{x_1, x_2, ..., x_n, y\}$;
    Dimension of low dimensional space **m**
**Process:**
    1: Centralization of all the samples $Y_1, Y_2, ..., Y_m$
    2: Calculated covariance matrix of the sample $XX^T$;
    3: Eigenvalue decomposition for covariance matrix $XX^T$
    4: Feature vectors corresponding to the maximum **m** eigenvalues $w_1, w_2, ..., w_m$.
**Output:** projection matrix $W = (w_1, w_2, ..., w_m)$

---

After the dimensionality reduction, the dimension m of low dimensional space is usually specified beforehand, or selected by cross validation of the k nearest neighbor classifier in a low dimension space with different m values. You can also set a reconstruction threshold from the perspective of reconfiguration, and then select the minimum m value for the lower set:

$$\frac{\sum_{i=1}^m \lambda_i}{\sum_{i=1}^m \lambda_i} \geq t \tag{1}$$

## 2.2   Support Vector Machine

The samples after PCA dimension reduction were divided into two groups: training set and test set. Suppose that $T = \{(x_1, y_1), (x_2, y_2), ..., (x_N, y_N)\}$, $y_i \in \{-1, +1\}$ is a set of training samples, where $X_i$ is the input vector, and $Y_i$ is the output vector, that is, the winner and the negative of the main team. N is the total number of training sample data points. SVM uses the next formula to estimate the function:

$$f(x) = \omega \cdot \phi(x) + b \tag{2}$$

where $\phi(x)$ is a nonlinear mapping from the input space to the high dimensional feature space. For a given set of sample sets T, and $\varepsilon \geq 0$ makes $|y_i - f(x)| \leq \varepsilon$, then f (x) is $\varepsilon$- linear regression of the sample set T, and the coefficients w and b are estimated by minimizing J:

$$J = c\frac{1}{n} \sum_{i=1}^{n} L_\varepsilon(y_i, f(x_i)) + \frac{1}{2}\|\omega\|^2 \tag{3}$$

where $c\frac{1}{n} \sum_{i=1}^{n} L_\varepsilon(y_i, f(x_i))$ is an empirical risk, $L_\varepsilon$ is a loss function, $\frac{1}{2}\|\omega\|^2$ is a regularized item to improve generalization ability of estimation function. c is a penalty factor, which determines the balance between empirical risk and regularized items. In order to find coefficients of and b, we introduce relaxation variables $\xi_i$ and $\xi_i^*$. The value of the relaxation variable actually indicates how far the corresponding point is, and the larger the value of the relaxation variable, the farther the corresponding point is from the classification plane, Thus obtained:

$$J = \min\left\{\frac{1}{2}\|\omega\|^2 + c\sum_{i=0}^{n}(\xi_i + \xi_i^*)\right\}$$
$$s.t. \begin{cases} \omega \cdot \phi(x_i) + b - y_i \leq \varepsilon + \xi_i^* \\ -\omega \cdot \phi(x_i) + b - y_i \leq \varepsilon + \xi_i^* \\ \xi_i, \xi_i^* \geq 0 \end{cases} \tag{4}$$

By using duality theory, the upper form can be transformed into a two programming problem and establishing Lagrange equation and making $K(x, y) = \phi(x) \cdot \phi(y)$, the duality of optimization problem can be obtained:

$$J = \min\left\{\frac{1}{2}\sum_{i,j=1}^{n}(a_i^* - a_i)(a_j^* - a_j)K(x_i, x_j) - \sum_{i=1}^{n}[a_i^*(y_i - \varepsilon) - a_i(y_i + \varepsilon)]\right\}$$
$$s.t. \sum_{i=1}^{n}(a_i^* - a_i) = 0, a_i^*, a_i \in [0, c]$$

$$\tag{5}$$

where $a_i^*, a_i$ is the Lagrange coefficient, the solution can be obtained. Also because of $b = \begin{cases} y_i + \varepsilon - \sum_{i,j=1}^{n}(a_i - a_i^*)K(x_i, x_j), a_i \in (0, c) \\ y_i - \varepsilon - \sum_{i,j=1}^{n}(a_i - a_i^*)K(x_i, x_j), a_i \in (0, c) \end{cases}$ , finally, we can get the classification result, that is, the expression of ice hockey results is:

$$f(x) = \omega \cdot \phi(x)$$
$$= \sum_{i=1}^{n} (a_i - a_i^*)\phi(x_i) \cdot \phi(x) + b \tag{6}$$
$$= \sum_{i=1}^{n} (a_i - a_i^*)K(x_i, x) + b$$

where $K(x_i, x)$ is a kernel function, it can be any positive definite function satisfying the Mercer condition. In this paper, the radial basis function is selected for the kernel function, the expression can be expressed as:

$$K(x, y) = \exp\left(-g\|x - y\|^2\right) \tag{7}$$

# 3    Improved Hybrid GAPSO Algorithm and Process Design of Ice Hockey Condition Prediction Model

The classification accuracy of SVM is related to training parameters [15]. Therefore when using the prediction model of the hockey condition based on support vector machine, if the selection of the parameters of the kernel function or the penalty factor C is unreasonable, the accuracy of the prediction model of the hockey condition based on support vector machine will be reduced The optimal is C changes with the difference of the characteristic subspace. When the C exceeds a certain value, the complexity of the SVM reaches the maximum allowable value of the sample space, and the empirical risk and generalization ability hardly change.

For the radial basis kernel function, the larger the parameter g is, the wider the sample's output response interval is, the smaller the structure risk of the optimal classification surface is, while the experience risk will increase; However the smaller the g is, the narrower the output response interval of the sample is, and the smaller the empirical risk will be, while the structural risk will increase, and it will easily lead to the over fitting phenomenon and reduce the performance of the SVM. Thus, a support vector machine classifier with superior performance can be obtained by selecting the appropriate support vector machine parameters, which can improve the accuracy of the prediction model based on the support vector machine (SVM) [16].

## 3.1    Hybrid GAPSO Algorithm

Compared with GA, PSO algorithm is simple because of no selection, crossover and mutation operation. However, the particle position in the algorithm mainly evolute by comparing its own position and the current optimal position in the surrounding group particles. The pattern is single, so it makes the convergence speed is not high in the later period of calculation and also it is easy to fall into the local extremum. The diversity of the solution is improved because of the existence of the selection, cross and mutation of GA which often leads to a large number of useless iterations, and lower time of calculation, when the GA is solved to a certain extent. Therefore, hybrid GA-PSO algorithm is proposed to optimize SVM parameters and improve the efficiency of the ice hockey prediction model based on support vector machine. PSO converges fast and has few adjustable parameters, but it is easy to fall into local optimal solution [17]. GA is relatively complex, with many adjustable parameters and binary encoding. However, due to the mutation operation, the probability of getting into the local optimal solution is smaller than that of PSO [18].

Considering that both PSO and GA have commonalities on optimizing by iterative, and to make use of the fast convergence ability of PSO and the local search ability of GA, we proposes a hybrid GAPSO algorithm to optimize the parameters C and G of SVM and train the training samples in this paper. The algorithm searches the overall optimal solution by comparing the optimal solutions of each group of two algorithms.

The concrete implementation steps of the hybrid GAPSO algorithm are described as follows:

**Step 1.** Initialize the relevant parameters of population GA and PSO. They contain maximum evolutionary algebra, population maximum number, cross probability in GA, mutation probability, particle local search capability (C1) in PSO, global search capability (C2), cross validation times, and output parameters C, G and so on.

According to the variation range of the initialized C and G, we can determine the number of chromosomes, and then R = unidrnd (N, m, n) is used to produce a group of discrete, uniform random integers with only 0 or 1, that is we generate a GA population by doing binary coding for the population; Besides the GA population is decoded to generate PSO population, and the PSO particle velocity is initialized.

**Step 2.** Taking SVM training classification accuracy rate as individual fitness, the GA population chooses the best solution [C, G], and PSO population updates the individual optimal solution pbest, and global optimal solution gbest.

**Step 3.** According to the optimal solution of the two populations, the overall optimal solution is generated. If the termination condition is satisfied, turn to Step5, and end parameter optimization utilize the assignment rules proposed in this paper to assign between the two algorithm.

Assignment rules: If the fitness of the PSO optimal solution is higher than that of GA, then the fitness of PSO will be taken as the overall optimal solution and assigned to the worst chromosome in the genetic algorithm. If the fitness of the PSO optimal solution is lower than that of GA, then the chromosome with the highest fitness in GA will be used as the overall optimal solution and assigned to the worst particle in the PSO algorithm.

$$v_i = w \times v_i + c_1 \times rand() \times (pbest_i - x_i) + c_2 \times rand() \times (gbest_i - x_i) \quad (8)$$

$$x_i = x_i + v_i \quad (9)$$

where $i = 1, 2, ..., N$, N is the total number of particles in the population. $w$ is the inertia factor, its value is non-negative; $v_i$ is the velocity of the particle; rand() is a random number between 0 and 1; $x_i$ is the particle's position; $c_1$ is the particle's own cognitive coefficient; $c_2$ is the particle's cognitive coefficient to the population;

**Step 4.** GA population chooses parent population to perform crossover and mutation operations; PSO population updates particle velocity and location, Return to Step2. Particle velocity and position update rules:

**Step 5.** Output the overall optimal solution of C and G.

Where the algorithm's terminating conditions include: 1. Reaching the objective function to achieve the convergence precision; 2. The number of iterations is up to N which is the maximum number we set.

Finally we get C and G which are optimized parameters of SVM classifier.

### 3.2 Prediction Process of Ice Hockey Condition Prediction Model Based on Support Vector Machine

The process of predicting race conditions is to establish a nonlinear relationship between competition results and representation indicators. To sum up, the prediction model of ice hockey condition based on support vector machine is set up, and this model is used to predict the situation of ice hockey.

The first stage is data preprocessing, as shown in Fig. 1, is to reduce the input index by principal component analysis. Then in the second stage, the hybrid GAPSO algorithm is used to optimize the parameters of the support vector machine. And in the third stage, SVM is used to train and predict the results. Some details are shown in Fig. 1:

## 4    Case Experiment

### 4.1 Evaluation Standard

In this part, we utilize the effect of the confusion matrix of classification results to measure the prediction, as Table 2 showing. In order to verify the effectiveness of the proposed method, and consider that most of the error optimization is completed by GA. The PSO is used to complete stability of the prediction results. To prove the running time is reduced by using hybrid GAPSO algorithm, we compare the hybrid GAPSO with GA-PCA-SVM, and PSO-PCA-SVM. Also we compare hybrid GAPSO-PCA-SVM. Finally we do simulation experiment to verify it.

### 4.2 Experimental Results and Analysis

In our experiment, the largest evolution algebra of the hybrid GAPSO population is defaulted set as 200, which is limited in the general value range [100,500]; The

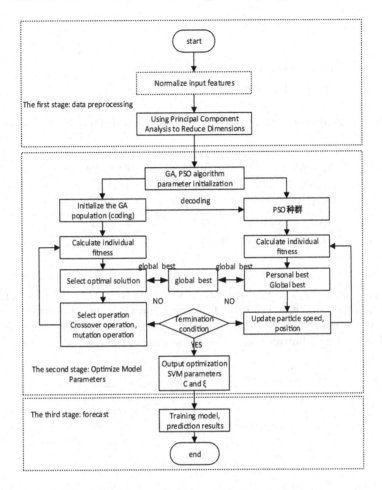

**Fig. 1.** The frame diagram of prediction algorithm

maximum number of population is defaulted set as 20 which is also limited in the value [20,100]; PSO's parameter local search capability which is defined as c1 is initially set as 1.5, The initial global search capability of the parameter which is denoted as c2 is 1.7; Besides, we set the number of cross validation as 5 and the run 20 times. Then the average accuracy and running time of each algorithm are shown in Table 3.

$$Accuracy = \frac{TP + TN}{TP + FP + FN + TN} \tag{10}$$

The following Fig. 2 (the abscissa of the figure below represents the number of runs and the ordinate represents the accuracy) is found, The stability of the prediction of SVM parameter selection by using Grid Search and the prediction result of SVM parameter optimization by PSO have little difference, but the

**Table 2.** Confusion matrix

|  | Actual positive | Actual negative |
|---|---|---|
| Predicted positive | TP | FP |
| Predicted negative | FN | TN |

**Table 3.** The prediction accuracy and running time of different algorithms are compared

| Algorithm | Accuracy rate | Running time |
|---|---|---|
| GA-SVM | 79.6 | 152.635 |
| PSO-SVM | 75.8 | 262.408 |
| GAPSO-SVM | 81.5 | 526.103 |
| GAPSO-N-PCA-SVM | 83.2 | 471.627 |

accuracy of prediction is very different. At the same time, GA-SVM has a higher prediction accuracy than PSO-SVM. After improving the optimized parameter algorithm, using GAPSO to select SVM parameters not only improves the prediction accuracy, but also improves the stability of the single GA prediction.

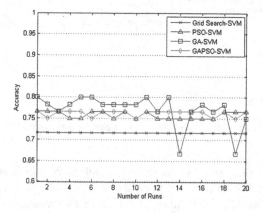

**Fig. 2.** Comparison of prediction accuracy of different parameter optimization algorithms

Figure 3 shows that the accuracy of different parameter optimization algorithm varies with the number of operations and Fig. 4 presents the run time of these four algorithms. As we observed, running time of the algorithm is shortened after using the PCA algorithm. However, it has no influence on the prediction results and stability of the model.

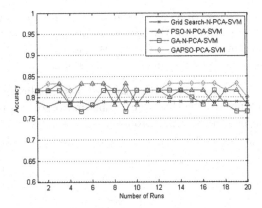

**Fig. 3.** Comparison of accuracy of mixed GAPSO and different parameter optimization algorithms

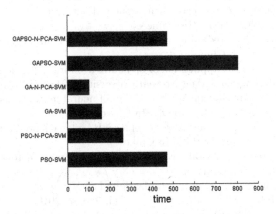

**Fig. 4.** Algorithm run time bar chart

## 5   Conclusion

In this paper, we focus on the selection of input variables and model parameters of the prediction model. To predict the situation of ice hockey match. To predict the situation of ice hockey match, we propose support vector machine prediction method based on principal component analysis and hybrid GAPSO algorithm. The result are showed as follow:

(1) Using the normalization and principal component analysis, the factors of the squadron affecting the competition condition are reduced, and the input of the simplified model is used to get some important variables covering the initial data. The results show that the input of the optimized prediction model can achieve the effect of reducing the running time.

(2) Comprehensive utilization of PSO's fast convergence ability and GA's local search capability. In this paper, hybrid GAPSO algorithm is proposed to

optimize the parameters of SVM. The measurement proves that the prediction accuracy and stability of the prediction model are improved.

(3) Compared with the traditional GA-SVM and PSO-SVM, the prediction accuracy is increased by 3.6% and 7.4% respectively.

The method proposed in this paper is a relatively new application in the field of support vector machines, so the theoretical contribution of this paper is limited. In other cases, whether the method proposed in this paper can work well is a problem to be solved in the future.

# References

1. Ning, W.J., Wang, S.: Restricted factors and countermeasures for the development of Chinese ice hockey players. China Winter Sports **6**, 23–25 (2013). (In Chinese)
2. Kong, Q.Y., Li, X.Q.: A method of predicting table tennis match based on the combination of condition number and genetic neural network. J. Guilin Univ. Areosp. Technol. **4**, 528–532 (2015). (In Chinese)
3. Wen, X.H., Wang, S.: Prediction method of time series neural network. J. Electron. **6**, 456–462 (1994). (In Chinese)
4. Xia, F.: Research on the application of bp neural network in predicting the outcome of football matches. Chongqing Normal University (2017). (In Chinese)
5. Li, J., Han, Z.Z.: The learning error function of neural network and its generalization. Control Decis. **1**, 95–97 (2000). (In Chinese)
6. Zeng, P., Zhu, A.M.: A SVM-based model for NBA playoffs prediction. J. Shenzhen Univ. Sci. Eng. **1**, 62–71 (2016). (In Chinese)
7. Gao, H.: The prediction of sailing speed based on support vector machine. Fudan University (2014). (In Chinese)
8. Ardjani, F., Sadouni, K., et al.: Optimization of SVM multiclass by particle swarm (PSO-SVM). Int. J. Mod. Educ. Comput. Sci. **2**, 1–4 (2010)
9. Yu, E., Cho, S.: GA-SVM wrapper approach for feature subset selection in keystroke dynamics identity verification. In: International Joint Conference on Neural Networks, USA, pp. 2253–2257. IEEE (2003)
10. Leung, F.H.F., Lam, H.K., et al.: Tuning of the structure and parameters of a neural network using an improved genetic algorithm. In: The 27th Annual Conference of the IEEE Industrial Electronics Society, pp. 79–88. IEEE (2003)
11. Kennedy, J., Eberhart, R.: Particle swarm optimization. In: Gass, S.I., Fu, M.C. (eds.) Encyclopedia of Operations Research and Management Science, pp. 1942–1948. Springer, Boston (2002). https://doi.org/10.1007/978-1-4419-1153-7_200581
12. Zhao, Z., Guo, H.: Visualization study of high-dimensional data classification based on PCA-SVM. In: IEEE Second International Conference on Data Science in Cyberspace, pp. 346–349. IEEE (2017)
13. Li, J., Zhao, B., et al.: Face recognition system using SVM classifier and feature extraction by PCA and LDA combination. In: International Conference on Computational Intelligence and Software Engineering, pp. 1–4. IEEE (2009)
14. Liang, S.J., Zhang, Z.H., et al.: Dimensionality reduction method based on PCA and KICAV. Syst. Eng. Electron. **9**, 2144–2148 (2011). (In Chinese)
15. Vapnik, V.N., et al.: The Essence of Statistical Learning Theory. Tsinghua University Press, Beijing (2000)

16. Alba, E., Garcanieto, J., et al.: Gene selection in cancer classification using PSO/SVM and GA/SVM hybrid algorithms. In: Evolutionary Computation, CEC, pp. 284–290. IEEE (2007)
17. Li, M.: Convergence analysis and improvement reseach of standard particle swawrm optimization algorithm. Bohai University (2017). (In Chinese)
18. Shi, Y.: Parameter estimation of nonlinear system based on SAPSO-MSFLA algorithm. Southwest JiaoTong University (2013). (In Chinese)

# Deep Structure of Gaussian Kernel Function Networks for Predicting Daily Peak Power Demands

Dae Hyeon Kim, Ye Jin Lee, Rhee Man Kil$^{(\boxtimes)}$, and Hee Yong Youn

College of Software, Sungkyunkwan Univesity, 2066, Seobu-ro, Jangan-gu, Suwon, Gyeonggi-do 440-746, Korea
{kdh92,yejini824,rmkil,youn7147}@skku.edu

**Abstract.** This paper proposes a novel method of predicting daily peak power demands using the deep structure of Gaussian kernel function networks (GKFNs). For the prediction model, the whole time series is divided into multiple parts and each part is trained using a GKFN. Then, the trained GKFNs are combined using the deep structure of GKFNs to minimize the mean square errors (MSEs) of prediction model. As a consequence, the proposed deep structure of GKFNs provides an improved performance of prediction accuracy compared with canonical GKFNs. The simulation for predicting daily peak power demands in Korea reveals that the proposed prediction model has the merits in prediction performances compared with the GKFN model and also other prediction models such as the $k$-NN and SVR.

**Keywords:** Daily peak power demand · Prediction model
Gaussian kernel function network · Deep structure

## 1 Introduction

In managing electric power plants, producing more power than necessary produces waste, and producing less than necessary power results in catastrophic consequences such as power failure or emergency power supply. In this respect, accurate prediction of short-term power demand is an essential factor for ensuring stability and efficient operation of the power system. It has been known that electric power load series can be treated as the nonlinear and non-stationary time series dependent upon the seasonal trend and also economic situation. According to [1], power load forecasting is classified into three types according to the period: they are up to 1 day for short-term load forecasting (STLF), 1 day to 1 year for medium-term load forecasting (MTLF), and 1–10 years for long-term load forecasting (LTLF). In this paper, a new prediction model for SLTF is investigated; that is, predicting peak power demands in the next day.

The SLTF method can be classified into four types [2]: they are statistical techniques, artificial intelligence (AI) techniques, knowledge based expert systems, and hybrid techniques. In the case of statistical techniques, the regression

© Springer Nature Switzerland AG 2018
L. Cheng et al. (Eds.): ICONIP 2018, LNCS 11305, pp. 116–126, 2018.
https://doi.org/10.1007/978-3-030-04221-9_11

model [3] or stochastic time series method such as ARIMA model [4] is usually used. The artificial neural networks (ANNs) are usually used as an AI technique [5,6]. The applications of ANNs began in the early 90s. Since then, a considerable amount of research has been done in this area [2]. Knowledge based expert systems make decisions based on expert experiences in a rule-based way and predict power demands through decision trees [7]. Finally, hybrid approaches combine two or more approaches to overcome the drawbacks of the original method [8,9]. In the majority of these cases, they use environmental features such as the weather information for power load prediction [3–9]. When predicting data that have a large impact on power usage or a large area of load forecasting, the complexity of predictions increases due to various weather information and the prediction accuracy is significantly influenced by the settings of environmental features. From this context, this paper proposes a novel method of predicting the maximum power demands in the next day only by using the maximum power demand series without using environmental features. For this purpose, a Gaussian Kernel Function Network (GKFN) [10] is selected since this model can be applied to various function approximation problems. In this model, the input structure for the given prediction model of daily peak power demands is searched by the analysis of daily peak power demand series. Then, in order to reduce the variance of the prediction model, the deep structure of GKFNs is investigated. As a consequence, the proposed model provides an improved performance of prediction accuracy compared with canonical GKFNs and also other prediction models such as the $k$-NN and support vector regression (SVR). To demonstrate the effectiveness of the proposed method, the daily peak power series from June 2014 to August 2017 in Korea is collected and used for the training of prediction model. The proposed approach of deep learning model is applicable to various problems of time series prediction or regression problems.

## 2    Analysis of Daily Peak Power Demand Series

For the analysis of daily peak power demand series, the dynamic structure of generating time series should be investigated. In this analysis, the delay coordinate embedding theory [11,12] is usually used. However, these methods only consider the attractor in the phase space, not the prediction model. In this context, this paper presents a method of determining the input structure of time series prediction models. Let the daily peak power demand series in discrete time be $x(i)$, where $i$ represents the time index. Then, the previous $E$ data including the current data $x(i)$ can be collected using the delay time $\tau$; that is, the $E$-dimensional vector is described as

$$\boldsymbol{x}_{\tau,\,E}(i) = (x(i),\, x(i-\tau),\ldots,x(i-(E-1)\tau)),\quad i=(E-1)\tau,\cdots \quad (1)$$

Here, the prediction model $f$ is described as

$$x(t_i + P) = f(\boldsymbol{x}_{\tau,\,E}(i)), \quad (2)$$

where $P$ represents the prediction step.

For the estimation of prediction model, an estimation function $\hat{f}$ should be trained for the target function $f$. In this case, the proper input structure (or the embedding parameters $\tau$ and $E$) of $f$ should be determined because the performance of the prediction model is significantly influenced by the choice of embedding parameters. For the determination of the embedding dimension $E$, the estimation of correlation dimension of the dynamical system [13] can be used. However, this method of determining the embedding dimension $E$ can be applied only to the attractor in the phase space, not the prediction model. In this context, a new method of determining the embedding parameters is suggested using the smoothness measure for predicting daily peak power demands.

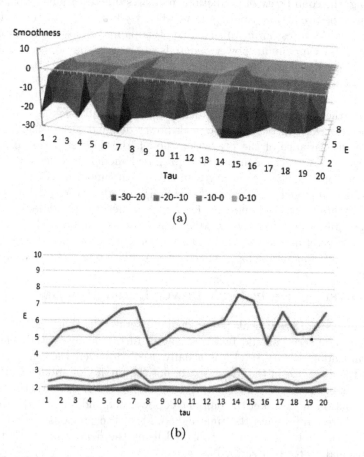

**Fig. 1.** The plot of smoothness measure for predicting daily peak power demands: (a) and (b) represent the 3D plot and contour map of smoothness measure. The selected embedding dimension $E = 5$ and the delay time $\tau = 19$ is indicated as a dot.

First, let us represent the nearest neighbor of the vector as $\boldsymbol{x}_{\tau, E}(k)$ by $\boldsymbol{x}_{\tau, E}^1(k)$. Then, the gradient of $f$ at each point $\boldsymbol{x}_{\tau, E}(k)$ is described by

$$\Delta f(\boldsymbol{x}_{\tau, E}^1) = \frac{\left|f(\boldsymbol{x}_{\tau, E}(i)) - f(\boldsymbol{x}_{\tau, E}^1(i))\right|}{||\boldsymbol{x}_{\tau, E}(i) - \boldsymbol{x}_{\tau, E}^1(i)||}, \tag{3}$$

where the norm in the denominator represents the Euclidian distance in $\mathbb{R}^E$. From this definition of gradient, the smoothness measure $S(\tau, E)$ [14] of a target function $f$ is described by

$$S(\tau, E) = 1 - \frac{1}{n - (E-1)\tau} \sum_{i=(E-1)\tau}^{n-1} \Delta f(\boldsymbol{x}_{\tau, E}^1(i)). \tag{4}$$

Here, the optimal the embedding parameters $\tau$ and $E$ can be determined by searching the region with higher values of smoothness measure (usually, near the smoothness value of 0). In this region, the smallest embedding dimension is preferred to reduce the sample complexity of training models. For the analysis of predicting daily peak power demands, the smoothness measure of (4) is calculated for every delay time $\tau$ between 1 and 20, and every embedding dimension $E$ between 2 and 10. Here, the smoothness measure for predicting daily peak power demand series is calculated and shown in Fig. 1. For this prediction problem, the smallest embedding dimension $E$ with positive value of smoothness measure is selected first; that is, the average gradient value in the phase space is less than 1. Then, for the selected $E$, the value of delay time $\tau$ is selected when smoothness measure has the maximum value. As a result, for the daily peak power demand series, the one step prediction model can be described by

$$x(t+1) = f(x(t), x(t-19), x(t-38), x(t-57), x(t-76)). \tag{5}$$

From this time series analysis, it is demonstrated that the input structure of 5 data with the delay time of 19 provides a good prediction model for training the data of daily peak power demands.

## 3   Deep Structure of GKFNs

For the prediction of peak power demands, a network with Gaussian kernel function is selected since this network is good for function approximation problems. Here, the prediction model $\hat{f}$ is given by

$$\hat{f}(\boldsymbol{x}) = \sum_{i=1}^{m} w_i \psi_i(\boldsymbol{x}), \quad \psi_i(\boldsymbol{x}) = e^{-||\boldsymbol{x} - \boldsymbol{\mu}_i||^2 / 2\sigma_i^2}, \tag{6}$$

where $m$ represents the number of kernel functions, $w_i$ represents the connection weight between the output and the $i$th kernel function $\psi_i$, and $\boldsymbol{\mu}_i$, and $\sigma_i$ represent the mean and standard deviation of the $i$th kernel function, respectively.

In (6), the number of kernels $m$ and also kernel parameters $w_i$, $\boldsymbol{\mu}_i$, and $\sigma_i$ should determined. For kernel parameters, [10] suggested an efficient estimation method to minimize the mean square error (MSE). The optimal number of kernels [15] can be determined by comparing the trained MSE with the estimated noise variance. In this prediction problem, the future (or target) value $x(t_k + P)$ is described by

$$x(t_i + P) = f(\boldsymbol{x}_{\tau, E}(i)) + \epsilon, \tag{7}$$

where $f$ represent a prediction function for the given input $\boldsymbol{x}_{\tau, E}(i)$ and $\epsilon$ represents a random noise with mean 0 and variance $\sigma^2$.

Then, for the target value $x(t_i + P)$, the predicted value $\hat{x}(t_i + P)$ is described by

$$\hat{x}(t_i + P) = \hat{f}(\boldsymbol{x}_{\tau, E}(i)). \tag{8}$$

The expected risk (or MSE) between the target and predicted values is determined by

$$E[(\hat{x} - x)^2] = E[(\hat{f} - f)^2] + Var(\epsilon), \tag{9}$$

where $E[(\hat{f} - f)^2]$ represents the regression error and $Var(\epsilon)$ represents the noise variance.

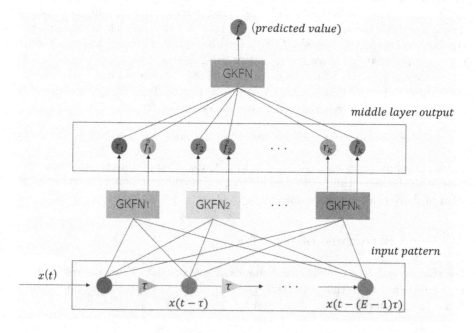

**Fig. 2.** Deep structure of GKFNs: each GKFN in the first layer is combined to reduce the variance of prediction model.

Then, the regression error is further decomposed by

$$
\begin{aligned}
E[(\hat{f} - f)^2] &= E[(\hat{f} - E[\hat{f}] + E[\hat{f}] - f)^2] \\
&= E[(\hat{f} - E[\hat{f}])^2] + E[(\hat{f} - f)^2] \\
&= Var(\hat{f}) + Bias^2(\hat{f}).
\end{aligned}
\tag{10}
$$

Suppose there are $k$ similar prediction models $\hat{f}_1, \hat{f}_2, \cdots, \hat{f}_k$; that is, the bias terms of $k$ prediction models are small and similar. Then, the optimal combination of these prediction models to minimize the MSE is described by

$$
\tilde{f} = \frac{1}{\sum_{i=1}^{k} 1/Var(\hat{f}_i)} \sum_{i=1}^{k} \frac{\hat{f}_i}{Var(\hat{f}_i)}.
\tag{11}
$$

In this case, the $Var(\tilde{f})$ is always less than the minimum value of $Var(\hat{f}_i)$, $i = 1, 2, \cdots, K$; that is,

$$
Var(\tilde{f}) = \frac{1}{\sum_{i=1}^{k} 1/Var(\hat{f}_i)} < \min_{i=1,2,\cdots,k} Var(\hat{f}_i).
\tag{12}
$$

However, in the regression model, $Var(\hat{f}_i)$ is changed according to the input pattern $\boldsymbol{x}$; that is, $Var(\hat{f}_i)$ is small near the center of input data distribution and vice versa. From this context, the average distance between the input pattern and center points of kernel functions of the $j$th prediction model $r_j$ is defined as the representative of input position:

$$
r_j(\boldsymbol{x}) = \left( \frac{1}{m_j} \sum_{i=1}^{m_j} ||\boldsymbol{x} - \boldsymbol{\mu}_{ij}||^2 \right)^{1/2},
\tag{13}
$$

where $m_i$ and $\boldsymbol{\mu}_{ij}$ represent the number of kernel functions and the mean of the $j$th prediction model, respectively. Then, for $k$ prediction models, the representative of input position $\boldsymbol{r}$ is described by

$$
\boldsymbol{r}(\boldsymbol{x}) = (r_1(\boldsymbol{x}), r_2(\boldsymbol{x}), \cdots, r_k(\boldsymbol{x}))
\tag{14}
$$

and a vector $\boldsymbol{y}$ as $k$ predicted values is described by

$$
\boldsymbol{y}(\boldsymbol{x}) = (\hat{f}_1(\boldsymbol{x}), \hat{f}_2(\boldsymbol{x}), \cdots, \hat{f}_k(\boldsymbol{x})).
\tag{15}
$$

From these vectors of $\boldsymbol{r}$ and $\boldsymbol{y}$, an unified vector $\boldsymbol{z}$ is defined by

$$
\boldsymbol{z}(\boldsymbol{x}) = (\boldsymbol{r}(\boldsymbol{x}), \boldsymbol{y}(\boldsymbol{x})).
\tag{16}
$$

Then, for the combination of $k$ prediction models of (11), another nonlinear model is suggested using $\tilde{m}$ Gaussian kernel functions as

$$
\tilde{f}(\boldsymbol{x}) = \sum_{i=1}^{\tilde{m}} w_i \psi_i(\boldsymbol{z}(\boldsymbol{x})).
\tag{17}
$$

As a result, the deep structure of GKFNs is constructed as illustrated in Fig. 2. For the training of the combination model of (17), the training patterns of $z(x) = (r(x), y(x))$ are applied after the training of $k$ prediction models for the purpose of minimizing the variance of prediction model. From this context, the deep structure of GKFNs is constructed using the following learning algorithm:

Learning Algorithm of Deep Structure of GKFNs

**Step 1.** For the given time series $x(t)$ and the prediction step $P$, the optimal embedding parameters $\tau$ and $E$ are determined by the smoothness measure of (4). Then, the target function is described by

$$x(t + P) = f(x(t)), \quad x(t) = (x(t), x(t - \tau), \cdots, x(t - (E - 1)\tau)).$$

**Step 2.** For the selected values of $E$, $\tau$, and $P$, make training patterns as

$$(x(t_i), x(t_i + P)), \quad i = 0, 1, \cdots, n - 1$$

and divide the training patterns evenly into $k$ sets.

**Step 3.** Determine the GKFN model as (6) and each set of training patterns is applied for the training of each GKFN using the learning algorithm of [10]. Then, $k$ GKFNs are obtained after the training of $k$ sets of training patterns.

**Step 4.** For the whole training patterns, obtain the input training vectors of

$$z(x(t_i)) = (r(x(t_i)), y(x(t_i))), \quad i = 0, 1, \cdots, n - 1$$

of (16) using $k$ trained GKFNs.

**Step 5.** Determine the GKFN model in the upper layer as (17) and the GKFN is trained for the training patterns of

$$(z(x(t_i)), x(t_i + P)), \quad i = 0, 1, \cdots, n - 1$$

using the learning algorithm of [10].

The proposed learning algorithm provides the reduction of the variance of the prediction model by combining $k$ GKFNs and it contributes the improvement of prediction accuracy (or the reduction of MSE).

## 4    Simulation

For the simulation for predicting daily peak power demands, the series of peak power data from June 2014 to August 2017 in Korea were collected for the prediction model. Among them, the data from June 2014 to May 2017 were used as the training data set, whereas the rest of data from May 2017 to June 2017 were used as test data set. The training data set was equally divided into three parts in which each part was divided into training and validation data sets; that is, the first 11 month data were used as the training data set, whereas the last one month data were used as validation data set as described in Table 1. These data were also normalized to a value between 0 and 1. For the prediction

**Table 1.** Division of daily peak power demand series into the training, validation, and test data sets.

| Name | Period |
|---|---|
| Train data set 1 | June 2014 ~April 2015 |
| Validation data set 1 | May 2015 |
| Train data set 2 | June 2015 ~April 2016 |
| Validation data set 2 | May 2016 |
| Train data set 3 | June 2016 ~April 2017 |
| Validation data set 3 | May 2017 |
| Test data set | June 2017 ~August 2017 |

model, the proper input structure (embedding dimension E and delay time $\tau$) were determined by the smoothness measure of (4) for the amount of prediction time $P = 1$; that is, one-step prediction. As a consequence, the prediction model has the form of (5).

For the evaluation of prediction models, the following performance measures were used: the root mean square error (RMSE) defined by

$$RMSE = \left( \frac{1}{l} \sum_{i=0}^{l-1} (y(t_i) - \hat{y}(t_i))^2 \right)^{1/2}, \tag{18}$$

where $y(t_i)$ and $\hat{y}(t_i)$ respectively represent the target and estimated values, and $l$ represents the number of test data, and the coefficient of determination $R^2$ defined by

$$R^2 = 1 - \left( \sum_{l=0}^{l-1} (y(t_i) - \hat{y}(t_i))^2 / \sum_{l=0}^{l-1} (y(t_i) - \bar{y})^2 \right), \tag{19}$$

where $\bar{y}$ represent the sample mean of $y(t_i)$. Here, the $R^2$ is a normalized measure between 0 and 1 indicating how well the regression model explains the variation of time series data.

**Table 2.** Comparison of prediction performances using the k-NN, SVR-RBF, and GKFN prediction models.

| | k-NN | SVR-RBF | GKFN |
|---|---|---|---|
| $RMSE$ | 0.104822 | 0.101392 | **0.099322** |
| $R^2$ | 0.694397 | 0.714068 | **0.725625** |

For the comparison of the proposed method, the prediction models using the k-nearest neighbor (k-NN), support vector regression with radial basis function

**Table 3.** Comparison of prediction performances using the GKFN prediction models with the deep GKFN (DGKFN) prediction model.

|  | GKFN | GKFN$_1$ | GKFN$_2$ | GKFN$_3$ | DGKFN |
|---|---|---|---|---|---|
| $RMSE$ | 0.099322 | 0.169867 | 0.099125 | 0.118739 | **0.093130** |
| $R^2$ | 0.725625 | 0.197452 | 0.726716 | 0.607863 | **0.758772** |

kernel (SVR-RBF) and GKFN were also trained for the same data. In the case of $k$-NN method, the nearest neighborhood $k$ was searched using the validation set in such a way of minimizing the coefficient of determination $R^2$. As a result, the value of $k$ was selected as 11 for this data set. In the case of SVR, the kernel functions were selected as Gaussian functions and the penalty parameter $C$ and kernel parameter $\gamma$ were also searched using the validation set in such a way of minimizing the coefficient of determination $R^2$. Here, the SVR is trained using the Scikit-learn toolkit [16]. As a result, the parameter values were determined by $C = 100$ and $\gamma = 2$.

In the case of GKFN, the proper number of Gaussian kernel functions was searched using the validation set in such a way of minimizing the coefficient of determination $R^2$ and the parameters of GKFN were trained for the training data set [10]. As a result, the number of kernel functions was determined by 67. After training these prediction models, the prediction performances of the RMSE and $R^2$ were collected and listed in Table 2. These simulation results showed that the GKFN provided the merits in prediction performances of $RMSE$ and also $R^2$. This is mainly due to the fact that the GKFN training includes the procedure of adjusting kernel parameters as well as weight parameters through fine tuning method. Then, the GKFN was compared with the proposed deep GKFN (DGKFN). For this purpose, three GKFNs (GKFN$_1$, GKFN$_2$, and GKFN$_3$) were trained for the train data set 1, 2, and 3, respectively. In this training of each GKFN, the number of kernel functions was selected using the assigned validation data set. Then, three GKFNs were combined using the deep structure of GKFNs. For this combination, the GKFN of the upper layer was trained for the whole training data set. After training GKFNs, the prediction performances of the RMSE and $R^2$ were collected and listed in Table 3. These simulation results showed that the prediction performances of GKFN$_1$, GKFN$_2$, and GKFN$_3$ were similar or less than the prediction performance of the single layer GKFN trained for the whole training data set, whereas the proposed DGKFN provided the better prediction performances than the single layer GKFN by combining the partially trained GKFNs (GKFN$_1$, GKFN$_2$, and GKFN$_3$). This implies that the proposed combination method using the deep structure of GKFNs is effective in improving the prediction performances of prediction models with Gaussian kernel functions. This is mainly from the fact that the variance of the proposed prediction model (DGKFN) is reduced by combining various GKFNs and it contributes to reduce prediction errors.

# 5    Conclusion

In managing electric power plants, accurate prediction of short-term power demand is an essential factor for ensuring stability and efficient operation of the power system. From this context, this paper proposes a novel method of predicting daily peak power demands using the deep structure of Gaussian kernel function networks (GKFNs). For the prediction model, the whole time series is divided into multiple parts and each part is trained using a GKFN. Then, the trained GKFNs are combined using the deep structure of GKFNs to minimize the mean square errors of prediction model. As a consequence, the proposed deep structure of GKFNs has an improved performance of prediction accuracy compared with canonical GKFNs. The simulation for predicting daily peak power demands in Korea reveals that the proposed prediction model has the merits in prediction performances compared with the GKFN model and also other prediction models such as the $k$-NN and SVR. The proposed deep structure of GKFNs is applicable to various problems of time series prediction or regression problems.

**Acknowledgment.** This work was partly supported by Institute for Information & communications Technology Promotion (IITP) grant funded by the Korea government (MSIT) (No. 2016-0-00133, Research on Edge computing via collective intelligence of hyperconnection IoT nodes).

# References

1. Srinivasan, D., Lee, M.A.: Survey of hybrid fuzzy neural approaches to electric load forecasting. In: IEEE International Conference on Systems, Man and Cybernetics, pp. 4004–4008. IEEE Press, New York (1995)
2. Srivastava, A.K., Pandey, A.S., Singh, D.: Short-term load forecasting methods: a review. In: IEEE International Conference on Emerging Trends in Electrical Electronics & Sustainable Energy Systems, pp. 130–138. IEEE Press, New York (2016)
3. Papalexopoulos, A.D., Hesterberg, T.C.: A regression-based approach to short-term system load forecasting. IEEE Trans. Power Syst. **5**(4), 1535–1547 (1990)
4. Amjady, N.: Short-term hourly load forecasting using time-series modeling with peak load estimation capability. IEEE Trans. Power Syst. **16**(4), 798–805 (2001)
5. Park, D.C., El-Sharkawi, M.A., Marks, R.J., Atlas, L.E., Damborg, M.J.: Electric load forecasting using an artificial neural network. IEEE Trans. Power Syst. **6**(2), 442–449 (1991)
6. Lee, K.Y., Cha, Y.T., Park, J.H.: Short-term load forecasting using an artificial neural network. IEEE Trans. Power Syst. **7**(1), 124–132 (1992)
7. Rahman, S., Bhatnagar, R.: An expert system based algorithm for short term load forecast. IEEE Trans. Power Syst. **3**(2), 392–399 (1988)
8. Fan, S., Chen, L.: Short-term load forecasting based on an adaptive hybrid method. IEEE Trans. Power Syst. **21**(1), 392–401 (2006)
9. Kim, K.H., Park, J.K., Hwang, K.J., Kim, S.H.: Implementation of hybrid short-term load forecasting system using artificial neural networks and fuzzy expert systems. IEEE Trans. Power Syst. **10**(3), 1534–1539 (1995)

10. Kil, R.: Function approximation based on a network with kernel functions of bounds and locality. ETRI J. **15**, 35–51 (1993)
11. Packard, N.H., Crutchfield, J.P., Farmer, J.D., Shaw, R.S.: Geometry from a time series. Phys. Rev. Lett. **45**, 712–716 (1980)
12. Takens, F.: Detecting strange attractors in turbulence. In: Rand, D., Young, L.-S. (eds.) Dynamical Systems and Turbulence, Warwick 1980. LNM, vol. 898, pp. 366–381. Springer, Heidelberg (1981). https://doi.org/10.1007/BFb0091924
13. Havstad, J.W., Ehlers, C.L.: Attractor dimension of nonstationary dynamical systems from small data sets. Phys. Rev. A **39**(2), 845–853 (1989)
14. Kil, R., Park, S., Kim, S.: Time series analysis based on the smoothness measure of mapping in the phase space of attractors. In: International Joint Conference on Neural Networks, vol. 4, pp. 2584–2589. IEEE Press, New York (1999)
15. Kim, D.K., Kil, R.M.: Stock price prediction based on a network with gaussian kernel functions. In: Lee, M., Hirose, A., Hou, Z.-G., Kil, R.M. (eds.) ICONIP 2013. LNCS, vol. 8227, pp. 705–712. Springer, Heidelberg (2013). https://doi.org/10.1007/978-3-642-42042-9_87
16. Pedregosa, F., et al.: Scikit-learn: machine learning in python. J. Mach. Learn. Res. **12**, 2825–2830 (2011)

# Convolutional Model for Predicting SNP Interactions

Suneetha Uppu$^{(\boxtimes)}$ and Aneesh Krishna

School of Electrical Engineering, Computing and Mathematical Sciences,
Curtin University, Bentley, Perth, Australia
suneetha.uppu@postgrad.curtin.edu.au,
A.Krishna@curtin.edu.au

**Abstract.** Single-nucleotide polymorphisms (SNPs) are genetic markers that empower researchers to examine for genes associated with complex diseases. Several efforts have been contributed by researchers to study the interaction effects between multi-locus SNPs for discerning the status of complex diseases. However, the current conventional machine learning techniques are still left with several caveats. Deep learning is a new breed of machine learning technique that elucidates the hidden structure of the raw data by transforming it into multiple high levels of abstractions, using the power of parallel and distributed computing. It promises empirical success in the number of applications including bioinformatics to drive insights of biological complexities. The deep learning approach in the multi-locus interaction studies is yet to meet its potential achievements. In this paper, a convolutional neural network is trained to identify true causative two-locus SNP interactions. The performance of the method is evaluated on hypertension data. Highly ranked two-locus SNP interactions are identified for the manifestation of hypertension.

**Keywords:** Convolutional neural network · SNP-SNP interactions
Deep learning · Multi-locus · Epistasis · Gene-gene interactions

## 1 Introduction

In this era of rapid development of high-throughput sequencing data, researchers gradually shifted the focus on genome-wide association studies (GWAS) for identifying the genetic variants associated with a complex disease. A Single nucleotide polymorphism (SNP) is a commonly occurring (>1%) genetic variation in a single nucleotide (A, T, G, and C) of a DNA sequence [1, 2]. SNPs are functionally insignificant compared with mutants (rare variants), however, recent studies in GWA explores a considerable effect in changing the functionality of proteins [2]. Predominantly, these studies are univariate, which leads to missing heritability problem in genetic epidemiology [3]. In reality, complex disease occurs due to the influence of individual or combined effect of genetic variants. Studying these combined effects of genetic variants is extremely complex due to high dimensionality problem, population stratification, biological complexities, presence of genetic heterogeneity and phenocopy. The previous study reviewed the current approaches in the literature, and

© Springer Nature Switzerland AG 2018
L. Cheng et al. (Eds.): ICONIP 2018, LNCS 11305, pp. 127–137, 2018.
https://doi.org/10.1007/978-3-030-04221-9_12

addressed number of challenges to be considered while designing new approaches [4]. Some of the well-established approaches are Multifactor Dimensionality Reduction (MDR) [5], LogicFS [6], AntEpiSeeker [7], Epistatic module detection (epiMODE) [8], Bayesian Epistasis Association Mapping (BEAM) [9], Boolean Operation-based Screening and Testing (BOOST) [10], Genetic Programming optimized Neural Network (GPNN) [11], SVM based Generalized MDR [12], PILINK [13], and Random Jungle [14]. However, these approaches are yet to produce remarkable results in searching for subset of informative SNPs.

In this new era of machine learning, application of deep learning into bioinformatics have been achieving great success [15, 16]. The deep learning in identifying SNP interactions is yet to meet its potential achievements. In the previous study, DNN is trained to identify highly ranked multi-locus SNP interactions, and compared with the previous approaches [17, 18]. The experimental results over simulated and real datasets showed potential results compared with other conventional methods. However, it was observed that training DNN for the multi-locus genome data became tedious and challenging due to huge number of network parameters. In this study, a convolutional neural network (CNN) has been explored to overcome the inherent problem [19]. CNNs uses convolution, a mathematical linear operation, instead of matrix multiplication at least in one of their hidden layers. The main features of CNN are to share the weights, and extract useful features with its trained weights. Hence, CNNs are considered to be less complex and use less memory compared with DNNs. These features of CNN motivated authors to incorporate it into this work for improving learning efficiently.

A CNN is trained to detect informative two-locus SNP interactions associated to a complex disease. The trained network is validated under different simulated scenarios. Further, the method is evaluated on hypertension data. The predictive performance of the optimal models is observed in terms of accuracy, and performance metrics of the models. The experimental evaluations show remarkable results for predicting the subset of informative SNP interactions over some of the previous approaches, such as, MDR, RF, support vector machine (SVM), neural networks (NN), Naïve Bayes', classification based on predictive association rules (CPAR), logistic regression (LR), PART, lazy associative classifier (LAC), generalised linear model (GLM), and Gradient Boosted Machines (GBM). Further experimentations are conducted by tuning the model's parameters to improve the accuracy of the optimal model. Top 20 highly ranked subset of two-locus causative SNP interactions are identified for the manifestation of hypertension in human.

Rest of the paper is organized as follows: Sect. 2 presents the method and a basic background of CNN. This section further includes data preparation and evaluation of the methods for various simulated and a real dataset. Experimental evaluations and discussions are provided in Sect. 3. Finally, the conclusion and future work is presented in Sect. 4.

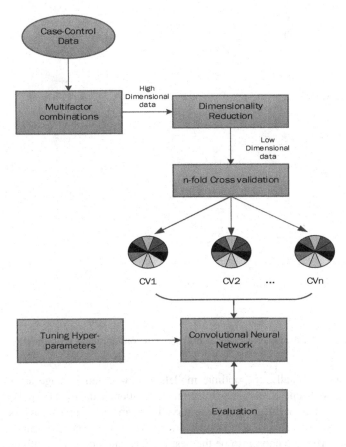

**Fig. 1.** Overview of the CNN model for predicting SNP interactions

## 2  Model Design

The main goal of the proposed method is to search for subset of informative interacting SNPs in high-dimensional genome by incorporating the capabilities of Convolutional neural networks. The workflow diagram of the proposed method is presented in Fig. 1 as ready reference (updated from [20]). There are six stages in the workflow of the proposed method. In the first stage, the case-control based datasets are represented by $n$-factors, whose examples determine their exposure to a disease. These $n$-factors are combined in $n$-dimensional space in the second stage. For example, four-locus combinations (each SNP have three genotyping factors due to bi-allelic in nature) will have 64 possible four-locus genotyping combinations. In stage three, the high-dimensional combinational data is reduced to low-dimensional data using PCA as a data prepro- cessing step. The case-control datasets are split, using 10-fold cross validation for training and testing purposes in step four. A deep multilayered convolutional stage is included, in which, a convolutional neural network is trained to obtain the useful knowledge from the data. Further in this stage, the models are optimized by tuning the

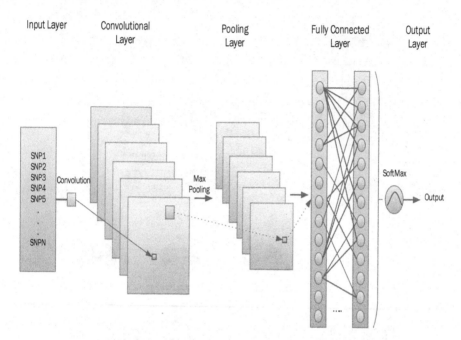

**Fig. 2.** An example of a basic structure of a convolutional neural network

hyper-parameters. Finally, the optimal models are evaluated in stage six for various simulated datasets, and a real data (hypertension patients' data [21]) application. The performance metrics of the models are observed in terms of accuracy, logloss, mse, and cross-validation consistency. The models with the highest accuracy and the lowest classification errors are chosen to be the best models, and compared with the previous approaches.

## 2.1  Convolutional Neural Networks

Convolutional neural networks (CNNs) are inspired by visual cortex of the brain [19]. In the visual cortex, there are simple neurons that respond to primitive patterns in the visual field, and complex neurons that respond to large intricate forms. CNNs leverages the idea of sparse interactions, parameter sharing and equivariant representations to improve machine learning [19]. Figure 2 represents an example of a basic structure of CNN that consists of an input layer, convolution layers, nonlinear layers, pooling/sub sampling layers, fully connected layers for the classification, and an output layer. CNN uses the notion of convolution, a mathematical linear operation, instead of matrix multiplication at least in one of their hidden layers. The input data is modeled as multidimensional arrays and Kernel with multidimensional array of learnable parameters [22]. Convolutional layer consists of several convolutional kernels that are used to determine the feature maps. Each neuron of the previous layers is connected to the neurons of a feature map. Consider a three-dimensional SNPs $S$ as input with three-dimensional kernal $K$. The convolution output $y$, which is commutative, and is expressed as [19]:

$$y[p,q,r] = (S * K)[p,q,r] = \sum_h \sum_i \sum_j S[p-h, q-i, r-j] K[h,i,j] \qquad (1)$$

The new feature map is obtained by convolving the input with the learned kernel. A non-linear activation function is applied on the convolved output $y$.

$$y[p,q,r] = \sigma(y[p,q,r]) \qquad (2)$$

where $\sigma(.)$ is a non-linear activation function. Activation function *tanh* is used in this method and is represented as:

$$\tanh(y[p,q,r]) = \frac{2}{1 + e^{-2y[p,q,r]}} - 1 \qquad (3)$$

The output of the activation function is given as the input to the pooling layer. It achieves shift-invariance by reducing the factors of the feature map and minimizes the calculations of the network. Each feature map is operated by a pooling layer independently by using max pooling. The fully connected layers take the input from pooling layers, and compute the output $y$ of $c$ dimensional vector of class by using softmax activation [19]. Fully connected layers are traditional multilayered neural networks, where every neuron in the previous layer is connected to all the neurons in the next layer. Softmax activation function computes probability for each class by interpreting its confidence value. The total error $e$ of the output layer is calculated by using cross-entropy function as follows:

$$e = -\frac{1}{N} \sum_s (\ln y * y' + \ln(1-y) * (1-y')) \qquad (4)$$

Where $y$ is the predicted output obtained from softmax, and $y'$ is the desired output. Gradients of the error with respect to weights in the network are computed by using backpropagation [22]. Stochastic gradient descent (SGD) is used to compute the partial derivative of each network parameters with respect to cross entropy loss function. It minimizes the loss function by optimizing the best fitting parameters using mini-batch strategy. The parameters of the CNN model are updated for every epoch from time t to t + 1.

$$w' \leftarrow w - \eta \frac{\partial e}{\partial w} \qquad (5)$$

Where, $\frac{\partial e}{\partial w}$ is the partial derivative of loss function with respect to $w$ and $\eta$ is the learning rate, $\eta > 0$.

## 2.2   Data Preparation

Case-control datasets consists of $s$ samples with $n$ factors along with a class label. The class label can be either 0 for controls or 1 for cases. Each factor is considered to be a

SNP at a locus. As SNPs are bi-allelic in nature, there are three genotyping factors at each locus. For example, SNP at locus Z will have genotypes ZZ (common homozygous), Zz/zZ (heterozygous), and zz (variant homozygous). Their corresponding numerical representations are 0, 1, and 2. Various two-locus simulated datasets and a real dataset (Hypertension patient data) are prepared for the validation of the models.

**Simulated Datasets.** Various case-control based simulated datasets are generated for six two-locus epistasis models that exhibits combined effects [23] for different penetrance values and allele frequencies ($p$ and $q$). Where $p$ is a frequency of a minor allele and $q$ is a frequency of an alternative allele [24]. These datasets are generated using GAMETES tool in the absence of main effects [24, 25]. All the simulated datasets are generated according to Hardy-Weinberg proportions, and they are replicated in the previous studies [26].

**Real World Data.** Hypertension data is obtained from the National Taiwan University hospital between July 1995 to June 2002 from the outpatient clinic [21]. The data consist of 443 Taiwanese residents' samples, in which there are 313 cases and 130 controls. The study considered eight SNPs in four genes, rs5050, rs5051, rs11568020, and rs5049 at AGT 5', rs4762, and rs699 at AGT, rs5186 at AT1-R, and rs4646994 at ACE. The genotypes are numerically represented by 0, 1, and 2 respectively, and hypertensives are represented by 1 and non-hypertensives as 0.

### 2.3    Data Evaluation

The datasets are split into equal parts of $m$ for training and independent testing without losing the data. That is, in $m$ fold cross validation, $m - 1$ splits are used for training and remaining one split is used in testing. The method runs $m$ times on training data by excluding different split each time for testing. In the proposed method, 10-fold cross validation is used for the validation of the proposed method. The performances of the models are validated by observing the metrics of the models. Further, the models are evaluated by varying parameters and changing non-linear activation functions. The optimal model is selected with the highest prediction accuracy and cross validation consistency (CVC), and the lowest prediction error. The final experimental results are evaluated statistically by determining the statistical significance of the findings, whose $p$ values are less than 0.05.

## 3    Experiments and Discussions

The proposed CNN method is implemented and trained in R [27, 28]. Several experimental results are studied over various simulated scenarios. Consistently, the accuracy of the CNN based method on simulated studies is much encouraging over the existing conventional methods. The method is further validated on a real world data application [21]. Hypertension patients' dataset is evaluated on the proposed CNN based method, and the performance of the optimal models is observed by determining the subset of informative SNP interactions. PCA computes the singular value decomposition of the gram matrix by using power method [29]. Figure 3 represents the eigenvectors of PCA over the multi-factor combinational data.

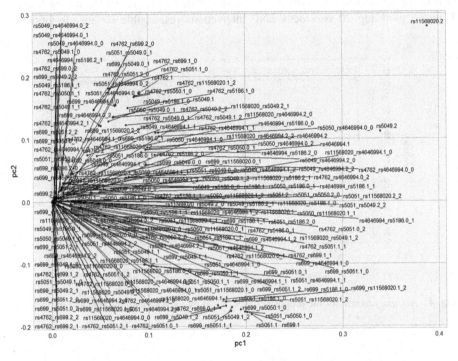

**Fig. 3.** PCA on two-locus combinations data

Preliminary studies of the method are evaluated for various non-linear activation functions, such as, rectifier, tanh, softmax, and maxout, and optimising hyper-parameters (using grid and random grid search). Further, the performance of the proposed method is validated for various non-linear activation functions with and without dropouts. These experimental results showed that *tanh* has highest prediction accuracy with low classification error. Various experiments are performed by tuning parameters, such as, epochs, momentum, learning rate, and hidden layers, using grid and random grid search methods. Even though, the grid search performed well in the large hyper-parameter space, it is sensitive in few parametric combinations. It exhaustively searches for all the models in the grid. However, random grid performed better than grid search by reducing the execution time. This is due to random selection of hyper-parameters rather than best guess. Figure 4 illustrates the accuracy metric of the proposed CNN model by varying momentum and learning rate. The best model with the optimal hyper-parameters is chosen for improving the model's predictive accuracy. The experimental observations showed that the accuracy is high for learning rate $\eta = 0.02$ and *momentum* $= 0.9$. The highest prediction accuracy attains saturation as number of epochs/iterations is increased for almost all $\eta$ and momentum values.

Figure 5 illustrates the performance of metrics of the optimal model with respect to *accuracy*, mean square error (*mse*), root mean square error (*rmse*), root mean square logarithmic error (*rmsle*), and mean absolute error (*mae*), during training. It is observed that *rmse* and *mae* decreased as number of epochs increased with lowest error rate of 0.684. This figure further illustrates roc curve (sensitivity vs specificity) during testing. Figure 6 plots top 20 two-locus SNP interactions responsible for manifestation of

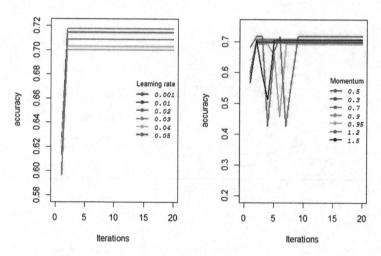

**Fig. 4.** Accuracy metric for the proposed CNN model by varying learning rate and momentum.

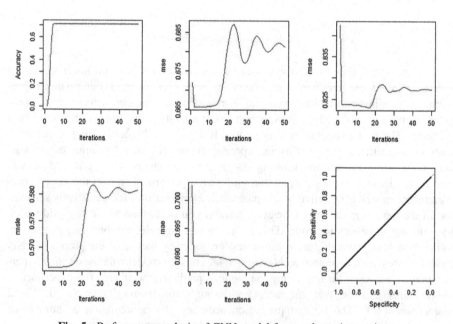

**Fig. 5.** Performance analysis of CNN model for two-locus interactions

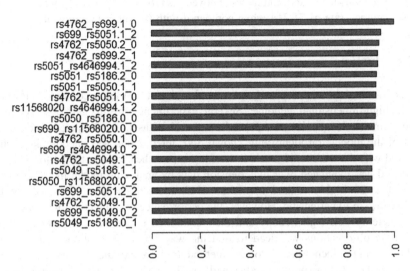

**Fig. 6.** Top 20 SNP interactions identified by CNN model on hypertension data

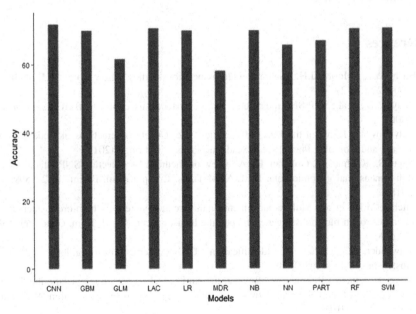

**Fig. 7.** Graphical representation of CNN model compared with some of the previous models

Hypertension. Figure 7 illustrates the accuracy bar chart of CNN model along with the previous methods in the area of machine learning. The proposed CNN based method is compared with some of the existing machine learning approaches, such as, MDR, RF,

GBM, GLM, LR, LAC, NB, SVM, PART, and NN. The CNN method identifies the highly ranked two-locus SNP interaction between rs4762 (1) – rs699 (0), which has the highest accuracy rate of 71.714 when compared with the other previous algorithms.

## 4  Conclusions

A CNN based method is proposed and trained to detect two-locus combined effects of SNP in high-dimensional genome. Several experimental results are observed over various simulated epistasis models and on a real dataset. In simulated studies, the power of the models on identifying known SNP interactions was encouraging, when compared with the traditional machine learning approaches. Hence, the performance of the method was evaluated over hypertension data by analyzing model metrics of each model. Top 20 highly ranked two-locus SNP interactions were identified for the manifestation of hypertension. However, the performance of the method was low when genetic heterogeneity, phenocopy and their combined effects were introduced in the data. These observations will lead to find the way for implementing random forest variable selection techniques into the method to improve the performance of the models. Further, optimizing techniques and parallel algorithms can be incorporated to improve the power of the method in searching for subset of informative SNPs from high-dimensional genome.

## References

1. Bush, W.S., Moore, J.H.: Genome-wide association studies. PLoS Comput. Biol. **8**(12), e1002822 (2012)
2. Onay, V.Ü., et al.: SNP-SNP interactions in breast cancer susceptibility. BMC Cancer **6**, 114 (2006)
3. Padyukov, L.: Between the Lines of Genetic Code: Genetic Interactions in Understanding Disease and Complex Phenotypes. Academic Press, Cambridge (2013)
4. Uppu, S., Krishna, A., Gopalan, R.: A review on methods for detecting SNP interactions in high-dimensional genomic data. IEEE/ACM Trans. Comput. Biol. Bioinf. **15**(2), 599–612 (2018)
5. Ritchie, M.D., et al.: Multifactor-dimensionality reduction reveals high-order interactions among estrogen-metabolism genes in sporadic breast cancer. Am. J. Hum. Genet. **69**, 138–147 (2001)
6. Schwender, H., Ickstadt, K.: Identification of SNP interactions using logic regression. Biostatistics **9**, 187–198 (2008)
7. Wang, Y., Liu, X., Robbins, K., Rekaya, R.: AntEpiSeeker: detecting epistatic interactions for case-control studies using a two-stage ant colony optimization algorithm. BMC Res. Notes **3**, 117 (2010)
8. Tang, W., Wu, X., Jiang, R., Li, Y.: Epistatic module detection for case-control studies: a Bayesian model with a Gibbs sampling strategy. PLoS Genet. **5**, e1000464 (2009)
9. Zhang, Y., Liu, J.S.: Bayesian inference of epistatic interactions in case-control studies. Nat. Genet. **39**, 1167–1173 (2007)
10. Wan, X., et al.: BOOST: a fast approach to detecting gene-gene interactions in genome-wide case-control studies. Am. J. Hum. Genet. **87**, 325–340 (2010)

11. Motsinger, A.A., Lee, S.L., Mellick, G., Ritchie, M.D.: GPNN: power studies and applications of a neural network method for detecting gene-gene interactions in studies of human disease. BMC Bioinform. **7**, 39 (2006)
12. Fang, Y.H., Chiu, Y.F.: SVM-based generalized multifactor dimensionality reduction approaches for detecting gene-gene interactions in family studies. Genet. Epidemiol. **36**, 88–98 (2012)
13. Purcell, S., et al.: PLINK: a tool set for whole-genome association and population-based linkage analyses. Am. J. Hum. Genet. **81**, 559–575 (2007)
14. Schwarz, D.F., König, I.R., Ziegler, A.: On safari to Random Jungle: a fast implementation of Random Forests for high-dimensional data. Bioinformatics **26**, 1752–1758 (2010)
15. LeCun, Y., Bengio, Y., Hinton, G.: Deep learning. Nature **521**, 436–444 (2015)
16. Min, S., Lee, B., Yoon, S.: Deep learning in bioinformatics. Brief. Bioinform. **18**(5), 851–869 (2016)
17. Uppu, S., Krishna, A., Gopalan, R.P.: A deep learning approach to detect SNP interactions. JSW **11**, 965–975 (2016)
18. Uppu, S., Krishna, A.: Improving strategy for discovering interacting genetic variants in association studies. In: Hirose, A., Ozawa, S., Doya, K., Ikeda, K., Lee, M., Liu, D. (eds.) ICONIP 2016. LNCS, vol. 9947, pp. 461–469. Springer, Cham (2016). https://doi.org/10.1007/978-3-319-46687-3_51
19. Bengio, Y., Goodfellow, I.J., Courville, A.: Deep learning. An MIT Press book in Preparation (2015). http://www.iro.umontreal.ca/~bengioy/dlbook
20. Uppu, S., Krishna, A.: Tuning hyperparameters for gene interaction models in genome-wide association studies. In: Liu, D., Xie, S., Li, Y., Zhao, D., El-Alfy, El-Sayed M. (eds.) ICONIP 2017. LNCS, vol. 10638, pp. 791–801. Springer, Cham (2017). https://doi.org/10.1007/978-3-319-70139-4_80
21. Wu, S.J., Chiang, F.T., Chen, W. J., Liu, P.H., Hsu, K.L., Hwang, J.J., Lai, L.P., Lin, J.L., Tseng, C.D., Tseng, Y.Z.: Three single-nucleotide polymorphisms of the angiotensinogen gene and susceptibility to hypertension: single locus genotype vs. haplotype analysis. Physiol. Genomics **17**, 79–86 (2004)
22. Wu, J.: Introduction to convolutional neural networks. National Key Lab for Novel Software Technology, Nanjing University, China (2017)
23. Moore, J.H., Hahn, L.W., Ritchie, M.D., Thornton, T.A., White, B.C.: Application of genetic algorithms to the discovery of complex models for simulation studies in human genetics. In Proceedings of the Genetic and Evolutionary Computation Conference/GECCO, p. 1150 (2002)
24. Ritchie, M.D., Hahn, L.W., Moore, J.H.: Power of multifactor dimensionality reduction for detecting gene-gene interactions in the presence of genotyping error, missing data, phenocopy, and genetic heterogeneity. Genet. Epidemiol. **24**, 150–157 (2003)
25. Urbanowicz, R.J., Kiralis, J., Sinnott-Armstrong, N.A., Heberling, T., Fisher, J.M., Moore, J. H.: GAMETES: a fast, direct algorithm for generating pure, strict, epistatic models with random architectures. BioData Min. **5**, 1–14 (2012)
26. Uppu, S., Krishna, A., Gopalan, R.P.: Rule-based analysis for detecting epistasis using associative classification mining. Netw. Model. Anal. Health Inform. Bioinform. **4**, 1–19 (2015)
27. Candel, A., Parmar, V., LeDell, E., Arora, A.: Deep Learning with H2O (2015)
28. Chen, T., et al.: MXNet: a flexible and efficient machine learning library for heterogeneous distributed systems. arXiv preprint arXiv:1512.01274 (2015)
29. Glander, S.: Building deep neural nets with H2O and rsparkling that predict arrhythmia of the heart (2017). https://shiring.github.io/machine_learning/2017/02/27/h2o

# Financial Data Forecasting Using Optimized Echo State Network

Junxiu Liu[1], Tiening Sun[1], Yuling Luo[1(✉)], Qiang Fu[1], Yi Cao[2],
Jia Zhai[3], and Xuemei Ding[4,5]

[1] Faculty of Electronic Engineering, Guangxi Normal University,
Guilin 541004, China
yuling0616@mailbox.gxnu.edu.cn
[2] Department of Business Transformation and Sustainable Enterprise,
Surrey Business School, University of Surrey, Surrey GU2 7XH, UK
[3] Salford Business School, University of Salford, 43 Crescent,
Salford M5 4WT, UK
[4] College of Mathematics and Informatics,
Fujian Normal University, Fuzhou 350108, China
[5] School of Computing, Engineering and Intelligent Systems, Ulster University,
Londonderry BT48 7JL, UK

**Abstract.** The echo state network (ESN) is a dynamic neural network, which simplifies the training process in the conventional neural network. Due to its powerful non-linear computing ability, it has been applied to predict the time series. However, the parameters of the ESN need to be set experimentally, which can lead to instable performance and there is space to further improve its performance. In order to address this challenge, an improved fruit fly optimization algorithm (IFOA) is proposed in this work to optimize four key parameters of the ESN. Compared to the original fruit fly optimization algorithm (FOA), the proposed IFOA improves the optimization efficiency, where two novel particles are proposed in the fruit flies swarm, and the search process of the swarm is transformed from two-dimensional to three-dimensional space. The proposed approach is applied to financial data sets. Experimental results show that the proposed FOA-ESN and IFOA-ESN models are more effective ($\sim$50% improvement) than others, and the IFOA-ESN can obtain the best prediction accuracy.

**Keywords:** Echo state network · Fruit fly algorithm · Time series
Algorithm optimization

## 1 Introduction

The reservoir computing, one computation framework, is an extension of recurrent neural networks, and has been successfully applied to the time series modeling tasks [1, 2]. The echo state network (ESN) [3], liquid state machine [4] and backpropagation decorrelation [5] are the common used reservoir computing methods. Comparing to the traditional neural networks, the ESN has the advantages of the excellent convergence speed, and the ability of avoiding local minimization etc. The core structure of the ESN

© Springer Nature Switzerland AG 2018
L. Cheng et al. (Eds.): ICONIP 2018, LNCS 11305, pp. 138–149, 2018.
https://doi.org/10.1007/978-3-030-04221-9_13

is a reservoir which remains unchanged once it is randomly generated. Meanwhile, the output weights are the only part to be adjusted, and the entire network only needs a linear regression algorithm for training. The ESN has been applied to many applications, e.g. nonlinear time series prediction [6], voice processing [7], power load forecast [8], short-term traffic flow forecast [9], pattern recognition [10], and financial stock data predication [11] etc. Although the ESN has been used in these domains, it still has some drawbacks, which seriously hinder the developments and applications of the ESN. Firstly, a randomly initialized reservoir can only achieve good performance for some tasks and it performs poorly on other specific data sets due to the randomness of the reservoir initialization [12]. In the meantime, the performance of ESNs is greatly affected by the parameters of the dynamic reservoir. However, these parameters are generally determined based on the experimental tuning process and the experiences of the researchers. In addition, due to the randomness of the reservoir and the black box structure, it is difficult to clearly understand the dynamic characteristics of the ESN. Hence, to determine the ESN parameters and the optimal reservoir structure are still one challenge of the ESN research [2].

In order to address this challenge, this paper employs the improved fruit fly optimization algorithm (IFOA) to optimize four key parameters of the dynamic reservoir. Firstly on the basis of typical fruit fly optimization algorithm (FOA) [13], the IFOA is proposed to improve the optimization efficiency, where two novel individuals are injected into the fruit fly population, and the search path of the population is changed from two-dimensional to three-dimensional. Then the IFOA is used to optimize the ESN parameters in order to achieve a better performance than the original ESN. The financial data sets, i.e. shanghai stock composite index and stock index option, are used as the experimental applications for the aim of forecasting. The trends of these datasets are predicted by using the proposed ESN with the IFOA model. Compared to the back-propagation (BP) and Elman neural networks, the proposed model achieves a better prediction accuracy. The rest is organized as follows. Section 2 introduces the ESN model. Section 3 proposes the IFOA and uses it for the ESN parameters optimization. Section 4 provides experimental results and Sect. 5 gives a conclusion.

## 2   ESN

The ESN is a special type of recurrent neural networks. It has three units, i.e. input, internal and output units, as illustrated in Fig. 1, where the numbers of neurons in the input, internal and output units are $M$, $N$ and $L$, respectively.

In the ESN, the input units $u(n)$, internal state $x(n)$, and output units $y(n)$ at time $n$ are calculated by

$$\begin{cases} u(n) = [u_1(n), u_2(n), \cdots, u_M(n)]^T \\ x(n) = [x_1(n), x_2(n), \cdots, x_N(n)]^T \\ y(n) = [y_1(n), y_2(n), \cdots, y_L(n)]^T \end{cases}. \tag{1}$$

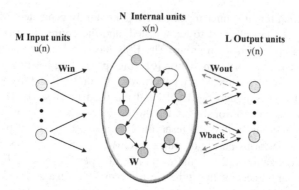

**Fig. 1.** The ESN structure. Black arrows represent the process of data-driven direction. The gray arrows represent the feedbacks from the output units to the reservoir.

The internal state $x(n)$ and output units $y(n)$ of the ESN are calculated by

$$x(n+1) = f(W(n) + W_{in}u(n) + W_{back}y(n)),  \qquad (2)$$

and

$$y(n+1) = f_{out}\left(W_{out}[x(n+1), u(n+1), y(n)] + W_{bias}^{out}\right),  \qquad (3)$$

where $W$, $W_{in}$, $W_{out}$ are the internal state matrix, the connection weight matrixs of input and output, respectively. $W_{back}$ represents the feedback matrix, $W_{bias}^{out}$ is the bias term. $W_{out}$ is the only variable that needs to be trained, and others (e.g. $W$, $W_{in}$, $W_{back}$) remain unchanged. $f$ and $f_{out}$ are the activation functions of the reservoir and output units, respectively, where the hyperbolic tangent functions are commonly used. The reservoir of the ESN is equivalent to a complicated nonlinear dynamic filter and changes with the input.

The number of the reservoir neurons, $N$, is one of the most important parameters of ESNs, and it determines the ability to simulate the complexity of the nonlinear system. In general, the larger the reservoir size, the more complex the nonlinear high-dimensional space the reservoir can generate, and the stronger the nonlinear simulation capability. However, increasing the reservoir size leads to more computing. Moreover, once the reservoir size reaches a certain level, it will cause the network over-training. Thus the reservoir size should be carefully designed. In addition, the parameters described below also should be designed efficiently.

The sparse degree (SD) is the proportion of the neurons which are connected to other neurons in the reservoir, which equals to the number of non-zero elements in the internal state matrix $W$. The SD is defined by

$$SD = n/N,  \qquad (4)$$

where $n$ is the number of neurons which connect to other neurons, and $N$ is the number of all the neurons in the reservoir. Using the sparse matrix as the internal reservoir, it is possible to speed up the calculation and increase the update speed of the reservoir status.

The spectral radius (SR) of the internal weight is another important parameter. It is defined by

$$SR = max\{abs(E(W))\}, \tag{5}$$

where $E(W)$ denotes the eigenvalue of the internal state matrix $W$. Only when $SR < 1$, the ESN can demonstrate the echo state. If the SR is too large, it will cause internal confusion and instable reservoir, and destroy the nature of the echo state. Thus In order to avoid this, the internal state matrix $W$ is usually randomly generated in the range of $[-1, 1]$, and follows the sparse matrix of the equally distributed distribution. The SR is obtained from the internal state matrix $W$ to ensure the stability of the network.

The final important parameter is the input scale (IS) which is a scale factor. Generally, the input scale is a single value for the purpose of scaling the whole input weight matrix $W_{in}$. However if the channels of $u(n)$ contribute differently to the task, different scale factors will be used for the columns of input weight matrix $W_{in}$ separately [14]. Because the activation function of the ESNs is usually a sigmoid function, it is necessary to scale the input data into the range of $[-1, 1]$ to ensure that it is within the scope of the activation function.

These aforementioned parameters are crucial for the ESNs. If these parameters are not optimized, it will affect the network performance. Therefore, in this paper, an improved fruit fly optimization algorithm is proposed which is used to optimize these key parameters of the ESN and improve the network performance. The details of the IFOA and the method to optimize the ESN parameters are presented in next section, and Sect. 4 provides the experimental results and analyses the network performance using the financial dataset as the benchmarking applications.

## 3   IFOA-ESN Model

Fruit fly is superior to other species in sensory perception, especially in the sense of smell and vision [13]. It can easily smell the odor in the air, and even smell the food from 40 km away [13]. Figure 2 shows the iterative food searching process of fruit flies. It can be seen that fruit fly $a$ initially smells the food by using its olfactory system. It also sends and receives information from other individuals (e.g. fruit fly $b$, $c$ and so on), and compares the smell concentration (fitness value) of food with the current best location. Fruit fly $a$ identifies the smell concentration, and moves to the location with the best smell concentration, and others have similar behaviors. Finally all individuals will be around the best location in every iteration, and use the vision systems to fly towards the target of food.

The FOA is based on the food searching behavior of the fruit fly. In the FOA, the smell concentration $C_i$ of the $i$-$th$ fruit fly is calculated by loading the smell

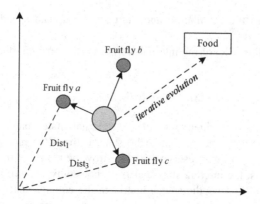

**Fig. 2.** The food search process of fruit fly swarm.

concentration value $S_i$ into the smell concentration judgment function $F$ (fitness function $F$), which is given by

$$C_i = F(S_i). \tag{6}$$

During the iteration process, the current optimal location is constantly adjusted until the FOA obtains the best optimal solution. The FOA is efficient for the computing and is easier for implementation than other optimization algorithms. More details about the FOA can be found in the approach of [13]. In this paper, four key parameters of the ESN dynamic reservoir need to be optimized which has the potential to improve the performance of the ESN model. However, the FOA has some limitations and cannot be directly applied for the ESN optimization. These limitations include it can easily fall into local optima and the searching path is too coarse. In order to address these problems, an improved FOA is proposed in this approach and the improvements include the following two-fold.

(a) **Three-dimensional space search.** The search space of the original FOA is in two dimensions, which limits the search range and flexibility in the search process [15]. In this work, the initial location of population is reset in three-dimensional coordinates. Correspondingly, the FOA is also converted into three-dimensional patterns. Hence, the flying range of fruit flies is calculated by

$$\begin{cases} X_i = X_o + \alpha \cdot X_o \cdot R \\ Y_i = Y_o + \alpha \cdot Y_o \cdot R, \\ Z_i = Z_o + \alpha \cdot Z_o \cdot R \end{cases} \tag{7}$$

where $(X_o, Y_o, Z_o)$ is the initial location of the fruit fly, $(X_i, Y_i, Z_i)$ is the updated location of the $i$-$th$ fruit fly, $\alpha$ is an adjustable parameter to control the flying distance, and $R$ is a random number in [0, 1]. So, the distance $D_i$ of the $i$-$th$ fruit fly to the origin is calculated by

$$D_i = \sqrt{X_i^2 + Y_i^2 + Z_i^2}. \tag{8}$$

In addition, when calculating smell concentration judgment value $S$, the perturbation factor $\beta$ is added to improve the local search performance of the FOA. The $\beta$ and $S_i$ of the $i$-th fruit fly is calculated by

$$\beta = \gamma \cdot D_i \cdot (0.5 - R), \tag{9}$$

$$S_i = 1/D_i + \beta, \tag{10}$$

where $\gamma$ is a real number. Since the $\beta$ only has a minimal interference, the value of $\gamma$ is usually small.

Then the smell concentration is calculated by (6), and the fruit flies fly toward the best smell concentration location, which can be described by

$$[b_s, b_i] = \min(C), \tag{11}$$

$$\begin{cases} X_o = X(b_i) \\ Y_o = Y(b_i) \\ Z_o = Z(b_i) \end{cases}, \tag{12}$$

where $b_s$ is the best smell concentration of the population, and $b_i$ is the index of the fruit flies with the best smell concentration values, and $X_o$, $Y_o$, $Z_o$ denote the locations of the fruit fly having the best smell concentration values obtained after iterations.

(b) **Population diversification.** In the original FOA, when a fruit fly finds the location and direction of food, it passes the information to other individuals. Then the population travels along the route which is provided by this fruit fly, i.e. all the other fruit flies follow the first fruit fly (which found the food). Therefore, the search process is mainly controlled by a few optimal individual fruit flies. However if the objective function is multimodal, the optimal individual is not able to obtain the global optimal solution directly in the long run, where a local optimum is always obtained [16]. This is different to the real world. As in nature, there are always some individuals which have different search pathways to the majority of the population, and some new findings can probably be found by these individuals. Therefore in this paper, inspired by the biology, two new novel fruit flies are added by considering and simulating the actual biological phenomenon to make other individuals have the opportunities to lead the search direction of the population. They are the "reverse" fruit flies and "radical" fruit flies in this approach, and their smell concentration judgment values are calculated by

$$\begin{cases} G_r = 1/(M_D + (U_{max} - M_D) \cdot R) \\ G_i = 1/(U_{min} + (M_D - U_{min}) \cdot R) \end{cases}, \tag{13}$$

where $G_r$ and $G_i$ are the smell concentration judgment values of the "reverse" and "radical" fruit flies, respectively, $U_{max}$ and $U_{min}$ are the upper and lower limits of the search range, and $M_D$ is the mean distance of all the current fruit flies to the origin. The "reverse" fruit flies search reversely and hover in the back of population invariably, and the "radical" fruit flies are flying in the front of the population and searching randomly. In every iteration, their flying ranges will change as the mean distance $M_D$ changes, but it is within the range of $U_{max}$ and $U_{min}$. These individuals have the probabilities to increase the population diversity and enhance the capability of global searching.

The improved FOA (IFOA) is used to optimize the key parameters of the ESN, and the optimization process is given by Fig. 3 which includes the following steps.

(a) Initialize the parameters of IFOA (i.e. population location, iterations, population size and flying range). The population size of fruit flies is one of the key parameters, which determines the final prediction accuracy. If the population size is too large, the prediction accuracy is higher, but it also consumes more computing power. If the size is too small, the prediction accuracy decreases and the convergence speed increases. In this paper, the parameters of the ESN (i.e. reservoir size, sparse degree, spectral radius, and input scale) are used as the smell concentration judgment values of fruit flies, the standard root mean square error (NRMSE) of each of the predicted value and the actual value is used as the smell concentration judgment function, and the minimum value of the function is defined as the search target.

(b) Calculate the smell concentrations of the normal and novel individuals using (6). Identify the individuals with the best smell concentration of the population. Keep the best smell concentration judgment value and its coordinates, and fruit flies fly to this location using vision.

(c) Iterative optimization is performed. The smell concentration judgment values and the locations of these fruit flies are updated and adjusted repeatedly by IFOA, so as to approach target values.

(d) When the conditions (e.g. reaching the maximum iterations or the target accuracy) are met, the optimization process stops. The current optimal smell concentration judgment values (which represent the ESN parameters) are obtained and will be used for the ESN.

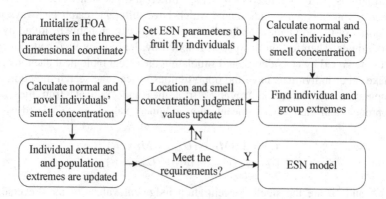

**Fig. 3.** The flow chart of the IFOA-ESN model.

# 4   Experimental Results

In this paper, the financial data sets of the shanghai composite index and stock index option, are used as the experimental applications for the testing. In the meantime, the FOA-ESN and IFOA-ESN are compared with other approaches. The standard root mean square error (NRMSE) and coefficient of determination ($R^2$) are used as the evaluation metric, and they are calculated by

$$NRMSE = \sqrt{\sum_{t=1}^{N} \frac{(o(t) - \hat{o}(t))^2}{N \cdot \sigma^2}}, \tag{14}$$

and

$$R^2 = 1 - \frac{\sum_{t=1}^{N}(o(t) - \hat{o}(t))^2}{\sum_{t=1}^{N}(o(t) - \bar{o})^2} \tag{15}$$

where $N$ is the total number of samples, $\hat{o}(t)$ and $o(t)$ are the actual and expected values, $\sigma^2$ is the variance of $o(t)$, $\bar{o}$ is the mean of the expected values and is calculated by $\bar{o} = \left(\sum_{t=1}^{N} o(t)\right)/N$.

## 4.1   Forecasting of Shanghai Composite Index

The shanghai stock composite index is a statistical indicator and reflects the overall trend of stocks listed on the shanghai stock exchange. It is a dynamic non-linear system with a variety of noise [17]. The data set comes from shanghai great wisdom and it has 4,579 sets where 80% is used for training and 20% is used for testing. The experiment is a one-step prediction, where six financial indicators (i.e. open, high, low, close, daily trading volume, and daily turnover) are used as input features, and the opening price of the next day is the output. Hence, the input and output layer of the model have six neurons and one neuron, respectively. In this work, the internal activation function is *tanh*, and the output activation function is a linear function. In the IFOA, the maximum iterations is set to 100, and the population size is set to 25, where the first 20 fruit flies are normal individuals, and the last are "reverse" and "radical" fruit flies. The four key parameters (i.e. reservoir size, sparse degree, spectral radius, and input scale) of the dynamic reservoir are automatically generated by the IFOA, and the optimization result is shown by Fig. 4.

When the system is iterated to 4 times, the NRMSE drops to 0.0227, which means that a better smell concentration is found. The location of the current best smell concentration is retained, and then the fruit flies fly to this location and continue to search around. When iteration goes to 53 times, a new best smell concentration is obtained, and the NRMSE drops to 0.0212. After 95 iterations, the NRMSE converges to 0.0209. It should be noted that at the beginning of the iteration, the NRMSE is small, i.e. 0.0233. This is mainly because a relatively good result has been obtained when the best smell concentration of the fruit fly population is first calculated. In other words, a

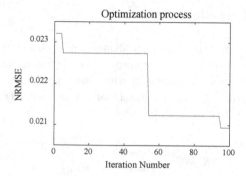

**Fig. 4.** IFOA parameter optimization for the shanghai stock composite index forecasting.

relatively good set of the ESN parameters has already been obtained at the initial optimization. This is random event, as it cannot be guaranteed that the target is approaching as much as possible before every iteration.

Based on the outputs of IFOA, the ESN is configured by the following parameters to forecast the opening price of the shanghai stock composite index: the number of the reservoir neurons is set to 56, the spectral radius (SR) of the internal weight is 0.3502, the sparse degree (SD) is 0.0280, and the input scales (IS) of the input weight matrix are set to 0.4816, 0.4646, 0.7067, 0.8900, 0.4912, and 0.1644, respectively. To verify the superiority of the FOA-ESN and IFOA-ESN, this work uses three models (i.e. BP, Elman, ESN) for the benchmarking. The comparison result is shown in Table 1.

**Table 1.** NRMSEs and $R^2$ of different methods for the shanghai stock composite index forecasting.

| Method | BP | Elman | ESN | FOA-ESN | IFOA-ESN (this work) |
|---|---|---|---|---|---|
| **NRMSE** | 0.4461 | 0.4667 | 0.1974 | 0.0482 | **0.0213** |
| $R^2$ | 0.8815 | 0.8566 | 0.9895 | 0.9976 | **0.9996** |

The NRMSE of the ESN is reduced by 55% and 58% compared to the BP and Elman models, respectively. In addition, the prediction accuracy of FOA-ESN and IFOA-ESN is improved by 76% and 89% compared to the original ESN, where the IFOA-ESN has an improvement of 56% than the FOA-ESN. The $R^2$ of IFOA-ESN is close to 1, which also indicates that the model has the best fitting accuracy. This is mainly because the key parameters of the ESN are optimized by the IFOA and the performance of the ESN model is significantly improved.

## 4.2    Forecasting of Stock Index Option

In the stock market, stock index option is an important basic financial derivative. According to the nature of the option, the options can be divided into call options and put options. Depending on the execution time of contracts, the options can be divided into American and European options [18]. According to the Black-Scholes option pricing formula [19], the price of the call option is affected by the following factors: the strike price of the option, the current price of the index, the maturity time, the expected volatility of the stock index, and the risk-free interest rate. Hence, these five indicators are selected as the input features for the proposed model, and the price of the call option is used as the target output. In this paper, the first 10,000 data samples are used for training, and the rest 2,000 rows are used for testing. In the IFOA, the maximum number of iterations is set to 300, and the population size is set to 20, where the first 15 fruit flies are normal individuals, and the others are "reverse" and "radical" fruit flies.

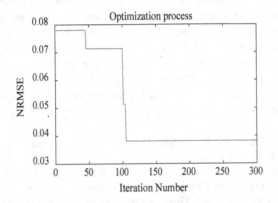

**Fig. 5.** IFOA parameter optimization for the stock index option forecasting.

The parameters of the ESN are optimized by IFOA, and the optimization process is shown in Fig. 5. When the system is iterated to 49 and 97 times, the NRMSE drops to 0.071 and 0.051, respectively. After the interactions of 100 times, the NRMSE converges to 0.037. According to this result, the following ESN parameters are used to forecast the price of the call option: the reservoir size is set to 264, the spectral radius is 0.4818, the sparse degree is 0.4848, and the input scale matrix are set to 0.3819, 0.3922, 0.4282, 0.5461, and 0.4215, respectively.

The FOA-ESN and IFOA-ESN are compared with the BP, Elman and ESN models. The comparison result is shown in Table 2. The ESN, BP and Elman models have similar NRMSEs, where the Elman model gets the lowest prediction accuracy. However, the FOA-ESN has 47%, 40% and 51% increases in accuracy over the original ESN, BP and Elman models, respectively. Compared to FOA-ESN, the IFOA-ESN achieves a better prediction accuracy, which is improved by 11%. In addition, the $R^2$ of IFOA-ESN also shows that the model has better fitting ability. Therefore, the IFOA-ESN gets a better performance than these approaches due to the lowest NRMSE and largest $R^2$.

**Table 2.** NRMSEs and $R^2$ of different methods for the stock index option forecasting.

| Method | BP | Elman | ESN | FOA-ESN | IFOA-ESN (this work) |
|--------|------|-------|------|---------|----------------------|
| **NRMSE** | 0.0713 | 0.0872 | 0.0797 | 0.0425 | **0.0379** |
| $R^2$ | 0.9948 | 0.9927 | 0.9937 | 0.9982 | **0.9986** |

## 5   Conclusions

In this work, a novel approach that combines the ESN and IFOA is proposed. Firstly, the IFOA is proposed to improve the optimization efficiency of the FOA, where two novel particles are designed in the fruit fly swarm, and the search process is performed in three-dimensional space. Then the IFOA is employed to search four key parameters of the dynamic reservoir in the ESN. The optimized ESN is used to predict financial time series. Results show that both the FOA-ESN and IFOA-ESN have improved the prediction accuracy than the BP, Elman and original ESN models. Therefore the FOA-ESN and IFOA-ESN models are more suitable for the prediction of financial time series data such as the shanghai stock composite index and stock index options.

**Acknowledgement.** This research is supported by the National Natural Science Foundation of China under Grant 61603104, the Guangxi Natural Science Foundation under Grants 2016GXNSFCA380017, 2015GXNSFBA139256 and 2017GXNSFAA198180, the funding of Overseas 100 Talents Program of Guangxi Higher Education, the Doctoral Research Foundation of Guangxi Normal University under Grant 2016BQ005, the Scientific Research Funds for the Returned Overseas Chinese Scholars from State Education Ministry, the Funds for Young Key Program of Education Department from Fujian Province, China (Grant No. JZ160425), and Program of Education Department of Fujian Province, China (Grant No. I201501005).

## References

1. Jaeger, H.: Survey: reservoir computing approaches to recurrent neural network training. Comput. Sci. Rev. **3**, 127–149 (2009)
2. Wang, H., Yan, X.: Improved simple deterministically constructed cycle reservoir network with sensitive iterative pruning algorithm. Neurocomputing **145**, 353–362 (2014)
3. Jaeger, H.: The "echo state" approach to analysing and training recurrent neural networks. Technology GMD Technical report 148, German National Research Center Information, German (2001)
4. Maass, W., Natschl, T., Markram, H.: Real-time computing without stable states: a new framework for neural computation based on perturbations. Neural Comput. **14**, 2531–2560 (2014)
5. Steil, J.J.: Backpropagation-decorrelation: online recurrent learning with O(N) complexity. In: IEEE International Joint Conference on Neural Networks, pp. 843–848 (2004)
6. Shi, Z., Han, M.: Support vector echo-state machine for chaotic time series prediction. IEEE Trans. Neural Netw. **18**, 359–372 (2007)
7. Skowronski, M.D., Harris, J.G.: Automatic speech recognition using a predictive echo state network classifier. Neural Netw. **20**, 414–423 (2007)

8. Song, Q., Zhao, X., Feng, Z., An, Y., Song, B.: Hourly electric load forecasting algorithm based on echo state neural network. In: IEEE Control and Decision Conference, pp. 3893–3897 (2011)
9. An, Y., Zhao, X., Song, Q.: Short-term traffic flow forecasting via echo state neural networks. In: IEEE Seventh International Conference on Natural Computation, pp. 844–847 (2011)
10. Embrechts, M., Alexandre, L., Linton, J.: Reservoir computing for static pattern recognition. In: European Symposium on Artificial Neural Networks, pp. 101–124 (2009)
11. Zhai, F., Lin, X., Yang, Z., Song, Y.: Notice of retraction financial time series prediction based on echo state network. In: IEEE Sixth International Conference on Natural Computation, pp. 3983–3987 (2010)
12. Ozturk, M.C., Xu, D.: Analysis and design of echo state networks for function approximation. Neural Comput. 19, 111–138 (2007)
13. Pan, W.: A new fruit fly optimization algorithm: taking the financial distress model as an example. Knowl.-Based Syst. 26, 69–74 (2012)
14. Lukoševičius, M.: A practical guide to applying echo state networks. In: Montavon, G., Orr, Geneviève B., Müller, K.-R. (eds.) Neural Networks: Tricks of the Trade. LNCS, vol. 7700, pp. 659–686. Springer, Heidelberg (2012). https://doi.org/10.1007/978-3-642-35289-8_36
15. Lin, W.: A novel 3D fruit fly optimization algorithm and its applications in economics. Neural Comput. Appl. 27, 1391–1413 (2016)
16. Wang, F.: A hybrid particle swarm algorithm with roulette selection operator. J. Southwest China Norm. Univ. 31, 93–96 (2006)
17. Shen, W., Han, Y.: Short term forecasting of Shanghai composite index based on GARCH and data mining technique. In: IEEE Circuits, Communications and System, pp. 63–66 (2010)
18. Phani, B.V., Chandra, B., Raghav, V.: Quest for efficient option pricing prediction model using machine learning techniques. In: International Joint Conference, pp. 654–657 (2011)
19. Black, F., Scholes, M.: The pricing of options and corporate liabilities. Polit. Econ. 81, 637–654 (2010)

# Application of SMOTE and LSSVM with Various Kernels for Predicting Refactoring at Method Level

Lov Kumar[1]([⊠]), Shashank Mouli Satapathy[2]([⊠]), and Aneesh Krishna[3]

[1] Department of Computer Science and Information Systems,
BITS Pilani Hyderabad, Hyderabad, India
lovkumar505@gmail.com
[2] School of Computer Science and Engineering, Vellore Institute of Technology,
Vellore, India
shashankamouli@gmail.com
[3] School of Electrical Engineering, Computing and Mathematical Sciences,
Curtin University, Bentley, Australia
a.krishna@curtin.edu.au

**Abstract.** Improving maintainability by refactoring is essentially being considered as one of the important aspect of software development. For large and complex systems, identification of code segments, which require re-factorization is a compelling task for software developers. Development of recommendation systems for suggesting methods, which require refactoring are achieved using this research work. *Materials and Methods:* Literature works considered source code metrics for object-oriented software systems in order to measure the complexity of a software. In order to predict the need of refactoring, the proposed system computes twenty-five different source code metrics at the method level and utilize them as features in a machine learning framework. An open source dataset consisting of five different software systems is being considered for conducting a series of experiments in order to assess the performance of proposed approach. LSSVM with SMOTE data imbalance technique are being utilized in order to overcome the class imbalance problem. *Conclusion:* Analysis of the results reveals that LS-SVM with RBF kernel using SMOTE results in the best performance.

**Keywords:** LSSVM · Kernels · Refactoring
Source code metrics · Smote

## 1 Research Motivation and Aim

Software refactoring is a disciplined technique for clarifying and simplifying the internal structure of an existing source code without affecting its external behavior [5,12]. Source code refactoring ruthlessly prevents rot, which in-turn helps in improving code maintainability and reducing complexity [5,12], without affecting semantics of the code. A series of refactoring techniques have been presented

© Springer Nature Switzerland AG 2018
L. Cheng et al. (Eds.): ICONIP 2018, LNCS 11305, pp. 150–161, 2018.
https://doi.org/10.1007/978-3-030-04221-9_14

by Martin Flower on a website called catalog[1] of refactoring. For systems having methods with identical results or subclasses, Pull Up Method refactoring technique is being considered. Detection of elements or regions of large and complex systems in need of refactoring, is technically challenging for software developers.

Identification of appropriate method(s) in a class, which require refactoring is the one of the prime objective of this research work. The proposed research is motivated by the need for developing recommendation systems, which can be integrated in the development environment and development processes of software engineers keeping in mind the end goal to recommend methods in need of refactoring. Previous research works emphasize on identification of classes and regions in a source code in need of refactoring (refer to the Section on Related Work). However, a deep study on the application of Machine Learning (ML) techniques considering several open source object-oriented based Java projects and furthermore utilizing software metrics as features or predictors is relatively unexplored.

**Research Contributions:** In the context of previous work, the proposed work exhibits several novel, unique and remarkable research contributions. The proposed work is an extension to our previous work on class-level refactoring prediction [11]. Computation of class-level metrics and prediction of the need of refactoring at class-level has been presented in the previous work [11]. The present research work emphasizes on the need of refactoring and prediction using software metrics at method-level. To the best of our knowledge, the work carried out in the paper is the first study on method-level refactoring prediction using 5 open-source Java based projects (antlr4, junit, mct, oryx, titan).

## 2   Related Work

For identification of refactoring candidates, a class-based approach has been introduced by Zhao et al. [15] considering a chosen set of static source code metrics and furthermore, utilizing a weighted ranking method for predicting a list of classes, which require refactoring. Al Dallal [2] presented a measure and a prescient model to decisively identify, whether method(s) in a class needing move method refactoring and achieved prediction accuracy of more than 90%. An in-depth analysis on the effects of refactoring over different internal quality attributes such as inheritance, complexity etc. have been presented by Chávez et al. [4]. They considered the history about different versions of twenty-three open source projects with more than 29,000 operations related to refactoring process. Kosker et al. [9] proposed an intelligent system by analyzing the code complexity in order to identify the class that require refactoring. Their approach is based on creating a machine learning model using Weighted Nave Bayes with InfoGain heuristic as the learner and conducted experiments on real world software system. Bavota et al. [3] conducted an empirical evaluation about the relationship between code smells and refactoring activities by mining more than 12000 refactoring operations.

---

[1] https://refactoring.com/catalog/.

## 3   Experimental Dataset

The experimental dataset used in our study is freely and publicly available at tera-PROMISE Repository[2]. The tera-PROMISE online resource is a well-known software engineering research dataset repository consisting of experimental datasets on several engineering topics such as source code analysis, faults, effort estimation, refactoring, source code metrics and test generation [1]. The tera-PROMISE repository dataset has been considered for experimental analysis purpose. This makes our work easily replicable and makes it easy for other researchers to compare or benchmark their approaches with our proposed method on the same dataset. The dataset used in our experiments is manually validated by the authors Kadar et al. [6,7], who shared the dataset on tera-PROMISE. The source code metrics have been created and dataset for two subsequent releases of 7 well-known OSS (open source software) Java applications have been refactored by them. [6,7]. The 7 Java-based OSS systems are available on GitHub[3] repository. Kadar et al. used the RefFinder tool [8] for identification of refactoring between two subsequent of releases of the source code. They compute the source code metrics using the SourceMeter tool[4]. We use the method level metrics in the work presented in this paper and not the system level metrics. The class level metrics have been considered in our previous work (refer to [10]) and the method level metrics of the same dataset have been considered in proposed solution. Table 1 shows the list of Java projects, Number of Methods (# NOM), Number of Refactored Methods (# NOMR), Number of Non-Refactored Methods (# NONMR),and Percentage of Refactored Methods (% RM). Table 1 reveals that the dataset is highly imbalanced as the % RM values are 0.13%, 0.19%, 0.52%, 0.75% and 1.21% respectively.

**Table 1.** Experimental data set description

| Project | NOM | NORM | NONRM | %RM |
| --- | --- | --- | --- | --- |
| antlr4 | 3298 | 40 | 3258 | 1.213 |
| junit | 2280 | 12 | 2268 | 0.526 |
| mct | 11683 | 16 | 11667 | 0.137 |
| oryx | 2507 | 19 | 2488 | 0.758 |
| titan | 8558 | 17 | 8541 | 0.199 |

## 4   Experimental Results

The approach for this experiment is a multi-step process. Initially, significant features using significance test is selected, followed by feature scaling using min-max

---

[2] http://openscience.us/repo/.
[3] https://github.com/.
[4] https://www.sourcemeter.com/.

normalization technique and handling imbalanced data using SMOTE approach. Then, LS-SVM learning algorithm with three different kernel techniques is considered in order to train the proposed model. Finally, with the help of AUC and ROC curves, the proposed model is compared. 25 source code metrics are considered as dataset for experimental evaluation process, which are computed through the SourceMeter[5] tool. The identification of a class as refactored or not in the successive releases, can be obtained from the target class represented as binary variable.

The proposed approach is initiated with a pre-processing phase, for identification of subset of source code metrics affecting refactoring through a statistical hypothesis test. The test dataset utilized in this proposed approach isn't equitably dispersed and exceptionally imbalanced. Trained imbalanced dataset can prompt predisposition for the minority class. Hence, a notable procedure called as SMOTE is considered in the proposed solution in order to deal with imbalanced dataset. SMOTE approach tends to oversample minority class. The experimental dataset characteristics is depicted in Table 1. Through the proposed research study, the effect of SMOTE technique over the classifier performance is investigated considering the results obtained from AUC values and ROC curves. As feature ranges are varied in nature, hence in order to rescale and standardize the features, a feature scaling method has been applied.

In order to scale down the all the features with in the range 0 to 1, the min-max normalization approach has been considered. LS-SVM learning algorithm, which is considered as one of the very popular method in software engineering domain in order to handle problems related to classification and pattern recognition [13, 14], is applied in the proposed study to develop a model for predicting refactoring. LS-SVM is a variation of SVM and can be seen as a least squares adaptation of SVM. SVM and LS-SVM are non-linear machine learning techniques, which maps the input feature space to a higher dimensional feature space. A 10-fold cross validation is applied over the dataset in order to measure the mean accuracy. The entire dataset is split into ten equal sizes, out of which nine sets are used for training and tenth one for evaluation purpose. The computation is performed for ten number of times and the final outcome of the experiment is computed by taking the average of those ten different experimental results.

### 4.1 Source Code Metrics

We use the source code metrics computed using the SourceMeter[6] for Java tool. We make use of 25 different source code metrics such as McCabes Cyclomatic Complexity (McCC), Clone Classes (CCL), Clone Complexity (CCO), Total Lines of Code (TLOC), Total Number of Statements (TNOS), Number of Incoming Invocations (NII) and Number of Outgoing Invocations (NOI). The list of all the metrics shown in Fig. 1 is provided on the SourceMeter tool website. The source code metrics are used as features or independent variables for the

---

[5] https://www.sourcemeter.com/.

[6] https://www.sourcemeter.com/resources/java/.

machine learning algorithms. The input to the machine learning algorithms are the source code metrics and the output is a binary class (whether a method is need of refactoring or now). We conduct a metrics selection using Wilcoxon rank-sum test and ULR (Univariate Logistic Regression). The graphs in Fig. 1 are represented using different symbols: filled circle: source code metrics selected using Wilcoxon rank-sum test, black circle with bold circle: source code metrics selected using Wilcoxon rank-sum test and ULR analysis. We perform metrics selection to remove metrics which are not relevant. Through metrics selection, we identify relevant metrics for the task of refactoring prediction.

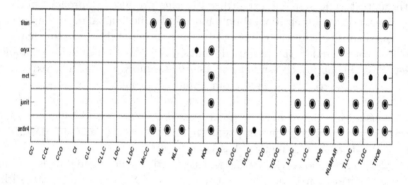

**Fig. 1.** Selected set of metrics using Wilcoxon rank-sum test and ULR analysis

**ROC Curve Analysis:** In order to validate the developed prediction models, some performance parameters need to compute and analyzed which indicate the usefulness of the trained models for predicting refactoring at class level. In this experiment, three different performance parameters such as Area Under the Receiver Operating Characteristics (ROC) Curve, accuracy, and F-Measure have been analyzed to select best model. The model having high value of these performance parameters denotes a best model for prediction. Figures 2 and 3 depict the ROC Curves for the trained model using different kernels with SMOTE and without SMOTE dataset for different datasets. The ROC Curve is very popular performance measure used for determining the accurateness of the trained models. It is used to find the performance of the model on the validated data. ROC Curve is a pictorial representation of 1-specificity and sensitivity at various cutoff points. In Figs. 2 and 3, 1-specificity and sensitivity are plotted on x-axis and y-axis respectively. Thus, area under curve (AUC) measure computed using ROC analysis has been considered to determine the performance of the kernels methods and imbalance techniques.

The technical challenges for this experiment to build a classifier for refactoring prediction is that the dataset is imbalanced. In this work, this issue has been resolved using Synthetic Minority Over-sampling Technique (SMOTE). One of the aspects of SMOTE is oversampling of the minority class. The objective

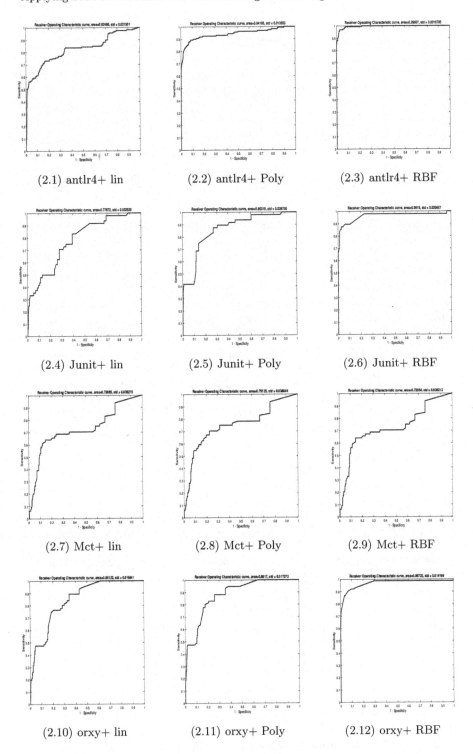

(2.1) antlr4+ lin          (2.2) antlr4+ Poly          (2.3) antlr4+ RBF

(2.4) Junit+ lin          (2.5) Junit+ Poly          (2.6) Junit+ RBF

(2.7) Mct+ lin          (2.8) Mct+ Poly          (2.9) Mct+ RBF

(2.10) orxy+ lin          (2.11) orxy+ Poly          (2.12) orxy+ RBF

**Fig. 2.** Sensitivity and specificity values for without SMOTE

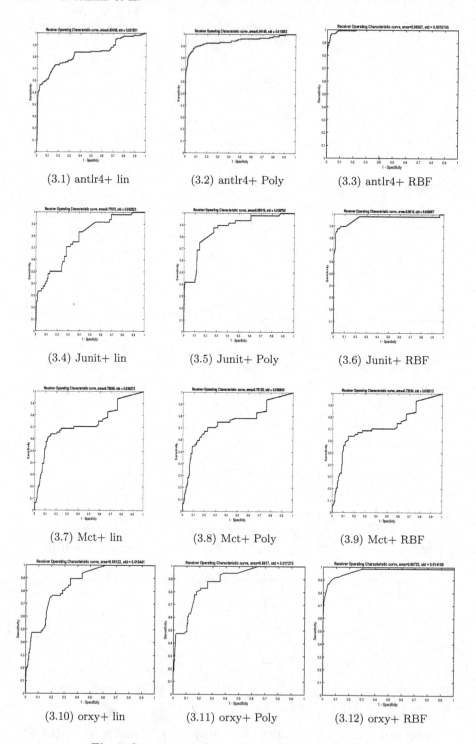

**Fig. 3.** Sensitivity and specificity values with SMOTE

to create the ROC curves for analyzing the performance of various kernel and imbalance technique through visualization in addition to determine performance value like AUC, accuracy, precision, and recall. The ROC curve shown in Figs. 2 and 3 depict the variation of 1-specificity value with sensitivity for different threshold value. It can be observed from the Figs. 2 and 3 that ROC curves for SMOTE having high upper left corner i.e., high values of 1-specificity value with sensitivity as compare to without smote as shown in Fig. 2.

**Performance Parameters Value:** Tables 2, and 3 present the performance results obtained after applying SMOTE and Without SMOTE. The tables show that the AUC, F-Measure, and accuracy for different kernels, imbalance techniques with different experimental dataset. The performance values shown in Tables 2, and 3 depict that the performance values with the SMOTE technique showing the best performance as compare to without smote. The experiment results in Tables 2, and 3 also displays the variations in the performance depending on the classifier. For example, SMOTE results in better performance in combination with LSSVM with RBF kerenl in comparison to other kernels.

**Table 2.** Performance results (without SMOTE)

|        | Accuracy |       |       | F-Measure |       |       | AUC   |       |       |
|--------|-------|-------|-------|-------|-------|-------|-------|-------|-------|
|        | Lin   | Poly  | RBF   | Lin   | Poly  | RBF   | Lin   | Poly  | RBF   |
| antlr4 | 98.79 | 98.88 | 99.85 | 0.977 | 0.991 | 0.996 | 0.811 | 0.876 | 1.000 |
| junit  | 99.47 | 99.47 | 99.47 | 0.989 | 0.991 | 0.996 | 0.757 | 0.734 | 0.720 |
| mct    | 99.86 | 99.86 | 99.86 | 0.997 | 0.997 | 0.997 | 0.719 | 0.752 | 0.718 |
| orxy   | 99.24 | 99.40 | 99.40 | 0.985 | 0.987 | 0.994 | 0.840 | 0.859 | 0.936 |
| titan  | 99.80 | 99.83 | 99.80 | 0.996 | 0.997 | 0.996 | 0.550 | 0.654 | 0.739 |

**Table 3.** Performance results (SMOTE)

|        | Accuracy |       |       | F-Measure |       |       | AUC   |       |       |
|--------|-------|-------|-------|-------|-------|-------|-------|-------|-------|
|        | Lin   | Poly  | RBF   | Lin   | Poly  | RBF   | Lin   | Poly  | RBF   |
| antlr4 | 95.57 | 98.20 | 99.14 | 0.977 | 0.991 | 0.996 | 0.825 | 0.942 | 0.995 |
| junit  | 97.91 | 98.17 | 99.26 | 0.989 | 0.991 | 0.996 | 0.779 | 0.863 | 0.962 |
| mct    | 99.45 | 99.45 | 99.45 | 0.997 | 0.997 | 0.997 | 0.727 | 0.751 | 0.726 |
| orxy   | 97.03 | 97.54 | 98.79 | 0.985 | 0.987 | 0.994 | 0.851 | 0.882 | 0.967 |
| titan  | 99.26 | 99.34 | 99.21 | 0.996 | 0.997 | 0.996 | 0.578 | 0.693 | 0.735 |

## 4.2  Performance Visualization Using Boxplots

Figures 5 and 4 show the multiple boxplots for analyzing the degree of spread
or dispersion, outliers, skewness, interquartile range in the accuracy, F-Measure
and AUC performance metrics for the classifiers and data sampling techniques.
The red line in the boxplot of Figs. 5 and 4 marks the mid-point (median value)
dividing the box into two segments. From Figs. 4, it has been observed that the
median value for the AUC, F-Measure and Accuracy of RBF kernel is higher that
all other kernels. Similarly from Figs. 5, it has been observed that the median
value for the AUC for SMOTE is higher than without SMOTE and accuracy
and F-Measure are similar in both cases.

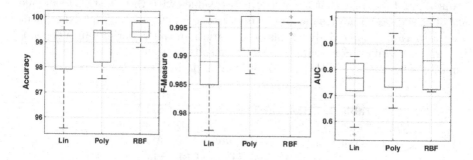

**Fig. 4.** Performance visualization using boxplots for kernels

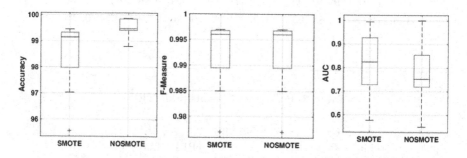

**Fig. 5.** Performance visualization using boxplots for imbalance technique

## 4.3  Descriptive Statistics In-Terms of Accuracy, F-Measure and AUC

Tables 4, 5, and 6 show the descriptive statistics of the overall performance for
LSSVM with linear kernel, polynomial kernel, and RBF kernel, SMOTE and

without SMOTE techniques in-terms of accuracy, F-Measure and AUC. For each kernel, we generate $5 Datasets * 2 imbalance techniques = 10$ data points consisting of 10 accuracy, F-Measure, and AUC values. For with and without SMOTE, we generate $5 Datasets * 3 kernels = 15$ data points consisting of 15 accuracy, F-Measure, and AUC values. Every value consist of the average performance for the 5 dataset. Table 4 reveals that the mean accuracy for the RBF kerenel is 99.423% and mean accuracy for the SMOTE is 98.51 respectively. From Tables 4, 5, and 6, we observed that SMOTE and LSSVM with RBF kerenl gives the best performance as compare to other techniques.

**Table 4.** Descriptive statistics of the overall performance of a technique across all datasets in term of Accuracy

| Accuracy | | | | | | | |
|---|---|---|---|---|---|---|---|
| | Min | Median | Q1 | Mean | Std Dev | Q3 | Max |
| SMOTE | 95.570 | 99.140 | 97.975 | 98.518 | 1.127 | 99.320 | 99.450 |
| NOSMOTE | 98.790 | 99.470 | 99.400 | 99.532 | 0.355 | 99.845 | 99.860 |
| LIN | 95.570 | 99.250 | 97.910 | 98.638 | 1.394 | 99.470 | 99.860 |
| POLY | 97.540 | 99.370 | 98.200 | 99.014 | 0.789 | 99.470 | 99.860 |
| RBF | 98.790 | 99.425 | 99.210 | 99.423 | 0.345 | 99.800 | 99.860 |

**Table 5.** Descriptive statistics of the overall performance of a technique across all datasets in-terms of F-Measure

| F-Measure | | | | | | | |
|---|---|---|---|---|---|---|---|
| | Min | Median | Q1 | Mean | Std Dev | Q3 | Max |
| SMOTE | 0.977 | 0.996 | 0.990 | 0.992 | 0.006 | 0.997 | 0.997 |
| NOSMOTE | 0.977 | 0.996 | 0.990 | 0.992 | 0.006 | 0.997 | 0.997 |
| LIN | 0.977 | 0.989 | 0.985 | 0.989 | 0.008 | 0.996 | 0.997 |
| POLY | 0.987 | 0.991 | 0.991 | 0.993 | 0.004 | 0.997 | 0.997 |
| RBF | 0.994 | 0.996 | 0.996 | 0.996 | 0.001 | 0.996 | 0.997 |

**Table 6.** Descriptive statistics of the overall performance of a technique across all datasets in-term of AUC

| AUC | | | | | | | |
|---|---|---|---|---|---|---|---|
| | Min | Median | Q1 | Mean | Std Dev | Q3 | Max |
| SMOTE | 0.578 | 0.825 | 0.729 | 0.818 | 0.119 | 0.927 | 0.995 |
| NOSMOTE | 0.550 | 0.752 | 0.719 | 0.778 | 0.113 | 0.854 | 1.000 |
| LIN | 0.550 | 0.768 | 0.719 | 0.744 | 0.105 | 0.825 | 0.851 |
| POLY | 0.654 | 0.806 | 0.734 | 0.801 | 0.095 | 0.876 | 0.942 |
| RBF | 0.718 | 0.838 | 0.726 | 0.850 | 0.130 | 0.967 | 1.000 |

## 5    Conclusion

In this work, the source code metrics computed using SourceMeter[7] are used as input to develop a model for predicting refactoring of software at method level. We make use of 25 different source code metrics such as McCabes Cyclomatic Complexity (McCC), Clone Classes (CCL), Clone Complexity (CCO), Total Lines of Code (TLOC), Total Number of Statements (TNOS), Number of Incoming Invocations (NII) and Number of Outgoing Invocations (NOI). The list of all the metrics shown in Fig. 1 is provided on the SourceMeter tool website. Wilcoxon rank-sum test and ULR (Univariate Logistic Regression) are used to remove irrelevant metrics and select best set of metrics for refactoring prediction. The graphs in Fig. 1 are represented using different symbols: filled circle: source code metrics selected using Wilcoxon rank-sum test, black circle with bold circle: source code metrics selected using Wilcoxon rank-sum test and ULR analysis. The experimental results of the comparison of different kernels evaluated using box-plots and descriptive statistics show the superiority of RBF kernel in code refactoring predictions. The experimental results also confirms the superiority of SMOTE as compare to without SMOTE for code refactoring predictions. Thus, we conclude that the LSSVM with different kernels and SMOTE have predictive capability and these methods can be applied on future releases of java projects.

## References

1. The promise repository of empirical software engineering data (2015)
2. Al Dallal, J.: Predicting move method refactoring opportunities in object-oriented code. Inf. Softw. Technol. **92**, 105–120 (2017)
3. Bavota, G., De Lucia, A., Di Penta, M., Oliveto, R., Palomba, F.: An experimental investigation on the innate relationship between quality and refactoring. J. Syst.Softw. **107**, 1–14 (2015)
4. Chávez, A., Ferreira, I., Fernandes, E., Cedrim, D., Garcia, A.: How does refactoring affect internal quality attributes?: A multi-project study. In: Proceedings of the 31st Brazilian Symposium on Software Engineering, pp. 74–83. ACM (2017)
5. Fowler, M., Beck, K.: Refactoring: Improving the Design of Existing Code. Addison-Wesley Professional, Boston (1999)
6. Kádár, I., Hegedus, P., Ferenc, R., Gyimóthy, T.: A code refactoring dataset and its assessment regarding software maintainability. In: 2016 IEEE 23rd International Conference on Software Analysis, Evolution, and Reengineering (SANER), vol. 1, pp. 599–603. IEEE (2016)
7. Kádár, I., Hegedűs, P., Ferenc, R., Gyimóthy, T.: A manually validated code refactoring dataset and its assessment regarding software maintainability. In: Proceedings of the The 12th International Conference on Predictive Models and Data Analytics in Software Engineering, p. 10. ACM (2016)
8. Kim, M., Gee, M., Loh, A., Rachatasumrit, N.: Ref-finder: a refactoring reconstruction tool based on logic query templates. In: Proceedings of the Eighteenth ACM SIGSOFT International Symposium on Foundations of Software Engineering, pp. 371–372. ACM (2010)

---

[7] https://www.sourcemeter.com/resources/java/.

9. Kosker, Y., Turhan, B., Bener, A.: An expert system for determining candidate software classes for refactoring. Expert Syst. Appl. **36**(6), 10000–10003 (2009)
10. Kumar, L., Sureka, A.: Application of LSSVM and SMOTE on seven open source projects for predicting refactoring at class level. In: 2017 24th Asia-Pacific Software Engineering Conference (APSEC), pp. 90–99 (2017)
11. Kumar, L., Sureka, A.: Application of LSSVM and SMOTE on seven open source projects for predicting refactoring at class level. In: 2017 24th Asia-Pacific Software Engineering Conference (APSEC), pp. 90–99. IEEE (2017)
12. Mens, T., Tourwé, T.: A survey of software refactoring. IEEE Trans. Softw. Eng. **30**(2), 126–139 (2004)
13. Suykens, J., Lukas, L., Van Dooren, P., De Moor, B., Vandewalle, J., et al.: Least squares support vector machine classifiers: a large scale algorithm. In: European Conference on Circuit Theory and Design, ECCTD, vol. 99, pp. 839–842 (1999)
14. Suykens, J.A., Lukas, L., Vandewalle, J.: Sparse approximation using least squares support vector machines. In: Proceedings of the IEEE International Symposium on Circuits and Systems, ISCAS 2000, vol. 2, pp. 757–760. IEEE, Geneva (2000)
15. Zhao, L., Hayes, J.: Predicting classes in need of refactoring: an application of static metrics. In: Proceedings of the 2nd International PROMISE Workshop, Philadelphia, Pennsylvania, USA (2006)

# Deep Ensemble Model with the Fusion of Character, Word and Lexicon Level Information for Emotion and Sentiment Prediction

Deepanway Ghosal, Md Shad Akhtar[✉], Asif Ekbal,
and Pushpak Bhattacharyya

Department of Computer Science and Engineering,
Indian Institute of Technology Patna, Patna, India
{deepanway.me14,shad.pcs15,asif,pb}@iitp.ac.in

**Abstract.** In this paper, we propose a novel neural network based architecture which incorporates character, word and lexicon level information to predict the degree of intensity for sentiment and emotion. At first we develop two deep learning models based on Long Short Term Memory (LSTM) & Convolutional Neural Network (CNN), and a feature based model. Each of these models takes as input a fusion of various representations obtained from the characters, words and lexicons. A Multi-Layer Perceptron (MLP) network based ensemble model is then constructed by combining the outputs of these three models. Evaluation on the benchmark datasets related to sentiment and emotion shows that our proposed model attains the state-of-the-art performance.

**Keywords:** Sentiment analysis · Emotion analysis
Intensity Prediction · Financial domain · Ensemble
Deep learning

## 1 Introduction

We live in a time where the access to information has never been so free. Online platforms like Twitter, Facebook etc. give a sense of power where an user can express his/her views, opinions and get to know about others' ideas. All this is possible in mere 280 characters that Twitter limits per tweet. This short piece of text has the potential to shape peoples' outlook towards any situation or product or a service. Companies and service providers can utilize this dynamic textual information, and infer the public opinions about a newly launched product or any service or market conditions.

Emotion analysis [27] deals with the automatic extraction of emotions expressed in a user written text. There are six basic emotions as categorized by Ekman [10]: *joy, sadness, surprise, fear, disgust* and *anger*. In comparison sentiment analysis [25] tries to automatically extract the subjective information

© Springer Nature Switzerland AG 2018
L. Cheng et al. (Eds.): ICONIP 2018, LNCS 11305, pp. 162–174, 2018.
https://doi.org/10.1007/978-3-030-04221-9_15

from user written textual contents and classify it into one of the predefined set of categories, for e.g. *positive, negative, neutral* or *conflict.* Emotions are usually shorter in duration whereas sentiments are more stable and valid for a longer period of time [8]. Also, sentiments are normally expressed towards a target entity, whereas emotions are not always target-centric [28].

Finding only the emotion or sentiment class does not always reflect the exact state of mood or opinion of any user. Level or intensity often differs on a case-to-case basis within a single emotion or sentiment class. Some emotions are gentle (e.g '*not good*') while the others can be very severe (e.g. '*terrible*'). Similarly, both the phrases '*its fine*' and '*its awesome*' carry positive sentiment but express different levels of intensities. Sentiment of the later is strong, whereas it is comparatively mild for the former. This kind of analysis of emotions and sentiments has a diversified set of real-world applications such as feedback systems for an organization or to an end-user *w.r.t.* a product or service, stock market prediction, policy making etc.

In this paper, we propose an effective deep learning architecture for determining the intensity of emotion and sentiment[1]. For emotion analysis we employ generic tweets, whereas for sentiment analysis our target domain is financial text. We first develop two deep learning models based on Convolutional Neural Network (CNN) and Long Short Term Memory (LSTM), and a Feature driven model based on Multi-Layer Perceptron (MLP). These models take as input a fusion of different representations obtained from the various sources including character-level embeddings, pre-trained word embeddings and the proposed lexicon embeddings. Next, we combine the outputs of these systems *via* an ensemble based MLP network.

We summarize the contributions of our proposed work as follows: **(a)** We propose lexicon embeddings, that when used along with character embeddings and the pre-trained word embeddings, improve the performance of both sentiment and emotion analysis; and **(b)** we propose an effective ensemble model by combining CNN, LSTM and the feature based MLP. It shows state-of-the-art performance for both fine-grained emotion and sentiment analysis (i.e. intensity prediction).

## 2 Methodology

### 2.1 Problem Definition

Our work focuses on determining the degree of intensities of emotion and sentiment.

**Emotion Analysis:** For a given tweet instance we aim to determine the degree of emotion (a score between 0 and 1) felt by the writer. Intensity value close to '1' reflects high-degree of emotion whereas value close to 0 reflects the low-degree of emotion. In our work we consider four different emotions i.e. '*anger*', '*fear*', '*joy*'

---

[1] We also call this problem as fine-grained emotion & sentiment analysis.

and *'sadness'*. Table 1 depicts four example scenarios with the emotion types and the intensity values.

**Sentiment Analysis:** Here sentiment analysis is carried out for two different kinds of financial texts, *viz. 'microblog messages'* and *'news headlines'*. The objective is to predict the sentiment score for each of the mentioned companies or stocks in the range of -1 (bearish) to 1 (bullish), with 0 implying neutral sentiment. A couple of examples are shown in Table 1.

## 2.2 Embeddings

We train our deep learning models on top of two distributed embeddings. We denote these as (a). *fused word embeddings* and (b). *char (character) convoluted embeddings*.

**(a). *Fused word embeddings*:** We generate the fused embedding representation of a word by concatenating the embeddings obtained from the pre-trained GloVe [26], pre-trained Word2Vec [19] & a lexicon embedding. In particular we use 300 dimensional *common crawl 840 billion* version of GloVe & 300 dimensional *skip gram with negative sampling* version of Word2Vec. Lexicon embedding is 20-dimensional, which is created from several sentiment and emotion lexicons. For any word ($W$) we obtain its lexicon embedding as follows,

- A single dimensional boolean value (1/0) denoting whether $W$ appears in the opinion lexicon [9].
- A single dimensional value denoting the sentiment score of $W$ in MPQA subjectivity lexicon [29]. We use the value of $\{-1/-0.5/0.5/1\}$ depending upon the appearance of $W$ in $\{strong\ negative/weak\ negative/weak\ positive/strong\ positive\}$.
- A five dimensional value denoting the sentiment scores of $W$ in NRC Hashtag Sentiment, NRC Hashtag Context, NRC Sentiment 140, NRC Sentiment 140 Context & NRC SemEval Twitter lexicons [5,20,22].
- An eight dimensional value denoting the emotion scores of {anticipation, fear, anger, trust, surprise, sadness, joy & disgust} of $W$ in NRC Emotion lexicon [23].
- A five dimensional value denoting the scores of {polarity, aptitude, attention, pleasantness, sensitivity} of $W$ in SenticNet lexicon [6].

We assign zero values for words that do not appear in the lexicon. Finally, the GloVe, Word2Vec and lexicon embedding are concatenated to generate the 620 dimensional *fused word embedding*.

**(b). *Char (character) convoluted embeddings*:** Our character convoluted embedding model is illustrated in Fig. 1. This is inspired from the character based neural language models [14]. Each word is represented as the concatenation of its constituent character embeddings ($C^k$). For e.g. in Fig. 1 'amazing' is represented as the concatenation of character embeddings of 'a', 'm', 'a', 'z', 'i', 'n' & 'g'. We represent the character embeddings as one-hot vector of dimension 38 (we had 38 unique characters in our corpus). This is a key difference of our model with that of [14], where the authors used a 4-dimensional character embedding.

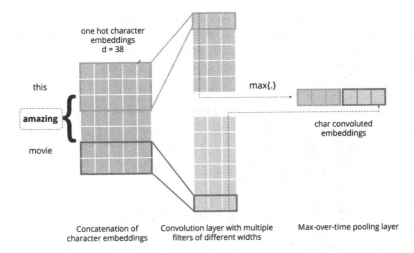

**Fig. 1.** Computation of *char convoluted embeddings*. Convolution operation is performed between the concatenated one-hot embeddings of {'a', 'm', 'a', 'z', 'n', 'i', 'g'} with 3 filters of width 3 (yellow) & 3 filters of width 2 (cyan). A max-over-time pooling is then performed to obtain a fixed dimensional vector (dim. equal to the total no. of filters) which is the *char convoluted embedding* of the word *amazing*. (Color figure online)

A convolution operation between $C^k$ and multiple filters with different widths is then performed. In particular we use 10 filters of width 2 (blue) and 10 filters of width 3 (yellow). Note that in Fig. 1, we have 3 filters of width 2 (blue) and 3 filters of width 3 (yellow). Finally, a max-over-time pooling operation is performed to obtain a fixed-sized (equal to the total no. of filters, here 20) representation of the word. We denote this 20-dimensional representation as the *char convoluted embedding* of the word. These *char convoluted embeddings* are incorporated in our LSTM & CNN model (which we describe next). Note that, the *char convoluted embeddings* are obtained from the trainable filter parameters during the convolution process. These parameters are thus trained jointly with our LSTM & CNN models.

### 2.3   LSTM Model

The overall architecture of the LSTM model is shown in Fig. 2. It consists of two LSTM sub-networks: first sub-network is applied over the *fused word embeddings*, whereas the second is applied over the *char convoluted embeddings*. Each of the sub-networks consists of two bidirectional LSTM layers stacked one upon another. We use 128 neurons in the first bidirectional LSTM layer of sub-network 1 (by concatenating 64 neurons from forward state & 64 neurons from backward state) and 64 neurons (32 + 32) in the first bidirectional LSTM layer of sub-network 2. The second bidirectional LSTM layer consists of 128 (64 + 64) neurons for sub-network 1 and 64 (32 + 32) neurons for sub-network 2. Finally, all the forward and backward states (of the second bidirectional LSTM layer) from the two sub-networks are concatenated to obtain a 192-dimensional vector. This is then passed through a hidden layer (50 neurons, fully connected) to the final output layer for the sentiment/emotion intensity prediction.

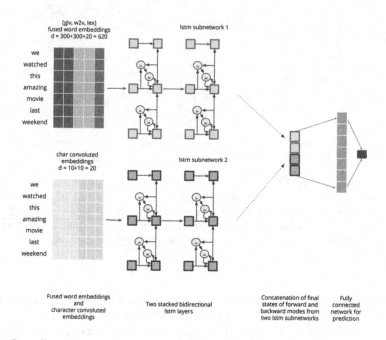

**Fig. 2.** Overall architecture of our LSTM model. The char convoluted embeddings are obtained from Fig. 1 and are trained jointly with this LSTM model.

## 2.4   CNN Model

We have two sub-networks in our CNN model. The first sub-network performs successive 1D convolution and pooling over the *fused word embeddings*, whereas the second sub-network performs the same operations over the *char convoluted embeddings*. We perform the following sequence of operations in each of the sub-networks: {*conv - pool - conv - pool*}. Both the *conv* operations correspond to 1D convolution with multiple filters of different widths. In particular we use 100 filters of width 2, 100 filters of width 3 & 100 filters of width 4, all with *stride* 1 and *same boundary* mode (i.e. zeros are padded at the boundary so that the size after convolution is equal to the size before convolution). The convoluted outputs from these 300 filters are then concatenated to obtain the final convoluted feature map. The *pool* operation corresponds to 1D max-pooling of the convoluted feature map (from the previous *conv* operation) with *factor* 2 and *stride* 2. The final pooled outputs from the two sub-networks are concatenated and flattened, which is then passed through a hidden layer (50 neurons, fully connected) to the final output layer for the sentiment/emotion intensity prediction.

## 2.5   Feature Based Model

In order to exploit the diversities of both deep learning and classical supervised model we develop a feature based model using MLP for both the tasks. We use the following set of features:

**1. *N-grams:*** We use Tf-Idf word n-grams (n = 1, 2, 3, 4) and character n-grams (n = 2, 3, 4).

**2. *Lexicon features:*** Lexicons are widely used in sentiment and emotion analysis. We extract the following features for each instance of text:

- count of positive and negative words using the MPQA subjectivity lexicon [29] and Opinion lexicon [9].
- positive and negative scores from: Sentiment140 [20], AFINN [24] and SentiWordnet [3] lexicons. It calculates the sum of positive and negative scores obtained from the lexicons.
- aggregate scores of hashtags from NRC Hashtag Sentiment lexicon [20].
- count of the number of words matching each emotion from the NRC Word-Emotion Association Lexicon [23].
- sum of emotion associations from NRC-10 Expanded lexicon [5].
- sum of emotion associations of tweet hashtags from the NRC Hashtag Emotion Association Lexicon [22].
- sum of the polarity and emotion scores from the SenticNet lexicon [6].

**3. *Average embedding:*** We scale the 620 dimensional fused embedding of words according to their Tf-Idf weights, and then compute the average of these weighted embeddings of all the words in the text to create a feature vector.

All these features are concatenated to generate the final feature vector, and subjected as input to the MLP network, which has 3 fully connected hidden layers (512, 128, 100 neurons) and the output layer (1 neuron).

## 2.6   Ensemble Model

We use a MLP based ensemble technique which learns on top of the predictions of the three base models (LSTM, CNN & Feature model). This network has a three-neuron (for the three base models) input layer and a single-neuron output layer. The weights corresponding to the three neurons are constrained to have non-negative values (in order to analyze relative importance of the models), however the bias can attain any value. Output of this network is the final prediction value.

## 2.7   Hyperparameter and Training Details

The *fused word embeddings* and the 38-dimensional one-hot character embeddings are dynamically updated through backpropagation during the training process. All the base models are trained independently. We use the *Rectified Linear (ReLU)* [11] activation function for the fully connected hidden layers in our LSTM, CNN and feature based model. 30% *Dropout* [12] is used in the fully connected hidden layers as a regularizer. We use *sigmoid* (emotion intensity) & *tanh* (sentiment score) activation for the output layer. The *Adam* optimizer [15] with learning rate of 0.001 and *mean squared error* loss function is used for the gradient based training. We keep the *batch size* equal to 32 during the training process. We train each model for 50 epochs with *Early Stopping* having patience of 10. The ensemble model is trained with the same output activation, optimizer, loss function & epoch configuration as described above. We run each base model 5 times (with different seeds) and after averaging the predictions (for each base model) we pass it to the ensemble network.

## 3   Datasets and Experiments

▶ **Datasets:** We perform experiments on the benchmark datasets of WASSA-2017 shared task on emotion intensity (EmoInt-2017) [21] for emotion analysis, and SemEval 2017 shared task on 'Fine Grained Sentiment Analysis on Financial Microblogs and News' [7] for sentiment analysis.

Datasets of EmoInt-2017 contain generic tweets representing four emotions i.e. *anger, fear, joy* & *sadness*. Datasets of SemEval 2017 consist of financial texts from microblog messages (StockTwits & Twitter) & news headlines.

Table 1 shows a few example scenarios for the problems of emotion analysis and sentiment analysis. In the first example, high intensity of emotion *'joy'* is

**Table 1.** Emotion & sentiment analysis examples from the benchmark datasets. Given text and emotion/target the goal is to determine the intensity.

| Emotion Analysis | | |
|---|---|---|
| Text | Emotion | Intensity |
| *Just died from laughter after seeing that* | Joy | 0.92 |
| *I am so gloomy today* | Sadness | 0.73 |
| *I genuinely think I have anger issues* | Anger | 0.60 |
| *What an actual nightmare!* | Fear | 0.73 |
| Sentiment Analysis | | |
| Text | Target | Intensity |
| *Best stock: $WTS +15%* | WTS | 0.86 |
| *HSBC chief hit with tax-avoidance scandal* | HSBC | −0.53 |

derived from the very intense phrase *'died from laughter'*. Similarly, the intensity of sadness is rather high because of the word *'gloomy'*. In the first example of sentiment analysis, a high positive sentiment is expressed towards the target *WTS*, whereas in the second example a moderately negative intensity is expressed towards *HSBC*.

▶ **Preprocessing:** We use the NLTK [4] for preprocessing and tokenization. Numbers, twitter user-names and urls are replaced with $<number>$, $<user>$ and $<url>$. Finally, we perform normalization of noisy text by employing a set of heuristics, as depicted in Table 2, in line with [2]. Elongation was handled by iteratively removing a character from the repeated sequence and verifying the new word against a dictionary[2].

**Table 2.** Examples of noisy text normalization.

| Normalization | Noisy | Corrected |
|---|---|---|
| Elongation | *'joooyyyyy'* | *'joy'* |
| Verb present participle | *'goin'* & *'gong'* | *'going'* |
| Frequent noisy term | *'g8'*, *'grt'* | *'great'* |
| Expand contractions | *'i'm'* | *'i am'* |
| Hashtag segmentation | *'#GreatDayEver'* | *'Great Day Ever'* |

**Table 3.** Cosine similarity (financial sentiment) & pearson correlation (emotion analysis) score of various models on test data. All scores are average of 5 runs with different seeds.

| Models | Financial sentiment | | Emotion analysis | | | | |
|---|---|---|---|---|---|---|---|
| | Microblogs | News | Anger | Joy | Sadness | Fear | Average |
| LSTM model | 0.749 | 0.747 | 0.712 | 0.702 | 0.738 | 0.753 | 0.726 |
| CNN model | 0.778 | 0.759 | 0.729 | 0.697 | 0.748 | 0.759 | 0.733 |
| Feature based model | 0.777 | 0.763 | 0.715 | 0.707 | 0.730 | 0.743 | 0.724 |

▶ **Experiments:** For evaluation we compute *Pearson correlation coefficient* and *Cosine similarity score* for emotion and sentiment intensity prediction, respectively. We use these metrics to remain consistent with the shared tasks of EmoInt-2017 [21] and SemEval-2017 [7].

Table 3 shows the cosine similarity and pearson scores of the models for the test data. Although the performance of these models are quite similar numerically, we found these to be quite contrasting on a qualitative side as shown in Fig. 3. Motivated by this contrasting nature, we combine these predictions through an ensemble network.

---

[2] Spell Checker Oriented Word Lists (SCOWL): http://wordlist.aspell.net/.

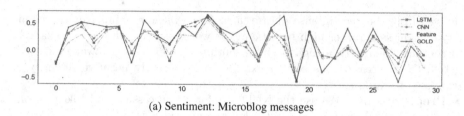

(a) Sentiment: Microblog messages

**Fig. 3.** Contrasting nature of different models *w.r.t.* the gold values in the microblog dataset; Sample size - 30. Y-axis denotes predicted values of different models & the gold value. X-axis denotes sample indices.

**Table 4.** Test result of the proposed ensemble model and comparison with the state-of-the-art systems. The ensemble model is the ensemble of the three models in Table 3. Evaluation metric is *cosine similarity* for financial sentiment & *pearson correlation* for emotion analysis. Financial sentiment: ECNU, Fortia-FBK & [1] are the top systems at SemEval-2017 task 5. Emotion analysis: Prayas & IMS are the top two systems at EmoInt-2017.

| Models | Financial sentiment | | Emotion analysis | | | | |
|---|---|---|---|---|---|---|---|
| | Microblogs | News | Anger | Joy | Sadness | Fear | Average |
| ECNU [17] | 0.778 | 0.710 | - | - | - | - | - |
| Fortia-FBK [18] | - | 0.745 | - | - | - | - | - |
| Akhtar et al.[1] | 0.797 | 0.786 | - | - | - | - | - |
| Prayas [13] | - | - | 0.732 | **0.732** | 0.765 | 0.762 | 0.747 |
| IMS [16] | - | - | 0.705 | 0.690 | **0.767** | 0.726 | 0.722 |
| Proposed ensemble | **0.813** | **0.801** | **0.754** | 0.724 | **0.767** | **0.785** | **0.757** |

For sentiment analysis, the proposed ensemble network (c.f. Table 4) shows the cosine similarity scores of 0.813 and 0.801 for microblog and news headlines, respectively. For emotion analysis, the proposed model demonstrates the pearson scores of 0.754, 0.724, 0.767 & 0.785 for '*anger*', '*joy*', '*sadness*' & '*fear*', respectively (resulting in an average pearson score of 0.757). Our predicted results are statistically significant (*t-test*) with *p*-values *0.0035*, *0.032* & *0.015* for microblog, news & average emotion, respectively with respect to the current state-of-the-art results.

▶ **Comparison to the other systems:** We compare our proposed system with the state-of-the-art systems in Table 4. For emotion analysis task, Prayas [13] and IMS [16] are the two best performing systems with average pearson scores of 0.747 and 0.722.

Prayas used an ensemble of five different neural network models (a feed-forward model, a multitasking feed-forward model and three joint CNN-LSTM models). The final predictions were obtained by a weighted average of these base models. IMS employed a random forest model on concatenated lexicon and CNN-LSTM features. The system made use of an external lexicon source (ACVH-Lexicons) containing the ratings for arousal, concreteness, valency and happiness, which were not part of the original baseline model [21]. They also used a Twitter corpus containing 800 million tokens to train their embeddings. Compared to these models, our proposed system attains the improved (average) pearson score of 0.757 without using such external resources. For microblog sentiment analysis, state-of-the-art models [1,17] reported cosine similarity of 0.778 & 0.797 compared to 0.813 of ours. For news headline, our model achieves 0.801 in comparison to [1,18] which reported 0.745 & 0.786, respectively. ECNU [17] used multiple regressors on top of the optimized feature set obtained from the *hill climbing* algorithm. Fortia-FBK [18] trained a CNN with sentiment lexicons. Akhtar et al. [1] trained 15 different deep learning models & 6 different SVR models with *financial word embeddings* trained on 126,000 news articles. Statistical *t-test* confirms these improvements over the state-of-the-art systems to be significant for both sentiment and emotion.

▶ **Ensemble analysis:** We analyze the *heatmap* (Fig. 4) of the weights & the bias of our ensemble model to understand the contribution of the base models. The first three rows represent the relative importance of the three base models, whereas the last row represents the bias. Darker *Reddish* cells signify higher contribution (higher weight) of the base model. It is evident that each model has a role to contribute in the overall ensemble.

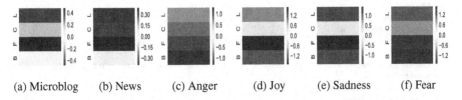

(a) Microblog     (b) News     (c) Anger     (d) Joy     (e) Sadness     (f) Fear

**Fig. 4.** Heatmap of weights & bias of the MLP ensemble network. L, C & F are weights corresponding to the LSTM, CNN & feature model. B represents the bias. Color coding (*Red: positive values; Blue: negative values; White: zero value.*) of cells signify the contribution of respective models in ensemble. (Color figure online)

▶ **Error analysis :** We analyze the top 25 error cases from the test dataset. We present the frequently occurring error cases along with their possible reasons in Table 5.

▶ **Ablation analysis:** The ablation analysis reported in Table 6 shows that all the sources of information (word, character, lexicon) are crucial. Any of these components, when omitted, causes a drop in the overall performance.

**Table 5.** Error analysis: frequent error cases with their possible reasons.

| Text | Dataset | Actual | Predicted | Possible reason |
|---|---|---|---|---|
| *I'm such a shy person, oh my lord* | Fear | 0.833 | 0.384 | Implicit sentiment/emotion |
| *Tesco abandons video-streaming ambitions in blinkbox sale.* | News | −0.335 | 0.332 | |
| *Verified $98.95 loss in $ENDP trades* | Microblogs | −0.146 | −0.441 | Numbers & symbols |
| *Best stock: $WTS +15%* | | 0.857 | 0.308 | |
| *Cannot wait to see you honey!* | Joy | 0.770 | 0.462 | Implicit emotion with negation |
| *I see things in the clouds that others cannot see so i can be late* | | 0.620 | 0.155 | |
| *Always so happy to support you brother, keep that fire burning* | Anger | 0.132 | 0.527 | Metaphoric sentence |

**Table 6.** Ablation study of different models. *w/o* stands for without. Average cosine similarity (financial sentiment) & pearson correlation (emotion analysis) score of 5 runs of various models on test data.

| Models | Financial sentiment | | Emotion analysis | | | | |
|---|---|---|---|---|---|---|---|
| | Microblogs | News | Anger | Joy | Sadness | Fear | Average |
| LSTM model w/o fused word embeddings | 0.512 | 0.487 | 0.401 | 0.365 | 0.409 | 0.429 | 0.400 |
| LSTM model w/o char convoluted embeddings | 0.735 | 0.730 | 0.699 | 0.701 | 0.735 | 0.740 | 0.719 |
| CNN model w/o fused word embeddings | 0.523 | 0.489 | 0.424 | 0.397 | 0.411 | 0.468 | 0.425 |
| CNN model w/o char convoluted embeddings | 0.753 | 0.733 | 0.726 | 0.693 | 0.743 | 0.751 | 0.728 |
| Feature based model w/o N-grams | 0.716 | 0.713 | 0.682 | 0.673 | 0.713 | 0.720 | 0.697 |
| Feature based model w/o lexicon Features | 0.748 | 0.754 | 0.708 | 0.689 | 0.730 | 0.726 | 0.713 |
| Feature based model w/o averaged embeddings | 0.756 | 0.733 | 0.689 | 0.677 | 0.724 | 0.711 | 0.700 |
| Ensemble w/o LSTM model | 0.791 | 0.780 | 0.745 | 0.718 | 0.760 | 0.775 | 0.750 |
| Ensemble w/o CNN model | 0.786 | 0.772 | 0.736 | 0.717 | 0.752 | 0.768 | 0.743 |
| Ensemble w/o feature model | 0.788 | 0.769 | 0.735 | 0.708 | 0.750 | 0.763 | 0.739 |

## 4    Conclusion

In this paper, we have presented an ensemble framework for intensity prediction of sentiment and emotion. The individual models are based on LSTM, CNN and feature based MLP, each of which incorporates word, character and lexicon level information. These models are finally combined together using a MLP based ensemble network. Our empirical evaluation shows that all these three sources of information are very important for final prediction. We have established that our proposed method is generic and adaptable to different domains and applications. Our model shows state-of-the-art performance for sentiment intensity prediction in financial microblogs, sentiment intensity prediction in financial news headlines and emotion intensity prediction in generic tweets.

**Acknowledgments.** Asif Ekbal acknowledges Young Faculty Research Fellowship (YFRF), supported by Visvesvaraya PhD scheme for Electronics and IT, Ministry of Electronics and Information Technology (MeitY), Government of India, being implemented by Digital India Corporation (formerly Media Lab Asia).

# References

1. Akhtar, M.S., Kumar, A., Ghosal, D., Ekbal, A., Bhattacharyya, P.: A multilayer perceptron based ensemble technique for fine-grained financial sentiment analysis. In: Proceedings of the 2017 Conference on Empirical Methods in Natural Language Processing, pp. 540–546 (2017)
2. Akhtar, M.S., Sikdar, U.K., Ekbal, A.: IITP: hybrid approach for text normalization in Twitter. In: Proceedings of the ACL 2015 Workshop on Noisy User-generated Text (WNUT-2015), Beijing, China, pp. 106–110 (2015)
3. Baccianella, S., Esuli, A., Sebastiani, F.: Sentiwordnet 3.0: an enhanced lexical resource for sentiment analysis and opinion mining. In: LREC, vol. 10, pp. 2200–2204 (2010)
4. Bird, S., Klein, E., Loper, E.: Natural Language Processing with Python: Analyzing Text with the Natural Language Toolkit. O'Reilly Media, Inc., Newton (2009)
5. Bravo-Marquez, F., Frank, E., Mohammad, S.M., Pfahringer, B.: Determining word-emotion associations from tweets by multi-label classification. In: WI 2016, pp. 536–539. IEEE Computer Society (2016)
6. Cambria, E., Poria, S., Bajpai, R., Schuller, B.: Senticnet 4: a semantic resource for sentiment analysis based on conceptual primitives. In: Proceedings of COLING 2016, the 26th International Conference on Computational Linguistics: Technical Papers, pp. 2666–2677 (2016)
7. Cortis, K., Freitas, A., Daudert, T., Huerlimann, M., Zarrouk, M., Davis, B.: Semeval-2017 task 5: fine-grained sentiment analysis on financial microblogs and news. In: Proceedings of the 11th International Workshop on SemEval-2017. ACL, Vancouver (2017)
8. Davidson, R.J., Sherer, K.R., Goldsmith, H.H.: Handbook of Affective Sciences. Oxford University Press, Oxford (2009)
9. Ding, X., Liu, B., Yu, P.S.: A holistic lexicon-based approach to opinion mining. In: Proceedings of the 2008 International Conference on Web Search and Data Mining, pp. 231–240. ACM (2008)
10. Ekman, P.: An argument for basic emotions. Cogn. Emot. **6**, 169–200 (1992)
11. Glorot, X., Bordes, A., Bengio, Y.: Deep sparse rectifier neural networks. In: Aistats, vol. 15, p. 275 (2011)
12. Hinton, G.E., Srivastava, N., Krizhevsky, A., Sutskever, I., Salakhutdinov, R.R.: Improving neural networks by preventing co-adaptation of feature detectors. arXiv preprint arXiv:1207.0580 (2012)
13. Jain, P., Goel, P., Kulshreshtha, D., Shukla, K.K.: Prayas at EmoInt 2017: an ensemble of deep neural architectures for emotion intensity prediction in tweets. In: Proceedings of the 8th Workshop on Computational Approaches to Subjectivity, Sentiment and Social Media Analysis, pp. 58–65. ACL, Copenhagen, September 2017
14. Kim, Y., Jernite, Y., Sontag, D., Rush, A.M.: Character-aware neural language models (2016)
15. Kingma, D.P., Ba, J.: Adam: a method for stochastic optimization. CoRR arxiv:abs/1412.6980 (2014)

16. Köper, M., Kim, E., Klinger, R.: IMS at EmoInt-2017: emotion intensity prediction with affective norms, automatically extended resources and deep learning. In: Proceedings of the 8th Workshop on Computational Approaches to Subjectivity, Sentiment and Social Media Analysis, pp. 50–57. ACL, Copenhagen, September 2017

17. Lan, M., Jiang, M., Wu, Y.: ECNU at SemEval-2017 task 5: an ensemble of regression algorithms with effective features for fine-grained sentiment analysis in financial domain. In: Proceedings of the 11th International Workshop on SemEval-2017, ACL, Vancouver (2017)

18. Mansar, Y., Gatti, L., Ferradans, S., Guerini, M., Staiano, J.: Fortia-FBK at SemEval-2017 task 5: bullish or bearish? Inferring sentiment towards brands from financial news headlines. In: Proceedings of the 11th International Workshop on SemEval-2017. ACL, Vancouver, Canada (2017)

19. Mikolov, T., Sutskever, I., Chen, K., Corrado, G.S., Dean, J.: Distributed representations of words and phrases and their compositionality. In: Advances in Neural Information Processing Systems, pp. 3111–3119 (2013)

20. Mohammad, S., Kiritchenko, S., Zhu, X.: NRC-Canada: building the state-of-the-art in sentiment analysis of tweets. In: Proceedings of the Seventh International Workshop on Semantic Evaluation Exercises (SemEval-2013), Atlanta, Georgia, USA, June 2013

21. Mohammad, S.M., Bravo-Marquez, F.: WASSA-2017 shared task on emotion intensity. In: Proceedings of the Workshop on Computational Approaches to Subjectivity, Sentiment and Social Media Analysis (WASSA), Copenhagen, Denmark (2017)

22. Mohammad, S.M., Kiritchenko, S.: Using hashtags to capture fine emotion categories from tweets. Comput. Intell. **31**(2), 301–326 (2015)

23. Mohammad, S.M., Turney, P.D.: Crowdsourcing a word-emotion association lexicon. Comput. Intell. **29**(3), 436–465 (2013)

24. Nielsen, F.Å.: A new ANEW: evaluation of a word list for sentiment analysis in microblogs. arXiv preprint arXiv:1103.2903 (2011)

25. Pang, B., Lee, L.: Opinion mining and sentiment analysis. Found. Trends Inf. Retrieval **2**(1–2), 1–135 (2008)

26. Pennington, J., Socher, R., Manning, C.D.: GloVe: global vectors for word representation. In: Empirical Methods in Natural Language Processing (EMNLP), pp. 1532–1543 (2014)

27. Picard, R.W.: Affective Computing. MIT Press, Cambridge (1997)

28. Russell, J.A., Barrett, L.F.: Core affect, prototypical emotional episodes, and other things called emotion: dissecting the elephant. J. Pers. Soc. Psychol. **76**(5), 805 (1999)

29. Wiebe, J., Mihalcea, R.: Word sense and subjectivity. In: Proceedings of the COLING/ACL, Australia, pp. 1065–1072 (2006)

# Research on Usage Prediction Methods for O2O Coupons

Jie Wu, Yulai Zhang[(⊠)], and Jianfen Wang

School of Information and Electronics Engineering,
Zhejiang University of Science and Technology,
Hangzhou 310023, People's Republic of China
zhangyulai@zust.edu.cn

**Abstract.** Activating old customers or attracting new ones with coupons is a frequently used and very effective marketing tool in O2O (Online to Offline) businesses. But without careful analysis, large amount of coupons can be wasted because of inappropriate delivery strategies. In this era of big data, O2O coupons can be more precisely delivered by using history usage records of customers. By implementing the mainstream data mining and machine learning models, customers' behaviors on O2O coupons can be predicted. Then as a result, individualized delivery can be performed. Coupons with particular discounts can be delivered to those customers who are more likely to use them. So coupon usage rate can be greatly increased. In this paper, multiple classification models are used to achieve this target. Experiments on real coupons' usage data show that compared with other methods, the Random Forest model has better classification performance, and its accuracy rate of coupon usage prediction is the highest.

**Keywords:** Coupon usage prediction · O2O marketing
Classification model performance

## 1 Introduction

In recent years, with the development of the mobile internet, the focus of enterprise marketing [1] has developed from offline to online. However, many local living services still have to be completed offline. So the integration of offline and online has become the center of many businesses. O2O (Online to Offline) businesses provide consumers with plenty of information online and lead them to complete their consumption offline [2]. There are at least 10 start-ups with a billion valuation in the O2O industry in China [3], and they are also very attractive to investors. On the other hand, the O2O industry is naturally linked to hundreds of millions of consumers, and various apps record over 10 billion users' behaviors and locations every day. Therefore, it has become one of the best combination points of big data researches and commercial marketing operations.

Activating old customers or attracting new ones with coupons is a traditional marketing tool. It is also very effective for O2O businesses in the age of mobile internet. However, random coupons [4] cause meaningless interruptions to most users.

© Springer Nature Switzerland AG 2018
L. Cheng et al. (Eds.): ICONIP 2018, LNCS 11305, pp. 175–183, 2018.
https://doi.org/10.1007/978-3-030-04221-9_16

For many businesses, spamming coupons can damage brand reputation, and make them difficult to estimate marketing costs. Personalized advertising with coupons [5] can give consumers certain preference to get real benefits, and give businesses a stronger marketing capability at the same time.

The most exciting thing is that such tasks can be well performed by analyzing the history data of consumers. And we can take advantage of the recent development in data mining and machine learning community to solve these problems. So in this article, the existing consumption data are used to predict the future usage of O2O coupons. Results of this paper can help improve marketing efficiency for O2O businesses. Coupons with more usage probabilities will be actually delivered to consumers.

## 2    Feature Extraction

The coupon usage data for O2O from Tianchi big data competition [6] is investigated in this paper. It is collected from real online and offline consumption behavior records of users on an e-commerce platform from January 1st, 2016 to June 30th, 2016. There are 1048576 records in historical data, including 629895 records of coupons that users get. There are 44967 positive record samples that users obtain consumption coupons and make consumption, and 584858 negative record samples with non-consumption.

**Table 1.** Data exploration and analysis.

|  | Count | Unique | Top | Freq |
|---|---|---|---|---|
| User_id | 1048576 | 323088 | 5054119 | 264 |
| Merchant_id | 1048576 | 12088 | 3381 | 74420 |
| Coupon_id | 1048576 | 9281 | Null | 418751 |
| Discount_rate | 1048576 | 80 | Null | 418751 |
| Distance | 1048576 | 13 | 0 | 493573 |
| Date_received | 1048576 | 169 | Null | 418751 |
| Date | 1048576 | 184 | Null | 584858 |

a. Count means the non-null value.
b. Unique means the unique value.
c. Top means value which has highest frequency.
d. Freq means the highest frequency.

As shown in Table 1, the number of data in each column is consistent with total samples, so it can be seen that there is no null value in the original data.

Through data exploration and analysis, it is found that there are records which have zero coupon discount rate and records which lack distance between users and merchants, etc. Due to the large amount of original data, such data accounts for a small proportion and has little impact on the overall data, thus we choose to discard it. Specific treatment methods are as follows:

a. Discard records without coupons which are delivered to users.
b. Discard records with zero-discount-rate coupons.
c. Discard records which lack distance between users and merchants.

Since the original data is not suitable for direct indicators, more indicators need to be obtained through the extraction of the original data. Coupons' value is not only connected with the distance between users and merchants' offline stores or discount rate, but also relates to users' own consumption behaviors and the popularity of merchants. Therefore, relevant features of users, merchants, coupons and user-merchants are extracted.

User's related features include FUser1, FUser2, FUser3, FUser4, FUser5, FUser6, FUser7, FUser8, FUser9, FUser10, FUser11, FUser12, FUser13 and FUser14. FUser1 is set as the number of coupons which are received by each user. FUser2 is set as offline consumption times for each user after they receive the coupon. FUser3 is set as the total offline consumption times of each user. Set FUser4 as offline non-consumption times after each user receives the coupon. Set FUser5 as consumption times without a coupon for each user. Set FUser6 as the verification rate after each user receives the coupon. Set FUser7 as the unwritten off rate after each user receives the coupon. FUser8 is set as times that each user receives a coupon with a minimum consumption above 50 yuan. FUser9 is set as times each user receives a coupon with a minimum consumption above 100 yuan. FUser10 is set as times each user receives a coupon with a minimum consumption above 150 yuan. Set FUser11 as times each user receives a coupon with a maximum discount as 5 yuan. Set FUser12 as times each user receives a coupon with a maximum discount as 10 yuan. Set FUser13 as times each user receives a coupon with a maximum discount as 20 yuan. Set FUser14 as times each user receives a coupon with a maximum discount as 30 yuan.

Merchant related features include FMer1, FMer2, FMer3, FMer4, FMer5, FMer6. FMer1 is the number of times each merchant issues coupons. Set FMer2 as times that each merchant is consumed offline after issuing coupons. Set FMer3 as times that each merchant is consumed offline. Set FMer4 as times that each merchant issues coupons but without being consumed offline. Set FMer5 as the verification rate of coupons issued by each merchant. Set FMer6 as the unwritten off rate of coupons issued by each merchant.

Coupon related features include Discount_min, Discount_cut and Cou1. Set Disount_min as the minimum consumption of each coupon requires. Let Disount_cut be the amount price deducted from each coupon. Let Cou1 be the discount rate for each coupon.

The user-merchant feature is Distance. Set Distance as the distance between each user and the merchant.

In the properties of the original data, we removed attributes that are not relevant or redundant to the system model, such as user number (User_id), merchant number (merchant ant_id) and coupon number (Coupon_id). Besides, we verified the correlation between each feature and the label item, as shown in Table 2.

**Table 2.** Feature correlation analysis.

| Feature | Corr | Feature | Corr |
|---------|------|---------|------|
| FUser1 | 0.186436 | FUser13 | 0.011215 |
| FUser2 | 0.33434 | FUser14 | −0.011368 |
| FUser3 | 0.319342 | FMer1 | −0.119655 |
| FUser4 | −0.040121 | FMer2 | 0.032431 |
| FUser5 | 0.23958 | FMer3 | 0.012732 |
| FUser6 | 0.743202 | FMer4 | −0.12297 |
| FUser7 | −0.743202 | FMer5 | 0.396969 |
| FUser8 | 0.212866 | FMer6 | −0.396969 |
| FUser9 | −0.019314 | Discount_min | −0.132686 |
| FUser10 | 0.011215 | Discount_cut | −0.132686 |
| FUser11 | 0.212866 | Cou1 | 0.041764 |
| FUser12 | −0.019314 | Distance | −0.160368 |

a. Corr means the correlation coefficient of each feature and flag.

## 3   Classification Method

Classification [7] are used to construct a classification model, input the attribute value of the sample, output the corresponding category, and map each sample to a predefined category. The classification model is based on the data set of existing class tags. The accuracy of the model on the existing samples can be easily calculated and it belongs to supervised learning [8].

### 3.1   Naive Bayes

The theory of Bayes [9] algorithm is based on Bayes formula

$$P(B|A) = \frac{P(A|B)P(B)}{P(A)}, \tag{1}$$

where P(A|B) is called conditional probability, P(B) is the prior probability, and P(B|A) is the posterior probability.

The naive Bayesian classifier is based on a simple assumption that the given target value attributes are independent of each other.

### 3.2   K-Nearest Neighbor

The idea of K-Nearest Neighbor [10] (KNN) algorithm is that if most of the k samples with the most similar sample in the feature space (i.e. the nearest one in the feature space) belong to a certain category, the sample also belongs to this category.

The optimal K value was selected by means of cross validation. Most voting rules are adopted and Euclidean distance is also adopted as the distance measurement.

### 3.3  Logistic Regression

Logistic Regression [11] is a probabilistic nonlinear regression method. For binary logistic regression, the dependent variable y only has the values of "yes or no", which is denoted as 1 and 0. Assume that under the action of independent variable x, the probability of y taking "yes" is p, and the probability of "no" is 1 − p. The relationship between the probability of y taking "yes" and the independent variable x is studied.

Since the number of features is not very large, L2 regularization [12] is chosen. The class weight parameter is "balanced", and the class library will calculate the weight according to the training sample size.

### 3.4  Neural Network

Artificial Neural Network [13] (ANN) is a mathematical model which simulates biological neural network for information processing. It is based on the results of physiological research on the brain, and it simulates some brain mechanisms to achieve some specific functions as its purpose.

Since we are doing binary classification, the loss function as binary cross entropy and the pattern as binary are specified. To add the input layer (24 nodes) to the hidden layer (30 nodes). The hidden layer is used "relu" function as the activation function, which can greatly provide accuracy. To add the connection from the hidden layer (30 nodes) to the output layer (1 node). Since the output layer is 0–1 model, the sigmoid function is used as the activation function of the output layer.

### 3.5  Decision Tree

The CART decision tree [14] (classification and regression tree) uses the Gini coefficient minimization criteria [15] to select partitioning attributes. Suppose the proportion of class k samples in the current sample set D is $p_k(\mathrm{k} = 1,2,\ldots,|y|)$, the purity of data set D can be measured by Gini value:

$$Gini(D) = \sum_{k=1}^{|y|} \sum_{k' \neq k} p_k p_{k'} = 1 - \sum_{k=1}^{|y|} p_k^2. \tag{2}$$

The smaller the value of Gini(D), the higher the purity of data set D.

Because the sample size is not particularly large, according to splitter's standard, feature points are selected splitter "best" to find the optimal partition point among all feature points. The number of sample features used in this paper is not more than 24. The default value of "none" is adopted to take into account the maximum number of features considered when partitioning. Because this sample is not balanced enough, in order to prevent some categories of training set samples too much or lead to the decision tree of training too biased towards these categories, the sample weight as "balanced" is specified. The corresponding sample weight of categories with less sample size will be higher.

### 3.6   Random Forest

Random Forest [16] (RF) introduces random attributes selection in the training of decision tree further. For each node of the base decision tree, a random subset containing k attributes is selected from the attribute set of the node, and then an optimal attribute is selected from the subset for partition.

Its purity is measured by Gini impurity and the number of trees in the forest is determined by grid search.

## 4   Experiments and Result Analysis

### 4.1   Experimental Data

This article uses the built expert sample to make training and then predict the probability of coupons' usage by users from July 1st to July 15th, 2016.

Expert samples are randomly divided into 10 groups on average. Then 10 group control experiments will be done respectively on RF (random forest), CART (CART decision tree), NN (neural network), BNB (Bernoulli naive Bayes), KNN (k-nearest neighbor), LR (logistic regression). Besides, 80% data of each group of experiments is randomly selected as the training set and other 20% data as the test set. The numbers of training sets and test sets of 10 subsets are 36770 and 9192 respectively.

### 4.2   Model Evaluation Criteria

Confusion Matrix [17] is a common expression in pattern recognition field. It describes the relationship between the real attribute of sample data and the recognition result type.

In this paper, TP is set to predict when coupons will be used and are actually used. Set TN to predict when coupons are not used and are not actually used. Set FP to predict when coupons are used but not actually used. Let FN be used to predict when coupons are not used but actually used.

Based on the calculation of Precision and Recall rate, then take the F1-measure, Accuracy, ROC (Receiver Operating Characteristic) curve as a model of evaluation index for O2O coupons' usage prediction performance.

Accuracy is used to predict the proportion of correct positive samples and negative samples in all samples. Precision is the ratio of the correct positive samples to all the predicted positive samples. Recall rate refers to the ratio of correct positive samples to all actual positive samples. The higher the recall rate, the more accurate the prediction of a positive sample of upcoming coupons is. Accuracy rate and Recall rate are a pair of quantities that interact with each other. Thus, F-measure is selected as the weighted harmonic average of Accuracy rate and Recall rate to evaluate the overall performance of the model. The formulation of these metrics can be expressed as:

$$Accuracy = \frac{TP + TN}{TP + TN + FP + FN} \times 100\%, \tag{3}$$

$$Precision = \frac{TP}{TP + FP} \times 100\%, \tag{4}$$

$$Recall = \frac{TP}{TP + FN} \times 100\%, \tag{5}$$

$$F = \frac{(a^2 + 1)Precision * Recall}{a^2(Precision + Recall)} \times 100\%. \tag{6}$$

When the parameter a equals to 1, the most common formula F1 that adopted in this paper can be expressed as

$$F = \frac{2 * Precision * Recall}{Precision + Recall} \times 100\%. \tag{7}$$

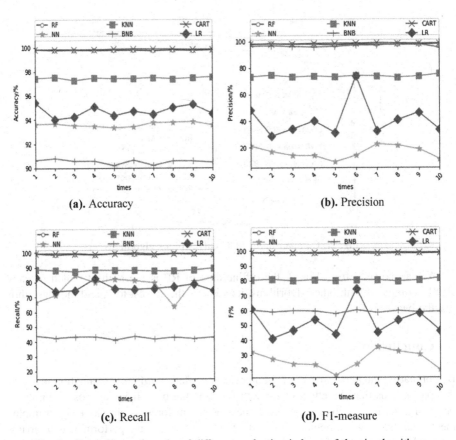

**(a).** Accuracy

**(b).** Precision

**(c).** Recall

**(d).** F1-measure

**Fig. 1.** Comparison of results of different evaluation indexes of the six algorithms

### 4.3    Experimental Results

Figure 1 is the comparison of the Accuracy, Precision, Recall rate, and F1-measure of the six algorithms on different test sets. It can be seen from (a) that CART and RF have the highest Accuracy rate and BNB has the lowest Accuracy rate. It can be seen from (b) that the Precision rate of CART, RF and BNB is the highest, and that of NN is the lowest as well. It also can be seen from (c) that the Recall rate of CART and RF is the highest and BNB's is the lowest. And it can be seen from (d) that the F1-measure of CART and RF is the highest and the F1-measure of NN is the lowest. In summary, the RF and CART indicators are high and relatively close, it is indicated that RF and CART algorithms have better performance.

Receiver Operating Characteristic (ROC) curve is a very effective model evaluation method, which can give quantitative hints for the selected threshold value. ROC curve can be obtained by setting Sensitivity on the vertical axis and 1-Specificity on the horizontal axis. The area under the curve is closely related to the advantages and disadvantages of each method, reflecting the statistical probability of classifier classi-fication. The closer the value to 1 is, the better the algorithm is.

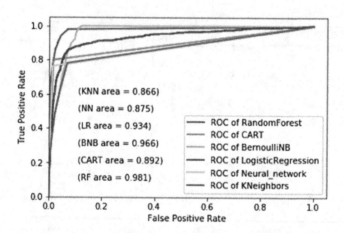

**Fig. 2.**  Comparison of ROC curves of six algorithms

As can be seen from Fig. 2, the offline area of RF is the highest which reaches 0.981. Compared with other algorithms, its value is the closest to 1, and the algorithm has the best effect.

## 5    Conclusion

By comparing the performance of six algorithm models in different evaluation indexes, this paper finds that the effect of random forest is the best. It is the most accurate in predicting the use of coupons by customers. Random forest is simple, easy to imple-ment, and has low computational cost, which shows strong performance in many practical tasks. Random forest's diversity of base learner is not only from the sample's disturbance, but also from the attribute's disturbance. It makes the final integrated generalization performance can get further improvement by the increase of the difference degree between individual learners.

Effective prediction on the actual use of coupons can improve the ability of merchants to deliver personalized coupons, strengthen their ability to retain old customers and attract new customers. So it is very helpful to improve the competitiveness of merchants and e-commerce platforms in marketing.

**Acknowledgments.** This work is supported by ZJSTF-LGF18F020011, ZJSTF-2017C31038, and NSFC 61803337.

# References

1. Grönroos, C., Ravald, A.: Service as business logic: implications for value creation and marketing. J. Serv. Manag. **22**(1), 5–22 (2013)
2. Phang, C.W., Tan, C.H., Sutanto, J., Magagna, F., Lu, X.: Leveraging O2O commerce for product promotion: an empirical investigation in Mainland China. IEEE Trans. Eng. Manag. **61**(4), 623–632 (2014)
3. Ji, S.W., Sun, X.Y., Liu, D.: Research on core competitiveness of Chinese retail industry based on O2O. In: Advanced Materials Research, vol. 834–836, pp. 2017–2020 (2014)
4. Zhao, D., Ma, H., Tang, S.: COUPON: cooperatively building sensing maps in mobile opportunistic networks. IEEE Trans. Parallel Distrib. Syst. **411**(2), 295–303 (2013)
5. Tucker, C.E.: Social networks, personalized advertising, and privacy controls. J. Mark. Res. **51**(5), 546–562 (2014)
6. Coupon Usage Data for O2O. https://tianchi.aliyun.com/datalab/dataSet.html?spm=5176.100073.0.0.746455b9zX7JO6&dataId=59
7. Ahmed, A.B.E.D., Elaraby, I.S.: Data Mining: a prediction for student's performance using classification method. World J. Comput. Appl. Technol. **2**(2), 43–47 (2014)
8. Rasmus, A., Valpola, H., Honkala, M., Berglund, M., Raiko, T.: Semi-supervised learning with Ladder networks. Comput. Sci. **9**(Suppl 1), 1–9 (2015)
9. Jiang, L., Cai, Z., Zhang, H., Wang, D.: Naive Bayes text classifiers: a locally weighted learning approach. J. Exp. Theor. Artif. Intell. **25**(2), 273–286 (2013)
10. Safar, M.: K nearest neighbor search in navigation systems. Mob. Inf. Syst. **1**(3), 207–224 (2014)
11. Namdari, M., Yoon, J.H., Abadi, A., Taheri, S.M., Choi, S.H.: Fuzzy logistic regression with least absolute deviations estimators. Soft. Comput. **19**(4), 909–917 (2015)
12. Sysoev, O., Burdakov, O.: A smoothed monotonic regression via L2 regularization. Knowl. Inf. Syst. **2018**, 1–22 (2018)
13. Floyd, C.E., Lo, J.Y., Yun, A.J., Sullivan, D.C., Kornguth, P.J.: Prediction of breast cancer malignancy using an artificial neural network. Cancer **74**(11), 2944–2948 (2015)
14. Rutkowski, L., Jaworski, M., Pietruczuk, L., Duda, P.: The CART decision tree for mining data streams. Inf. Sci. **266**(5), 1–15 (2014)
15. Rodríguez, J.G., Salas, R.: The Gini coefficient: majority voting and social welfare. J. Econ. Theory **152**(152), 214–223 (2014)
16. Belgiu, M., Drăguţ, L.: Random forest in remote sensing: a review of applications and future directions. ISPRS J. Photogramm. Remote. Sens. **114**, 24–31 (2016)
17. Beauxis- Deng, X., Liu, Q., Deng, Y., Mahadevan, S.: An improved method to construct basic probability assignment based on the confusion matrix for classification problem. Inf. Sci. **340–341**, 250–261 (2016)

# Prediction Based on Online Extreme Learning Machine in WWTP Application

Weiwei Cao and Qinmin Yang[✉]

College of Control Science and Engineering, Zhejiang University,
Zheda Rd. 38, Hangzhou 310027, China
{cww,qmyang}@zju.edu.cn

**Abstract.** Predicting the plant process performance is essential for controlling in wastewater treatment plant (WWTP), which is a complex nonlinear time-variant system. Extreme learning machine (ELM) is a single-hidden layer feed-forward neural network (SLFN), which randomly generates the feed-forward parameters without tuning the parameters from the input to the output layer. The output weights are calculated via the theory of Moore-Penrose generalized inverse and the minimum norm least-squares. In this paper, online extreme learning machine (Online ELM) is proposed as a predictor in WWTP, which trains the output weights and predicts the next outputs according to the real-time data collected from the process in an online manner. Furthermore, extensive comparison studies have been conducted by using other four neural network structures, including extreme learning machine, ELM with kernel, online sequential ELM (OSELM) and back propagation (BP) neural network.

**Keywords:** Online ELM algorithm
Wastewater treatment plant
Benchmark simulation model No. 1 (BSM1)
Dissolved oxygen (DO) concentration
Nitrate concentration

## 1 Introduction

Wastewater treatment plant (WWTP) is a very complex time-variant dynamic system, which contains plenty of physic and biochemistry reactions. Modeling and controlling of the WWTP is still a challenging problem due to its nonlinearity, long delay, strong coupling, great energy consumption, and large scale disturbances. In recent two decades, many researchers contributes lots of excellent works to overcome such issues within WWTP.

Most researches focus on controlling the target variables to save energy consumption under a stable condition. Conventional control strategies, such as PID

Q. Yang—This work is supported by the National Natural Science Foundation of China (61673347, U1609214, 61751205).

L. Cheng et al. (Eds.): ICONIP 2018, LNCS 11305, pp. 184–195, 2018.
https://doi.org/10.1007/978-3-030-04221-9_17

control, feed-forward PI control and so on, have been extensively discussed to control dissolved oxygen (DO) concentration or nitrate concentration [1–4]. But the control performance will be highly deteriorated if the inflow is influenced by a strong disturbance, due to the fact that the parameters of PID controller can't adapt to the changing condition online. To address this concern, Ferrer [5] has proposed a fuzzy logic algorithm to control DO concentration, which saves about 40% energy consumption compared with traditional on-off control. Traore [6], Punal [7], Zhu [8] also have introduced fuzzy logic control into WWTP with different purposes and demonstrated great performance improvement on control effect and energy consumption. However, they highly rely on the expert experience, which is difficult to identify easily.

Furthermore, neural networks have been recognized as a useful predictor widely, and Qiao [10] has proposed a multi-variables control based on PI algorithm with feed-forward neural network model. Han [11] introduces adaptive DO control based on dynamic structure neural network. Based on self-organizing recurrent radial basis function (RBF) neural network, Han et al. [12–15] proposes a novel nonlinear model predictive control strategy to WWTP. Essentially, these methods are model-free or neural network based parametric control, and have shown great performance on stability, convergence and cost. However, considering the intricate characteristics of wastewater treatment plants, more complex structure should be adopted. Moreover, gradient descent algorithm has been widely utilized in these applications, which is criticized by relatively high training time and the low efficiency [9]. Neural network is also apt to be captured in local minima and suffers from fitting problem if the parameters' setting is inappropriate.

In the meantime, extreme learning machine (ELM) was proposed by Huang [18] et al. in 2004, which has been widely applied in plenty of fields as a classifier or predictor, such as electricity market price prediction [19], wind turbine control [20], traffic signal recognition [21], and so on. ELM is one kind of single-hidden layer feed-forward neural network (SLFN). But different from most neural networks. ELM randomly generates the feed-forward weights and biases and doesn't need tune the initial parameters. This method not only can ensure the regression performance but also achieves rapid prediction calculation. Because of these great merits of ELM, some researchers have introduced it into the control WWTP as well. Han [16] proposes hierarchical extreme learning machine to forecast biochemical oxygen demand (BOD) value and sludge volume index (SVI) value. Lin [17] introduces basic differential evolution (BDE) ELM, self-adaptive differential evolution (SaDE) ELM, and trigonometric mutation differential evolution (TDE) ELM into prediction of effluent from WWTP and compares three methods to obtain the result that TDE-ELM is most effective among three candidates. These methods do not change the output weights after finishing the training process in the predicting system, so the predict performance could be reduced in the presence of disturbance and drift of the process parameters. Therefore, this paper is proposed to discuss a novel method for online prediction algorithm – Online ELM for the control of WWTP. After initialization, the output weights of the ELM are updated according to the real-time collected data in an online manner, and thus the output can be predicted with the new weights to accommodate the time-variant characteristics of the plant.

## 2  Preliminary Description About the Benchmark Simulation Model

Considering the difficulty of building the practical math model and comparing the efficiency with different control strategies in WWTP, benchmark simulation model No. 1 (BSM1) has been proposed with the framework of COST 624 and 628 [23]. It contains a series of simulation features including the process model, influent load, and testing and evaluating standards.

BSM1 (Fig. 1) is composed of two anoxic zones, three aerobic zones and a secondary settler. The anoxic/aerobic zone is followed by the secondary settler. All anoxic zones and aerobic zones are based on the activated sludge model no. 1 (ASM1) proposed by the international water association (IWA) [23]. 13 state variables and 8 processes are in each aerobic/anoxic tank. ASM1 not only reflects the relationship between influent components and effluent components in treatment process, but also considers the relevance between microorganism and substrate concentration. So it strictly simulates the carbon and nitrogen removal. The double exponential model based on the solids flux concept is to be chosen as the model of the secondary settler [23], which has 10 non-reactive layers and each layer is composed by 8 main components.

There are two internal recycles in this platform, which are the nitrate internal recycle from the fifth aerobic zone to the first anoxic zone, and the return activated sludge (RAS) recycle from the secondary underflow to the end of the plant [24]. And in BSM1, basic control strategy is proposed with two control loops, two control loops aim at the level of DO in the effluent of the fifth aerobic zone and nitrate in the second anoxic zone, and the manipulators are the oxygen transfer coefficient ($K_La_5$) of the fifth aerobic zone and the internal recycle flow rate ($Q_{int}$) respectively.

## 3  Online ELM

### 3.1  Review of Extreme Learning Machine

ELM has been proposed by Huang et al. [18–22], which is a kind of single-hidden layer feedforward neural network (SLFN). Conventional neural networks are slower than required because of their structures. In many applications, traditional methods need to train the different layer's parameters by massive data set before classification and prediction. Gradient descend-based method is widely used in various algorithms of feedforward neural networks to tuning or training the parameters, and improper step or converging to local minima would make gradient descent method slow down seriously [22].

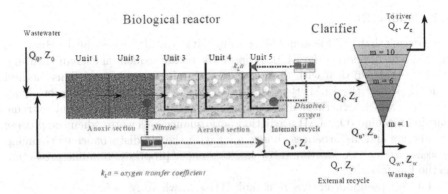

**Fig. 1.** The structure of benchmark simulation model no. 1 (BSM1)

Different from conventional feedforward neural networks, the weights and biases from the input layer to the hidden layer generate randomly in ELM. The Moore-Penrose generalized inverse and the minimum norm least-squares methods are applied to choose the proper weights from the hidden layer to the output layer. A brief description is given as follows.

For N arbitrary distinct training samples $(x_i, t_i) \in \mathbf{R}^n \times \mathbf{R}^m$. $\mathbf{x}_j$ is a $n \times 1$ input vector, and $\mathbf{t}_j$ is $m \times 1$ target vector. $G(\bullet)$ is the activation function with L nodes in hidden layer. A SLFN approximating these N training samples with zero error means that there exist $\boldsymbol{\omega}_i, b_i \text{ and } \boldsymbol{\beta}_i$ such that

$$\mathbf{t}_j = \sum_{i=1}^{L} \boldsymbol{\beta}_i G(\boldsymbol{\omega}_i, b_i, \mathbf{x}_j), \qquad j = 1, ..., N. \tag{1}$$

which can be summarized as

$$\mathbf{H}\boldsymbol{\beta} = \mathbf{T}. \tag{2}$$

where, $\mathbf{H}, \boldsymbol{\beta} \text{ and } \mathbf{T}$ can be expressed as

$$\mathbf{H} = \begin{bmatrix} \mathbf{h}(\mathbf{x}_1) \\ \vdots \\ \mathbf{h}(\mathbf{x}_N) \end{bmatrix} = \begin{bmatrix} G(\boldsymbol{\omega}_1, b_1, \mathbf{x}_1) \dots G(\boldsymbol{\omega}_L, b_L, \mathbf{x}_1) \\ \vdots \quad \dots \quad \vdots \\ G(\boldsymbol{\omega}_1, b_1, \mathbf{x}_N) \dots G(\boldsymbol{\omega}_L, b_L, \mathbf{x}_N) \end{bmatrix}_{N \times L}. \tag{3}$$

$$\boldsymbol{\beta} = \begin{bmatrix} \boldsymbol{\beta}_1^T \\ \vdots \\ \boldsymbol{\beta}_L^T \end{bmatrix}_{L \times m} \quad and \quad \mathbf{T} = \begin{bmatrix} \mathbf{t}_1^T \\ \vdots \\ \mathbf{t}_N^T \end{bmatrix}_{N \times m}. \tag{4}$$

and

$$\boldsymbol{\beta} = \mathbf{H}^{\dagger}\mathbf{T}. \tag{5}$$

## 3.2   Online ELM

Considering that ELM assumes all training data is available previously, Huang et al. [25], who developed the ELM, propose the online sequential extreme learning machine that can deal with data chunk by chunk or one by one. This method is developed by the ELM. Because ELM can't train data one by one or chunk by chunk which generates like that in reality. But online sequential extreme learning machine (OSELM) prediction performance gets worse when the process parameters and disturbances change a lot with the big disturbance or changing process parameters. Motivated by this, this paper proposes an online predicting method as follows.

The left pseudoinverse of $H$ if rank $(\mathbf{H}) = L$, where $N > L$.

$$\mathbf{H}^{\dagger} = (\mathbf{H}^T\mathbf{H})^{-1}\mathbf{H}^T. \tag{6}$$

Once the left pseudoinverse of H tends to singular, the proper action is to choose smaller hidden nodes or more initial training data to make the pseudoinverse nonsingular.

The initial training data $\mathfrak{N}_0 = (x_i, t_i)_{i=1}^{N_0}$ and $N_0 \geq L$, and the initial output of the hidden layer and target matrix as follows

$$\mathbf{H}_0 = \begin{bmatrix} G(\boldsymbol{\omega}_1, b_1, \mathbf{x}_1) & \cdots & G(\boldsymbol{\omega}_L, b_L, \mathbf{x}_1) \\ \vdots & \cdots & \vdots \\ G(\boldsymbol{\omega}_1, b_1, \mathbf{x}_{N_0}) & \cdots & G(\boldsymbol{\omega}_L, b_L, \mathbf{x}_{N_0}) \end{bmatrix}_{N_0 \times L}. \tag{7}$$

and

$$\mathbf{T}_0 = \begin{bmatrix} t_1^T \\ \vdots \\ t_{N_0}^T \end{bmatrix}_{N_0 \times m}. \tag{8}$$

Considering the training with zero error, minimizing the $\|\mathbf{H}_0\boldsymbol{\beta} - \mathbf{T}_0\|$, the answer is

$$\boldsymbol{\beta}^{(0)} = \mathbf{K}_0^{-1}\mathbf{H}_0^T\mathbf{T}_0. \tag{9}$$

Where   $\mathbf{K}_0 = \mathbf{H}_0^T\mathbf{H}_0$.

when the new chunk of data $\mathfrak{N}_1 = (x_i, t_i)_{i=N_0+1}^{N_0+N_1}$ is given, the minimizing problem becomes

$$\left\| \begin{bmatrix} \mathbf{H}_0 \\ \mathbf{H}_1 \end{bmatrix} \boldsymbol{\beta} - \begin{bmatrix} \mathbf{T}_0 \\ \mathbf{T}_1 \end{bmatrix} \right\|. \tag{10}$$

where

$$\mathbf{H}_1 = \begin{bmatrix} G(\boldsymbol{\omega}_1, b_1, \mathbf{x}_{N_0+1}) & \cdots & G(\boldsymbol{\omega}_L, b_L, \mathbf{x}_{N_0+1}) \\ \vdots & \cdots & \vdots \\ G(\boldsymbol{\omega}_1, b_1, \mathbf{x}_{N_0+N_1}) & \cdots & G(\boldsymbol{\omega}_L, b_L, \mathbf{x}_{N_0+N_1}) \end{bmatrix}_{N_1 \times L}. \tag{11}$$

and

$$\mathbf{T}_1 = \begin{bmatrix} \mathbf{t}_{N_0+1}^T \\ \vdots \\ \mathbf{t}_{N_0+N_1}^T \end{bmatrix}_{N_1 \times m} . \tag{12}$$

considering the former and new training data, the updating weight from the hidden layer to the output layer derives as

$$\boldsymbol{\beta}^{(1)} = \mathbf{K}_1^{-1} \begin{bmatrix} \mathbf{H}_0 \\ \mathbf{H}_1 \end{bmatrix}^T \begin{bmatrix} \mathbf{T}_0 \\ \mathbf{T}_1 \end{bmatrix} = \boldsymbol{\beta}^{(0)} + \mathbf{K}_1^{-1} \mathbf{H}_1^T (\mathbf{T}_1 - \mathbf{H}_1 \boldsymbol{\beta}^{(0)}). \tag{13}$$

where

$$\mathbf{K}_1 = \begin{bmatrix} \mathbf{H}_0 \\ \mathbf{H}_1 \end{bmatrix}^T \begin{bmatrix} \mathbf{H}_0 \\ \mathbf{H}_1 \end{bmatrix} = \mathbf{K}_0 + \mathbf{H}_1^T \mathbf{H}_1. \tag{14}$$

Similarly, when the $(k+1)$th new batch training data $\mathfrak{N}_{k+1} = (x_i, t_i)_{i=N_k+1}^{N_k+N_{k+1}}$ generates, the Eq. (13) derives to

$$\boldsymbol{\beta}^{(k+1)} = \boldsymbol{\beta}^{(k)} + \mathbf{K}_{k+1}^{-1} \mathbf{H}_{k+1}^T (\mathbf{T}_{k+1} - \mathbf{H}_{k+1} \boldsymbol{\beta}^{(k)}). \tag{15}$$

where

$$\mathbf{K}_{k+1} = \mathbf{K}_k + \mathbf{H}_k^T \mathbf{H}_k. \tag{16}$$

$$\mathbf{T}_{k+1} = \begin{bmatrix} \mathbf{t}_{\sum_k^{j=0} N_j+1}^T \\ \cdots \\ \mathbf{t}_{\sum_{k+1}^{j=0} N_j}^T \end{bmatrix}_{N_{k+1} \times m} . \tag{17}$$

$$\mathbf{H}_{(k+1)} = \begin{bmatrix} G(\boldsymbol{\omega}_1, b_1, \mathbf{x}_{\sum_{j=0}^k N_j+1}) & \cdots & G(\boldsymbol{\omega}_L, b_L, \mathbf{x}_{\sum_{j=0}^k N_j+1}) \\ \vdots & \cdots & \vdots \\ G(\boldsymbol{\omega}_1, b_1, \mathbf{x}_{\sum_{j=0}^{k+1} N_j}) & \cdots & G(\boldsymbol{\omega}_L, b_L, \mathbf{x}_{\sum_{j=0}^{k+1} N_j}) \end{bmatrix}_{N_{k+1} \times L} . \tag{18}$$

where define $\mathbf{P}_{k+1} = \mathbf{K}_{k+1}^{-1}$, the updating formulas of $\boldsymbol{\beta}^{(k+1)}$ change as follows

$$\mathbf{P}_{k+1} = \mathbf{P}_k - \mathbf{P}_k \mathbf{H}_{k+1}^T (\mathbf{I} + \mathbf{H}_{k+1} \mathbf{P}_k \mathbf{H}_{k+1}^T)^{-1} \mathbf{H}_{k+1} \mathbf{P}_k. \tag{19}$$

$$\boldsymbol{\beta}^{(k+1)} = \boldsymbol{\beta}^{(k)} + \mathbf{P}_{k+1} \mathbf{H}_{k+1}^T (\mathbf{T}_{k+1} - \mathbf{H}_{k+1} \boldsymbol{\beta}^{(k)}). \tag{20}$$

The predicting value

$$f_L^{(k+2)}(\mathbf{x}) = \sum_{i=1}^L \beta_i^{(k+1)} G(\omega_i, \mathbf{b}_i, x_i^{(k+2)}). \tag{21}$$

The Online ELM is shown in Algorithm 1.

---

**Algorithm 1.** Online Extreme Learning Machine algorithm
___
1: **Training**
2:     Prepare the initial samples and key parameters;
3:     Generate the $\boldsymbol{\omega}$ and $\mathbf{b}$ vector in a presetting scale.
4:     Calculate the output of the hidden layer $\mathbf{H}_0$;
5:     Calculate the initial output weights $\boldsymbol{\beta}^{(0)}$, $\boldsymbol{\beta}^{(0)} = \mathbf{K}_0^{-1}\mathbf{H}_0^T\mathbf{T}_0$.
6: **repeat**
7:     Get the kth batch data training data;
8:     Calculate the output of the hidden layer $\mathbf{H}_k$;
9:     Update the output weights
10:    $\boldsymbol{\beta}^{(k)} = \boldsymbol{\beta}^{(k-1)} + \mathbf{P}_k\mathbf{H}_k^T(\mathbf{T}_k - \mathbf{H}_k\boldsymbol{\beta}^{(k-1)})$;
11:    where
12:    $\mathbf{P}_k = (\mathbf{K}_{k-1} + \mathbf{H}_{k-1}^T\mathbf{H}_{k-1})^{-1}$
13:    the new predicting value for the next step
14:    $f_L^{(k+1)}(\mathbf{x}) = \sum_{i=1}^{L}\beta_i^{(k)}G(\boldsymbol{\omega}_i, b_i, \mathbf{x}^{(k+1)})$.
15: **until** (the process stop)

---

## 4    Experiment

In this section, Online ELM is used to predict the DO concentration of 5th aerobic zone and the nitrate concentration of second anoxic zone with data from the BSM1. Firstly, data processing before predicting is necessary to offset the influence of the dimension. Then, eliminating the components which don't impact on DO concentration and nitrate concentration. Finally, predicting the DO concentration from the output of the 5th aerobic zone with Online ELM. Because of the stochastic of the parameter initialization, the performance of this method is unstable. So the final results are the appropriate choice of statistics in 50 times' simulation. All simulations are programmed with Matlab, 2017b, and executed on a station with 2.3 GHz CPU and 64 GB RAM, under a Microsoft Windows service 2008 R2 Enterprise environment.

In the following simulations, the performance of the methods measured using the root mean-square-error (RMSE) function and R-square ($R^2$) function are defined as

$$RMSE = \sqrt{\frac{1}{n}\sum_{i=1}^{n}(\hat{y}_i - y_i)}. \tag{22}$$

$$R^2 = \frac{\sum_{i=1}^{n}\omega_i(\hat{y}_i - \bar{y}_i)^2}{\sum_{i=1}^{n}(y_i - \bar{y}_i)^2}. \tag{23}$$

respectively.

### 4.1    Online ELM Based on Off-Line Data

In order to evaluate the performance of Online ELM regression, before this method is applied in WWTP platform, prediction based on off-line data from the BSM1 platform is necessary. DO concentration and nitrate concentration are concerned as the output $\hat{y}(t)$ of the prediction model. Trying to make the prediction is closer to real process, 13 components of the inflow, inflow rate value, total suspended solids (TSS) and 2 control variables, oxygen transfer coefficient of the fifth aerobic zone $(K_La_5)$ and the internal recycle flow rate $(Q_{int})$, are considered as the u(t-5), and 5 sample time delay in the input value is concerned. u(t-5) and the real target y(t) from BSM1 conform the input of the model.

There are three different weather cases in BSM1 platform, dry, rain and storm weather. Different case indicates different inflow data. This paper chooses the dry day data as the inflow of BSM1. Data set includes 53018 group samples, each group data contains 13 components, inflow rate, TSS, $K_La_5$ and $Q_{int}$. 15000 group samples are chosen as the training data or initialization data, rest samples will be used to predict the DO concentration and nitrate concentration. All samples are normalized by min-max normalization. Based on Online ELM predicting model is made up by 18 input nodes, 200 hidden nodes and one output node. Radial basis function (RBF) is chosen as activation function. The weights from input layer to hidden layer are randomly generate under the scale from $-1$ to 1, in turn biases are chosen from 0 to 1 randomly.

Considering the clarity of the properties in Online ELM, this paper introduces ELM, ELM with kernel, online sequential ELM and BP neural network as reference. Figure 2 shows the $SO_5$ as the predicting model's output with four algorithms. Same as the DO concentration in Fig. 2, Fig. 3 is about the nitrate concentration prediction. As shows in two figures, the performances of predicting nitrate concentration are better than the DO concentration, generally. The reason is that the nitrate in the second zone is closer to the inflow than the DO in the fifth zone. The longer distance is, the more disturbance involves, and this can be verified by the different sample points. Four algorithms all track the real line in general, and ELMs behaviors is better than others. Furthermore, the difference of four methods is illustrated in Table 1. Because of the ELM, as data shows, the training time and testing time of ELM with 150 hidden nodes are smaller than BP with 40 hidden nodes a lot, and accuracy of prediction is approximate to BP neural network. Online ELM's training RMSE is the lowest among four methods in two cases, and the $R^2$ of Online ELM is still higher than others in nitrate concentration and DO concentration.

## 4.2    Online ELM Prediction Model in BSM1 Platform

Different from the off-line testing, this subsection discusses the performance about based on Online ELM predicting model on the BSM1 platform. See in Fig. 4, the outflow of the forth aerobic zone (u(t-5), TSS(t-5), Q(t-5)), the manipulators value $K_La$(t-5), $Q_{int}$(t-5) and the former controlled value $SO_5(t-1)$ (DO) are chosen as the input of Online ELM predictor. Obviously, the output is the $SO_5(t)$ concentration in the outflow of the fifth tank. Platform sample time is set to 0.0001. The Online ELM is composed by 18 input nodes, 300 hidden nodes and one output node. RBF is chosen as activation function, the initialization

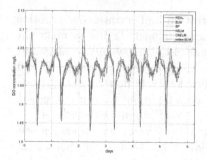

Fig. 2. SO5 prediction with Online ELM

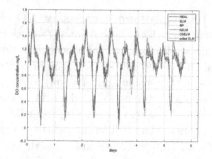

Fig. 3. SNO2 prediction with Online ELM

Table 1. Comparison with different algorithm

|  | Algorithm | Training | | | Testing | | |
|---|---|---|---|---|---|---|---|
|  |  | RMSE | $R_2$ | Time | RMSE | $R_2$ | Time |
| SO5 | Online ELM | 0.0044 | 0.9766 | 1.6722 | 0.0129 | 0.9357 | – |
|  | ELM | 0.0048 | 0.9717 | 0.2868 | 0.0216 | 0.8453 | 0.0486 |
|  | OSELM | 0.0044 | 0.9761 | 11.9295 | 0.0221 | 0.7611 | 0.5142 |
|  | KELM | 0.0108 | 0.8566 | 75.3056 | 0.0230 | 0.8563 | 14.6169 |
|  | BP | 0.0837 | 0.94548 | 183 | 0.0243 | 0.8352 | 0.0383 |
| SNO2 | Qnline ELM | 0.0375 | 0.9839 | 1.6391 | 0.0430 | 0.9768 | – |
|  | ELM | 0.0481 | 0.9797 | 0.2816 | 0.0857 | 0.9206 | 0.0471 |
|  | OSELM | 0.0395 | 0.9822 | 11.0451 | 0.0861 | 0.9162 | 0.5359 |
|  | KELM | 0.0844 | 0.9187 | 80.9478 | 0.0676 | 0.9423 | 17.8024 |
|  | BP | 0.0864 | 0.9730 | 182 | 0.0701 | 0.9390 | 0.0794 |

of weights and bias is same as the off-line experiment. The first six days' data serves as the initialization set.

The result of prediction test is shown in Fig. 5, which illustrates the performance of the PI control strategy with Online ELM forecasting compared with only PI control. The blue line is the controlled value $SO_5$ with only conventional PI control algorithm, and the red line is the result of PI control With Online ELM. Compared with the conventional PI controller, PI controller with Online ELM shows the equivalent tracking performance, that indicates the Online ELM prediction can reflect the inner structure and discriminate the parameters about the fifth zone. So it's a novel strategy to achieve more precise prediction with Online ELM, because of the characteristic of more convenient iteration, faster initialization and better generalization performance.

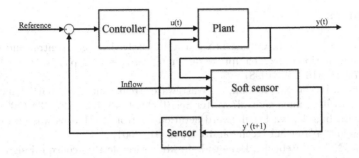

**Fig. 4.** PI control strategy with Online ELM predicting model in BSM1

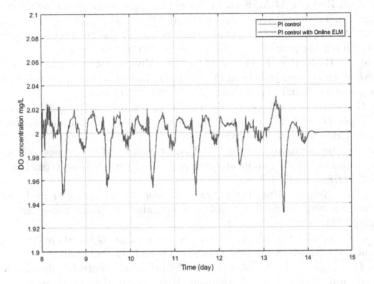

**Fig. 5.** The online prediction of the $SO_5$ (Color figure online)

## 5   Conclusion

This paper introduces Online ELM algorithm into wastewater treatment process as a predictor. First predicting the controlled variables DO concentration and nitrate concentration in Online ELM with the off-line from the BSM1 platform, which verifies the performance of tracking the actual output of the tank based on real input. Then, Trying to using Online ELM algorithm in BSM1 platform as predictor. Result shows that this method tacks the real output, analyzes the structure and recognises the parameters of the fifth tank very well. And in the future work, the main task is to propose a proper control algorithm to make the controlled variables track set point with lower cost and smaller errors based on Online ELM predictor. Future work may include robust design for faulty conditions and optimal control [26].

# References

1. Suescun, J., Irizar, I., Ostolaza, X., et al.: Dissolved oxygen control and simultaneous estimation of oxygen uptake rate in activated-sludge plants. Water Environ. Res. **70**(3), 316–322 (1998)
2. Vreko, D., Hvala, N., Kocijan, J.: Wastewater treatment benchmark: what can be achieved with simple control? Water Sci. Technol. **45**(4–5), 127–134 (2002)
3. Wett, B., Ingerie, K.: Feedforward aeration control of a biocos wastewater treatment plant. Water Sci. Technol. **43**(3), 85–91 (2001)
4. Wahab, N.A., Katebi, R., Balderud, J.: Multivariable PID control design for activated sludge process with nitrification and denitrification. Biochem. Eng. J. **45**(3), 239–248 (2009)
5. Ferrer, J., Rodrigo, M.A., Seco, A., et al.: Energy saving in the aeration process by fuzzy logic control. Water Sci. Technol. **38**(3), 209–217 (1998)
6. Traore, A., Grieu, S., Puig, S., et al.: Fuzzy control of dissolved oxygen in a sequencing batch reactor pilot plant. Chem. Eng. J. **111**(1), 13–19 (2005)
7. Punal, A., Rodriguez, J., Franco, A., et al.: Advanced monitoring and control of anaerobic wastewater treatment plants: diagnosis and supervision by a fuzzy-based expert system. Water Sci. Technol. **43**(7), 191–198 (2001)
8. Zhu, G., Peng, Y., Ma, B., et al.: Optimization of anoxic/oxic step feeding activated sludge process with fuzzy control model for improving nitrogen removal. Chem. Eng. J. **151**(1–3), 195–201 (2009)
9. Yang, Q., Jagannathan, S., Sun, Y.: Robust integral of neural network and error sign control of MIMO nonlinear systems. IEEE Trans. Neural Netw. Learn. Syst. **26**(12), 3278–3286 (2015)
10. Qiao, J.F., Han, G., Han, H.G.: Neural network online modeling and controlling method for multi variable control of wastewater treatment processes. Asian J. Control **16**(4), 1213–1223 (2014)
11. Han, H.G., Qiao, J.F.: Adaptive dissolved oxygen control based on dynamic structure neural network. Appl. Soft Comput. **11**(4), 3812–3820 (2011)
12. Han, H.G., Qiao, J.F., Chen, Q.L.: Model predictive control of dissolved oxygen concentration based on a self-organizing RBF neural network. Control Eng. Practice **20**(4), 465–476 (2012)
13. Han, H.G., Wu, X.L., Qiao, J.F.: Real-time model predictive control using a self-organizing neural network. IEEE Trans. Neural Netw. Learn. Syst. **24**(9), 1425–1436 (2013)
14. Han, H.G., Qiao, J.F.: Nonlinear model-predictive control for industrial processes: an application to wastewater treatment process. IEEE Trans. Ind. Electron. **61**(4), 1970–1982 (2014)
15. Han, H.G., Zhang, L., Hou, Y., et al.: Nonlinear model predictive control based on a self-organizing recurrent neural network. IEEE Trans. Neural Netw. Learn. Syst. **27**(2), 402–415 (2016)
16. Han, H.G., Wang, L.D., Qiao, J.F.: Hierarchical extreme learning machine for feedforward neural network. Neurocomputing **128**, 128–135 (2014)
17. Lin, M., Zhang, C., Su, C.: Prediction of effluent from WWTPS using differential evolutionary extreme learning machines. In: 2016 35th Chinese Control Conference (CCC), pp. 2034–2038. IEEE (2016)
18. Huang, G.B., Zhu, Q.Y., Siew, C.K.: Extreme learning machine: a new learning scheme of feedforward neural networks. In: Proceedings of 2004 IEEE International Joint Conference on Neural Networks, vol. 2, pp. 985–990. IEEE (2004)

19. Shrivastava, N.A., Panigrahi, B.K.: A hybrid wavelet-ELM based short term price forecasting for electricity markets. Int. J. Electr. Power Energy Syst. **55**, 41–50 (2014)
20. Mahmoud, T.K., Dong, Z.Y., Ma, J.: A developed integrated scheme based approach for wind turbine intelligent control. IEEE Trans. Sustain. Energy **8**(3), 927–937 (2017)
21. Zeng, Y., Xu, X., Shen, D., et al.: Traffic sign recognition using kernel extreme learning machines with deep perceptual features. IEEE Trans. Intell. Transp. Syst. **18**(6), 1647–1653 (2017)
22. Huang, G.B., Zhu, Q.Y., Siew, C.K.: Extreme learning machine: theory and applications. Neurocomputing **70**(1–3), 489–501 (2006)
23. Jeppsson, U., Pons, M.N.: The COST benchmark simulation model current state and future perspective. Control Eng. Practice **12**(3), 299–304 (2004)
24. Copp, J.B.: The COST Simulation Benchmark: Description and Simulator Manual: a Product of COST Action 624 and COST Action 628. EUP-OP (2002)
25. Liang, N.Y., Huang, G.B., Saratchandran, P., et al.: A fast and accurate online sequential learning algorithm for feedforward networks. IEEE Trans. Neural Netw. **17**(6), 1411–1423 (2006)
26. Yang, Q., Ge, S.S., Sun, Y.: Adaptive actuator fault tolerant control for uncertain nonlinear systems with multiple actuators. Automatica **60**, 92–99 (2015)

# Pattern Recognition

# Learning a Joint Representation for Classification of Networked Documents

Zhenni You and Tieyun Qian[✉]

School of Computer Science, Wuhan University, Wuhan, Hubei, China
{znyou,qty}@whu.edu.cn

**Abstract.** Recently, several researchers have incorporated network information to enhance document classification. However, these methods are tied to some specific network representations and are unable to exploit different representations to take advantage of data specific properties. Moreover, they do not utilize the complementary information from one source to the other, and do not fully leverage the label information. In this paper, we propose CrossTL, a novel representation model, to find better representations for classification. CrossTL improves the learning at three levels: (1) at the input level, it is a general framework which can accommodate any useful text or graph embeddings, (2) at the structure level, it learns a text-to-link and link-to-text representation to comprehensively describe the data; (3) at the objective level, it bounds the error rate by incorporating four types of losses, i.e., text, link, and the combination and disagreement of text and link, into the loss function. Extensive experimental results demonstrate that CrossTL significantly outperforms the state-of-the-art representations on datasets with either rich or poor texts and links.

**Keywords:** Representation learning · Networked documents
Document classification

## 1 Introduction

Networked documents, such as referenced papers and hyper-linked webpages, are becoming pervasive nowadays. The links between documents (nodes, or vertices) often encode useful information for document classification because they convey human knowledge beyond document contents. Hence the objects in a document network are often classified not just based on their own text attributes but also based on the related nodes.

Traditionally, links are used to pass messages or labels between neighbors [8,11,20], and are represented as binary variables or network regularization [4,12] under the assumption that linked documents should share similar topic distribution. More recently, two studies investigated to combine distributed network

© Springer Nature Switzerland AG 2018
L. Cheng et al. (Eds.): ICONIP 2018, LNCS 11305, pp. 199–209, 2018.
https://doi.org/10.1007/978-3-030-04221-9_18

representations with the content information. By proving that the typical network embedding method DeepWalk [16] was equivalent to matrix factorization (MF), Yang et al. [22] presented a text-associated deep walk (TADW) to introduce text features generated by SVD decomposition of tf-idf matrix into network embedding through MF. Pan et al. [15] proposed to combine DeepWalk [16] with a document embedding, Doc2Vec [10], to learn a tri-party (node-node, node-content, and content-label) deep network representation TriDNR.

While TADW and TriDNR exploited distributed representations for enhancing document classification, they both have the drawback that their graph embedding is tied to DeepWalk [16], a specific network representation method. This prevents TADW and TriDNR from benefiting from other embedding approaches. Different graphs can have different semantics, and different graph densities can have different implications on embeddings too. For example, for denser graphs, LINE [18] can better leverage structural properties than Deep-Walk. Furthermore, the architecture of TADW and TriDNR lacks a deep interaction between texts and links, thus one representation can not utilize the complementary information in the other. Finally, neither TADW nor TriDNR fully explores the valuable label information, since TADW is totally unsupervised and TriDNR associates the label with contents and is partially supervised.

In this paper, we present a novel model, which is cross the text and link (CrossTL), to learn the joint representation from texts and links. CrossTL improves the learning performance at the following three levels.

- At the input level, CrossTL can accommodate any types of graph or text embedding. We can choose the text or network representations which well capture the data properties as the inputs. To the best of our knowledge, this is the first such framework.
- At the structure level, CrossTL introduces a novel neural network architecture which recurrently learns a text-to-link and link-to-text representation to comprehensively and accurately describe the data.
- At the objective level, CrossTL bounds the error rate by incorporating four types of losses, i.e., text, link, and the combination and disagreement of text and link, into the loss function.

We conduct extensive experiments on two real world datasets. Results demonstrate that our framework can better leverage and combine text and structure information in the networked documents and achieve the state-of-the-art performance.

## 2 Related Work

**Text Embedding:** Word embedding has shed lights on many NLP tasks. Typical techniques include NNLM [1], LBL [14], CBOW and SkipGram [13]. Skip-Gram significantly speeds up the training process of NNLM and LBL and performs better than CBOW. Once getting the embedding for every word in the corpus, one can use the average of the embeddings of all words, i.e., WAvg [17], to represent the embedding for a document. Instead of computing word embedding first, the Doc2Vec technique [10] directly embeds any piece of text like a paragraph or an entire document in a distributed vector. We will use both WAvg and Doc2Vec as baselines to show the performance on the single text representation in our experiments.

**Graph Embedding:** In recent years, inspired by the SkipGram model [13], a number of graph embedding approaches are proposed to model the network structures [3,5,7,16,21]. For example, researchers proposed DeepWalk using random walk [16], and LINE using edge sampling [18] for large scale network embedding, and Node2Vec [7] using a biased random walk procedure to explore diverse neighborhoods, and SDNE using a deep architecture to optimize the first and second order proximity [21]. Node2Vec and SDNE are much more computationally expensive than DeepWalk and LINE. Hence we adopt DeepWalk [16] and LINE [18] as the basic blocks for the single link representation.

## 3    Our CrossTL Model

In this section, we first give the problem definition and then present our model for classifying networked documents.

**Definition 1.** (Networked Documents) can be viewed as a data graph $G = (T, L)$ composed of a set of documents $T$ connected to each other via a set of links $L$. In a classification task, $G$ contains a set of labeled $G^S$ and unlabeled documents $G^U$, i.e., $G = G^U \bigcup G^S$,

**Definition 2.** (Classification of Networked Documents) is to label the unlabeled documents $G^U \in G$ from a predefined set of categorical values $C$, given a set of training nodes $G^S \in G$.

### 3.1    CrossTL Learning Architecture

In order to utilize the complementary information in the other representation, we carefully design a three-layer cross-view learning architecture. Figure 1 shows the architecture of our CrossTL method.

The bottom of the architecture is the embedding layer, and the interaction layer lies in the middle. At the top we provide a decision layer to transform the representation into class distribution.

*Embedding Layer:* The embedding layer contains the text embedding for the text source and the graph embedding for the link source. Moreover, these two representations can be pre-trained using any existing methods since our CrossTL model is a general framework. In this paper, we adopt DeepWalk [16], LINE [18] and WAvg [17] to get the link and text representations as they are more efficient and often achieve better performance than other methods. We denote the text and link embedding of an input example $s$ as $T_1(s)$ and $L_1(s)$, or $T_1$ and $L_1$ for short, respectively.

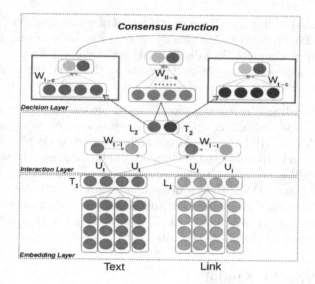

Text          Link

**Fig. 1.** The architecture for CrossTL

*Interaction Layer:* Previous works in learning joint representations mainly concatenate two embeddings $T_1$ and $L_1$ directly. However, the text and link have their own semantics, which are hard to be mixed together through a simple concatenating. We aim to learn a joint representation with the following characteristics. (1) It can reduce the semantic gap between different representations. (2) It can lend the complement information from one source to the other.

To this end, we design the interactive layer as follows. First, we add a hidden layer to further reduce the semantic gap between text and link. This step transforms the original text and link representations $T_1$ and $L_1$ into $T_1^t = U_t \cdot T_1$ and $L_1^t = U_l \cdot L_1$, where $U_t$ and $U_l$ are the weights of the hidden text and link layer. Second, the transformed text representation $T_1^t$ will be supplemented with a link-to-text representation $L_1^{l \to t} = W_{l \to t} L_1^t$, where $W_{l \to t}$ is the weight matrix for the mapping of link-to-text, and we have the second layer text representation $T_2$ with the interactive information from its counterpart. Similarly we can get the second layer link representation $L_2$. The process can be represented as follows.

$$T_2 = tanh(T_1^t + L_1^{l \to t}), \quad L_2 = tanh(L_1^t + T_1^{t \to l}), \tag{1}$$

where *tanh* is the activation function.

*Decision Layer:* Upon the interaction layer, we can get for each node three higher level representations including $T_2$, $L_2$ and $T_2 \oplus L_2$, where $\oplus$ is a concatenation of $T_2$ and $L_2$. Although these representations all contain fusion information from link and text, each representation still has its distinct focus. For example, $T_2$ is mainly a text representation with additional link information. Hence we employ all three representations for the classification of networked documents. To this end, we first apply a linear layer to mapping three representations, $T_2$, $L_2$ and $T_2 \oplus L_2$, into their target class space $C^t$, $C^l$, and $C^{tl}$, respectively. We then use a softmax layer to obtain the probability distribution for class $c$. The linear mapping and softmax layer can be defined as follows.

$$C^t = W_{t \to c} \cdot T_2 + b_{t \to c}, \quad C^l = W_{l \to c} \cdot L_2 + b_{l \to c}, \tag{2}$$

$$C^{tl} = W_{tl \to c} \cdot (T_2 \oplus L_2) + b_{tl \to c}, \tag{3}$$

$$p_c^t = \frac{exp(C_c^t)}{\sum_{m=1}^{M} exp(C_m^t)}, \quad p_c^l = \frac{exp(C_c^l)}{\sum_{m=1}^{M} exp(C_m^l)}, \tag{4}$$

$$p_c^{tl} = \frac{exp(C_c^{tl})}{\sum_{m=1}^{M} exp(C_m^{tl})}, \tag{5}$$

where $W_{t \to c}$, $W_{l \to c}$, $W_{tl \to c}$ are the weight matrices for the linear layer of $T_2$, $L_2$ and $T_2 \oplus L_2$, $b_{t \to c}$, $b_{l \to c}$, $b_{tl \to c}$ the corresponding biases, $C_c^t$, $C_c^l$, $C_c^{tl}$ the $c^{th}$ component in the class space, $M$ the number of classes, and $p_c^t$, $p_c^l$, $p_c^{tl}$ the probability of class $c$ predicted by $T_2$, $L_2$, and $T_2 \oplus L_2$.

### 3.2   Objective Function

In our CrossTL architecture, each representation can get supplementary information from the other. The learning process will benefit from such an architecture if the objective function is consistent with the representations. We hence design an objective function composed of four types of losses, i.e., the link ($J_l$) and text ($J_t$), the combination ($J_{tl}$) and the disagreement ($J_{dif}$) of link and text, so as to fully utilize the CrossTL representations. Here $J_l$ and $J_t$ reflect the contribution from the enriched text and link. Meanwhile, $J_{tl}$ emphasizes their joint effects and $J_{dif}$ is used to balance the scale of each part in the loss function.

We use the cross-entropy as the loss, and four objective functions can be defined as:

$$J_l = - \sum_{s \in G^S} \sum_{c=1}^{C} p_c^g(s) \cdot log(p_c^t(T_2(s))), \quad J_t = - \sum_{s \in G^S} \sum_{c=1}^{C} p_c^g(s) \cdot log(p_c^l(L_2(s)))$$

$$\tag{6}$$

$$J_{tl} = - \sum_{s \in G^S} \sum_{c=1}^{C} p_c^g(s) \cdot log(p_c^{tl}(T_2(s) \oplus L_2(s))), \tag{7}$$

$$J_{dif} = - \sum_{s \in G^S} \sum_{c=1}^{C} (p_c^t(T_2(s)) p_c^l(L_2(s))) \cdot log(p_c^t(T_2(s)) p_c^l(L_2(s))), \tag{8}$$

where $s$ is a node in the labeled data $G^S$, $p_c^g$ the gold probability of class $c$. Finally, the total loss $J_{com}$ is a linear combination of four functions.

$$J_{com} = \alpha \cdot J_l + \beta \cdot J_t + \gamma \cdot J_{dif} + J_{tl}, \tag{9}$$

where $\alpha$, $\beta$, and $\gamma$ are the weights to balance the text, link, the combination and disagreement of text and link representations. For simplicity, we set $\alpha$, $\beta$, and $\gamma$ to 1. Note that they can also be chosen by grid search on validation set for fine tuning, which will further improve our performance.

The parameter space in our CrossTL model would be $\Theta = \{U_t, U_l, W_{l \to t}, W_{t \to l}, W_{t \to c}, W_{l \to c}, W_{tl \to c}, b_{t \to c}, b_{l \to c}, b_{tl \to c}\}$. We adopt the widely-used stochastic gradient descent (SGD) [2] to train the model.

# 4    Experimental Evaluation

## 4.1    Experimental Setup

**Datasets:** We conduct experiments on two well known and publicly available datasets. One is DBLP [19] containing bibliography data in computer science. The other is Cora [23] which comes from an online achieve of computer science research papers.

The statistics for two datasets are summarized in Table 1.

**Table 1.** The statistics for DBLP and Cora datasets

|      | #of labels | # of nodes | #of edges | avg. degree | avg. doclen | # null docs | # null edges |
|------|-----------|-----------|-----------|-------------|-------------|-------------|--------------|
| DBLP | 4         | 60,744    | 52,890    | 0.87        | 8.25        | 0           | 43,019       |
| Cora | 7         | 4,263     | 89,819    | 21.07       | 133.22      | 0           | 78           |

Overall, DBLP is not as good for graph embedding as Cora, because it contains many outliers which do not connect with any other nodes. However, the number of documents in DBLP is 60,744, almost 15 times that in Cora. By evaluating on such two datasets with varied characteristics, we pursue to investigate the applicability of our model to different types of data.

**Settings:** We use linear SVM implemented by Liblinear [6] as our classifier to make a fair comparison with existing methods [15,22]. We train a one-vs.-rest classifier for each class and select the class with the maximum score as the label. We take representations of vertices as features to train classifiers, and evaluate classification performance with different training ratios (ratio r = 10%, 30%, 50%, 70%, respectively). The documents in training and testing splits are all randomly selected. We repeat our experiments for 10 times and report the average macro-$F_1$ as the evaluation metric.

We follow the parameter settings in [15,22]. Specifically, we set the dimension $k = 300$ and $k = 100$ on DBLP and Cora, and set window size = 8, batch size = 128 and learning rate = 0.1 on both datasets.

**Baselines.** We conduct extensive experiments to compare our methods with the following 9 baselines.

**DeepWalk (DW)** [16] learns node representations based on the combination of SkipGram model and random walks.

**LINE (LE)** [18] is a state-of-the-art algorithm which exploits the first-order and second-order proximity, i.e., the local and global structure, in the learning process.

**Word Average (WAvg)** [17] uses SkipGram [13] to obtain the embedding of words in corpus and takes the average of word vectors as the feature for document.

**Doc2Vec (D2V)** [10] extends vector representations for words to that for arbitrary piece of text by adding paragraph matrix.

**DW+D2V** concatenates the vectors from DeepWalk and Doc2Vec into a long vector.

**DW+WAvg** concatenates the vectors from DeepWalk and WAvg into a long vector.

**LE+WAvg** concatenates the vectors from LINE and WAvg into a long vector.

**TADW** [22] is based on the matrix-factorization formalization of DeepWalk and brings the SVD decomposition of tf-idf matrix into the process of matrix factorization.

**TriDNR** [15] integrates DeepWalk with Doc2Vec to learn representations from three parties: network, contents, and labels.

## 4.2  Experimental Results and Analysis

We now report our results on two datasets. Note that in the following tables and figures, the symbols ** and * indicate that the difference between our CrossTL model and other baselines is significant according to the paired-sample T-test at the level of $p < 0.01$ and $p < 0.05$, respectively.

**Comparison Results on DBLP.** We first present classification results on DBLP dataset in Table 2.

From Table 2, we have the following important observations.

**Table 2.** Average macro-$F_1$ score and significance on DBLP dataset

| r(%) | DeepWalk | LINE | WAvg | Doc2Vec | DW+D2V | DW+WAvg | LE+WAvg | TADW | TriDNR | CrossTL |
|---|---|---|---|---|---|---|---|---|---|---|
| 10 | 0.400** | 0.390** | 0.703** | 0.652** | 0.653** | 0.718** | 0.704** | 0.459** | 0.692** | **0.738** |
| 30 | 0.424** | 0.394** | 0.712** | 0.679** | 0.681** | 0.744** | 0.724** | 0.542** | 0.727** | **0.762** |
| 50 | 0.427** | 0.394** | 0.713** | 0.686** | 0.686** | 0.751** | 0.728** | 0.564** | 0.740** | **0.770** |
| 70 | 0.428** | 0.394** | 0.715** | 0.690** | 0.690** | 0.754** | 0.731** | 0.574** | 0.747** | **0.775** |

**Table 3.** Average macro-$F_1$ score and significance on Cora dataset

| r(%) | DeepWalk | LINE | WAvg | Doc2Vec | DW+D2V | DW+WAvg | LE+WAvg | TADW | TriDNR | CrossTL |
|---|---|---|---|---|---|---|---|---|---|---|
| 10 | 0.625** | 0.841** | 0.683** | 0.587** | 0.712** | 0.658** | 0.845** | 0.468** | 0.651** | **0.861** |
| 30 | 0.717** | 0.859** | 0.727** | 0.684** | 0.745** | 0.763** | 0.866** | 0.659** | 0.735** | **0.875** |
| 50 | 0.753** | 0.868** | 0.734** | 0.719** | 0.775** | 0.797** | 0.870** | 0.709** | 0.773** | **0.877** |
| 70 | 0.767** | 0.870** | 0.741** | 0.733** | 0.801** | 0.817** | 0.876* | 0.740** | 0.804** | **0.880** |

– Our proposed CrossTL model achieves the best performance among all approaches. Its improvements over other baselines are all significant at the level of $p < 0.01$. For example, when training on 10% dataset, the macro-$F_1$ score for TriDNR is 0.692. In contrast, CrossTL reaches a score of 0.738, showing a 6.65% increase over TriDNR.
– Two state-of-the-art approaches, TriDNR and TADW, perform worse than the naive combination DW+WAvg. This can be arisen from the fact that the single text based approach WAvg has already gotten good enough macro-$F_1$ and its combination with DeepWalk further enhances performance. This shows the limitation of TriDNR and TADW: they use the same DeepWalk-style graph representations but their performances are bounded by the Doc2Vec and SVD decomposed text representations.
– Two link based methods DeepWalk and LINE perform much worse than WAvg and Doc2Vec which use texts. The reason is that a large number of nodes (43019 among 60744) in DBLP do not have any links. This clearly damages the performance of graph embeddings.

**Comparison Results on Cora.** We now give in Table 3 the classification results for Cora. Once again, our CrossTL is the best among all. Its enhancements over other methods are significant at the level of $p < 0.01$ in nearly all cases. The only exception is the one for LE+WAvg on 70% training data. However, it is still significant at the level of $p < 0.05$. More importantly, CrossTL outperforms TADW and TriDNR by 77.16% and 32.26% on 10% training data, respectively. This clearly demonstrates the superiority of CrossTL. It works well on either sparse or dense graph, rich or poor text information.

In contrast to the findings on DBLP, the graph embedding approaches LINE and DeepWalk outperforms WAvg and Doc2Vec in most cases on Cora. In particular, LINE performs much better than others. This is consistent with the characteristic of the dataset. Note from Table 1 that Cora is a much denser dataset than DBLP. Its average degree is 21.07 while that for DBLP is only 0.87.

The naive combination approach LE+WAvg beats other four combination-based baselines on Cora. We further see that even the single graph based method LINE outperforms others. The reason is that other four methods all use the style of DeepWalk, which deploys a truncated random walk for graph embedding. For a denser network, LINE learns better representations which preserve both the first and second order proximities [18]. This, from another point of view, shows the advantage of our CrossTL model, since it does not rely on a specific graph or text embedding technique.

The poor performance of TADW is because the Cora dataset used in our experiment is not a closed graph. It contains a lot of out-going edges which have to be removed before matrix factorization. This makes its performance worse than DeepWalk.

**Effects of Parameters.** We show the effects of dimensionality ($d$) and training ratio ($r$) on DBLP and Cora in Fig. 2(a) and (b), respectively.

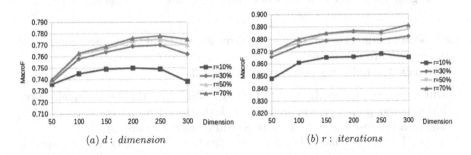

(a) $d$ : *dimension*                         (b) $r$ : *iterations*

**Fig. 2.** Effects of training ratio and dimensionality

It can be seen from Fig. 2 that, when the number of dimensions increases, the overall trends for CrossTL model are upward. This infers that a large value of dimensionality is beneficial for the model. In general, not much difference is observed with various dimensions, showing that CrossTL is stable with different number of features.

As for the training ratio ($r$), it is clear that the macro $F_1$ scores increase steadily when there are more training data. The relative performance on Cora is not as obvious as that on DBLP. The reason is that DBLP has an order of magnitude more nodes than Cora. Hence increasing training ratio has greater impacts on DBLP than on Cora.

### 4.3    Visualization

We visualize the learned representations of the Cora network. We use the embeddings learned by different approaches as the input to the visualization tool t-SNE [9]. As a result, the embedding of each node is mapped as a two-dimensional vector and then visualized as a point on a two dimensional space. Nodes in different categories are shown in different colors. The visualization results for CrossTL and other TADW and TriDNR baselines are shown in Fig. 3. We do not present the visualization of other baselines because TriDNR and TADW are two state-of-the-art baselines which combine two sources of information with carefully designed architecture or objective function. In contrast, other baselines either use only

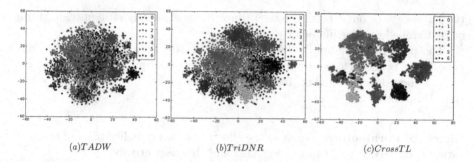

(a)TADW          (b)TriDNR          (c)CrossTL

**Fig. 3.** Visualizations of the Cora citation network. (Color figure online)

one type of information, or are a simple concatenating of two embeddings whose performances are not stable.

We have the following observations for Fig. 3.

- Our proposed CrossTL approach in Fig. 3(c) has the best visualization, because the points of the same color in CrossTL are close to each other, and also because the borders between different colors are quite clear.
- TriDNR is the second, as it has a relatively exclusive area for one specific color except a detached red block and a large area mixing black, red, and purple on mid-right in Fig. 3(b).
- TADW in Fig. 3(a) is the worst since it mixes the points in all colors into a blurred image. For example, the cyan and black dots scatter over the entire image.

## 5    Conclusion

In this paper, we present CrossTL, a novel representation method, to jointly learn the representations from two different sources. Specifically, with a general framework, our CrossTL model can utilize more sophisticated input representations like WAvg or LINE which well capture the content or structure properties. Furthermore, CrossTL can comprehensively describe the data by adding supplementary information from text to link or link to text. Finally, with a carefully designed objective function, CrossTL achieves the consensus between the text and link sources. Experimental results demonstrate that, compared to other state-of-the-art baselines, CrossTL is much more effective. It is also robust to various types of datasets.

**Acknowledgments.** The work described in this paper has been supported in part by the NSFC project (61572376).

## References

1. Bengio, Y., Ducharme, R., Vincent, P., Jauvin, C.: A neural probabilistic language model. JMLR **3**, 1137–1155 (2003)

2. Bottou, L.: Stochastic gradient learning in neural networks. In: Neuro-Nîmes (1991)
3. Bui, T.D., Ravi, S., Ramavajjala, V.: Neural graph machines: learning neural networks using graphs. In: ICIR (2017)
4. Chang, J., Blei, D.M.: Relational topic models for document networks. In: Proceedings of International Conference on Artificial Intelligence and Statistics, pp. 81–88 (2009)
5. Chen, J., Zhang, Q., Huang, X.: Incorporate group information to enhance network embedding. In: Proceedings of CIKM, pp. 1901–1904 (2016)
6. Fan, R.E., Chang, K.W., Hsieh, C.J., Wang, X.R., Lin, C.J.: Liblinear: a library for large linear classification. JMLR **9**, 1871–1874 (2008)
7. Grover, A., Leskovec, J.: node2vec: scalable feature learning for networks. In: Proceedings of SIGKDD, pp. 855–864 (2016)
8. Jensen, D., Neville, J., Gallagher, B.: Why collective inference improves relational classification. In: Proceedings of SIGKDD, pp. 593–598 (2004)
9. Laurens, V.D.M., Hinton, G.: Visualizing data using t-SNE. JMLR **9**(2605), 2579–2605 (2008)
10. Le, Q.V., Mikolov, T.: Distributed representations of sentences and documents. In: Proceedings of ICML, pp. 1188–1196 (2014)
11. Lu, Q., Getoor, L.: Link-based classification. In: Proceedings of ICML, pp. 496–503 (2003)
12. Mei, Q., Cai, D., Zhang, D., Zhai, C.: Topic modeling with network regularization. In: Proceedings of WWW, pp. 101–110 (2008)
13. Mikolov, T., Chen, K., Corrado, G., Dean, J.: Efficient estimation of word representations in vector space. In: Proceedings of ICLR (2013)
14. Mnih, A., Hinton, G.: Three new graphical models for statistical language modelling. In: Proceedings of ICML, pp. 641–648 (2007)
15. Pan, S., Wu, J., Zhu, X., Zhang, C., Wang, Y.: Tri-party deep network representation. In: Proceedings of 25th IJCAI, pp. 1895–1901 (2016)
16. Perozzi, B., Al-Rfou, R., Skiena, S.: Deepwalk: online learning of social representations. In: Proceedings of SIGKDD, pp. 701–710 (2014)
17. Tang, D., Qin, B., Liu, T.: Document modeling with gated recurrent neural network for sentiment classification. In: Proceedings of EMNLP, pp. 1422–1432 (2015)
18. Tang, J., Qu, M., Wang, M., Zhang, M., Yan, J., Mei, Q.: Line: large-scale information network embedding. In: Proceedings of WWW, pp. 1067–1077 (2015)
19. Tang, J., Zhang, J., Yao, L., Li, J., Zhang, L., Su, Z.: ArnetMiner: extraction and mining of academic social networks. In: Proceedings of SIGKDD, pp. 990–998 (2008)
20. Taskar, B., Abbeel, P., Koller, D.: Discriminative probabilistic models for relational data. In: Proceedings of the 18th UAI, pp. 485–492 (2002)
21. Wang, D., Cui, P., Zhu, W.: Structural deep network embedding. In: Proceedings of SIGKDD, pp. 1225–1234 (2016)
22. Yang, C., Liu, Z., Zhao, D., Sun, M., Chang, E.Y.: Network representation learning with rich text information. In: Proceedings of 24th IJCAI, pp. 2111–2117 (2015)
23. Zhang, X., Hu, X., Zhou, X.: A comparative evaluation of different link types on enhancing document clustering. In: Proceedings of SIGIR, pp. 555–562 (2008)

# A Comparison of Modeling Units in Sequence-to-Sequence Speech Recognition with the Transformer on Mandarin Chinese

Shiyu Zhou[1,2(✉)], Linhao Dong[1,2], Shuang Xu[1], and Bo Xu[1]

[1] Institute of Automation, Chinese Academy of Sciences, Beijing, China
{zhoushiyu2013,donglinhao2015,shuang.xu,xubo}@ia.ac.cn
[2] University of Chinese Academy of Sciences, Beijing, China

**Abstract.** The choice of modeling units is critical to automatic speech recognition (ASR) tasks. Conventional ASR systems typically choose context-dependent states (CD-states) or context-dependent phonemes (CD-phonemes) as their modeling units. However, it has been challenged by sequence-to-sequence attention-based models. On English ASR tasks, previous attempts have already shown that the modeling unit of graphemes can outperform that of phonemes by sequence-to-sequence attention-based model. In this paper, we are concerned with modeling units on Mandarin Chinese ASR tasks using sequence-to-sequence attention-based models with the Transformer. Five modeling units are explored including context-independent phonemes (CI-phonemes), syllables, words, sub-words and characters. Experiments on HKUST datasets demonstrate that the lexicon free modeling units can outperform lexicon related modeling units in terms of character error rate (CER). Among five modeling units, character based model performs best and establishes a new state-of-the-art CER of 26.64% on HKUST datasets.

**Keywords:** ASR · Multi-head attention · Modeling units
Sequence-to-sequence · Transformer

## 1 Introduction

Conventional ASR systems consist of three independent components: an acoustic model (AM), a pronunciation model (PM) and a language model (LM), all of which are trained independently. CD-states and CD-phonemes are dominant as their modeling units in such systems [3,11,13]. However, it recently has been challenged by sequence-to-sequence attention-based models. These models are commonly comprised of an *encoder*, which consists of multiple recurrent neural network (RNN) layers that model the acoustics, and a *decoder*, which consists of one or more RNN layers that predict the output sub-word sequence. An *attention* layer acts as the interface between the encoder and the decoder: it selects frames

© Springer Nature Switzerland AG 2018
L. Cheng et al. (Eds.): ICONIP 2018, LNCS 11305, pp. 210–220, 2018.
https://doi.org/10.1007/978-3-030-04221-9_19

in the encoder representation that the decoder should attend to in order to predict the next sub-word unit [9]. In [10], Tara et al. experimentally verified that the grapheme-based sequence-to-sequence attention-based model can outperform the corresponding phoneme-based model on English ASR tasks. This work is very interesting and amazing since a hand-designed lexicon might be removed from ASR systems. As we know, it is very laborious and time-consuming to generate a pronunciation lexicon. Furthermore, the latest work shows that attention-based encoder-decoder architecture achieves a new state-of-the-art WER on a 12500 h English voice search task using the word piece models (WPM), which are sub-word units ranging from graphemes all the way up to entire words [2].

Since the outstanding performance of grapheme-based modeling units on English ASR tasks, we conjecture that maybe there is no need for a hand-designed lexicon on Mandarin Chinese ASR tasks as well by sequence-to-sequence attention-based models. In Mandarin Chinese, if a hand-designed lexicon is removed, the modeling units can be words, sub-words and characters. Character-based sequence-to-sequence attention-based models have been investigated on Mandarin Chinese ASR tasks in [1,15], but the performance comparison with different modeling units are not explored before. Building on our work [19], which shows that syllable based model with the Transformer can perform better than CI-phoneme based counterpart, we investigate five modeling units on Mandarin Chinese ASR tasks, including CI-phonemes, syllables, words, sub-words and characters. The Transformer is chosen to be the basic architecture of sequence-to-sequence attention-based model in this paper [7,19]. Experiments on HKUST datasets confirm our hypothesis that the lexicon free modeling units, i.e. words, sub-words and characters, can outperform lexicon related modeling units, i.e. CI-phonemes and syllables. Among five modeling units, character based model with the Transformer achieves the best result and establishes a new state-of-the-art CER of 26.64% on HKUST datasets without a hand-designed lexicon and an extra language model integration, which is a 4.8% relative reduction in CER compared to the existing best CER of 28.0% by the joint CTC-attention based encoder-decoder network with a separate RNN-LM integration [4].

The rest of the paper is organized as follows. After an overview of the related work in Sect. 2, Sect. 3 describes the proposed method in detail. We then show experimental results in Sect. 4 and conclude this work in Sect. 5.

## 2   Related Work

Sequence-to-sequence attention-based models have achieved promising results on English ASR tasks and various modeling units have been studied recently, such as CI-phonemes, CD-phonemes, graphemes and WPM [2,9,10]. In [10], Tara et al. first explored sequence-to-sequence attention-based model trained with phonemes for ASR tasks and compared the modeling units of graphemes and phonemes. They experimentally verified that the grapheme-based sequence-to-sequence attention-based model can outperform the corresponding phoneme-based model on English ASR tasks. Furthermore, the modeling units of WPM have been explored in [2], which are sub-word units ranging from graphemes all the way up to entire words. It achieved a new state-of-the-art WER on a 12500 h English voice search task.

Although sequence-to-sequence attention-based models perform very well on English ASR tasks, related works are quite few on Mandarin Chinese ASR tasks. Chan et al. first proposed Character-Pinyin sequence-to-sequence attention-based model on Mandarin Chinese ASR tasks. The Pinyin information was used during training for improving the performance of the character model. Instead of using joint Character-Pinyin model, [15] directly used Chinese characters as network output by mapping the one-hot character representation to an embedding vector via a neural network layer. What's more, [20] compared the modeling units of characters and syllables by sequence-to-sequence attention-based models.

Besides the modeling unit of character, the modeling units of words and sub-words are investigated on Mandarin Chinese ASR tasks in this paper. Sub-word units encoded by byte pair encoding (BPE) have been explored on neural machine translation (NMT) tasks to address out-of-vocabulary (OOV) problem on open-vocabulary translation [14], which iteratively replace the most frequent pair of characters with a single, unused symbol.

## 3   System Overview

### 3.1   ASR Transformer Model Architecture

The Transformer model architecture is the same as sequence-to-sequence attention-based models except relying entirely on self-attention and position-wise, fully connected layers for both the encoder and decoder [17]. The encoder maps an input sequence of symbol representations $\mathbf{x} = (x_1, \ldots, x_n)$ to a sequence of continuous representations $\mathbf{z} = (z_1, \ldots, z_n)$. Given $\mathbf{z}$, the decoder then generates an output sequence $\mathbf{y} = (y_1, \ldots, y_m)$ of symbols one element at a time.

The ASR Transformer architecture used in this work is the same as our work [19] which is shown in Fig. 1. We describe it briefly here. It stacks multi-head attention (MHA) [17] and position-wise, fully connected layers for both the encode and decoder. The encoder is composed of a stack of $N$ identical layers. Each layer has two sub-layers. The first is a MHA, and the second is a position-wise fully connected feed-forward network. Residual connections are employed around each of the two sub-layers, followed by a layer normalization. The decoder is similar to the encoder except inserting a third sub-layer to perform a MHA over the output of the encoder stack. To prevent leftward information flow and preserve the auto-regressive property in the decoder, the self-attention sub-layers in the decoder mask out all values corresponding to illegal connections. In addition, positional encodings [17] are added to the input at the bottoms of these encoder and decoder stacks, which inject some information about the relative or absolute position of the tokens in the sequence.

The difference between the NMT Transformer [17] and the ASR Transformer is the input of the encoder. We add a linear transformation with a layer normalization to convert the log-Mel filterbank feature to the model dimension $d_{model}$ for dimension matching [19], which is marked out by a dotted line in Fig. 1.

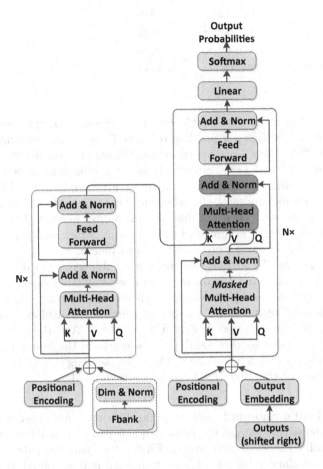

**Fig. 1.** The architecture of the ASR Transformer.

## 3.2   Modeling Units

Five modeling units are compared on Mandarin Chinese ASR tasks, including CI-phonemes, syllables, words, sub-words and characters. Table 1 summarizes the different number of output units investigated by this paper. We show an example of various modeling units in Table 2.

**CI-Phoneme and Syllable Units.** CI-phoneme and syllable units are compared in our work [19], which 118 CI-phonemes without silence (phonemes with tones) are employed in the CI-phoneme based experiments and 1384 syllables

**Table 1.** Different modeling units explored in this paper.

| Modeling units | Number of outputs |
| --- | --- |
| CI-phonemes | 122 |
| Syllables | 1388 |
| Characters | 3900 |
| Sub-words | 11039 |
| Words | 28444 |

(pinyins with tones) in the syllable based experiments. *Extra tokens* (i.e. an unknown token (<UNK>), a padding token (<PAD>), and sentence start and end tokens (<S>/<\S>)) are appended to the outputs, making the total number of outputs 122 and 1388 respectively in the CI-phoneme based model and syllable based model. Standard tied-state cross-word triphone GMM-HMMs are first trained with maximum likelihood estimation to generate CI-phoneme alignments on training set. Then syllable alignments are generated through these CI-phoneme alignments according to the lexicon, which can handle multiple pronunciations of the same word in Mandarin Chinese.

The outputs are CI-phoneme sequences or syllable sequences during decoding stage. In order to convert CI-phoneme sequences or syllable sequences into word sequences, a greedy cascading decoder with the Transformer [19] is proposed. First, the best CI-phoneme or syllable sequence $s$ is calculated by the ASR Transformer from observation $X$ with a beam size $\beta$. And then, the best word sequence $W$ is chosen by the NMT Transformer from the best CI-phoneme or syllable sequence $s$ with a beam size $\gamma$. Through cascading these two Transformer models, we assume that $Pr(W|X)$ can be approximated.

**Sub-word Units.** Sub-word units, using in this paper, are generated by BPE[1] [14], which iteratively merges the most frequent pair of characters or character sequences with a single, unused symbol. Firstly, the symbol vocabulary with the character vocabulary is initialized, and each word is represented as a sequence of characters plus a special end-of-word symbol '@@'[2], which allows to restore the original tokenization. Then, all symbol pairs are counted iteratively and each occurrence of the most frequent pair ('A', 'B') are replaced with a new symbol 'AB'. Each merge operation produces a new symbol which represents a character n-gram. Frequent character n-grams (or whole words) are eventually merged into a single symbol. Then the final symbol vocabulary size is equal to the size of the initial vocabulary, plus the *number of merge operations*, which is the hyperparameter of this algorithm [14].

BPE is capable of encoding an open vocabulary with a compact symbol vocabulary of variable-length sub-word units, which requires no shortlist. After

---

[1] https://github.com/rsennrich/subword-nmt.
[2] '@@' is a special end-of-word symbol to connect sub-words.

encoded by BPE, the sub-word units are ranging from characters all the way up to entire words. Thus there are no OOV words with BPE and high frequent sub-words can be preserved.

In our experiments, we choose the *number of merge operations* 5000, which generates the number of sub-words units 11035 from the training transcripts. After appended with 4 *extra tokens*, the total number of outputs is 11039.

**Word and Character Units.** For word units, we collect all words from the training transcripts. Appended with 4 *extra tokens*, the total number of outputs is 28444.

For character units, all Mandarin Chinese characters together with English words in training transcripts are collected, which are appended with 4 *extra tokens* to generate the total number of outputs 3900[3].

## 4 Experiment

### 4.1 Data

The HKUST corpus (LDC2005S15, LDC2005T32), a corpus of Mandarin Chinese conversational telephone speech, is collected and transcribed by Hong Kong University of Science and Technology (HKUST) [8], which contains 150-h speech, and 873 calls in the training set and 24 calls in the test set. All experiments are conducted using 80-dimensional log-Mel filterbank features, computed with a 25 ms window and shifted every 10 ms. The features are normalized via mean subtraction and variance normalization on the speaker basis[4]. Similar to [5,12], at the current frame $t$, these features are stacked with 3 frames to the left and downsampled to a 30 ms frame rate. As in [4], we generate more training data by linearly scaling the audio lengths by factors of 0.9 and 1.1 (speed perturb), which can improve the performance in our experiments.

**Table 2.** An example of various modeling units in this paper.

| Modeling units | Example |
|---|---|
| CI-phonemes | Y IY1 JH UH3 NG3 X IY4 N4 N IY4 AE4 N4 |
| Syllables | YI1 ZHONG3 XIN4 NIAN4 |
| Characters | 一 种 信 念 |
| Sub-words | 一种 信@@ 念 |
| Words | 一种 信念 |

---

[3] We manually delete two tokens · and +, which are not Mandarin Chinese characters.
[4] Experiment code: https://github.com/shiyuzh2007/ASR/tree/master/transformer.

## 4.2   Training

We perform our experiments on the *base model* and *big model* (i.e. D512-H8 and D1024-H16 respectively) of the Transformer from [17]. The basic architecture of these two models is the same but different parameters setting. Table 3 lists the experimental parameters between these two models. The Adam algorithm [6] with gradient clipping and warmup is used for optimization. During training, label smoothing of value $\epsilon_{ls} = 0.1$ is employed [16]. After trained, the last 20 checkpoints are averaged to make the performance more stable [17].

**Table 3.** Experimental parameters configuration.

| Model | $N$ | $d_{model}$ | $h$ | $d_k$ | $d_v$ | $warmup$ |
|-------|-----|-------------|-----|-------|-------|----------|
| D512-H8 | 6 | 512 | 8 | 64 | 64 | 4000 *steps* |
| D1024-H16 | 6 | 1024 | 16 | 64 | 64 | 12000 *steps* |

In the CI-phoneme and syllable based model, we cascade an ASR Transformer and a NMT Transformer to generate word sequences from observation $X$. However, we do not employ a NMT Transformer anymore in the word, sub-word and character based model, since the beam search results from the ASR Transformer are already the Chinese character level. The total parameters of different modeling units list in Table 4.

**Table 4.** Total parameters of different modeling units.

| Model | D512-H8 (ASR) | D1024-H16 (ASR) | D512-H8 (NMT) |
|-------|---------------|-----------------|---------------|
| CI-phonemes | $57M$ | $227M$ | $71M$ |
| Syllables | $58M$ | $228M$ | $72M$ |
| Words | $71M$ | $256M$ | – |
| Sub-words | $63M$ | $238M$ | – |
| Characters | $59M$ | $231M$ | – |

## 4.3   Results

According to the description from Sect. 3.2, we can see that the modeling units of words, sub-words and characters are lexicon free, which do not need a hand-designed lexicon. On the contrary, the modeling units of CI-phonemes and syllables need a hand-designed lexicon.

Our results are summarized in Table 5. It is clear to see that the lexicon free modeling units, i.e. words, sub-words and characters, can outperform corresponding lexicon related modeling units, i.e. CI-phonemes and syllables on HKUST datasets. It confirms our hypothesis that we can remove the need for a hand-designed lexicon on Mandarin Chinese ASR tasks by sequence-to-sequence attention-based models. What's more, we note here that the sub-word based model performs better than the word based counterpart. It represents that the modeling unit of sub-words is superior to that of words, since sub-word units encoded by BPE have fewer number of outputs and without OOV problems. However, the sub-word based model performs worse than the character based model. The possible reason is that the modeling unit of sub-words is bigger than that of characters which is difficult to train. We will conduct our experiments on larger datasets and compare the performance between the modeling units of sub-words and characters in future work. Finally, among five modeling units, character based model with the Transformer achieves the best result. It demonstrates that the modeling unit of character is suitable for Mandarin Chinese ASR tasks by sequence-to-sequence attention-based models.

**Table 5.** Comparison of different modeling units with the Transformer on HKUST datasets in CER (%).

| Modeling units | Model | CER |
| --- | --- | --- |
| CI-phonemes [19] | D512-H8 | 32.94 |
| | D1024-H16 | **30.65** |
| | D1024-H16 (speed perturb) | 30.72 |
| Syllables [19] | D512-H8 | 31.80 |
| | D1024-H16 | 29.87 |
| | D1024-H16 (speed perturb) | **28.77** |
| Words | D512-H8 | 31.98 |
| | D1024-H16 | 28.74 |
| | D1024-H16 (speed perturb) | **27.42** |
| Sub-words | D512-H8 | 30.22 |
| | D1024-H16 | 28.28 |
| | D1024-H16 (speed perturb) | **27.26** |
| Characters | D512-H8 | 29.00 |
| | D1024-H16 | 27.70 |
| | D1024-H16 (speed perturb) | **26.64** |

### 4.4    Comparison with Previous Works

In Table 6, we compare our experimental results to other model architectures from the literature on HKUST datasets. First, we can find that our best results of different modeling units are comparable or superior to the best result by the deep multidimensional residual learning with 9 LSTM layers [18], which is a hybrid LSTM-HMM system with the modeling unit of CD-states. We can observe that the best CER 26.64% of character based model with the Transformer on HKUST datasets achieves a 13.4% relative reduction compared to the best CER of 30.79% by the deep multidimensional residual learning with 9 LSTM layers. It shows the superiority of the sequence-to-sequence attention-based model compared to the hybrid LSTM-HMM system.

Moreover, we can note that our best results with the modeling units of words, sub-words and characters are superior to the existing best CER of 28.0% by the joint CTC-attention based encoder-decoder network [4], which is the state-of-the-art on HKUST datasets to the best of our knowledge. Character based model with the Transformer establishes a new state-of-the-art CER of 26.64% on HKUST datasets without a hand-designed lexicon and an extra language model integration, which is a 7.8% relative CER reduction compared to the CER of 28.9% of the joint CTC-attention based encoder-decoder network when no external language model is used, and a 4.8% relative CER reduction compared to the existing best CER of 28.0% by the joint CTC-attention based encoder-decoder network with separate RNN-LM [4].

**Table 6.** CER (%) on HKUST datasets compared to previous works.

| Model | CER |
| --- | --- |
| LSTMP-9×800P512-F444 [18] | 30.79 |
| CTC-attention+joint dec. (speed perturb., one-pass)+VGG net | 28.9 |
| +RNN-LM (separate) [4] | **28.0** |
| CI-phonemes-D1024-H16 [19] | 30.65 |
| Syllables-D1024-H16 (speed perturb) [19] | 28.77 |
| Words-D1024-H16 (speed perturb) | 27.42 |
| Sub-words-D1024-H16 (speed perturb) | 27.26 |
| Characters-D1024-H16 (speed perturb) | **26.64** |

## 5 Conclusions

In this paper we compared five modeling units on Mandarin Chinese ASR tasks by sequence-to-sequence attention-based model with the Transformer, including CI-phonemes, syllables, words, sub-words and characters. We experimentally verified that the lexicon free modeling units, i.e. words, sub-words and characters, can outperform lexicon related modeling units, i.e. CI-phonemes and syllables on HKUST datasets. It represents that maybe we can remove the need for a hand-designed lexicon on Mandarin Chinese ASR tasks by sequence-to-sequence attention-based models. Among five modeling units, character based model achieves the best result and establishes a new state-of-the-art CER of 26.64% on HKUST datasets. Moreover, we find that sub-word based model with the Transformer achieves a promising result, although it is slightly worse than character based counterpart.

**Acknowledgments.** The research work is supported by the National Key Research and Development Program of China under Grant No. 2016YFB1001404.

# References

1. Chan, W., Lane, I.: On online attention-based speech recognition and joint Mandarin character-pinyin training. In: Interspeech, pp. 3404–3408 (2016)
2. Chiu, C.C., et al.: State-of-the-art speech recognition with sequence-to-sequence models. arXiv preprint arXiv:1712.01769 (2017)
3. Dahl, G.E., Yu, D., Deng, L., Acero, A.: Context-dependent pre-trained deep neural networks for large-vocabulary speech recognition. IEEE Trans. Audio Speech Lang. Process. **20**(1), 30–42 (2012)
4. Hori, T., Watanabe, S., Zhang, Y., Chan, W.: Advances in joint CTC-attention based end-to-end speech recognition with a deep CNN encoder and RNN-LM. arXiv preprint arXiv:1706.02737 (2017)
5. Kannan, A., Wu, Y., Nguyen, P., Sainath, T.N., Chen, Z., Prabhavalkar, R.: An analysis of incorporating an external language model into a sequence-to-sequence model. arXiv preprint arXiv:1712.01996 (2017)
6. Kingma, D.P., Ba, J.: Adam: a method for stochastic optimization. arXiv preprint arXiv:1412.6980 (2014)
7. Dong, L., Xu, S., Xu, B.: Speech-transformer: a no-recurrence sequence-to-sequence model for speech recognition. In: 2018 IEEE International Conference on Acoustics, Speech and Signal Processing, ICASSP, pp. 5884–5888. IEEE (2018)
8. Liu, Y., Fung, P., Yang, Y., Cieri, C., Huang, S., Graff, D.: HKUST/MTS: a very large scale Mandarin telephone speech corpus. In: Huo, Q., Ma, B., Chng, E.-S., Li, H. (eds.) ISCSLP 2006. LNCS, vol. 4274, pp. 724–735. Springer, Heidelberg (2006). https://doi.org/10.1007/11939993_73
9. Prabhavalkar, R., Sainath, T.N., Li, B., Rao, K., Jaitly, N.: An analysis of attention in sequence-to-sequence models. In: Proceedings of Interspeech (2017)
10. Sainath, T.N., et al.: No need for a lexicon? Evaluating the value of the pronunciation lexica in end-to-end models. arXiv preprint arXiv:1712.01864 (2017)
11. Sak, H., Senior, A., Beaufays, F.: Long short-term memory recurrent neural network architectures for large scale acoustic modeling. In: Fifteenth Annual Conference of the International Speech Communication Association (2014)
12. Sak, H., Senior, A., Rao, K., Beaufays, F.: Fast and accurate recurrent neural network acoustic models for speech recognition. arXiv preprint arXiv:1507.06947 (2015)
13. Senior, A., Sak, H., Shafran, I.: Context dependent phone models for LSTM RNN acoustic modelling. In: 2015 IEEE International Conference on Acoustics, Speech and Signal Processing, ICASSP, pp. 4585–4589. IEEE (2015)
14. Sennrich, R., Haddow, B., Birch, A.: Neural machine translation of rare words with subword units. arXiv preprint arXiv:1508.07909 (2015)
15. Shan, C., Zhang, J., Wang, Y., Xie, L.: Attention-based end-to-end speech recognition on voice search. In: 2018 IEEE International Conference on Acoustics, Speech and Signal Processing (ICASSP), pp. 4764–4768 (2018)
16. Szegedy, C., Vanhoucke, V., Ioffe, S., Shlens, J., Wojna, Z.: Rethinking the inception architecture for computer vision. In: Proceedings of the IEEE Conference on Computer Vision and Pattern Recognition, pp. 2818–2826 (2016)
17. Vaswani, A., et al.: Attention is all you need. In: Advances in Neural Information Processing Systems, pp. 6000–6010 (2017)

18. Zhao, Y., Xu, S., Xu, B.: Multidimensional residual learning based on recurrent neural networks for acoustic modeling. In: Interspeech, pp. 3419–3423 (2016)
19. Zhou, S., Dong, L., Xu, S., Xu, B.: Syllable-based sequence-to-sequence speech recognition with the transformer in Mandarin Chinese. ArXiv e-prints, April 2018
20. Zou, W., Jiang, D., Zhao, S., Li, X.: A comparable study of modeling units for end-to-end Mandarin speech recognition. arXiv preprint arXiv:1805.03832 (2018)

# Multi-view Emotion Recognition Using Deep Canonical Correlation Analysis

Jie-Lin Qiu[1], Wei Liu[1], and Bao-Liang Lu[1,2,3]($\boxtimes$)

[1] Center for Brain-Like Computing and Machine Intelligence,
Department of Computer Science and Engineering, Shanghai Jiao Tong University,
Shanghai, China
{Qiu-Jielin,liuwei-albert,bllu}@sjtu.edu.cn
[2] Key Laboratory of Shanghai Education Commission for Intelligent Interaction
and Cognition Engineering, Shanghai Jiao Tong University, Shanghai, China
[3] Brain Science and Technology Research Center, Shanghai Jiao Tong University,
Shanghai, China

**Abstract.** Emotion is a subjective, conscious experience when people face different kinds of stimuli. In this paper, we adopt Deep Canonical Correlation Analysis (DCCA) for high-level coordinated representation to make feature extraction from EEG and eye movement data. Parameters of the two views' nonlinear transformations are learned jointly to maximize the correlation. We propose a multi-view emotion recognition framework and evaluate its effectiveness on three real world datasets. We found that DCCA efficiently learned representations with high correlation, which contributed to higher emotion classification accuracy. Our experiment results indicate that DCCA model is superior to the state-of-the-art methods with mean accuracies of 94.58% on SEED dataset, 87.45% on SEED IV dataset, and 88.51% and 84.98% for four classification and two dichotomies on DEAP dataset, respectively.

**Keywords:** Emotion recognition · EEG · Eye movement
Deep Canonical Correlation Analysis · Coordinated representation
Multi-view deep networks

## 1 Introduction

Emotion recognition is important for communication, decision making, and human-machine interface. Since emotions are complex psycho-physiological phenomena associated with many nonverbal cues, it is difficult to build robust emotion recognition models using only one single modality. Signals from different modalities can represent different aspects of the emotions, and the complementary supplemental information from different modalities can be integrated to build a more robust emotional recognition model. Emotion recognition based on electroencephalography (EEG) and eye movement data have attracted increasing interest. Integrating different features with fusion technologies is important

© Springer Nature Switzerland AG 2018
L. Cheng et al. (Eds.): ICONIP 2018, LNCS 11305, pp. 221–231, 2018.
https://doi.org/10.1007/978-3-030-04221-9_20

to construct robust emotion recognition models [1]. The combination of signals from the central nervous system, EEG, and external behaviors, eye movement, has been a remarkable method for utilizing the complementarity of different modes of features [1–3].

In recent years, various deep neural networks have been introduced to affective computing and their attractive results showed the superior performance of such networks compared with the conventional shallow methods [13]. And various multimodal deep architectures have been proposed to leverage the advantages of two modalities, which can be concluded into two categories of representation: joint and coordinated [4]. The joint representation combines the unimodal signals into the same representation space, while the coordination representation processes the unimodal signals separately, enforces some similarity constraints on them, and brings them to the coordination space. Multimodal emotion recognition intends to distinguish emotions using different forms of physiological data collected at the same time, where complementary features of different modalities can be employed [2,3,12]. Deep neural networks have also been used for multimodal emotion recognition in an end-to-end method. Lu *et al.* used both EEG data and eye movement data to classify three kinds of emotions [3]. Liu *et al.* furthermore used Bimodal Deep AutoEncoder to extract high level representation features [5]. Tang *et al.* adopted the Bimodal-LSTM model to recognize multimodal emotions [6], and achieved better results than [5]. However, all the achievements above are based on joint representations and few coordinated based methods have been studied.

Coordinated representations first enforced similarity between representations. For example, the similarity models try to minimize distance between different modalities. With the rapid development of neural networks, they have shown the ability to reconstruct coordinated representations when learning jointly in an end-to-end manner [7]. What's more, structure coordinated space added more constrains between the modality representations [8]. Order-embedding is another example of a structured coordinated representation, which was proposed by Vendrov *et al.*, enforcing a dissimilarity metric and implementing the notion of partial order in the multimodal space [9]. Canonical correlation analysis (CCA) based structured coordinated is another case, where CCA computes the linear projection and maximizes the correlation between two modalities. CCA based models have been widely used for cross-modal retrieval and signal analysis. Kernel canonical correlation analysis (KCCA) uses reproducing kernel Hilbert spaces for projection but shows poor performance on large real-world datasets [10]. Deep canonical correlation analysis (DCCA) was introduced with deep network extension to optimize the correlation over the representations and showed better performance [11].

In this paper, we adopt DCCA to extract multimodal features for emotion recognition and achieved remarkable results. DCCA is a deep network based extension of canonical correlation analysis. It can learn separate representations nonlinearly for each modality, and coordinate them through a constraint. In this paper, we use deep networks to learn the nonlinear transformation of two views into a highly correlated space.

The main contributions of this paper are as follow:

(1) We first took coordinated representation of multimodal signals to recognize emotions, which means extracting more correlated high-level representations.
(2) We proposed a multi-view framework to deal with multimodal emotion recognition problem and achieved better classification accuracy than the state-of-the-art methods.

## 2  Deep Canonical Correlation Analysis

### 2.1  Background

Canonical correlation analysis can learn linear transformation of two vectors in order to maximize the correlation between them, which is widely used in economics, medical studies, and meteorology [21]. Lai *et al.* designed Colin's CCA, which performed Canonical Correlation Analysis with Artificial Neural Network [21]. With rapid development of deep learning, Andrew *et al.* proposed Deep Canonical Correlation Analysis (DCCA) with deep networks extension, which is a non-linear version of Canonical Correlation Analysis (CCA) that uses neural networks as the mapping functions [15]. DCCA calculates the representation of two views by multiplying them through stacked layers that are non-linearly transformed. Hossain *et al.* proposed a novel FS method based on Network of Canonical Correlation Analysis, NCCA, which is a robust method to acquisition noise and ignores mutual information computation based on Colin's CCA [21]. CCA is a standard statistical technique to find linear projections of two random vectors that are maximally correlated, while in Colin's CCA network [21], activation is fed forward from each input to the corresponding output through the respective weights to maximise the correlation. In Deep Canonical Correlation Analysis, deep networks are used for feature extraction with back propagation applied to maximise the correlation between two views.

### 2.2  Our Model

We use DCCA for feature transformation, fuse the features after extraction, and apply SVM as the classifier. The model is shown in Fig. 1, and the model contains three parts: non-linear feature transformation ($L_2$ and $L_3$ in Fig. 1), CCA calculation ($CCA$ layer in Fig. 1), and feature fusion and classification. EEG features and eye movement features are separated into two views denoted as $L_1$ in Fig. 1, and we set two views' input features as $X_1$, and $X_2$.

**Nonlinear Feature Transformation.** In the deep networks, for simplicity, we assume that each intermediate layer in the network for the first view has $c_1$ units, and the output layer has $o$ units, as shown in Fig. 1 as 'View 1'. Let $x_1 \in R^{n_1}$ be an instance of the first view, and the outputs of the first layer in the hidden layers for the instance $x_1$ are $h_1 = s(W_1^1 x_1 + b_1^1) \in R^{c_1}$, where $W_1^1 \in R^{c_1 \times n_1}$ is a matrix of weights, $b_1^1 \in R^{c_1}$ is a vector of biases, and $s(\cdot)$ is a non-linear function applied componentwise. The outputs $h_1$ can then be used to compute the outputs of the next one in hidden layers as $h_2 = s(W_2^1 x_1 + b_2^1) \in R^{c_1}$, and so on until the final representation in the hidden layers $f_1(x_1) = s(W_d^1 h_d - 1 + b_d^1) \in R^o$ is computed, for a network with $d$ layers.

Given an instance $x_2$ of the second view, as shown in Fig. 1 as 'View 2', the representation $f_2(x_2)$ is computed the same way, with different parameters $W_l^2$ and $b_l^2$ (and potentially different architectural parameters $c_2$), here $l$ is the number of layers in the View 2 network, and the total network function is defined as $f_1$ and $f_2$, from $L_1$ to $L_2$, for building two neural networks to transform features non-linearly, respectively. The layer sizes of both views are the same, including input layer $L_1$, hidden layers $L_2$, and output layer $L_3$ with each layer's nodes fully connected. The two views' output features are defined as $H_1$ and $H_2$, respectively. We use back propagation to update parameters of each view to acquire higher correlation in the CCA layer.

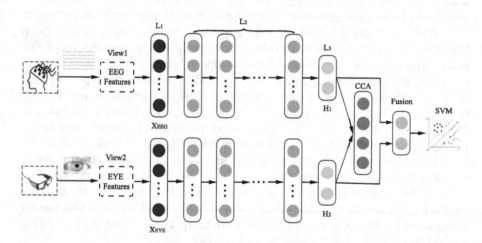

**Fig. 1.** Our Deep Canonical Correlation Analysis model, including deep networks (input layer, hidden layers and output layer), Canonical Correlation Analysis layer, and classifier SVM.

**CCA Calculation.** The goal is to jointly learn parameters for both views $W_l^i$ and $b_l^i$, where $i = \{1, 2\}$, such that $corr(f_1(X_1), f_2(X_2))$ is as high as possible. Let $\theta_1$ be the vector of all parameters $W_l^1$ and $b_l^1$ of the first view for $l = 1, \ldots, d$, where $d$ is the number of hidden layers, and similarly for $\theta_2$. The optimization function is:

$$(\theta_1^*, \theta_2^*) = \arg\max corr(H_1, H_2) = \arg\max_{\theta_1, \theta_2} corr(f_1(X_1; \theta_1), f_2(X_2; \theta_2)) \quad (1)$$

According to [15], the correlation of two views' transformed features ($H_1$ and $H_2$) can be calculated as follows:

$$corr(H_1, H_2) = corr(f_1(X_1), f_2(X_2)) = ||T||_{tr} = tr(T'T)^{1/2} \qquad (2)$$

where

$$T = \hat{\Sigma}_{11}^{-1/2} \hat{\Sigma}_{12} \hat{\Sigma}_{22}^{-1/2}$$

$$\hat{\Sigma}_{11} = \frac{1}{m-1} \overline{H}_1 \overline{H}_1' + r_1 I,$$

$$\hat{\Sigma}_{22} = \frac{1}{m-1} \overline{H}_2 \overline{H}_2' + r_2 I,$$

$$\hat{\Sigma}_{12} = \frac{1}{m-1} \overline{H}_1 \overline{H}_2'.$$

The $\overline{H}_1$ and $\overline{H}_2$ are the centered data matrixes:

$$\overline{H}_1 = H_1 - \frac{1}{m} H_1 1, \quad \overline{H}_2 = H_2 - \frac{1}{m} H_2 1 \qquad (3)$$

and $r_1, r_2$ are the regularization constants. To update the weighs of networks, we calculate the gradients. If the singular value decomposition of $T$ is $T = UDV'$, then

$$\frac{\partial corr(H_1, H_2)}{\partial H_1} = \frac{1}{m-1} (2\nabla_{11} \overline{H}_1 + \nabla_{12} \overline{H}_2), \qquad (4)$$

where

$$\nabla_{11} = -\frac{1}{2} \hat{\Sigma}_{11}^{-1/2} UDU' \hat{\Sigma}_{22}^{-1/2}, \quad \nabla_{12} = \hat{\Sigma}_{11}^{-1/2} UV' \hat{\Sigma}_{22}^{-1/2}.$$

**Feature Fusion and Classification.** We take weighted average of two views' extracted features. DCCA is used for feature extraction, and linear SVM is used as classifier to recognize emotions. The fusion function is defined as follows:

$$F_{fusion} = \alpha H_1 + \beta H_2 \qquad (5)$$

where $F_{fusion}$ is fusion features, $H_1$ and $H_2$ are extracted features of EEG and eye movement, respectively, and $\alpha$ and $\beta$ are the fusion weights. In our experiment, in order to balance the composition of features, we set $\alpha = \beta = 0.5$.

## 3   Experiment Settings

### 3.1   Dataset

We evaluate the performance of our approach on three real world datasets, the SEED[1] dataset, the SEED IV (See footnote 1) dataset, and the DEAP[2] dataset.

---

[1] http://bcmi.sjtu.edu.cn/~seed/.
[2] http://www.eecs.qmul.ac.uk/mmv/datasets/deap/.

- **SEED.** The SEED dataset contains EEG data with three emotions (happy, neutral, and sad) of 15 subjects, and all subjects' data were collected when they watching 15 four-minute-long emotional movie clips, where first 9 movie clips were used as training data and the rest were used as test data. The EEG signals were recorded with ESI NeuroScan System at a sampling rate of 1000 Hz with a 62-channel electrode cap. The eye movement signals were recorded with SMI ETG eye tracking glasses. To compare with the existing work, we used the same data, which contained 27 experiment results from 9 subjects.
- **SEED IV.** The SEED IV dataset contains EEG and eye movement features in total of four emotions (happy, sad, fear, and neutral) [16]. A total of 72 movie clips were used for the four emotions, and forty five experiments were taken by participants to evaluate their emotions while watching the movie clips with keywords of emotions and ratings out of ten points for two dimensions: valence and arousal. The valence scale ranges from sad to happy, and the arousal scale ranges from calm to excited.
- **DEAP.** The DEAP dataset contains EEG signals and other peripheral physiological signals from 32 subjects. These data were collected when participants were watching emotional music videos, which was one-minute-long each. We chose 5 as a threshold to divide the trials into two classes according to the rated levels of arousal and valence. We used 10-fold cross validation to compare our results with Liu *et al.* [5], Yin *et al.* [12], and Tang *et al.* [6] (Fig. 2).

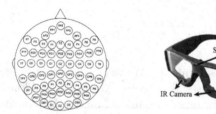

**Fig. 2.** The EEG electrode layout and SMI ETG eye tracking glasses.

## 3.2 Feature Extraction

For the SEED and SEED IV datasets, we extracted Differential Entropy (DE) features [19] from each EEG signal channel in five frequency bands: $\delta$ (1–4 Hz), $\theta$ (4–8 Hz), $\alpha$ (8–14 Hz), $\beta$ (14–31 Hz), and $\gamma$ (31–50 Hz). So at each time step, the dimension of EEG features is 310 (5 bands × 62 channels). As for eye movement features, we used the same features as Lu *et al.* 2015 [3], which were listed in Table 1. At each time step, there were 39 dimensions of pupil diameters in total, including both Power Spectral Density (PSD) and DE features. The extracted EEG features and eye movement features were scaled between 0 and 1.

For the DEAP dataset, because a 4–45 Hz bandpass frequency filter was applied during pre-processing, so we extracted DE features from EEG signals in four frequency bands: $\theta$ (4–8 Hz), $\alpha$ (8–14 Hz), $\beta$ (14–31 Hz), and $\gamma$ (31–45 Hz). Then in total, the dimension of extracted 32-channel EEG features is 128 (4 bands × 32 channels). As for peripheral physiological signals, six time-domain features were extracted to describe the signals in different perspective, including minimum value, maximum value, mean value, standard deviation, variance, and squared sum. So the dimension of peripheral physiological features is 48 (6 features × 8 channels).

**Table 1.** The details of the extracted eye movement features.

| Eye movements parameters | Extracted features |
|---|---|
| Pupil diameter (X and Y) | Mean, standard deviation, DE in four bands (0–0.2 Hz, 0.2–0.4 Hz, 0.4–0.6 Hz, 0.6–1 Hz) |
| Disperson (X and Y) | Mean, standard deviation |
| Fixation duration (ms) | Mean, standard deviation |
| Blink duration (ms) | Mean, standard deviation |
| Saccade | Mean, standard deviation of saccade duration (ms) and saccade amplitude |
| Event statistics | Blink frequency, fixation frequency, fixation dispersion total, fixation duration maximum, fixation dispersion maximum, saccade frequency, saccade duration average, saccade latency average, saccade amplitude average |

### 3.3 Parameter Details

In this paper, we build subject-specific models. We use grid search to find optimal hyperparameters, including learning rate, batch size, regulation parameters, and layer nodes. Taking several experiment results and time consuming into account, we choose learning rate as $1e^3$, batch size as 100, and regulation parameter as $1e^7$. The hidden units in our models are presented in Table 2.

**Table 2.** Layer's framework of different datasets in our experiments

| Dataset | Layers |
|---|---|
| SEED | $400 \pm 40, 200 \pm 20, 150 \pm 20, 120 \pm 10, 60 \pm 10, 20 \pm 2$ |
| SEEC IV | $400 \pm 40, 200 \pm 20, 150 \pm 20, 120 \pm 10, 90 \pm 10, 60 \pm 10, 20 \pm 2$ |
| DEAP | $1500 \pm 50, 750 \pm 50, 500 \pm 25, 375 \pm 25, 130 \pm 20, 65 \pm 20, 30 \pm 20$ |

## 4    Experimental Results

### 4.1    Results on Different Datasets

Table 3 shows the comparison results of different approaches on the SEED dataset, different feature extraction methods are listed in the first line and SVM is used as classifier for all methods. From Table 3, DE Feature fusion tested on SVM achieved higher classification accuracy and less std than the CCA method, which directly used CCA on EEG and eye movement features. BDAE used RBM pre-training to build a multimodal autoencoder model performed a better result of 93.19% [5]. Tang *et al.* used Bimodal-LSTM to make fusion by considering timing and classification layer parameters and achieved the state-of-the-art performance [6]. In our DCCA model, we extracted highly correlated features, bringing closer these high-level representations, and achieved better results with test classification accuracy of 94.58% and std of 6.16.

**Table 3.** Average accuracies (%) and standard deviation of different approaches for three emotions classification on the SEED dataset

|              | CCA   | DE features | BDAE [5] | Bimodal-LSTM [6] | DCCA  |
| ------------ | ----- | ----------- | -------- | ---------------- | ----- |
| Accuracy(%)  | 40.35 | 81.21       | 93.19    | 93.97            | **94.58** |
| Std          | 16.38 | 12.51       | 8.23     | 7.03             | **6.16**  |

Comparison results on the SEED IV dataset is shown in Table 4. We regard Zheng *et al.*'s deep learning results as our baseline [16]. We compare our DCCA model with different existing feature extraction methods. Table 4 presents that BDAE achieved better results than DE features. Compared with CCA based approach and other methods, we conclude that DCCA model coordinating high-level features achieves the best results.

**Table 4.** Average accuracies (%) and standard deviation of different approaches for four emotions classification on the SEED IV dataset

|               | CCA   | DE features | BDAE [16] | DCCA  |
| ------------- | ----- | ----------- | --------- | ----- |
| Accuracy (%)  | 49.56 | 75.88       | 85.11     | **87.45** |
| Std           | 19.24 | 16.14       | 11.79     | **9.23**  |

Tables 5 and 6 demonstrate comparison results of different feature extraction methods on the DEAP dataset, which are for two dichotomous classification and four categories classification, respectively, while SVM is used as classifier. For two dichotomous classification, Liu *et al.*'s multimodal autoencoder model achieved 2% higher than AutoEncoder. Yin *et al.* used an ensemble of deep classifiers, making higher-level abstractions of physiological features [12]. Then Tang *et al.* used Bimodal-LSTM and achieved the state-of-the-art accuracy for two dichotomous classification [6]. For four categories classification, Tripathi achieved accuracy of 81.41% [18]. As for our DCCA method, we learned high-level correlated features and achieved better results than the state-of-the-art method with mean test accuracies of 84.33% and 85.62% for arousal and valence classification and 88.51% for four categories classification.

**Table 5.** Comparison of average accuracies (%) of different approaches on the DEAP dataset for two dichotomous

|  | CCA | AutoEncoder [3] | Liu *et al.* [5] | Yin *et al.* [12] | Tang *et al.* [6] | DCCA |
|---|---|---|---|---|---|---|
| Arousal (%) | 61.25 | 74.49 | 80.5 | 84.18 | 83.23 | **84.33** |
| Valence (%) | 69.58 | 75.69 | 85.2 | 83.04 | 83.82 | **85.62** |

**Table 6.** Comparison of average accuracies (%) of different approaches on the DEAP dataset for four categories

| Method | CCA | KNN+RF [17] | Tripathi *et al.* [18] | DCCA |
|---|---|---|---|---|
| Accuracy (%) | 40.35 | 70.04 | 81.41 | **88.51** |

### 4.2 Discussion

The shortcoming of the existing feature-level fusion and multimodal deep learning methods is very difficult to relate the original features in one modality to features in other modality [14]. Moreover, the relations across various modalities are deep instead of shallow. In our DCCA model, we can learn coordinated representation from high-level features and make two views of features become more complementary, which in return improves the classification performance.

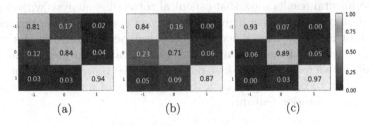

(a)          (b)          (c)

**Fig. 3.** Confusion matrices of DCCA outputs on the SEED dataset of single modality and feature fusion methods. Each row of the confusion matrices represents the target class and each column represents the predicted class. The element $(i, j)$ is the percentage of samples in class $i$ that is classified as class $j$. (a) EEG features; (b) Eye movement features; and (c) Fusion features.

Figure 3 shows the confusion matrices of SEED feature classification. The EEG features have classification accuracy of 0.86 while eye movement features' of 0.81, and the fusion feature has classification accuracy of 0.94. We can draw a conclusion from the confusion matrices that EEG features and eye movement features are complementary.

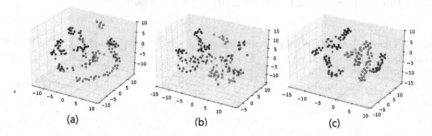

(a)    (b)    (c)

**Fig. 4.** t-SNE 3D visualization of extracted features on the SEED dataset, where blue for negative emotion, red for neutral emotion, and green for positive emotion. (a) EEG features; (b) Eye movement features; and (c) Fusion features. (Color figure online)

To find out the distribution of fusion features, we use t-SNE to make dimensionality reduction of the high-dimensional extracted features for visualization [20]. Figure 4 presents high-dimensional input features which are reduced to three dimensions for visualization. Comparing the EEG features, eye movement features, and fusion features, we can directly conclude that the fusion features are more reasonable and have better distribution than single-model of EEG and eye movement features, which are beneficial for classification.

## 5   Conclusion

In this paper, we have used Deep Canonical Correlation Analysis to extract highly correlated high-level features of two views on three real world datasets. The experimental results show that canonical correlation analysis with deep networks extension can achieve higher classification accuracy of emotion recognition when higher correlation is acquired. The deep networks with nodes' weights updated by back propagation can extract better features, which are more correlated of two views. Our work first put coordinated representation into multimodal emotion recognition and indicated a new way of multimodal representation in high-level fusion features.

**Acknowledgments.** This work was supported in part by the grants from the National Key Research and Development Program of China (Grant No. 2017YFB1002501), the National Natural Science Foundation of China (Grant No. 61673266), and the Fundamental Research Funds for the Central Universities.

# References

1. Soleymani, M., Pantic, M., Pun, T.: Multimodal emotion recognition in response to videos. IEEE Trans. Affect. Comput. **3**, 211–223 (2012)
2. Zheng, W.L., Dong, B.N., Lu, B.L.: Multimodal emotion recognition using EEG and eye tracking data. In: EMBS 2014, pp. 5040–5043 (2014)
3. Lu, Y., Zheng, W.L., Li, B., Lu, B.L.: Combining eye movements and EEG to enhance emotion recognition. In: IJCAI 2015, pp. 1170–1176 (2015)
4. Baltrusaitis, T., Ahuja C., Morency, L.P.: Multimodal machine learning: a survey and taxonomy. IEEE Trans. Pattern Anal. 1–20 (2018)
5. Liu, W., Zheng, W.-L., Lu, B.-L.: Emotion recognition using multimodal deep learning. In: Hirose, A., Ozawa, S., Doya, K., Ikeda, K., Lee, M., Liu, D. (eds.) ICONIP 2016. LNCS, vol. 9948, pp. 521–529. Springer, Cham (2016). https://doi.org/10.1007/978-3-319-46672-9_58
6. Tang, H., Liu, W., Zheng, W.L., Lu, B.L.: Multimodal emotion recognition using deep neural networks. In: Liu, D., Xie, S., Li, Y., Zhao, D., El-Alfy, E.S. (eds.) ICONIP 2017. LNCS, vol. 10637, pp. 811–819. Springer, Cham (2017). https://doi.org/10.1007/978-3-319-70093-9_86
7. Frome, A., et al.: Devise: a deep visual-semantic embedding model. In: NIPS 2013, pp. 2121–2129 (2013)
8. Bronstein, M., Michel, F., Paragios, N.: Data fusion through cross-modality metric learning using similarity-sensitive hashing. In: CVPR 2010, pp. 3594–3601 (2010)
9. Zhang, H., Hu, Z., Deng, Y., Sachan, M., Yan, Z., Xing, E.P.: Learning concept taxonomies from multimodal data. In: ACL 2016, pp. 1791–1801 (2016)
10. Lai, P.L., Fyfe, C.: Kernel and nonlinear canonical correlation analysis. Int. J. Neural Syst. **10**, 365–377 (2000)
11. Sohn, K., Lee, H., Yan, X.: Learning structured output representation using deep conditional generative models. In: NIPS 2015, pp. 3483–3491 (2015)
12. Yin, Z., Zhao, M., Wang, Y., Yang, J., Zhang, J.: Recognition of emotions using multimodal physiological signals and an ensemble deep learning model. Comput. Methods Progr. Biomed. **140**, 93–110 (2017)
13. Zheng, W.L., Zhu, J.Y., Peng, Y., Lu, B.L.: EEG-based emotion classification using deep belief networks. In: IEEE ICME 2014, pp. 1–6 (2014)
14. Ngiam, J., Khosla, A., Kim, M., Nam, J., Lee, H., Ng, A.Y.: Multimodal deep learning. In: ICML 2011, pp. 689–696 (2011)
15. Andrew, G., Arora R., Bilmes, J.A., Livescu, K.: Deep canonical correlation analysis. In: ICML 2013, pp. 1247–1255 (2013)
16. Zheng, W.L., Liu, W., Lu, Y., Lu, B.L., Cichocki, A.: EmotionMeter: a multimodal framework for recognizing human emotions. IEEE Trans. Cybern. **99**, 1–13 (2018)
17. Chen, J., Hu, B., Wang, Y., Dai, Y., Ya, Y., Zhao, S.: A three-stage decision framework for multi-subject emotion recognition using physiological signals. In: IEEE BIBM 2016, pp. 470–474 (2016)
18. Tripathi, S., Acharya, S., Sharma, R.D., Mittal, S., Bhattacharya, S.: Using deep and convolutional neural networks for accurate emotion classification on DEAP dataset. In: IAAI 2017 (2017)
19. Duan, R.N., Zhu, J.Y., Lu, B.L.: Differential entropy feature for EEG-based emotion classification. In: IEEE NER 2013, pp. 81–84 (2013)
20. Maaten, L., Hinton, G.E., Bengio, Y.: Visualizing data using t-SNE. J. Mach. Learn. Res. **9**, 2579–2605 (2008)
21. Hossain, M.Z., Kabir, M.M., Shahjahan, M.: A robust feature selection system with colins CCA network. Neurocomputing **173**, 855–863 (2016)

# Neural Machine Translation for Financial Listing Documents

Linkai Luo[1]([✉]), Haiqin Yang[1,2], Sai Cheong Siu[1], and Francis Yuk Lun Chin[1,3]

[1] Deep Learning Research and Application Centre, Hang Seng Management College,
Siu Lek Yuen, Hong Kong
llk1896@gmail.com, hqyang@ieee.org, scsiu@hsmc.edu.hk, chin@cs.hku.hk
[2] MTdata, Meitu, Xiamen, China
[3] Department of Computer Science, The University of Hong Kong,
Lung Fu Shan, Hong Kong

**Abstract.** In this paper, we focus on developing a Neural Machine Translation (NMT) system on English-to-Traditional-Chinese translation for financial prospectuses of companies which seek listing on the Hong Kong Stock Exchange. To the best of our knowledge, this is the first work on NMT for this specific domain. We propose a domain-specific NMT system by introducing a domain flag to indicate the target-side domain. By training the NMT model on the data from both the IPO corpus and the general domain corpus, we can expand the vocabulary while capturing the common writing styles and sentence structures. Our experimental results show that the proposed NMT system can achieve a significant improvement on translating the IPO documents. More significantly, through a blind assessment by a translator expert, our system outperforms two mainstream commercial tools, the Google translator and SDL Trado for some IPO documents.

**Keywords:** Neural machine translation · Financial listing documents
Domain flag

## 1 Introduction

Recently, neural machine translation (NMT) based on deep neural network architecture [1–4] has demonstrated its superior translation performance than traditional statistical machine translation methods such as phrase-based machine translation (PBMT) [5]. However, it is still awkward for specific domains because the sentences in each domain may contain individual writing styles, sentence structures, and terminology [6].

In this paper, we focus on developing a Neural Machine Translation (NMT) system for translating financial documents of companies which seek listing, i.e., initial public offering (IPO), on the Hong Kong Stock Exchange (HKSE). In this specific domain, the documents need to provide in two languages, English and Traditional Chinese. Its goal is to disclose business and financial information

© Springer Nature Switzerland AG 2018
L. Cheng et al. (Eds.): ICONIP 2018, LNCS 11305, pp. 232–243, 2018.
https://doi.org/10.1007/978-3-030-04221-9_21

**Table 1.** A translation example. *src*: an English source sentence; *ref*: a reference sentence translated by human; *nmt*: translated by our developed NMT system trained on data from the IPO corpus and the UN corpus without inserting the domain flag. The NMT result show that it can translate the source sentence meaningfully, but the style is bias to the UN corpus.

| | | | | | | |
|---|---|---|---|---|---|---|
| *src:* | *Certain* | *facts and statistics in this prospectus* | *may not* | *be reliable and accurate.* | | |
| *ref:* | 本招股章程內之 | 若干 | 事實及統計資料 | 未必 | 可靠及準確。 | |
| *nmt:* | 本招股章程中的 | 某些 | 事實和統計數據 | 可能不 | 可靠準確。 | |

about a financial security to potential buyers. Other than the financial listing documents, the listed public company is also required to issue interim and annual reports periodically or make disclosure announcements occasionally. All the documents are required to be written in both English and Traditional Chinese. Here, we start from the English-to-Traditional-Chinese translation. In the following, we interchangeably use source and En to denote English, target and Zh to denote Chinese, and En → Zh to denote English-Chinese translation. Without specified indication, Chinese refers to Traditional Chinese.

One significant problem in this domain is that more and more listing companies come from the "new economics". The previously paired translated prospectuses are not sufficient for translating the general content such as the industry overview, company history, and company business. To resolve this issue, we decide to include more bilingual resources because they may provide supplement information and demonstrate its effectiveness than using only the monolingual resource [6,7]. More specifically, we collect the United Nation (UN) parallel corpus because it consists of rich English and Simplified Chinese parallel official records and parliament documents. Though this dataset is not well suited for the task of En → Zh in the IPO domain, due to the scarcity of high-quality parallel corpus, we have to utilize them and omit the difference between Traditional Chinese and Simplified Chinese. We therefore define the IPO domain as the primary domain, namely the P domain, and the UN corpus as the secondary domain, referred as the S domain.

After incorporating data from two domains, one predominant problem is that we need to differentiate the writing style in the IPO corpus and the general corpus because data in each domain may be writing in a certain style with different sentence structures. For example, the word "we", normally translated as " 我們", is translated as " 本公司" (meaning "this Company") in the documents. Other cases include "you" (translated as " 閣下" instead of " 你" for the sake of a more respectful tone), "if" (" 倘若" instead of " 如果", a more formal tone) and "in other words" (" 換言之" instead of " 換句話說", an ancient Chinese style). Moreover, it is observed that without any appropriate processing, the writing style of translation will be bias toward one type when data from two domains are imbalanced; see Table 1 for an example.

To tackle this issue, we introduce a **domain flag** to indicate which domain the source sentence should be translated to. This setting is inspired by Google's multilingual translation system [8] due to its simplicity without modifying the NMT architecture. Our later experimental results show that this setting can capture the domain information and improve the quality of translation without bias toward the S domain.

This rest paper is organized as follows: in Sect. 2, we depict the related work. In Sect. 3, we present the proposed NMT architecture. In Sect. 4, we detail the experimental setup and results. Finally, we conclude the whole paper in Sect. 5.

## 2    Related Work

In the following, we first review several pieces of related work based on domain-specific Statistical Machine Translation (SMT) by including more domain features or domain data. In [9], a general translation engine with a phrase-based log-linear model is proposed to combine the domain dependent features and language models. The phrase-based SMT system and a hybrid MT is presented to achieve improvement with a combination of a small in-domain bilingual corpus and a larger out-of-domain corpus. More monolingual corpora are also included to improve the performance of domain-specific SMT systems [7,10].

Later, NMT attains significant improvement on translation than traditional SMT systems. For example, Google announces a deep learning powered multilingual translation system [8], based solely on a single NMT model to translate among corpora in multiple languages. The underlying idea is to introduce an artificial token to indicate the target language the model should be translated to. This work motivates several recent work [11,12] and a detailed analysis in [6]. In [6], two types of methods are test, one inserting a domain token at the end of each source sentence and the other expanding the word embedding with a domain embedding. Though both methods work equally, adding an extra domain flag is simpler and inspires us to adopt it in our design.

## 3    The NMT System

### 3.1    Neural Machine Translation

Recently, deep learning architecture has been actively developed to solve various real-world applications, such as text mining and knowledge tracing [13–15]. State-of-the-art NMT systems are generally trained a sequence-to-sequence (S2S) model based on an encoder-decoder and an attention mechanism to learn the mapping from a set of paired sentences $\{(\mathbf{x}^{(n)}, \mathbf{y}^{(n)})\}_{n=1}^{N}$, where a source sentence is $\mathbf{x}^{(n)} = (x_1, \ldots, x_{S_n})$ and a target sequence is $\mathbf{y}^{(n)} = (y_1, \ldots, y_{T_n})$. More concretely, it is applied a neural network to parameterize the conditional probability, $p(\mathbf{y}|\mathbf{x})$, governed by $\theta$. The parameter $\theta$ is learned by maximizing the log-likelihood on the parallel training set:

$$\mathcal{L}(\theta) = \sum_{n=1}^{N} \sum_{t=1}^{T_n} \log p(y_t^{(n)} | \mathbf{y}_{<t}^{(n)}, \mathbf{x}^{(n)}; \theta) \qquad (1)$$

Several new neural network architectures have been proposed to build the S2S model, including bidirectional RNNs [1,2,16], stacked convolutional S2S (ConvS2S) [3] and the Transformer [4]. The Transformer is built solely on self-attention, which eliminates the recurrence processing and positional dependency in the encoder. It has been proven to yield a remarkable improvement over RNNs as well as significant time reduction for training. Hence, we choose Transformer as the base translation architecture, which is shown in Fig. 1.

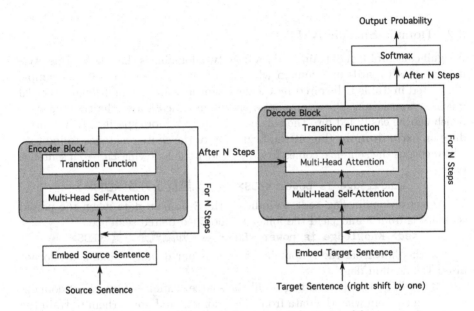

**Fig. 1.** Transformer architecture described in [4].

The Transformer performs a small, constant number of steps (chosen empirically) in combining the information. In each step, it applies a self-attention mechanism which directly models relationships between all words in a sentence, regardless of their respective position [4]. In our setting, both encoder and decoder are composed of $N$ (=6) stacked identical block, where the multi-head attention is calculated by:

$$\text{Attention}\,(Q, K, V) = \text{Softmax}\left(\frac{QK^T}{\sqrt{d_k}}\right) V, \tag{2}$$

$$\text{MultiHead}\,(Q, K, V) = \text{Concat}(\text{head}_1, \ldots, \text{head}_h)\, W^O, \tag{3}$$

$$\text{head}_i = \text{Attention}(QW_i^Q, KW_i^K, VW_i^V), \tag{4}$$

where $(Q, K, V)$ is a (query, key, value)-triple embedded the hidden units; $d_k$ is the dimension of the key $K$, $h$ is the number of heads, and $W^O, W_i^Q, W_i^K, W_i^V$ are weight matrices to be estimated.

The transition function is a fully-connected neural network which consists of a single rectified-linear activation function:

$$\text{FFN}(x) = \max(0, xW_1 + b_1)W_2 + b_2, \tag{5}$$

where $x$ is the input from a hidden layer, $W_1$, $b_1$, $W_2$, and $b_2$ are weights and biases to be estimated. Moreover, layer normalization, dropout and residual networks are also applied to enhance the model performance. More details can be referred to [4].

### 3.2 Domain-Specific NMT

A significant problem of training data from two domains is that the writing style may be bias towards one domain when it contains more data, see an example illustrated in Table 1. Inspired by the zero-shot learning for multilingual neural machine translation [8], we propose a simple modification to the training data which can be easily fed into the single NMT model. More specifically, we introduce an extra **domain flag** at the beginning of the source sentence to indicate which corpus the sentence comes from. Consider the following En → Zh sentence pair:

Knowledge is power <EOS> →    知識就是力量 <EOS>

We set a domain flag <sd> to indicate the S domain data when the paired sentence is from S and yield the following modified paired sentence:

<sd> Knowledge is power <EOS> →    知識就是力量 <EOS>

For the data from the P domain, since it is our default domain, we do not insert the domain flag.

After adding the domain flag to all the source sentences from the S domain, we then mix them with the data from the P domain and apply them to train the Transformer. After training, we sequentially output the word from the vocabulary $V_T$ with the most conditional probability on the given input sentence $\mathbf{x}$ from the test set and the produced output words:

$$y_t = \arg \max_{V_T} p(\mathbf{y}|\mathbf{y}_{<t}, \mathbf{x}). \tag{6}$$

Moreover, we apply a left-to-right beam search decoder to search for the most likely translations containing $B$ hypotheses. That is, at each time step we extend each partial hypothesis in the beam with every possible word in the vocabulary. Finally, the decoder will yield $B$ candidate sentences and the one with the highest probability will be chosen as the final translation output.

## 4    Experiments and Results

### 4.1    Data Pre-processing and Experimental Setup

The dataset consists of two corpus: (1) the IPO corpus (the P data) crawled from the Hong Kong Exchanges and Clearing Market Website[1] during June 2007 and

---

[1] https://www.hkex.com.hk/.

February 2017, totally 956 PDF files in both English and Chinese, and (2) the UN corpus (the S data) obtained from the training set of [17]. Moreover, we conduct the following steps to filter and annotate the raw data:

- In the UN corpus, the Simplified Chinese sentences are converted to the Traditional Chinese ones using OpenCC[2] and segmented by Jieba[3].
- All the English sentences are tokenized by the Moses script[4] and converted to lowercase.
- All sentences are restricted to 5 to 80 words and eliminated illegal symbols.
- Named entities such as person names, organizations and locations are identified with reference to Stanford CoreNLP [18] while regular expressions are conducted to determine other named entities such as date and number.
- wordpieces[5] is adopted to build a fixed vocabulary with 32K sub-word units for both languages.
- An artificial token <sd> is added at the beginning of each English sentence in the UN corpus if needed.

Finally, the IPO corpus yields 1,979,674 parallel sentences and the UN corpus consists of 13.3 million sentence pairs. Hence, in the training set, 15% of the data comes from the IPO corpus while 85% of the data coming from the UN corpus. To test the effect of bilingual domains, we select two test sets to evaluate the model performance:

- IPOtestset: consisting of 15K paired sentences from 6 company prospectuses released between March to September 2017 at HKSE. The selected companies are new to the market while covering quite diverse industry sectors, ranging from the publishing sector to the food sector.
- UNtestset: the test set for the WMT 2017 English-Chinese challenge, which consists of 4K paired sentences after conducting the Chinese conversion.

In the comparison, we train the proposed NMT system on three different training sets and name them as: (1) $NMT_P$: trained on only the P data; (2) $NMT_{PUS}$: trained on both the P data and the S data; and (3) $NMT_{PUS_+}$: trained on both domain data and an additional domain flag added to the data in S.

The NMT system is implemented by the library, Tensor2Tensor[6]. Without further specification, we adopt the default hyper-parameters defined in Transformer-big settings [4], i.e., a 6-layer transformer with the size of the hidden units being 2014, a feed forward network with the size being 4096, and 8-head attentions. During training, the batch size is set to 2048 and checkpoints are saved every one hour. Each model is trained on 2 Tesla P100 GPUs for 500K steps, or equivalently 4.5 days nonstop running under the above configurations.

---

[2] https://github.com/BYVoid/OpenCC.

[3] https://github.com/fxsjy/jieba.

[4] https://github.com/moses-smt/mosesdecoder/blob/master/scripts/tokenizer/tokenizer.perl.

[5] https://github.com/google/sentencepiece.

[6] https://github.com/tensorflow/tensor2tensor.

The final model is an ensemble model averaged on the parameters for the last 20 checkpoints. When decoding, the beam size is set to 4 and the length penalty is 1.0. In the test, the BLEU score [19], a standard metric, is applied to measure the translation quality for different systems.

**Table 2.** BLEU scores for the NMT system trained on different datasets. The baseline is the best result reported in [17], which is trained on S and an additional news dataset without conducting Chinese conversion. The marker * and † indicate that the domain flag is inserted or not inserted in the test data.

| System | IPOtestset | UNtestset |
|--------|-----------|-----------|
| Baseline | - | **37.68** |
| $NMT_P$ | 45.71 | 5.47 |
| $NMT_{PUS}$ | 46.40 | 32.34 |
| $NMT_{PUS_+}$ | **47.55** | 34.62*(22.49†) |

### 4.2  Experimental Results

Table 2 reports the experimental results after handling the name entities in the translation. From the results, we have the following observations:

- By examining the results on IPOtestset, we can see that $NMT_{PUS}$ and $NMT_{PUS_+}$ achieve 0.69 BLEU score and 1.84 BLEU score improvement over $NMT_P$. The results show that including more data indeed can improve the BLEU score on the IPO translation. By introducing the domain flag, the translation quality can be further improved by 1.15 BLEU score.
- On UNtestset, we observe that $NMT_P$ yields very poor translation quality, i.e., attaining only 5.47 BLEU score. This result makes sense because the paired sentences in the training set and the test set follow totally different writing styles. It can observe that $NMT_{PUS}$ attains significant improvement, +26.87 BLEU score, over $NMT_P$. The result again confirms that including more related data will help to train a model with better generalization. More-over, $NMT_{PUS_+}$ achieves a higher BLEU score (+1.15) than $NMT_{PUS}$ after inserting the domain token. However, the perform deteriorates, decreasing from 34.62 to 22.49, when the data is from UNtestset, but the model deems it as the IPO corpus. It is therefore evident that the flag effectively plays the role of specifying the information of which target domain the system being translated to.
- One exceptional case is that $NMT_{PUS}$ and $NMT_{PUS_+}$ attains relatively poor performance, up to −5.34 and −3.06 drops, respectively, comparing to the baseline result in [17]. We conjecture that two reasons may trigger the results. First, the best baseline result is trained in both the UN corpus and an addi-tional news dataset. The training set is larger and more relevant. Second, the

best baseline result is evaluated on the English-to-Simplified-Chinese task, while ours is on the English-to-Traditional-Chinese task. The additional Chinese conversion may also decay the performance a little bit.

- In sum, our proposed NMT system can attain the state-of-the-art performance.

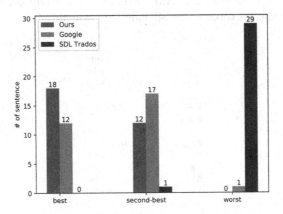

**Fig. 2.** Human evaluation results.

### 4.3   Human Evaluation Results

Though BLEU is widely used to evaluate the translation quality of MT systems, it is understood that BLEU can be misleading in its indications when it is used to compare some kinds of MT systems. Hence, we invite a human expert with over ten years' research and teaching experience on English-Traditional Chinese translation to conduct a blind test. The evaluation follows the rest steps:

1. Thirty English sentences are randomly selected from IPOtestset due to the limitation by the workload of the expert.
2. The English-Traditional Chinese translation is conducted on the selected sentences by three NMT systems: our $NMT_{PUS_+}$, Google Translator[7] and SDL Trados, where SDL Trados is a commercial translation tool developed by SDL Plc[8] and widely used by printing companies.
3. The translated results are permuted randomly and presented to the human expert.
4. The human expert then scores the translation quality of each English sentence according to the perceived quality.

---

[7]  https://translate.google.com.
[8]  https://www.freetranslation.com.

Figure 2 summarizes the evaluation result. Among the thirty sentences, our NMT system ranks the best translation quality for 18 sentences and the second best for 12 sentences while the Google translator rank the best translation quality for 17 sentences and the second best for 17 sentences. Trados achieves the poorest translation results, only one sentence attaining the second best translation quality and the rest 29 sentences yielding the worst translation quality. The evaluation results show that our NMT system successfully beats two popular commercial translation tools in the specific IPO domain.

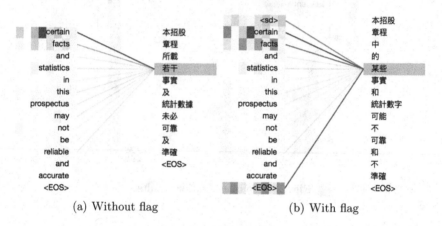

(a) Without flag                    (b) With flag

**Fig. 3.** An example visualizes the attention results on the effect of the domain flag.

## 4.4  Case Study

We first investigate a case to show the effect of the domain flag in word dependency. We adopt the encoder-decoder attention visualizer in Tensor2Tensor to illustrate the results. Because the second to the last layer, i.e., the 5th layer, of the decoder contains the most informative information of the encoder-decoder dependencies, we unveil the results in that layer.

Figure 3 shows the dependencies of a specific target word and the input source translated without and with the domain flag, respectively. It is noted that the colored blocks in the source word correspond to the weights of different attention heads. Hence, totally eight colored blocks are placed for each source word. In each block, the darker the color, the more attention the source word is paid to the target word. The lines connecting the paired words emphasize the largest attentions among the eight heads and apply the same color as the block with the largest weight.

In this case, without setting the domain flag, see Fig. 3(a), the source sentence is deemed as a sentence from the IPO domain, the target word " 若干" attends mainly to the source word "certain". After inserting the token, <sd>, see Fig. 3(b), the source sentence is deemed as one from the UN domain. The corresponding word " 若干" is then translated to " 某些". Without the domain flag, " 若干" is mainly connected to the source "certain" and weakly related to "facts". But after adding the domain flag, the translated word " 某些" not only establishes connection to the source word "certain" and "facts", but also connects significantly to other words such as the end-of-sentence token <EOS> and the domain flag <sd>. It seems that the flag not only rephrases the sentence in terms of the writing style, but also attends to nearly every target word during decoding.

We also provide one more example to show the advantage of our developed NMT system over a general translation system in Table 3. Obviously, our system can capture the IPO style and translate the phrase "The Group" in a more respectful tone to " 貴集團" rather than " 本集團" by the Google translator and SDL Trados. The word "strong" is translated to " 穩健", literally meaning "firm" or "steady" in a more modest tone for official documents, by our system. This is much better than a general translation " 強大" by the Google translator and SDL Trados, which literally means "powerful".

**Table 3.** An example of translated sentences from a prospectus. *src:* an English source sentence; *ref:* a reference translation by human; *ours:* translated by our developed NMT system; *Google:* translated by the Google translator; and *SDL Trados:* translated by the SDL Trados. The correspondingly emphasized word is highlighted with the same color.

| | |
|---|---|
| *src:* | *The Group 's policy is to maintain a* strong *capital base to maintain investor, creditor and market confidence and to sustain future developments of the business.* |
| *ref:* | 貴集團 的政策是保持 穩健 的資本基礎，從而維持投資者、債權人及市場信心，並維持業務的日後發展。 |
| *ours:* | 貴集團 的政策為維持 穩健 的資本基礎，以維持投資者，債權人及市場信心，並維持業務的未來發展。 |
| *Google:* | 本集團 的政策是維持 強大 的資本基礎，維持投資者，債權人及市場信心，並維持業務未來發展。 |
| *SDL Trados:* | 本集團 的政策是維持 強大 的資本基礎以維持投資者、債權人及市場信心和維持未來發展業務。 |

## 5   Conclusion

We present the first investigation of neural machine translation on English-Traditional-Chinese translation for financial listing documents. We have tackled two difficulties of NMT in this specific domain, i.e., lack of sufficient general domain data and being hard to differentiate the writing styles of multiple domains. Our solution by including more bilingual resources, i.e., the United Nation parallel corpus, and introducing the domain flag in the training demonstrate that the included secondary domain data expand the vocabulary and indeed can provide more supplement information while the simple setting of the domain flag can effectively capture the domain information such as the writing style and select favorite words from synonymous words based on the domain information. Though a blind test, our developed NMT system also outperforms two popular commercial translators, the Google translator and the SDL Trados, in this specific IPO domain.

There are still several challenging but promising directions to improve the translation quality in this domain. First, it is easy to face rare words in the translation. How to design effective mechanisms to tackle this problem is worthy of investigation for deploying the system in real-world. Second, it is worthwhile to explore how to simplify the model architecture which can be trained and decoded faster. Third, it is valuable to scale up our system, especially translating well not only in a specific domain, but also in general domains.

**Acknowledgments.** The work described in this paper was partially supported by the Research Grants Council of the Hong Kong Special Administrative Region, China (Project No. UGC/IDS14/16).

# References

1. Bahdanau, D., Cho, K., Bengio, Y.: Neural machine translation by jointly learning to align and translate. CoRR abs/1409.0473 (2014)
2. Wu, Y., et al.: Google's neural machine translation system: bridging the gap between human and machine translation. CoRR abs/1609.08144 (2016)
3. Gehring, J., Auli, M., Grangier, D., Yarats, D., Dauphin, Y.N.: Convolutional sequence to sequence learning. In: ICML, pp. 1243–1252 (2017)
4. Vaswani, A., et al.: Attention is all you need. In: Guyon, I., et al. (eds.) NIPS, pp. 6000–6010 (2017)
5. Koehn, P., Och, F.J., Marcu, D.: Statistical phrase-based translation. In: Hearst, M.A., Ostendorf, M. (eds.) HLT-NAACL. The Association for Computational Linguistics (2003)
6. Kobus, C., Crego, J.M., Senellart, J.: Domain control for neural machine translation. In: RANLP, pp. 372–378 (2017)
7. Bertoldi, N., Federico, M.: Domain adaptation for statistical machine translation with monolingual resources. In: WMT@EACL, pp. 182–189 (2009)
8. Johnson, M., et al.: Google's multi-lingual neural machine translation system: enabling zero-shot translation. TACL 5, 339–351 (2017)
9. Stajner, S., Querido, A., Rendeiro, N., Rodrigues, J.A., Branco, A.: Use of domain-specific language resources in machine translation. In: LREC (2016)
10. Wu, H., Wang, H., Zong, C.: Domain adaptation for statistical machine translation with domain dictionary and monolingual corpora. In: COLING, pp. 993–1000 (2008)
11. Tiedemann, J.: Emerging language spaces learned from massively multilingual corpora. CoRR abs/1802.00273 (2018)

12. Chu, C., Dabre, R., Kurohashi, S.: An empirical comparison of simple domain adaptation methods for neural machine translation. CoRR abs/1701.03214 (2017)
13. Hu, Z., Zhang, Z., Yang, H., Chen, Q., Zhu, R., Zuo, D.: Predicting the quality of online health expert question answering services with temporal features in a deep learning framework. Neurocomputing **275**, 2769–2782 (2018)
14. Yang, H., Cheung, L.P.: Implicit heterogeneous features embedding in deep knowledge tracing. Cognit. Comput. **10**(1), 314 (2018)
15. Cheung, L.P., Yang, H.: Heterogeneous features integration in deep knowledge tracing. In: Neural Information Processing - 24th International Conference, ICONIP 2017, Guangzhou, China, 14–18 November 2017, Proceedings, Part II, pp. 653–662 (2017)
16. Britz, D., Goldie, A., Luong, M., Le, Q.V.: Massive exploration of neural machine translation architectures. CoRR abs/1703.03906 (2017)
17. Ziemski, M., Junczys-Dowmunt, M., Pouliquen, B.: The united nations parallel corpus v1.0. In: LREC (2016)
18. Manning, C.D., Surdeanu, M., Bauer, J., Finkel, J.R., Bethard, S., McClosky, D.: The stanford CoreNLP natural language processing toolkit. In: ACL, pp. 55–60 (2014)
19. Papineni, K., Roukos, S., Ward, T., Zhu, W.: Bleu: a method for automatic evaluation of machine translation. In: ACL, pp. 311–318 (2002)

# Unfamiliar Dynamic Hand Gestures Recognition Based on Zero-Shot Learning

Jinting Wu[1,2(✉)], Kang Li[1,2], Xiaoguang Zhao[1,2], and Min Tan[1,2]

[1] The State Key Laboratory of Management and Control for Complex System, Institute of Automation, Chinese Academy of Sciences, Beijing, China
[2] University of Chinese Academy of Sciences, Beijing, China
{wujinting2016,likang2014,xiaoguang.zhao,min.tan}@ia.ac.cn

**Abstract.** Most existing robots can recognize trained hand gestures to interpret user's intent, while untrained dynamic hand gestures are hard to be understood correctly. This paper presents a dynamic hand gesture recognition approach based on Zero-Shot Learning (ZSL), which can recognize untrained hand gestures and predict user's intention. To this end, we utilize a Bidirectional Long-Short-Term Memory (BLSTM) network to extract hand gesture feature from skeletal joint data collected by Leap Motion Controller (LMC). Specifically, this data is used to construct a novel dynamic hand gesture dataset for human-robot interaction application. Twenty common hand gestures are included and fifteen concrete semantic attributes are condensed. Based on these features and semantic attributes, a Semantic Autoencoder (SAE) is employed to learn a mapping from feature space to semantic space. By matching the most similar semantic information, the unfamiliar hand gestures are recognized as correct as possible. Experimental results on our dataset indicate that the proposed approach can effectively identify unfamiliar hand gestures.

**Keywords:** Dynamic hand gesture recognition
Bidirectional Long-Short-Term Memory (BLSTM)
Zero-Shot Learning (ZSL) · Semantic Autoencoder (SAE)
Leap Motion Controller (LMC)

## 1 Introduction

Recently, hand gesture recognition has been widely applied in various fields, such as medical technology [14,24], sign language recognition [3,8], virtual reality and human-robot interaction [2,9,17]. Specially, in human-robot interaction, hand gesture, as one of the most intuitive and efficient interactive interfaces, can help a person with speech barrier communicate with the robot [17], remote control the robot [9] and express intention [2]. Therefore, a robot with the capability of recognizing hand gestures becomes more practical and valuable.

However, a limitation of existing hand gesture recognition algorithms is that they need to learn gestures from large amounts of labeled image data. Thus, they

© Springer Nature Switzerland AG 2018
L. Cheng et al. (Eds.): ICONIP 2018, LNCS 11305, pp. 244–254, 2018.
https://doi.org/10.1007/978-3-030-04221-9_22

can only classify familiar hand gestures based on training dataset. In real human-robot interaction, the robot may encounter some unfamiliar hand gestures. Under these circumstances, the robot needs to have the ability to guess what meanings the unseen hand gestures convey. Fortunately, ZSL methods provide a solution for identifying unseen categories.

ZSL relies on a labeled training set of seen classes and the semantic relationship between the seen and unseen classes. Seen and unseen classes are usually related in a semantic embedding space. The semantic relationships between classes can be measured by a distance in this space. In the recognition task, the class label of a test sample is assigned to the nearest unseen class prototype in the semantic space. Currently, ZSL is mainly applied in 2D image recognition for object classification [5, 10, 11, 23]. The studies about hand gesture recognition [18] are very rare.

In this paper, we present a novel unfamiliar dynamic hand gesture recognition method based on ZSL. First, hand and finger skeletal joint data is collected by a LMC and used to build a novel hand gesture dataset involving twenty common hand gestures and fifteen concrete semantic attributes. Then, we utilize a BLSTM network [7] to extract hand gesture features and employ SAE [10] to analyze the semantic information. By matching the predicted semantic representation and the semantic prototypes, the unfamiliar gestures can be inferred. Finally, experimental results on the novel hand gesture dataset demonstrate effectiveness of the proposed method.

The remainder of the paper is organized as follows. Related work on hand gesture recognition is reviewed in Sect. 2. Afterwards, the unfamiliar hand gesture recognition approach is elaborated in Sect. 3. Section 4 provides the experimental results and analysis. Finally, conclusions and future work are summarized in Sect. 5.

## 2  Related Work

Traditional work on hand gesture recognition mostly uses information captured by data gloves [4] or 2D digital cameras [22]. However, the data gloves are not user-friendly, and the 2D images include limited information for dynamic hand gesture recognition. In recent years, depth sensors, such as the Leap Motion controller (LMC) [6, 16, 19] and Microsoft Kinect [12, 20, 25], have widely used in hand gesture recognition because they can contribute rich 3D information to enhance the accuracy. Especially, LMC is cheaper and more portable, and has higher localization precision (which is about 0.2 mm [26]). Abundant 3D hand data, such as palm positions, hand directions and skeletal joint positions, can be easily collected by a LMC without extra computational work. For instance, Lu et al. used palm direction, palm normal and fingertip data captured from a LMC to recognize dynamic hand gestures, and reached an accuracy of 95.0% for the Handicraft-Gesture dataset [16]. Chen et al. utilized a LMC to acquire the motion trajectory of 36 hand gestures, and the accuracy of SVM approach is 98.24% [6].

In addition to determining the source of the data, the recognition method is also important for hand gesture recognition. Wang et al. utilized a Hidden Markov Model (HMM) to estimate motion trajectory of hand gesture in a service robot system [9]. Lu et al. first recognized dynamic hand gestures using Hidden Conditional Random Field (HCRF) which was only applied in speech recognition [16]. Tang et al. employed Deep Neural Networks (DNNs) to extract robust features and precisely recognize hand postures [25]. All of the aforementioned hand gesture recognition methods have a common drawback that they cannot identify unfamiliar hand gestures. How to do it? ZSL algorithms make it possible.

Since Lampert et al. first proposed the attribute-based classification approach, which was used to identify new classes based on attribute representation [11], a large amount of ZSL models have been proposed one after another to improve the performance of unseen class recognition. Paredes et al. proposed a ZSL approach which adopted two linear layers to construct relationships between features, attributes and categories [23]. However, that method left a large of dimensions of the semantic space unconstrained. To solve this problem, Morgado et al. combined two main strategies of ZSL, which are Recognition using independent semantics (RIS) and Recognition using semantic embeddings (RULE) [21]. However, the algorithm is limited to its model complexity and computational cost. In our proposed method, we use a linear SAE for ZSL [10], which achieved state-of-the-art performance and had lower computational cost.

Most ZSL methods in object recognition application extract features by Convolutional Neural Networks (CNNs). However, CNNs are not suitable for extracting dynamic spatio-temporal sequential hand gesture features. Considering that recurrent neural networks (RNNs) can encoder temporal information of dynamic hand gesture sequences [15], we use RNNs to pre-process our original skeletal joint data. In practice, Long Short-term Memory (LSTM), as a special RNN architecture which replaces traditional artificial neurons in the hidden layer with memory cells [15], can overcome the issue of gradient vanishing and error blowing up. The Bidirectional LSTM network involves two hidden LSTM layers (forwards and backwards) to store and process both past and future information [7]. Thus, we use a BLSTM network to extract features of hand data.

# 3   Unfamiliar Hand Gesture Recognition Approach Based on Zero-Shot Learning

## 3.1   Overview of Unfamiliar Hand Gesture Recognition Approach

The brief flow of our approach is shown in Fig. 1. Three modules are included: data collection, feature extraction and ZSL. First, hand gesture data are captured by LMC and pre-processed. Then, a BLSTM network is employed to extract hand gesture features from the pre-processed data. Finally, we utilize a SAE model for ZSL to learn a mapping from the feature space to the semantic space. By comparing the distances between the estimated semantic representations and the prototypes in the semantic space, hand gestures can be recognized.

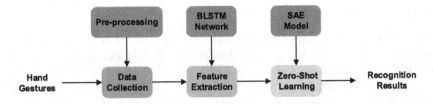

**Fig. 1.** Brief flow of the approach architecture.

## 3.2 Collecting Hand Gesture Data

We collect hand gesture data by a LMC. As shown in Fig. 2, the data frames of LMC involve much information, such as palm positions, skeletal joint positions, and so on. We choose the following information on a single right hand as the input of our recognition system:

1. Palm center position in 3D space.
2. The pitch, yaw and roll of the hand, which are calculated from the hand direction vector and palm normal vector.
3. The 3D positions of finger skeletal joints.

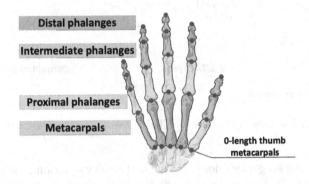

**Fig. 2.** Hand bones captured by the LMC. The red circles are finger skeletal joints. (Color figure online)

We record hand gesture data with 50 Hz sampling rate. Pre-processing mainly includes three steps. First, we eliminate the invalid data and select a fixed number of frames from each sequence. Then, to decrease the influence of different hand location, the skeletal joint positions are replaced by the positions in relation to the palm center position. Finally, these data are normalized to the interval [0, 1] based on z-score.

### 3.3   Feature Extraction

To better analyze the time correlation among sequential frame, a BLSTM network is used to extract spatio-temporal features from hand gesture data. The structure of this network is shown in Fig. 3. It is comprised of one input layer, one BLSTM layer and one output layer. The BLSTM layer concludes two LSTM layers (a forward one and a backward one), which can respectively deal with the past and future spatio-temporal context [13]. The BLSTM layer is fully connected to the input layer, and the outputs of the BLSTM layer are high-level feature expressions. The size of the output layer is equal to the number of labeled hand gestures. We use the softmax classifier to predict recognition results.

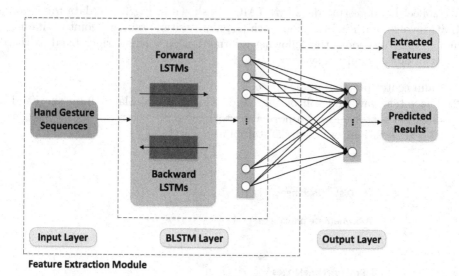

**Fig. 3.** The structure of the BLSTM network. (Color figure online)

In the training stage, we adopt the five-fold cross validation to select the best model. The labeled data is divided into five parts. Each part is chosen as the validation set without repetition. Meanwhile, the other four parts constitute the training set to train the network. In the feature extraction stage, we delete the output layer of the trained BLSTM model and put the labeled and unlabeled data into this network to extract the feature vectors (see the output of red dashed box in Fig. 3).

### 3.4   ZSL for Unfamiliar Hand Gesture Recognition

To recognize unfamiliar hand gesture, we need to learn high-level semantic representations from extracted feature vectors in Sect. 3.3. Our method is based on the SAE for ZSL [10], which achieved state-of-the-art performance currently. We employ the simplest antoencoder which is linear and only has one hidden layer. An antoencoder contains an encoder and a decoder. The encoder projects features into the hidden layer which represents the attribute space in our experiment, and the decoder projects the attribute vectors back to the feature space to reconstruct the original features.

During training, the input hand gesture features are denoted as $\mathbf{X}_Y = \{\mathbf{x}_i\} \in \mathbb{R}^{d \times N}$ and the semantic attributes are denoted as $\mathbf{S}_Y = \{\mathbf{s}_i\} \in \mathbb{R}^{k \times N}$, where $\mathbf{x}_i$ is a d-dimensional feature vector extracted from the i-th training sample, and $\mathbf{s}_i$ is a k-dimensional corresponding semantic attribute vector of the i-th training sample. The goal of ZSL algorithm is to obtain a projection matrix $\mathbf{W} \in \mathbb{R}^{k \times d}$ which can describe the mapping from the feature space to the semantic attribute space. The objective function is formulated as:

$$\min_{\mathbf{W}} \left\| \mathbf{X} - \mathbf{W}^T \mathbf{S} \right\|_F^2 + \lambda \left\| \mathbf{W}\mathbf{X} - \mathbf{S} \right\|_F^2 \tag{1}$$

This unconstrained optimization problem can be transformed into solving a Sylvester equation by using Bartels-Stewart algorithm [1].

During testing, the test hand gesture features $\mathbf{X}_Z = \{\mathbf{x}_i\}$ and the attribute prototypes of unseen classes $\mathbf{S}_Z$ are provided to predict the labels of untrained samples. Based on encoder projection matrix $\mathbf{W}$ obtained in training, we can project a new test sample $\mathbf{x}_i \in \mathbf{X}_Z$ to the semantic space by $\hat{\mathbf{s}}_i = \mathbf{W}\mathbf{x}_i$. Then, we predict an ideal hand gesture class label with minimum distance between estimated semantic representation $\hat{\mathbf{s}}_i$ and the projection prototypes $\mathbf{S}_Z$:

$$\Phi(\mathbf{x}_i) = \arg\min_j D\left(\hat{\mathbf{s}}_i, \mathbf{S}_{Z_j}\right) \tag{2}$$

where $\mathbf{S}_{Z_j}$ is the attribute vector of the j-th unseen class, $D$ is the L2 distance function, and $\Phi(\cdot)$ returns the class label of the sample. More theoretical derivation and proof details about SAE model can be found in [10].

## 4    Experimental Results and Analysis

### 4.1    Experimental Setting

**Dataset.** Because of the lack of open 3D dynamic hand gesture data captured by the LMC, we build a novel dataset for our recognition task. Sixteen training hand gestures and four test hand gestures contained in our dataset are shown in Fig. 4. For each hand gesture class, we collected 50 data sequences, and each sequence consists of 50 frames. Particularly, a frame contains hand direction, palm center and 25 skeletal joint positions. Therefore, each frame is described as an 81-dimensional vector.

**Parameter Settings.** For training the BLSTM network, the number of training epochs is set to 100, and both batch size and the number of forward and backward LSTM neurons are set to 64. We select the cross-entropy function as the loss function, and the Adam optimization algorithm is utilized to minimize the loss. After training, we extract a 128-dimensional feature vector from each sample before the output layer. The output features of training set and test set are included in our dataset for the subsequent ZSL.

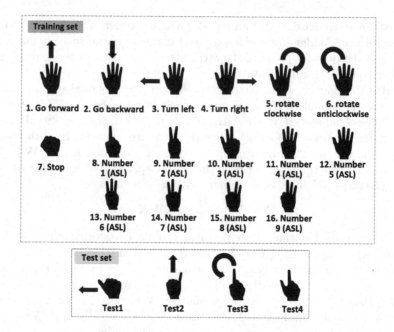

**Fig. 4.** Hand gestures in our dataset.

**Semantic Representation.** We condense fifteen semantic attributes about various hand gestures. The correspondence between hand gesture classes and semantic attributes is shown in Table 1.

**Table 1.** The semantic attribute description

| Lable | Hand gesture name | Dynamic state of hand | | Shape of trajectory | | Direction of trajectory | | | | | | Finger bending | | | | |
|---|---|---|---|---|---|---|---|---|---|---|---|---|---|---|---|---|
| | | Rest | Motion | Straight line | Circle | Clockwise | Anticlockwise | Forward | Backward | Left | Right | Thumb | Index finger | middle finger | Ring finger | Pinky finger |
| 0 | go forward | 0 | 1 | 1 | 0 | 0 | 0 | 1 | 0 | 0 | 0 | 0 | 0 | 0 | 0 | 0 |
| 1 | go backward | 0 | 1 | 1 | 0 | 0 | 0 | 0 | 1 | 0 | 0 | 0 | 0 | 0 | 0 | 0 |
| 2 | turn left | 0 | 1 | 1 | 0 | 0 | 0 | 0 | 0 | 1 | 0 | 0 | 0 | 0 | 0 | 0 |
| 3 | turn right | 0 | 1 | 1 | 0 | 0 | 0 | 0 | 0 | 0 | 1 | 0 | 0 | 0 | 0 | 0 |
| 4 | rotate clockwise | 0 | 1 | 0 | 1 | 1 | 0 | 0 | 0 | 0 | 0 | 0 | 0 | 0 | 0 | 0 |
| 5 | rotate anticlockwise | 0 | 1 | 0 | 1 | 0 | 1 | 0 | 0 | 0 | 0 | 0 | 0 | 0 | 0 | 0 |
| 6 | stop | 1 | 0 | 0 | 0 | 0 | 0 | 0 | 0 | 0 | 0 | 1 | 1 | 1 | 1 | 1 |
| 7 | number 1 | 1 | 0 | 0 | 0 | 0 | 0 | 0 | 0 | 0 | 0 | 1 | 0 | 1 | 1 | 1 |
| 8 | number 2 | 1 | 0 | 0 | 0 | 0 | 0 | 0 | 0 | 0 | 0 | 1 | 0 | 0 | 1 | 1 |
| 9 | number 3 | 1 | 0 | 0 | 0 | 0 | 0 | 0 | 0 | 0 | 0 | 0 | 0 | 0 | 1 | 1 |
| 10 | number 4 | 1 | 0 | 0 | 0 | 0 | 0 | 0 | 0 | 0 | 0 | 1 | 0 | 0 | 0 | 0 |
| 11 | number 5 | 1 | 0 | 0 | 0 | 0 | 0 | 0 | 0 | 0 | 0 | 0 | 0 | 0 | 0 | 0 |
| 12 | number 6 | 1 | 0 | 0 | 0 | 0 | 0 | 0 | 0 | 0 | 0 | 1 | 0 | 0 | 0 | 1 |
| 13 | number 7 | 1 | 0 | 0 | 0 | 0 | 0 | 0 | 0 | 0 | 0 | 1 | 0 | 0 | 1 | 0 |
| 14 | number 8 | 1 | 0 | 0 | 0 | 0 | 0 | 0 | 0 | 0 | 0 | 1 | 0 | 1 | 0 | 0 |
| 15 | number 9 | 1 | 0 | 0 | 0 | 0 | 0 | 0 | 0 | 0 | 0 | 1 | 1 | 0 | 0 | 0 |
| 17 | test1 | 0 | 1 | 1 | 0 | 0 | 0 | 0 | 0 | 0 | 0 | 1 | 0 | 1 | 1 | 1 |
| 18 | test2 | 0 | 1 | 1 | 0 | 0 | 0 | 1 | 0 | 0 | 0 | 1 | 1 | 1 | 1 | 0 |
| 19 | test3 | 0 | 1 | 0 | 1 | 0 | 1 | 0 | 0 | 0 | 0 | 1 | 0 | 1 | 1 | 1 |
| 20 | test4 | 1 | 0 | 0 | 0 | 0 | 0 | 0 | 0 | 0 | 0 | 0 | 0 | 1 | 1 | 1 |

## 4.2 Experimental Results and Analysis

In this section, we conducted extensive evaluation on unfamiliar hand gesture recognition task on our dataset. In the first experiment, we conducted comparisons with two state-of-the-art ZSL models, which are the Embarrassingly Simple Zero-Shot Learning (ESZSL) [23] and the Synthesized Classifiers (SYNC) [5], respectively. The qualitative evaluation results are shown in Fig. 5, and the confusion matrices are shown in Fig. 6. We can observe that our model can significantly outperform other methods. We also evaluate the computational cost of these three ZSL methods. Table 2 shows that for model training and testing, our method is the fastest.

In the second experiment, the effect of various size of training dataset is evaluated. We randomly delete fixed number classes from original sixteen classes in

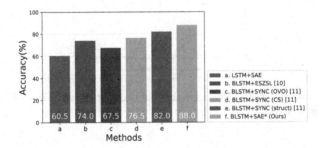

**Fig. 5.** The qualitative evaluation results of different recognition methods.

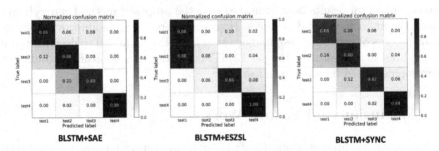

**Fig. 6.** The confusion matrices of different ZSL models.

**Table 2.** Comparative evaluation on computation cost

| Method | Training time (s) | Test time (s) |
|---|---|---|
| ESZSL | 1.1311 | 0.147546 |
| SYNC (CS) | 1.4941 | 0.43829 |
| SYNC (OVO) | 1.8176 | 0.41969 |
| SYNC (struct) | 1.0786 | 0.48054 |
| SAE | 0.437094 | 0.018566 |

training dataset, and calculate the average accuracies of 10 repeated experiments. Average accuracies with different numbers of deleted classes are shown in Fig. 7. From the trend of curve, we can see that the average accuracies will decrease when deleting more classes from the training dataset. Because more attributes cannot be learned, the more unlabeled classes will be not recognized. In despite of deleting 5 training classes, we achieve the recognition accuracy of 37.5% , which still demonstrates the effectiveness of our method.

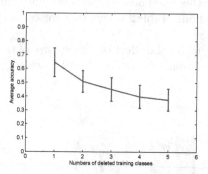

**Fig. 7.** Average accuracy of different numbers of training classes. Vertical bars indicate ±1 standard deviation.

### 4.3   Discussion

The experimental results have indicated that the proposed method can well recognize the unfamiliar hand gestures. However, there are some improving space. In the process of attribute design, more attributes can be added to describe more complicated hand gestures. Deeper BLSTM network can be utilized to extract more sophisticated hand gesture features.

## 5   Conclusion and Future Work

In this paper, we present a novel unfamiliar hand gesture recognition approach based on ZSL. We collect hand and finger joint data from the LMC and construct a hand gesture dataset with semantic information. The BLSTM network is used to extract features and the SAE model for ZSL is built to infer semantic description of unlabeled hand gestures. By matching the predicting semantic information and ground truth, the unfamiliar hand gestures can be inferred. Finally, the experimental results verify that our method achieves state-of-the-art performance and has lower computational cost than other methods.

In the future, the proposed method will be applied to the real-life human-robot interaction system. We plan to realize a real-time online hand gesture recognition interface which can make the robot correctly understand user's intention even though they use unfamiliar hand gestures.

**Acknowledgments.** This work is partially supported by the National Natural Science Foundation of China under Grants 61673378 and 61421004.

# References

1. Bartels, R.H., Stewart, G.W.: Solution of the matrix equation AX+ Xb = C [F4]. Commun. ACM **15**(9), 820–826 (1972)
2. Van den Bergh, M., et al.: Real-time 3D hand gesture interaction with a robot for understanding directions from humans. In: RO-MAN, pp. 357–362. IEEE (2011)
3. Bheda, V., Radpour, D.: Using deep convolutional networks for gesture recognition in American sign language. arXiv preprint arXiv:1710.06836 (2017)
4. Camastra, F., De Felice, D.: LVQ-based hand gesture recognition using a data glove. In: Apolloni, B., Bassis, S., Esposito, A., Morabito, F. (eds.) Neural Nets and Surroundings. Smart Innovation, Systems and Technologies, vol. 19, pp. 159–168. Springer, Heidelberg (2013). https://doi.org/10.1007/978-3-642-35467-0_17
5. Changpinyo, S., Chao, W.L., Gong, B., Sha, F.: Synthesized classifiers for zero-shot learning. In: Proceedings of the IEEE Conference on Computer Vision and Pattern Recognition, pp. 5327–5336 (2016)
6. Chen, Y., Ding, Z., Chen, Y.L., Wu, X.: Rapid recognition of dynamic hand gestures using leap motion. In: 2015 IEEE International Conference on Information and Automation, pp. 1419–1424. IEEE (2015)
7. Graves, A., Schmidhuber, J.: Framewise phoneme classification with bidirectional lstm and other neural network architectures. Neural Netw. **18**(5–6), 602–610 (2005)
8. Ji, Y., Liu, C., Gong, S., Cheng, W.: 3D hand gesture coding for sign language learning. In: 2016 International Conference on Virtual Reality and Visualization (ICVRV), pp. 407–410. IEEE (2016)
9. Ke, W., Li, W., Ruifeng, L., Lijun, Z.: Real-time hand gesture recognition for service robot. In: 2010 International Conference on Intelligent Computation Technology and Automation (ICICTA), vol. 2, pp. 976–979. IEEE (2010)
10. Kodirov, E., Xiang, T., Gong, S.: Semantic autoencoder for zero-shot learning. arXiv preprint arXiv:1704.08345 (2017)
11. Lampert, C.H., Nickisch, H., Harmeling, S.: Learning to detect unseen object classes by between-class attribute transfer. In: IEEE Conference on Computer Vision and Pattern Recognition, CVPR 2009, pp. 951–958. IEEE (2009)
12. Lefebvre, G., Berlemont, S., Mamalet, F., Garcia, C.: BLSTM-RNN based 3D gesture classification. In: Mladenov, V., Koprinkova-Hristova, P., Palm, G., Villa, A.E.P., Appollini, B., Kasabov, N. (eds.) ICANN 2013. LNCS, vol. 8131, pp. 381–388. Springer, Heidelberg (2013). https://doi.org/10.1007/978-3-642-40728-4_48
13. Lefebvre, G., Berlemont, S., Mamalet, F., Garcia, C.: Inertial gesture recognition with BLSTM-RNN. In: Koprinkova-Hristova, P., Mladenov, V., Kasabov, N.K. (eds.) Artificial Neural Networks. SSB, vol. 4, pp. 393–410. Springer, Cham (2015). https://doi.org/10.1007/978-3-319-09903-3_19
14. Li, W.J., Hsieh, C.Y., Lin, L.F., Chu, W.C.: Hand gesture recognition for post-stroke rehabilitation using leap motion. In: 2017 International Conference on Applied System Innovation (ICASI), pp. 386–388. IEEE (2017)
15. Lipton, Z.C., Berkowitz, J., Elkan, C.: A critical review of recurrent neural networks for sequence learning. arXiv preprint arXiv:1506.00019 (2015)
16. Lu, W., Tong, Z., Chu, J.: Dynamic hand gesture recognition with leap motion controller. IEEE Signal Process. Lett. **23**(9), 1188–1192 (2016)

17. Luo, R.C., Wu, Y.C.: Hand gesture recognition for human-robot interaction for service robot. In: 2012 IEEE Conference on Multisensor Fusion and Integration for Intelligent Systems (MFI), pp. 318–323. IEEE (2012)
18. Madapana, N., Wachs, J.P.: A semantical & analytical approach for zero shot gesture learning. In: 2017 12th IEEE International Conference on Automatic Face & Gesture Recognition (FG 2017), pp. 796–801. IEEE (2017)
19. Marin, G., Dominio, F., Zanuttigh, P.: Hand gesture recognition with leap motion and kinect devices. In: 2014 IEEE International Conference on Image Processing (ICIP), pp. 1565–1569. IEEE (2014)
20. Molchanov, P., Gupta, S., Kim, K., Kautz, J.: Hand gesture recognition with 3D convolutional neural networks. In: Proceedings of the IEEE Conference on Computer Vision and Pattern Recognition Workshops, pp. 1–7 (2015)
21. Morgado, P., Vasconcelos, N.: Semantically consistent regularization for zero-shot recognition. In: CVPR, vol. 9, p. 10 (2017)
22. Rautaray, S.S., Agrawal, A.: Vision based hand gesture recognition for human computer interaction: a survey. Artif. Intell. Rev. **43**(1), 1–54 (2015)
23. Romera-Paredes, B., Torr, P.: An embarrassingly simple approach to zero-shot learning. In: International Conference on Machine Learning, pp. 2152–2161 (2015)
24. Shen, J., Luo, Y., Wang, X., Wu, Z., Zhou, M.: GPU-based realtime hand gesture interaction and rendering for volume datasets using leap motion. In: 2014 International Conference on Cyberworlds (CW), pp. 85–92. IEEE (2014)
25. Tang, A., Lu, K., Wang, Y., Huang, J., Li, H.: A real-time hand posture recognition system using deep neural networks. ACM Trans. Intell. Syst. Technol. (TIST) **6**(2), 21 (2015)
26. Weichert, F., Bachmann, D., Rudak, B., Fisseler, D.: Analysis of the accuracy and robustness of the leap motion controller. Sensors **13**(5), 6380–6393 (2013)

# Chinese Event Recognition via Ensemble Model

Wei Liu[✉], Zhenyu Yang, and Zongtian Liu

School of Computer Engineering and Science, Shanghai University,
Shanghai 200444, China
liuw@shu.edu.cn

**Abstract.** Event recognition is one of the most fundamental and critical field in information extraction. In this paper, Event recognition task can be divided into two sub-problems containing candidate event triggers identification and the classification of candidate event trigger words. Firstly, we use trigger vocabulary generated by trigger expansion to identify candidate event trigger, and then input sequences are generated according to the following three features: word embedding, POS (part of speech) and DP (dependency parsing). Finally multiclass classifier based on joint neural networks is introduced in the step of candidate trigger classification. The experiments in CEC (Chinese Emergency Corpus) have shown the superiority of our proposal model with a maximum F-measure of 80.55%.

**Keywords:** Event recognition · Bi-RNN · Dependency parsing

## 1 Introduction

Natural-language organized texts express higher-level semantic information through events. Recognizing these events is the most fundamental task in event-oriented natural language process. Event recognition is also one of the most significant tasks in Information Extraction field. Event is defined as a thing happened in particular time and place and showed a number of characteristics of the movement [1]. Event trigger is a word which clearly indicates the occurring of event. Event trigger identification is a crucial step in most of event recognition approaches. Event recognition can be transformed in to the problem of trigger recognition in some situations. For example, "the earthquake happened yesterday caused 21 wounded". There is a trigger *wounded* indicates the occurrence of casualty. However, "the earthquake happened yesterday hasn't caused any wounded." There is also an event trigger wounded in the sentence, but the trigger "wounded" doesn't indicate the occurrence of any events.

Therefore, Event Recognition can be divided into two sub-problems containing candidate event trigger identification and the classification of candidate event trigger. In previous work, we counted all event triggers in CEC and expanded it by using Tongyici Cilin (Extended) developed by HIT IR-Lab which is a synonym dataset applied to Chinese words. Hence this paper mainly focuses on the classification of candidate event trigger.

© Springer Nature Switzerland AG 2018
L. Cheng et al. (Eds.): ICONIP 2018, LNCS 11305, pp. 255–264, 2018.
https://doi.org/10.1007/978-3-030-04221-9_23

Existing event recognition approaches can be generally classified into two categories: rule based approaches and machine learning based approaches. Rule based approaches refer to the recognition and extraction of certain types of events are conducted under the guidance of a number of templates, using variety of pattern matching algorithms to match the events to be extracted in texts. However, the effectiveness of rule based approaches depending on the quality of the templates researchers defined. These methods also have poor portability and robustness. Machine learning based approaches based on machine learning convert event recognition to classification of word-level information. The performances of traditional machine learning based methods depend on suitability of the features designed by researchers. Existing methods of event recognition achieved certain results, however there are some restrictions. Methods based on rules lack portability and robustness. Scientists have to design rules to maintain optimal performance for texts in new fields. Methods based on machine learning treat the event recognition as a word features classification problem, where many features are extracted using various NLP toolkit, these methods fail to mine deep semantic information.

In recent years, deep learning has achieved great success in speech recognition, image processing and natural language [10–12]. Deep learning creates more abstract high-level representations by combining lower-level features to discover distributed representations of the data. Currently, some researchers have made some attempts to apply deep learning in the field of event recognition. These methods achieved better results than traditional models. However, there are few researches on the event recognition applied in Chinese texts.

In this paper we propose a joint model constituted by RNN and CNN structures to extract events from Chinese texts. In this paper, the words in texts are converted into feature vectors containing word embedding and semantic features. Deep learning model is used to automatically mine deep semantic information. The result of experiment has shown to achieve better results compared to other approaches applied into Chinese text.

The remained of this paper is organized as follows: we sketch related works in Sect. 2. Section 3 introduces our model architecture in detail. Section 4 presents the experiment results on CEC, and the conclusion is drawn in Sect. 5.

## 2   Related Work

Existing event recognition approaches can be generally classified into two categories: rule based approaches and machine learning based approaches.

Methods based on rules mainly apply pre-defined rules to event recognition. Surdeanu et al. [2] proposed an event extraction system applied to open-domain. Liang et al. [3] proposed an approach combined manual definition and automatic filling to extract disaster events. Mcclosky et al. [4] proposed an approach for the extraction of structures by taking tree of event-argument relations and using it to capture event structures. Yankova et al. [5] defined FRET (Football Reports Extraction Templates) applied in event extraction of football match. Methods based on rules lack portability and robustness. Rules designed in these researches work in certain field. Researchers have to design new rules to adapt texts in other field.

Methods based on machine learning regards the task of event recognition as a classification problem, with the main emphasis on the construction of classifier and the discovery and selection of features. Relatively speaking, these approaches are more objective, doesn't require too much human intervention and domain knowledge, so the current event recognition research mostly adopts the method of machine learning. In 2002, Hai et al. [6] firstly applied maximum entropy classifier to event extraction for event recognition. Fu et al. [7] proposed approached event classifier based on the semantic features. Zhao et al. [8] proposed an approach based on trigger expansion and classification. Mccracken et al. [9] proposed an approach for event extraction which integrates a statistical approach using machine learning over Propbank semantic role labels. Machine learning based methods design different features extracted by various NLP toolkits for different tasks, thus not making them generalizable.

Deep learning has achieved great success in speech recognition, image processing and natural language [10–12]. Deep learning creates more abstract high-level representations by combining lower-level features to discover distributed representations of the data. Currently, some researchers have made some attempts to apply deep learning in the field of event recognition. Zhang et al. [13] proposed CEERM based on RBM to recognize events in Chinese Emergency Corpus. Patchigolla et al. [14] proposed a model based on RNN to identify biomedical event trigger. Chen et al. [15] proposed DMCNN model, which uses a dynamic multi-pooling layer according to event triggers and arguments. Ghaeini et al. [16] applied forward-backward recurrent neural networks to detect events.

These methods achieved better results than traditional models. However, there are few researches on the event recognition applied in Chinese texts. In this paper we propose a joint model constituted by RNN and CNN structures to extract events from Chinese texts. In this paper, the words in texts are converted into feature vectors containing word embedding and semantic features. Deep learning model is used to automatically mine deep semantic information. The result of experiment has shown to achieve better results compared to other approaches applied into Chinese text.

## 3 Proposed Model

In this section we will introduce the whole structure of event recognition model and the details of each procedure.

As shown in Fig. 1, our event recognition model can be divided into three parts, candidate trigger extraction, input sequence generation module and deep classifier. Candidate trigger extraction mainly extracts word sequences with candidate trigger by trigger vocabulary. Input sequence generation module converts word sequences into word feature sequences by word embedding trained by skip-gram and NLP toolkit analyze result. Deep Classifier receives word feature sequence as input and output the category of candidate trigger in input sequence.

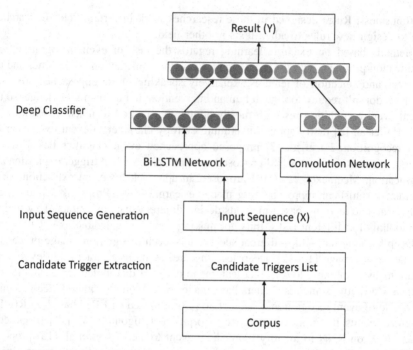

**Fig. 1.** The whole architecture of event recognition model using joint neural network

### 3.1 Candidate Trigger Extraction

The main function of Candidate Trigger Extraction procedure is extracting sentences with candidate triggers and then provide these sentence with their corresponding labels. In previous work, we have counted the event triggers in CEC. Due to the limited size of CEC, the trigger vocabulary constructed by triggers extracted from CEC is also limited, making it difficult to achieve large-scale coverage in open domain. This paper used Tongyici Cilin (Extended) developed by HIT IR-Lab to automatically expand the trigger vocabulary and cover triggers as many as possible. Then Preprocessing System calls LTP to process the sentences with candidate triggers and get POS and DP results of these sentences.

### 3.2 Input Sequence Generation

The input sequence generation module generates input sequence for each sentence processed by last procedure. In this module, we use the word feature around the candidate trigger word to seize the context information. So each sequence contains seventeen feature vectors including the features of eight words before the candidate trigger, the feature of trigger word itself and the features of eight words. Before feeding the data into the deep classifier, each word is represented by feature vector contains word embedding and semantic feature.

**Word Embedding.** Word embedding is the collective name for a set of language modeling and feature learning techniques in natural language processing where words or phrases from the vocabulary are mapped to vectors of real numbers. Several researches have raised methods for representing words as real-valued vectors in order to capture the hidden semantic and syntactic properties of words. In this paper, we use pre-trained 64-dimension word embedding as part of word feature.

**Semantic Feature.** According to words in sentences containing candidate trigger words, we define two features to describe their semantic features.

POS (parts of speech) is the most basic grammatical feature of a word in sentence. In sentence, POS as a generalization of the word plays an important role in language recognition, syntax analysis, information extraction and other tasks. For instance, nouns and verbs can act as trigger words in sentences, however using an adverb as trigger word is absolutely impossible. In CEC, trigger words have a concentrated distribution, 80% of trigger words are verbs and 14% are nouns. Therefore, using POS as part of semantic feature can improve recognition accuracy.

DP (dependency parsing) reveals its syntactic structure by analyzing the dependencies among the components of language units. Intuitively speaking, dependent syntactic analysis identifies the grammatical components such as "verb-object" and "subject-verb" in sentences, and analyzes the relationship among the components. DP exactly describes semantic role relationships between words. In CEC, according to results of DP, 62% of trigger words have subject-verb relationships and 18% have verb-object relationships. In most cases, certain dependencies between trigger words indicate the occurrences of the events.

The part-of-speech categories here can be nouns, verbs, adjectives, or others. Part of speech is the most basic grammatical feature of a word. DP (Dependency Parsing) reveals its syntactic structure by analyzing the dependencies among the components of language units. Popularly, DP is the relationships like verb-object and verdict-complement in sentence. We use one-hot vector to present these feature. The feature of DP can be divided into five classes contains SBV (subject-verb relationship), VOB (verb-object relationship), COO (coordinate relationship), ATT (attribute relationship) and default (other relationships).

### 3.3   Feature Extraction

In this section, we will discuss about details of our study model. Our study model structure is presented as shown in Fig. 2. In this paper, CNN and Bi-RNN Model are used to learn features from the input sequence simultaneously and the outputs from two models are concatenate to make classification. After the procedures introduced in last section, each tagged candidate trigger word was converted into the input sequence of word features.

**Extracting Context Features with Bi-RNN.** RNN is a powerful model for learning features from sequential data. RNN model is suitable for our inputs which are sequences of word feature. In this paper, we use LSTM unit [18] to handle the vanishing and exploding gradient problem and use bidirectional model to let the hidden state of event trigger to capture features from both past and future.

As shown in Fig. 2, we use Bi-RNN to extract the context information of the candidate trigger. The hidden state of trigger word is extracted as Bi-RNN feature. We choose the state of the middle moment as the output of Bi-RNN model.

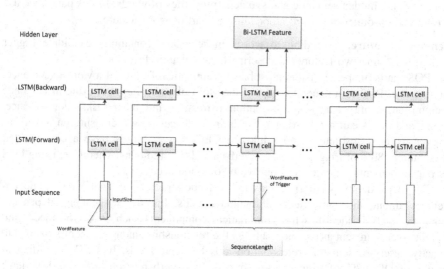

**Fig. 2.** Bi-RNN Layer expanded according to series

**Extracting Sentence-Level Features with CNN.** The CNN with max-pooling layers, is a good choice to capture the semantics of n-grams within a sentence. In this paper, various convolution filters are used to encode the semantic information in word sequence, the experimental result in Sect. 4 shows that the features extracted by CNN model helps to improve the effect of event recognition.

## 4    Experiment Results

### 4.1    Hyperparameters Setting and Corpus

Our experimental dataset is CEC 2.0. CEC 2.0 is an event-based Chinese natural language corpus developed by the Semantic Intelligence Laboratory of Shanghai University. It has collected 333 newspaper reports about earthquakes, fires, traffic accidents, terrorist attacks and food poisoning. We labeled event triggers, participants, objects, times, places and relationships between events by using a semi-automatic method. CEC is available at https://github.com/shijiebei2009/CEC-Corpus.

We extract texts in CEC as experiment Corpora, and find out candidate trigger with trigger vocabulary. These trigger were divided into 7 classes including 6 classes denoted in CEC and one more class which means this candidate trigger is not an event trigger. As shown in Table 1, the experiment Corpora contains 9224 candidate trigger, these triggers were divided into 7 classes.

Cross entropy is used as loss function and the model is trained using Adam optimization. We use pre-trained word embedding with 64 dimensions. In our final model, RNN hidden state dimension of 80 (for each direction), CNN with 3 filters whose sizes was 3, 4, 5. In the whole network, we use dropout of 0.3.

**Table 1.** Event statistics in train and test corpora

| Trigger class | Train count | Test count |
| --- | --- | --- |
| Non-event | 2882 | 720 |
| Perception | 212 | 53 |
| stateChange | 797 | 199 |
| Emergency | 534 | 133 |
| Statement | 688 | 171 |
| Action | 1893 | 473 |
| Movement | 376 | 93 |

### 4.2 Comparison Between Different Values of CNN Features

Our study model is a joint model combining RNN feature with CNN feature. It's necessary for us to decide threshold of CNN features. The comparison of ensemble model with different variable and single RNN or CNN model is shown in Table 2. In this table, parameter n is the feature amount for each filter in CNN. As shown in Table 2 and Fig. 3, the joint model with 11 as parameter performed best in our experiment. Compared to the result of single RNN model, F-measure was increased by 2.2%. However single CNN model performed worst in our experiment. The features amount of CNN features need to set according to specific issue.

**Table 2.** Comparison of results based on various amount of features extracted by CNN

| Model | Precision (%) | Recall (%) | F-measure (%) |
| --- | --- | --- | --- |
| RNN model | 79.11 | 77.33 | 78.04 |
| CNN model | 64.53 | 60.25 | 62.31 |
| Joint model (n = 6) | 79.62 | 75.30 | 77.16 |
| Joint model (n = 7) | 79.64 | 77.05 | 78.28 |
| Joint model (n = 8) | 81.03 | 77.84 | 79.20 |
| Joint model (n = 9) | 80.04 | 78.72 | 79.47 |
| Joint model (n = 10) | 80.82 | 79.57 | 80.08 |
| Joint model (n = 11) | 82.12 | 79.25 | **80.55** |
| Joint model (n = 12) | 79.69 | 79.85 | 79.80 |
| Joint model (n = 13) | 80.27 | 77.94 | 78.75 |
| Joint model (n = 14) | 80.95 | 78.91 | 79.71 |

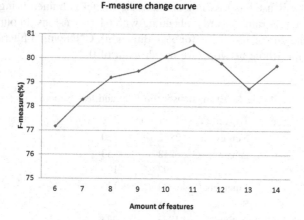

**Fig. 3.** The change curve of F-measure changes with features extracted by CNN

### 4.3    Comparison Between Various Word Representation

In this section we investigate the importance of various features shown in Table 3. Here Word, POS and DP refer to using word embedding, part-of-speech and dependency as word representation in event recognition. For example, Word + POS means using both word embedding and part-of-speech as input feature. As shown in Table 3, F-measure was increased when we use POS and DP to expand the word feature. We can say that POS and DP help to expand word semantic information and improve the model's performance.

**Table 3.** Comparison between various word representations

| Word representation | Precision (%) | Recall (%) | F-measure (%) |
|---|---|---|---|
| Word embedding | 77.61 | 79.53 | 78.40 |
| Word + POS | 79.27 | 79.20 | 79.19 |
| Word + DP | 80.76 | 78.86 | 79.69 |
| Word + POS + DP | 82.12 | 79.25 | 80.55 |

### 4.4    Comparison with Baseline Models in Chinese Event Recognition

As shown in Table 4, our method performed better f-measure than other traditional methods in Chinese. Different from the traditional method using manually engineered features, we use word embedding and a small amount of designed features as input of deep learning model, and let computer learn RNN and CNN feature automatically. The experimental result shows deep learning method can perform better without artificial features.

**Table 4.** Comparison with baseline models applied in Chinese texts

| Approach | Precision (%) | Recall (%) | F-measure (%) |
|---|---|---|---|
| Our method | 82.12 | 79.25 | 80.65 |
| Fu's method [7] | 71.60 | 67.20 | 69.30 |
| Zhao's method [8] | 57.14 | 64.22 | 60.48 |

# 5 Discussion and Conclusion

This paper examined event recognition in Chinese Emergency Corpus. We converted event recognition into two sub-problems containing candidate event trigger identification and the classification of the candidate event trigger. We used trigger vocabulary to identify event triggers and proposed a joint study model combining bi-directional RNN model with CNN model to solve the second sub-problem. The study model learned both RNN feature and CNN feature to make classification. Our experiments have shown to achieve better f-measure than other methods in Chinese text. We also tested the relationship between recognition performance of model and the number of CNN features. Experimental results show that recognition performance can be improved when the number of CNN features increases to a certain extent. In addition, the experimental results show that DP and POS feature of word help to improve the effect of event recognition.

**Acknowledgments.** This paper is supported by the Natural Science Foundation of China, No. 61305053 and No. 61273328.

# References

1. Liu, Z.T., et al.: Research on event-oriented ontology model. Comput. Sci. **36**(11), 189–192 (2009)
2. Surdeanu, M., Harabagiu, S.M.: Infrastructure for open-domain information extraction. In: Proceedings of the Second International Conference on Human Language Technology Research, pp. 325–330. Morgan Kaufmann Publishers Inc., San Francisco (2002)
3. Liang, H., Chen, X., Wu, P.B.: Information extraction system based on event frame. J. Chin. Inf. Process. **20**(2), 40–46 (2006)
4. McClosky, D., Surdeanu, M., Manning, C.D.: Event extraction as dependency parsing. In: Proceedings of the 49th Annual Meeting of the Association for Computational Linguistics: Human Language Technologies, vol. 1, pp. 1626–1635. ACL, Stroudsburg (2011)
5. Yankova, M., Boytcheva, S.: Focusing on scenario recognition in information extraction. In: Tenth Conference on European Chapter of the Association for Computational Linguistics, pp. 41–48. ACL, Stroudsburg (2003)
6. Hai, L.C., Ng, H.T.: A maximum entropy approach to information extraction from semi-structured and free text. In: Eighteenth National Conference on Artificial Intelligence, pp. 786–791. AAAI, Menlo Park (2002)
7. Fu, J.F., Liu, Z.T., Fu, X.F., Zhou, W., Zhong, Z.M.: Dependency parsing based event recognition. Comput. Sci. **36**(11), 217–219 (2009)

8. Zhao, Y., Qin, B., Che, W., Liu, T.: Research on chinese event extraction. J. Chin. Inf. Process. **22**(1), 3–8 (2008)

9. Mccracken, N., Ozgencil, N.E., Symonenko, S.: Combining techniques for event extraction in summary reports. In: Proceedings of AAAI Workshop Event Extraction and Synthesis, pp. 7–11. AAAI, Menlo Park (2006)

10. Kim, Y.: Convolutional neural networks for sentence classification. arXiv preprint arXiv: 1408.5882 (2014)

11. Graves, A.: Generating sequences with recurrent neural networks. arXiv preprint arXiv: 1308.0850 (2013)

12. Tai, K.S., Socher, R., Manning, C.D.: Improved semantic representations from tree-structured long short-term memory networks. arXiv preprint arXiv:1503.00075 (2015)

13. Zhang, Y.J., Liu, Z.T., Zhou, W.: Event recognition based on deep learning in Chinese texts. PLoS ONE **11**(8), e0160147 (2016)

14. Patchigolla, R.V.S.S., Sahu, S., Anand, A.: Biomedical event trigger identification using bidirectional recurrent neural network based models. arXiv preprint arXiv:1705.09516 (2017)

15. Chen, Y., Xu, L., Liu, K., Zeng, D., Zhao, J.: Event extraction via dynamic multi-pooling convolutional neural networks. In: Proceedings of the 53rd Annual Meeting of the Association for Computational Linguistics and the 7th International Joint Conference on Natural Language Processing, pp. 167–176. ACL, Stroudsburg (2015)

16. Ghaeini, R., Fern, X., Huang, L., Tadepalli, P.: Event nugget detection with forward-backward recurrent neural networks. In: Meeting of the Association for Computational Linguistics, pp. 369–373. ACL, Stroudsburg (2016)

17. Ananiadou, S., Thompson, P., Nawaz, R., Mcnaught, J., Kell, D.B.: Event-based text mining for biology and functional genomics. Brief. Funct. Genomics **14**(3), 213–230 (2015)

18. Hochreiter, S., Schmidhuber, J.: Long short-term memory. Neural Comput. **9**(8), 1735–1780 (1997)

19. Che, W.X., Li, Z.H., Liu, T.: LTP: a Chinese language technology platform. In: International Conference on Computational Linguistics: Demonstrations, pp. 13–16. ACL, Stroudsburg (2010)

# Convolutional Neural Network for Machine-Printed Traditional Mongolian Font Recognition

Hongxi Wei[1,2(✉)], Ya Wen[1,2], Weiyuan Wang[1,2], and Guanglai Gao[1,2]

[1] School of Computer Science, Inner Mongolia University,
Hohhot 010021, China
cswhx@imu.edu.cn

[2] Provincial Key Laboratory of Mongolian Information Processing Technology,
Hohhot, China

**Abstract.** Although font recognition is a fundamental issue in the field of document analysis and recognition, it was usually ignored in the past. With the development of optical character recognition (OCR), font recognition becomes more and more important. This paper proposed a well-designed convolutional neural network (CNN) architecture for traditional Mongolian font recognition by means of a single word. To be specific, the whole word image is regarded as input of CNN. Hence, the word images should be normalized into the same size before being inputted into CNN. By comparison, an appropriate aspect ratio for the traditional Mongolian word images has been determined. Experimental results demonstrate that the proposed CNN architecture outperforms three classic CNN architectures, including LeNet-5, AlexNet and GoogLeNet. Therefore, the proposed CNN is much more suitable for the task of the traditional Mongolian font recognition in the way of a single word.

**Keywords:** Traditional Mongolian · Font recognition
Convolutional neural network · Word image · Aspect ratio

## 1 Introduction

A formal document generally contains multiple parts such as title, main body and so forth. These parts are usually edited in some certain fonts. The existence of multiple fonts makes the character recognition difficult, which results in decreasing the accuracy of optical character recognition (OCR) systems considerably. If the fonts are known before character recognition, an individual recognizer can be constructed for per font. Such the mono-font character recognition strategy can achieve higher accuracy. Moreover, reproduction of a digitized document requires the identification of the characters and the fonts used in the original document. Font recognition is very useful for determining the logical entities of a document including title, subtitle and paragraphs. Therefore, font recognition is able to not only improve the accuracy of OCR system, but also recover the layouts of a document exactly.

In the literatures, many approaches have been proposed for solving the problem of font recognition. Most of these approaches were applied to handling font recognition

© Springer Nature Switzerland AG 2018
L. Cheng et al. (Eds.): ICONIP 2018, LNCS 11305, pp. 265–274, 2018.
https://doi.org/10.1007/978-3-030-04221-9_24

for the western and Arabic. Zramdini *et al.* [1] presented a statistical approach based on global typographic features for recognizing 10 English fonts. Jung *et al.* [2] used typographical attributes such as ascenders, descenders and serifs obtained from a word image for classifying 7 English fonts. Moussa *et al.* [3] proposed global texture features based on fractal geometry for categorizing 10 Arabic fonts. Lutf *et al.* [4] extracted rotation invariant features from diacritics of Arabic characters. For Chinese font recognition, Zhu *et al.* [5] firstly utilized a set of Gabor filters to extract textual features. Ding *et al.* [6] introduced an algorithm for context-independent font recognition on a single Chinese character and extracted wavelet features. Moreover, Song *et al.* [7] proposed a novel multi-scale sparse representation for font recognition on a single Chinese character as well.

The above approaches can be divided into the following two categories: typographical features based approaches [1, 2] and textual features based approaches [3–7]. In [8], Joshi *et al.* have investigated and reported that the textual features based approach is much more efficient than the former one. Although the existing approaches can realize font recognition for various languages, there is still room to improve the recognition precision.

In [9], font recognition was regarded as a special form of image classification problem. Therefore, the key issue of font recognition is how to represent or describe images. In recent years, convolutional neural network (CNN) has attracted much attention and shown superior performance in the domain of image classification [10–12] as well as font categorization [13, 14]. In [13], handwritten Chinese characters were classified into five kinds of calligraphy categories using deep features extracted from a CNN pretrained on natural images. In [14], a patch based classification framework has been presented for font classification. Specifically, two classic CNN architectures (i.e. AlexNet and ResNet-50) were used for classifying patches extracted from line image or page image. The advantage of CNN is able to learn discriminative feature representations from training samples.

Until now, there is no literature about Mongolian font recognition. In this paper, a designed CNN architecture has been used to accomplish Mongolian font recognition. To be specific, Mongolian word images (i.e. font recognition on a single word) are treated as inputs of a well-designed CNN. Therefore, Mongolian word images extracted from original document images need to be normalized into the same size before being inputted into CNN. In this study, several normalized sizes have been compared with each other and a suitable one has been confirmed. As far as we know, this study is the first time to realize Mongolian font recognition.

The remainder of the paper is organized as follows. The related work is presented in Sect. 2. The proposed CNN is described detailedly in Sect. 3. Experimental results are provided in Sect. 4. Section 5 gives the conclusions.

## 2 Related Work

In this study, the mentioned Mongolian is called *traditional Mongolian*, which is widely used in Inner Mongolia Autonomous Region of China. The traditional Mongolian is a kind of alphabetic script. All letters of one Mongolian word are conglutinated together in vertical direction, and letters have initial, medial or final visual forms according to their positions within a word. A blank space is used to separate two words. Meanwhile, the traditional Mongolian is also a kind of agglutinative language. Its word formation and inflection is built through connecting different suffixes to the roots or stems. Furthermore, the traditional Mongolian has a very special writing system, which is quite different from Chinese, Arabic, English and other Latin languages. Its writing order is from top to bottom and its column order is from left to right [15, 23]. A fragment of one document written in the traditional Mongolian is depicted in Fig. 1.

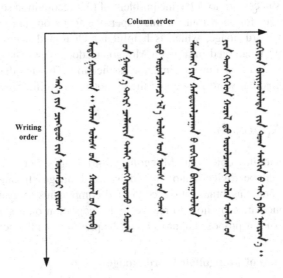

**Fig. 1.** A fragment of one machine-printed traditional Mongolian document.

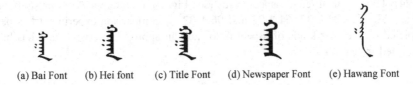

(a) Bai Font     (b) Hei font     (c) Title Font     (d) Newspaper Font     (e) Hawang Font

**Fig. 2.** The same Mongolian word in different fonts.

Regarding to OCR for the machine-printed traditional Mongolian, a segmentation-based scheme has been proposed in our previous work [16, 17]. One Mongolian word should be segmented into multiple glyphs, and then a classifier was designed to recognize these glyphs. A multi-layer perceptron model and a convolutional neural network have been treated as the classifier in [16] and [17], separately. However, the segmentation is quite costly and increases the chances of errors. Especially, such the segmentation-based scheme is incompetent for a certain Mongolian font (e.g. Hawang font in Fig. 2).

Recently, we have presented a segment-free traditional Mongolian OCR system in [18]. The traditional Mongolian word can be recognized directly, which avoids the problem of segmentation. Therein, the process of the machine-printed traditional Mongolian words recognition was taken as a sequence to sequence mapping problem. To be specific, the input word image is treated as a sequence of image frames and the output word as a sequence of letters. In this way, the model can be trained to obtain the relationship between letters and image frames. Nevertheless, the traditional Mongolian language has a very large vocabulary, and daily used vocabulary is about 0.1 to 1 million. Meanwhile, many fonts are frequently used in practice. As a result, the process for training the model is so slowly.

Therefore, it is necessary to solve the problem of font recognition so as to design an individual recognizer for each font. In this paper, we focus on font recognition on a single Mongolian word image, which is helpful for OCR and document analysis if more than one font is adopted in a given Mongolian document. The considered font recognition is formalized as a task of image classification. Through designing a CNN, the input is a Mongolian word image and the output is a specific category of font.

## 3   Proposed Approach

In our study, the handling objects are Mongolian word images. For a given Mongolian document, it should be divided into individual word images through several pre-processing steps, including binarization, connected components analysis and so forth. Here, we only concentrate on how to realize font recognition on a single Mongolian word. The details of our proposed approach will be presented in the next sub-sections.

### 3.1   Normalization of Mongolian Word Images

Because all letters of one Mongolian word are conglutinated together in vertical direction, the aspect ratios of the Mongolian word images are more than one in general [19, 24]. Thus, four different aspect ratios (see Fig. 3) are attempted in our study and they are 28 * 28, 32 * 32, 64 * 32 and 96 * 32. A comparison experiment has been completed on the four kinds of aspect ratios and an appropriate aspect ratio has been determined in Sect. 4.

28*28        32*32        64*32        96*32

**Fig. 3.**  Four kinds of aspect ratios.

## 3.2   The Framework of the Designed CNN

The proposed convolutional neural network is very simple and its structure consists of four convolutional layers (denoted by C1, C2, C3 and C4, severally), two fully-connected (denoted by FC1 and FC2) layers and one softmax output layer. The second, third and fourth convolutional layers are followed by a max-pooling layer, respectively. The detailed structure is shown in Fig. 4.

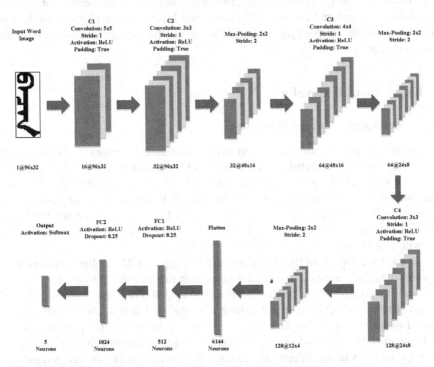

**Fig. 4.** The structure of the designed CNN.

The running procedure of the proposed CNN is presented as below. If the aspect ratio is set to 96 * 32, each Mongolian word is resized into 96 * 32 and then inputted into the first convolutional layer (denoted by C1), which contains 16 kernels of size 5 * 5 with a stride of one pixel. Next, the second convolutional layer (denoted by C2) takes as input the output of the first convolutional layer and filters it with 32 kernels of size 3 * 3 with a stride of one pixel. After that, one max-pooling layer follows the second convolutional layer, in which kernel size is 2 * 2 with a stride of two pixels. The max-pooling layer makes the output of C2 reduce to a half. Then, the third convolutional layer (denoted by C3) takes as input the output of the first max-pooling layer and filters it with 64 kernels of size 4 * 4 with a stride of one pixel. Another max-pooling layer follows the third convolutional layer. Its kernel size is also 2 * 2 with a stride of two pixels, which reduces the output of C3 into a half. The fourth convolutional layer (denoted by C4) takes as input the output of the second max-pooling layer and filters it with 128 kernels of size 3 * 3 with a stride of one pixel.

Afterwards, the third max-pooling layer follows the fourth convolutional layer. Its kernel size is still 2 * 2 with a stride of two pixels, which reduces the output of C4 into a half. The output of the third max-pooling layer will be converted into a vector by a flatten layer. Finally, two fully-connected layers (denoted by FC1 and FC2) are followed by an output layer. The amount of neurons in FC1 and FC2 is 512 and 1024, separately. Because the number of categories of fonts is 5, the number of neurons in the output layer equals 5. Specially, zero-padding scheme is utilized in per convolutional layer. In addition, ReLU [10] activation function is utilized for the neurons of the four convolutional layers and the two fully-connected layers. As an effective method to prevent over-fitting, dropout [20] is applied to the two fully connected layers. Both the probabilities are set to 0.25.

## 4 Experimental Results

### 4.1 Dataset and Experiment Settings

To evaluate the performance of the proposed approach, a dictionary of the traditional Mongolian has been collected and the number of vocabularies is more than 240,000. Here, five types of Mongolian fonts are involved, which includes *Bai font*, *Hei font*, *Title font*, *Newspaper font* and a handwriting-like font, named *Hawang font*. Several examples of the five fonts are given in Fig. 2. From Fig. 2, we can see that the words vary in width, height and shape from font to font, which results in the font recognition task with challenging.

For each vocabulary of the dictionary, the five types of Mongolian fonts are used for generating the corresponding word images with font size of 10 point, respectively. In the above process, three resolutions (i.e. 300 dpi, 400 dpi and 500 dpi) are considered. Thus, a collection with 1.2 million (240,000 × 5 fonts) word images is constructed under per resolution. Our dataset consists of 3.6 million samples in total.

In our experiment, *2-fold cross validation* is used for evaluating the performance of the proposed CNN under different image resolution, severally. In this way, samples per image resolution have been divided randomly into two parts with equal sizes, which results in 0.6 million samples in the training set and the testing set, separately. Evaluation metric is the accuracy of font recognition.

The proposed CNN was trained and tested on the training set and the testing set, severally. During the above process, a standard stochastic gradient descent (SGD) with momentum and weight decay is adopted. In this study, the momentum is set to 0.9 and weight decay is $10^{-6}$. Meanwhile, the batch size and learning rate is set to 350 and 0.01, respectively. The amount of training epochs is 500.

### 4.2 Performance of the Proposed CNN

In this experiment, we have tested the performance of the proposed CNN under the three different image resolutions. Meanwhile, the four types of the aspect ratios have been compared with each other. The detailed results are shown in Tables 1, 2 and 3.

**Table 1.** The performance of the proposed CNN using 2-fold cross validation with 300 dpi.

| Aspect ratio | 28 * 28 | 32 * 32 | 64 * 32 | 96 * 32 |
|---|---|---|---|---|
| Fold 1 | 99.75% | 99.76% | 99.92% | 99.95% |
| Fold 2 | 99.66% | 99.79% | 99.94% | 99.99% |
| Average | 99.62% | 99.78% | 99.93% | **99.97%** |

**Table 2.** The performance of the proposed CNN using 2-fold cross validation with 400 dpi.

| Aspect ratio | 28 * 28 | 32 * 32 | 64 * 32 | 96 * 32 |
|---|---|---|---|---|
| Fold 1 | 99.19% | 99.88% | 99.82% | 99.98% |
| Fold 2 | 99.87% | 99.67% | 99.88% | 99.98% |
| Average | 99.53% | 99.78% | 99.85% | **99.98%** |

**Table 3.** The performance of the proposed CNN using 2-fold cross validation with 500 dpi.

| Aspect ratio | 28 * 28 | 32 * 32 | 64 * 32 | 96 * 32 |
|---|---|---|---|---|
| Fold 1 | 99.97% | 99.98% | 99.99% | 99.99% |
| Fold 2 | 99.97% | 99.98% | 99.99% | 99.99% |
| Average | 99.97% | 99.98% | **99.99%** | **99.99%** |

From Tables 1, 2 and 3, we can see that the performance is better and better with increasing the resolution. It comes to a conclusion that the best aspect ratio is always 96 * 32 under the three different image resolutions. Among the four aspect ratios, 96 * 32 is the most suitable for the traditional Mongolian word images. In subsequent experiment, only the aspect ratio of 96 * 32 is considered.

### 4.3 Comparison Between Our Proposed CNN with Baseline Methods

In this experiment, three classic CNN architectures were chosen as baseline methods. They are LeNet-5 [21], AlexNet [10] and GoogLeNet [22], respectively. For the three classic CNNs, the experimental settings are the same as our proposed CNN. We have tested the corresponding performance of the three classic CNNs under the three different image resolutions.

The performance of the three classic CNNs is depicted in Fig. 5 and the comparison results between our proposed CNN and the baselines are listed in Table 4. From Fig. 5, the same conclusion can be drawn that the performance is also better and better with increasing the resolution. The GoogLeNet can obtain the best performance in various image resolutions. It demonstrates that the GoogLeNet is superior to the other CNN architectures.

From Table 4, we can see that the proposed CNN consistently outperforms than the three classic CNNs in various image resolutions. In particular, the performance of the proposed CNN is identical to the GoogLeNet, when the image resolution is 500 dpi. But, the architecture of the proposed CNN is much shallower compared with the GoogLeNet. Therefore, we can conclude that the proposed CNN with a shallow architecture is competent for the task of font recognition in the way of a single Mongolian word.

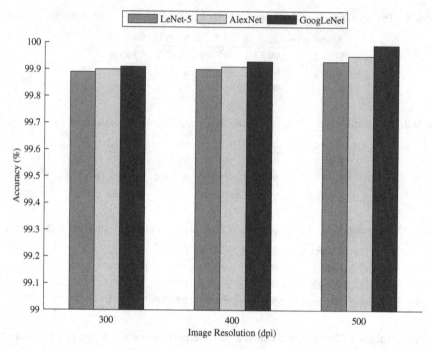

**Fig. 5.** The performance of the three classic CNNs.

**Table 4.** The comparison results between the proposed CNN and the baseline methods.

| Method | Resolution | | |
|---|---|---|---|
| | 300 dpi (96 * 32) | 400 dpi (96 * 32) | 500 dpi (96 * 32) |
| LeNet-5 | 99.89% | 99.90% | 99.93% |
| AlexNet | 99.90% | 99.91% | 99.95% |
| GoogLeNet | 99.91% | 99.93% | **99.99%** |
| The proposed CNN | **99.97%** | **99.98%** | **99.99%** |

## 5 Conclusion

In this paper, we proposed a novel CNN architecture for accomplishing the traditional Mongolian font recognition. On the one hand, a single Mongolian word is taken as input of CNN. Thereby, it is helpful for OCR and document analysis if more than one font is adopted in a given Mongolian document. On the other hand, an appropriate aspect ratio has been determined according to the word-formation characteristics of Mongolian language.

By analyzing the experimental results, the performance of the proposed CNN is superior to LeNet-5, AlexNet and GoogLeNet under the image resolutions of 300 dpi and 400 dpi. When the image resolution is set to 500 dpi, the performance of the proposed CNN is the same as the GoogLeNet. However, the architecture of the proposed CNN is much shallower than the GoogLeNet. Furthermore, the appropriate aspect ratio (i.e. 96 * 32) for the traditional Mongolian word images has been determined as well. By comparison, we find that the proposed CNN with a shallow architecture is much more suitable for the task of font recognition by means of a single Mongolian word.

**Acknowledgements.** This paper is supported by the National Natural Science Foundation of China under Grant 61463038.

# References

1. Zramdini, A., Ingold, R.: Optical font recognition using typographical features. IEEE Trans. Pattern Anal. Mach. Intell. **20**(8), 877–882 (1998)
2. Jung, M., Shin, Y., Srihari, N.: Multifont classification using typographical attributes. In: Proceedings of ICDAR 1999, pp. 353–356. IEEE Press, New York (1999)
3. Moussa, B., Zahour, A., Benabdelhafid, A., Alimi, M.: New features using fractal multi-dimensions for generalized Arabic font recognition. Pattern Recogn. Lett. **31**(5), 361–371 (2010)
4. Lutf, M., You, X., Cheung, Y., Chen, P.: Arabic font recognition based on diacritics features. Pattern Recogn. **47**(2), 672–684 (2014)
5. Zhu, Y., Tan, T., Wang, Y.: Font recognition based on global texture analysis. IEEE Trans. Pattern Anal. Mach. Intell. **23**(10), 1192–1200 (2001)
6. Ding, Q., Li, C., Tao, W.: Character independent font recognition on a single Chinese character. IEEE Trans. Pattern Anal. Mach. Intell. **29**(2), 195–204 (2007)
7. Song, W., Lian, Z., Tang, Y., Xiao, J.: Content-independent font recognition on a single Chinese character using sparse representation. In: Proceedings of ICDAR 2015, pp. 376–380. IEEE Press, New York (2015)
8. Joshi, G., Garg, S., Sivaswamy, J.: A generalized framework for script identification. Int. J. Doc. Anal. Recogn. **10**(2), 55–68 (2007)
9. Tao, D., Lin, X., Jin, L., Li, X.: Principal component 2-D long short-term memory for font recognition on single Chinese characters. IEEE Trans. Cybern. **46**(3), 756–765 (2016)
10. Krizhevsky, A., Sutskever, I., Hinton, G.: ImageNet classification with deep convolutional neural networks. In: Proceedings of NIPS 2012, pp. 1097–1105. Curran Associates Inc. (2012)
11. Ciresan, D., Meier, U., Schmidhuber, J.: Multi-column deep neural networks for image classification. In: Proceedings of CVPR 2012, pp. 3642–3649. IEEE Press, New York (2012)
12. Oquab, M., Bottou, L., Laptev, I., Sivic, J.: Learning and transferring mid-level image representations using convolutional neural networks. In: Proceedings of CVPR 2014, pp. 1717–1724. IEEE Press, New York (2014)
13. Gao, P., Gu, G., Wu, J., Wei, B.: Chinese calligraphic style representation for recognition. Int. J. Doc. Anal. Recogn. **20**(1), 59–68 (2017)
14. Tensmeyer, C., Saunders, D., Martinez, T.: Convolutional neural networks for font classification. In: Proceedings of ICDAR 2017, pp. 985–990. IEEE Press, New York (2017)
15. Wei, H., Gao, G.: A keyword retrieval system for historical Mongolian document images. Int. J. Doc. Anal. Recogn. **17**(1), 33–45 (2014)

16. Wei, H., Gao, G.: Machine-printed traditional Mongolian characters recognition using BP neural networks. In: Proceedings of CiSE 2009, pp. 1–7. IEEE Press, New York (2009)

17. Hu, H., Wei, H., Liu, Z.: The CNN based machine-printed traditional Mongolian characters recognition. In: Proceedings of CCC 2017, pp. 3937–3941. IEEE Press, New York (2017)

18. Zhang, H., Wei, H., Bao, F., Gao, G.: Segmentation-free printed traditional Mongolian OCR using sequence to sequence with attention model. In: Proceedings of ICDAR 2017, pp. 585–590. IEEE Press, New York (2017)

19. Ma, L., Liu, J., Wu, J.: A new database for online handwritten Mongolian word recognition. In: Proceedings of ICPR 2016, pp. 1131–1136. IEEE Press, New York (2016)

20. Hinton, G., Srivastava, N., Krizhevsky, A., Sutskever, I., Salakhutdinov, R.: Improving neural networks by preventing co-adaptation of feature detectors. arXiv preprint arXiv:1207.0580 (2012)

21. LeCun, Y., Bottou, L., Bengio, Y., Haffner, P.: Gradient-based learning applied to document recognition. Proc. IEEE 86(11), 2278–2324 (1998)

22. Szegedy, C., Liu, W., Jia, Y., Sermanet, P., Reed, S.: Going deeper with convolutions. In: Proceedings of CVPR 2015, pp. 1–9. IEEE Press, New York (2015)

23. Wei, H., Gao, G., Bao, Y.: A method for removing inflectional suffixes in word spotting of Mongolian Kanjur. In: Proceedings of ICDAR 2011, pp. 88–92. IEEE Press, New York (2011)

24. Wei, H., Zhang, H., Gao, G.: Representing word image using visual word embeddings and RNN for keyword spotting on historical document images. In: Proceedings of ICME 2017, pp. 1368–1373. IEEE Press, New York (2017)

# WGAN Domain Adaptation for EEG-Based Emotion Recognition

Yun Luo[1], Si-Yang Zhang[1], Wei-Long Zheng[1], and Bao-Liang Lu[1,2,3(✉)]

[1] Center for Brain-Like Computing and Machine Intelligence,
Department of Computer Science and Engineering, Shanghai Jiao Tong University,
800 Dong Chuan Road, Shanghai 200240, China
[2] Key Laboratory of Shanghai Education Commission for Intelligent Interaction
and Cognition Engineering, Shanghai Jiao Tong University,
800 Dong Chuan Road, Shanghai 200240, China
[3] Brain Science and Technology Research Center, Shanghai Jiao Tong University,
800 Dong Chuan Road, Shanghai 200240, China
{angeleader,zhangsiyang-sjtu,weilong,bllu}@sjtu.edu.cn

**Abstract.** In this paper, we propose a novel Wasserstein generative adversarial network domain adaptation (WGANDA) framework for building cross-subject electroencephalography (EEG)-based emotion recognition models. The proposed framework consists of GANs-like components and a two-step training procedure with pre-training and adversarial training. Pre-training is to map source domain and target domain to a common feature space, and adversarial-training is to narrow down the gap between the mappings of the source and target domains on the common feature space. A Wasserstein GAN gradient penalty loss is applied to adversarial-training to guarantee the stability and convergence of the framework. We evaluate the framework on two public EEG datasets for emotion recognition, SEED and DEAP. The experimental results demonstrate that our WGANDA framework successfully handles the domain shift problem in cross-subject EEG-based emotion recognition and significantly outperforms the state-of-the-art domain adaptation methods.

**Keywords:** EEG · Emotion recognition · Domain adaptation · GAN

## 1 Introduction

With rapid development of affective computing and emotional intelligence, affective brain-computer interfaces (aBCIs) have recently attracted widespread attention [13]. aBCIs aim to equip machines with the ability to detect users' affective states from neurophysiological signals and provide humanized interactions. Recently, many researchers have made significant progresses in EEG-based emotion recognition models, especially in subject-specific models [1,8,10,21]. However, due to domain shift [18] caused by the non-stationary nature of EEG signals

ⓒ Springer Nature Switzerland AG 2018
L. Cheng et al. (Eds.): ICONIP 2018, LNCS 11305, pp. 275–286, 2018.
https://doi.org/10.1007/978-3-030-04221-9_25

and structural variability between individuals [12,15], an EEG-based emotion recognition model trained with data from one specific subject usually does not generalize well to another. In practical aBCI applications, a cross-subject emotion recognition model which is capable of recognizing the emotions of a new subject with unlabeled data is required rather than a subject-specific one. To deal with the domain shift problem caused by inter-subject variability, we focus on developing cross-subject emotion recognition approach in this work.

A promising solution to the domain shift problem is to take advantage of the domain adaptation methods. The basic idea of these methods is to transfer knowledge from source domain to unlabeled target domain. Under the circumstance of domain shift, marginal probability distributions of source domain and target domain are different. Domain adaptation methods are able to handle this difference by mapping features of both domains into a common feature space, where the marginal probability distributions of the two mappings are similar.

Various domain adaptation methods have been developed to find the common feature space for source and target subjects. Most of them aim to minimize some metrics between two probability distributions, such as maximum mean discrepancy (MMD) [5]. For example, transfer component analysis (TCA) [14], a typical domain adaptation method, minimizes MMD between distributions of source and target domains by constructing kernel matrix. This method, along with kernel principle component analysis (KPCA) [17] and transductive parameter transfer (TPT) [16], has been successfully used for implementing personalized EEG-based emotion models [23].

An alternative way of finding the common space is to leverage the transferability of deep neural networks [9]. One of attractive approaches is to apply generative adversarial domain adaptation [19], which is closely related to generative adversarial networks (GANs) [4]. The adversarial training procedure of GANs can be formulated as a minimax problem. When the game achieves its equilibrium, the distribution of generated data is approximate to the distribution of real data. By taking advantage of the generative ability of GANs, generative adversarial domain adaptation methods have made considerable progresses in dealing with the domain shift problem in computer vision [19].

In this paper, we adopt the generative adversarial domain adaptation method to build a cross-subject EEG-based emotion recognition framework. Our work is based on Wasserstein GAN [2], which is an improved stable version of traditional GAN. Instead of using the source subject features, we consider their mappings in a new feature space as the real data distribution, which has been adopted in Adversarial Discriminative Domain Adaptation (ADDA) as well [19]. The features of the target subjects are mapped to the same feature space, in which their mappings are considered as the generated distribution. The distance between marginal probability distributions of the two mappings are reduced through adversarial training, and then the domain shift problem is fixed.

Our proposed Wasserstein GAN domain adaptation (WGANDA) framework aims to solve the domain shift problem in EEG-based emotion recognition caused by inter-subject variability. Compared with subject-specific models, the proposed

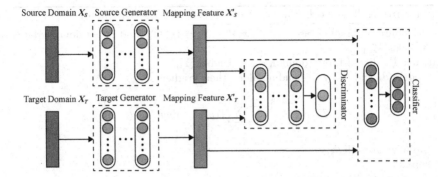

**Fig. 1.** Illustration of the proposed WGANDA framework, which consists of four parts: the source and target generators for mapping source domain and target domain to a common feature space, the discriminator for distinguishing source and target distribution in the common feature space, and the classifier for recognizing emotional states.

cross-subject framework makes better use of the EEG data collected from different subjects. The framework is also able to recognize the emotions of a new subject with unlabeled data more precisely. The application of Wasserstein GAN in this work overcomes the gradients vanish and instability problems of traditional GANs' training procedure. Besides, the implementation of the gradient-penalty Wasserstein GAN loss [6] speeds up the convergence process. According to experimental results on two public EEG datasets, the proposed WGANDA framework significantly outperforms the state-of-the-art domain adaptation methods.

## 2    Methods

### 2.1    Notations and Framework Structure

Our proposed framework consists of four components as shown in Fig. 1. Assume that a labeled dataset $X_s$ is collected from the source subjects, and an unlabeled dataset $X_t$ is collected from the target subjects:

$$X_s = \{x_s^i\}_{i=1}^m, \ Y_s = \{y_s^i\}_{i=1}^m, \ X_t = \{x_t^i\}_{i=1}^n \tag{1}$$

where $m$ and $n$ represent the numbers of data samples in source and target datasets, respectively.

The source generator $\psi_s$ and the target generator $\psi_t$ map source data $X_s$ and target data $X_t$ to a common feature space, respectively:

$$X_s' = \psi_s(X_s), \ X_t' = \psi_t(X_t) \tag{2}$$

where $X_s'$ and $X_t'$ are expected to have the same feature dimensions.

The classifier $C$ takes $X_s'$ and $X_t'$ as inputs and outputs emotion predictions, $Y_{sp}$ and $Y_{tp}$, as follows:

$$Y_{sp} = C(X_s'), \ and \ Y_{tp} = C(X_t'). \tag{3}$$

---

**Algorithm 1.** The work flow of the proposed WGANDA framework

---

**Input:** Source domain dataset $X_s = \{x_s^i\}_{i=1}^m, Y_s = \{y_s^i\}_{i=1}^m$ and target domain dataset
$X_t = \{x_t^i\}_{i=1}^n$
**Output:** Predicted target domain dataset labels $Y_{tp}$
1: Update $\theta_s$ and $\theta_c$ by descending along their gradients:

$$\nabla_{\theta_s, \theta_c} \left[ -\frac{1}{m} \sum_{i=1}^m \sum_{h=1}^H \mathbb{I}(y_s^i = h) log C(\psi_s(x_s^i)) \right]$$

2: Initialize $\theta_t$ with $\theta_s$;
3: **repeat**
4:     **for** *critic* iterations **do**
5:         Update $\theta_d$ by ascending along its gradient:

$$\nabla_{\theta_d} \left[ \frac{1}{m} \sum_{i=1}^m D(\psi_s(x_s^i)) - \frac{1}{n} \sum_{i=1}^n D(\psi_t(x_t^i)) - \frac{\lambda}{q} \sum_{i=1}^q (||\nabla_{\hat{x}^i} D(\hat{x}^i)||_2 - 1)^2 \right]$$

6:     **end for**
7:     Update $\theta_t$ by descending along its gradient:

$$\nabla_{\theta_t} \left[ -\frac{1}{n} \sum_{i=1}^n D(\psi_t(x_t^i)) \right]$$

8: **until** convergence
9: Predict target domain dataset labels with the target generator and the classifier:

$$Y_{tp} = C(\psi_t(X_t))$$

10: **return** $Y_{tp}$

---

Then a discriminator $D$ is applied to distinguish $X_s'$ and $X_t'$. Note that all four components are parameterized by feedforward neural networks. The parameters of the source generator, the target generator, the classifier, and the discriminator are represented with $\theta_s$, $\theta_t$, $\theta_c$ and $\theta_d$, respectively.

### 2.2   Training Procedure

The training procedure of the WGANDA framework consists of the following two steps:

(i) **Pre-training:** feed source data $X_s$ through the source generator $\psi_s$ to the classifier $C$, minimize cross-entropy loss with source dataset labels $Y_s$, and initialize target generator parameters $\theta_t$ with source generator parameters $\theta_s$.

(ii) **Adversarial-training:** train the network through an adversarial way and update discriminator parameters $\theta_d$ as well as target generator parameters $\theta_t$ alternatively with source data $X_s$ and target data $X_t$. Note that in each adversarial iteration, the discriminator is updated a certain number of times denoted with *critic*, while the target generator is updated only once.

After the two-step training procedure, recognition accuracy can be calculated by feeding $X_t'$ to the pre-trained classifier $C$. The whole work flow of the proposed WGANDA framework is described in Algorithm 1.

In the pre-training step, our goal is to minimize the cross-entropy loss by optimizing $\theta_s$ and $\theta_c$:

$$\min_{\theta_s, \theta_c} L_C(X_s, Y_s) = -\mathbb{E}_{(x_s, y_s) \sim (X_s, Y_s)} \left[ \sum_{h=1}^{H} \mathbb{I}(y_s = h) log C(\psi_s(x_s)) \right] \quad (4)$$

where $H$ is the number of emotion states. Then $\theta_s$ is fixed through the following adversarial-training step, and $\theta_c$ is fixed for the final target emotion prediction.

We initialize $\theta_t$ with $\theta_s$ when $L_C$ is minimized. Without this target generator parameter initialization step, the discriminator can easily distinguish samples from $X_s'$ and samples from $X_t'$, which makes it hard to optimize target generator. Initializing $\theta_t$ with $\theta_s$ ensures the distribution of $X_t$ is relatively close to $X_s$. The discriminator will thus not be able to distinguish the two distributions too easily, and target generator can be optimized faster.

In the adversarial-training step, the network is trained to narrow down the gap between marginal distributions $P(X_s)$ and $P(X_t)$. With fixed $\theta_s$ and $\theta_c$, the framework can be treated as a typical GAN model. However, the traditional training procedure of GANs is prone to fall into model collapse, and it is troubled with gradients vanish as well. To prevent these two drawbacks, we implement Wasserstein GAN loss with gradient-penalty rather than traditional GANs' adversarial loss, which is applied in ADDA.

The training procedure of traditional GANs can be viewed as minimizing the Jensen-Shannon divergence between the real and generated distributions. As a metric for the distance of two distributions, Jensen-Shannon divergence is discontinuous, which makes it difficult to provide useful gradients for optimizing the generator. It is also the main reason of the GANs' instability. The Wasserstein GAN adopts Earth-Mover distance (EMD, also called Wasserstein-1) to eliminate the instability problem [2]. The EMD between two distributions is:

$$W(X_r, X_g) = \inf_{\gamma \sim \Pi(X_r, X_g)} \mathbb{E}_{(x_r, x_g) \sim \gamma}[||x_r - x_g||] \quad (5)$$

where $\Pi(X_r, X_g)$ denotes all possible joint distributions of real distribution $X_r$ and generated distribution $X_g$ defined in traditional GANs. The EMD is almost continuous and differentiable almost everywhere, and thus overcomes the instability problem. Since the infimum of Eq. (5) is computationally highly intractable, its Kantorovich-Rubinstein duality form is usually utilized [20]:

$$W(X_r, X_g) = \frac{1}{K} \sup_{||f||_L \leq K} \mathbb{E}_{x_r \sim X_r}[f(x_r)] - \mathbb{E}_{x_g \sim X_g}[f(x_g)] \quad (6)$$

where $f$ denotes the set of 1-Lipschitz functions. In realistic implementations, $f$ is replaced by discriminator $D$ and $||f||_L \leq K$ is replaced by $||D||_L \leq 1$. The loss function of Wasserstein GAN is then formulated by:

$$\min_{\theta_G} \max_{\theta_D} L(X_r, X_g) = \mathbb{E}_{x_r \sim X_r}[(D(x_r))] - \mathbb{E}_{x_g \sim X_g}[(D(x_g))] \tag{7}$$

where $\theta_D$ and $\theta_G$ represent the parameters of discriminator and generator in traditional GANs, respectively. The discriminator realizes 1-Lipschitz function by clipping the weights and constraining them within a bounded range.

Gulrajani et al. enforced Lipschitz constraint with gradient penalty instead of weight clipping to directly constrain the gradient norm [6], which makes the training procedure more stable and make convergence faster. An extra penalty term is appended to the loss function in their approach:

$$\min_{\theta_G} \max_{\theta_D} L(X_r, X_g) = \mathbb{E}_{x_r \sim X_r}[(D(x_r))] - \mathbb{E}_{x_g \sim X_g}[(D(x_g))]$$
$$- \lambda \mathbb{E}_{\hat{x} \sim \hat{X}}[(||\nabla_{\hat{x}} D(\hat{x})||_2 - 1)^2] \tag{8}$$

where $\lambda$ is a hyperparameter controlling the trade-off between original objective and gradient penalty, and $\hat{x}$ denotes the data points sampled from the straight line between real distribution $X_r$ and generator distribution $X_g$:

$$\hat{x} = \alpha x + (1 - \alpha)\tilde{x}, \alpha \sim U[0,1], x \sim X_r, \tilde{x} \sim X_g \tag{9}$$

In Algorithm 1, the number of sampled data points is denoted with $q$.

Our WGANDA framework can be treated as a Wasserstein GAN when the source generator is fixed. In this case, $X_s'$ and $X_t'$ correspond to the real data $X_r$ and the generated data $X_g$ in traditional GANs, respectively. We present our adversarial loss in Wasserstein GAN gradient penalty form as follows. First, the discriminator is trained by maximizing the discriminator loss (D-Loss) with target generator fixed:

$$\max_{\theta_d} L_D(X_s, X_t) = \mathbb{E}_{x_s \sim X_s}[D(\psi_s(x_s))] - \mathbb{E}_{x_t \sim X_t}[D(\psi_t(x_t))]$$
$$- \lambda \mathbb{E}_{\hat{x} \sim \hat{X}}[(||\nabla_{\hat{x}} D(\hat{x})||_2 - 1)^2] \tag{10}$$

Then the target generator is trained by minimizing the generator loss (G-Loss) with discriminator fixed:

$$\min_{\theta_t} L_G(X_t) = -\mathbb{E}_{x_t \sim X_t}[D(\psi_t(x_t))] \tag{11}$$

The two losses are optimized in an alternating procedure, and the parameters of different components are updated in an interleaved manner. Note that in Wasserstein GANs, the discriminator aims to fit the 1-Lipschitz function. In each adversarial training iteration, the discriminator is fully trained to its optimization. Thus $\theta_d$ is updated for *critic* times and $\theta_t$ is updated only once in each adversarial training iteration. When the discriminator is fully trained, D-Loss represents the EMD between the marginal distribution of $X_s'$ and $X_t'$. In our experiments, D-Loss is used as an indicator of training process.

The assumption of most domain adaptation methods is that, the conditional probability distributions of source domain and target domain equal when the marginal probability distributions of source domain and target domain are the same. When D-Loss converges, the marginal distribution of $X_s'$ is approximate to the marginal distribution of $X_t'$:

$$P(X_s') \approx P(X_t') \tag{12}$$

According to the assumption mentioned above, the conditional distribution of $X_s'$ and the conditional distribution of $X_t'$ are also similar:

$$P(Y_s'|X_s') \approx P(Y_t'|X_t') \tag{13}$$

where $Y_t'$ denotes the true labels of the dataset collected from target subject. Under this circumstance, the classifier pre-trained with $X_s'$ is able to recognize the emotions of the target subject from $X_t'$. Thus after the adversarial-training procedure, we feed $X_t$ to the pre-trained classifier and compare the output $Y_{tp}$ with its true label to get the recognition accuracy.

## 3    Experiment Settings

### 3.1    EEG Datasets

We evaluate our framework on two public EEG datasets, SEED[1] [22] and DEAP[2] [7]. The SEED dataset consists of 15 participants. Each of them was required to watch 15 emotional film clips to elicit three emotions: positive, neutral, and negative. The EEG signals were recorded at a sampling rate of 1000 Hz with ESI NeuroScan System, which had a 62 electrode cap. The data in DEAP are formed with 8-channel peripheral physiological signals and 32-channel EEG signals. 32 participants watched 40 music videos and their EEG signals were collected by an international 10–20 system. The level of each video was rated 1–9 by the participants in terms of arousal, valence, like, and dislike.

The EEG signals of both datasets are preprocessed before feeding to the framework. Differential entropy (DE) features are extracted per second from five frequency bands for SEED dataset: $\delta$: 1–3 Hz, $\theta$: 4–7 Hz, $\alpha$: 8–13 Hz, $\beta$: 14–30 Hz, and $\gamma$: 31–50 Hz [3,22]. The feature dimension is 310 (62 channels $\times$ 5 frequency bands) and the number of samples for each subject is 3394. For DEAP dataset, the DE features are extracted per second except for $\delta$ frequency since the low frequency band is filtered in this dataset. The feature dimension is 128 (32 channels $\times$ 4 frequency bands) and the number of samples for each subject is 2400. Valence model (high valence: level > 5, low valence: level $\leq$ 5) and arousal model (high arousal: level > 5, low arousal: level $\leq$ 5) are adopted in this work.

---

[1] http://bcmi.sjtu.edu.cn/~seed/index.html.
[2] http://www.eecs.qmul.ac.uk/mmv/datasets/deap/.

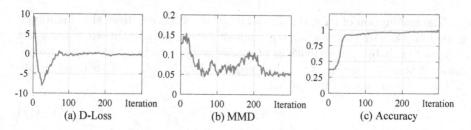

**Fig. 2.** Discriminator loss (D-Loss) (a), MMD (b) and accuracy (c) tendency along with training steps of SEED dataset.

## 3.2   Evaluation Details

To demonstrate the effectiveness of the proposed framework, a leave-one-subject-out cross validation is conducted. We chose one subject as the target subject and leave the others (14 for SEED, and 31 for DEAP) as source subjects.

To optimize the network structure, we perform grid search on the number of network layers. The numbers of layers are searched from 3 to 6 for both generator and discriminator. Each hidden layer of both the source generator network and the target generator network has 512 nodes for SEED dataset and 256 nodes for DEAP dataset. The outputs of the two generators have the same dimension as the input data, which is 310 for SEED dataset and 128 for DEAP dataset. Each hidden layer of discriminator network has the same number of nodes as the hidden layers of the generators, and the output has only one dimension. For the classifier, the numbers of network layers are searched from 1 to 3. The output dimension is 3 and 2 for SEED and DEAP datasets, respectively. Each hidden layer of the classifier network has 64 nodes. The ReLU activation function is used for all hidden layers.

In our experiments, we observe that the loss of discriminator is fluctuating with less discriminator training iterations in each round. And the discriminator should be fully optimized to ensure the convergence in each adversarial training iteration according to the theory of Wasserstein GANs. So the *critic* value is set to 20 to ensure the convergence and training speed. It means that we update discriminator 20 times and update target generator once in each adversarial-training iteration. Besides, Adam optimizer is more likely to cause fluctuation than RMSProp optimizer. Thus we use RMSProp optimizer during adversarial-training and Adam optimizer during pre-training. To speed up the training procedure, we use mini-batch instead of full batch shown in Algorithm 1. The size of mini-batch is set to 256. And the hyperparameter $\lambda$ is set to 10.

MMD is frequently used as a measurement of the distance between two distributions [9,14], thus we adopt it in this work to evaluate the distance between the probability distributions of $X'_s$ and $X'_t$, and demonstrate the effectiveness of our framework.

We use the recognition results before adversarial-training as baseline to show the ability of adversarial domain adaptation. In order to evaluate the effective-

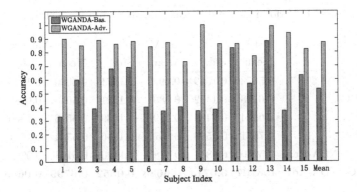

**Fig. 3.** Accuracy comparison between the strategy using adversarial-training and the baseline without using adversarial-training on SEED dataset.

ness of our framework, we compare it with the state-of-the-art methods including KPCA, TCA and TPT on SEED dataset [23]. We also implement these methods and evaluate their performances on DEAP dataset. All the hyperparameters are adjusted following the strategies used in [23].

## 4    Experimental Results

In this section, we demonstrate the effectiveness of our proposed WGANDA framework. Figure 2 depicts the training process of the adversarial-training procedure. The discriminator loss (D-loss) converges to a small value along with the training epoch as illustrated in Fig. 2(a). As the EMD between the distributions of source and target mappings, D-Loss converging to a small value demonstrates that the two marginal distributions are approximate to each other. The MMD curve in Fig. 2(b) has a similar converged tendency with D-Loss, which also implies that adversarial-training has reduced the distance between the two mapping distributions. Moreover, the recognition accuracy shown in Fig. 2(c) increases while MMD decreases. This phenomenon confirms the domain adaptation assumption. Since the classifier is optimized according to the conditional distribution of $X'_s$, only when the two conditional distributions are similar, $X'_t$ can achieve high recognition accuracy with the same classifier.

We first compare our proposed framework with its baseline. Figure 3 shows the accuracy comparison of using adversarial-training (WGANDA-Adv.) and without using adversarial-training (WGANDA-Bas.) on SEED dataset. The recognition accuracy of the baseline WGANDA-Bas. is calculated with the target mappings $X'_t$ directly fed into the classifier after target generator initialization. WGANDA-Bas. performs poorly due to the fact that domain shift exists when neglecting inter-subject variability. Without adversarial-training, the source mappings and the target mappings share no common marginal distributions as well as conditional distributions. The classifier trained with $X'_s$ hence can not predict the emotion states of the target subject precisely according to $X'_t$. By

**Table 1.** Performance of different domain adaptation methods

| Methods | SEED | | DEAP-Arousal | | DEAP-Valence | |
|---|---|---|---|---|---|---|
| | Mean | Std. | Mean | Std. | Mean | Std. |
| SVM | 0.5673 | 0.1629 | 0.4922 | 0.1571 | 0.5036 | 0.1125 |
| KPCA | 0.6128 | 0.1462 | 0.5891 | 0.1521 | 0.5658 | 0.0980 |
| TCA | 0.6364 | 0.1488 | 0.5193 | 0.1539 | 0.5516 | 0.1069 |
| TPT | 0.7631 | 0.1589 | 0.5577 | 0.1496 | 0.5564 | 0.1221 |
| WGANDA-Bas. | 0.5260 | 0.1831 | 0.5183 | 0.1406 | 0.5164 | 0.0929 |
| WGANDA-Adv. | **0.8707** | **0.0714** | **0.6685** | **0.0552** | **0.6799** | **0.0656** |

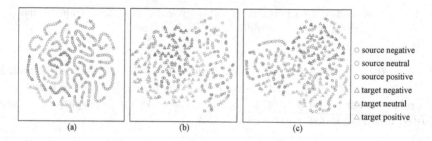

(a)                    (b)                    (c)

○ source negative
○ source neutral
○ source positive
△ target negative
△ target neutral
△ target positive

**Fig. 4.** Two-dimension visualization of source and target domain distributions in different training stages: (a) original distribution; (b) distribution after pre-training procedure; and (c) distribution after adversarial-training procedure. Small circles represent source data samples of three classes and small triangles represent target data samples of three classes.

using adversarial-training, the accuracy of WGANDA-Adv. shows a significant improvement for each subject compared with the baseline result.

Next, we compare our proposed framework with three state-of-the-art domain adaptation methods. Table 1 presents mean accuracies and standard deviations of our proposed framework WGANDA-Adv., the baseline WGANDA-Bas., and other three domain adaptation methods, KPCA, TCA, and TPT. The experimental results of KPCA, TCA and TPT on SEED dataset are referenced from [23]. From Table 1, we see that domain adaptation methods are effective when handling domain shift problem in EEG-based emotion recognition. Our framework significantly outperforms the state-of-the-art methods with mean accuracy of 87.07% and standard deviation of 0.0714 on SEED dataset. On DEAP dataset, our framework achieves mean accuracies of 66.85% and 67.99% and standard deviations of 0.0552 and 0.0656 on arousal and valence classifications, respectively, which is also superior to other methods.

In order to have a better view of the effectiveness of our proposed framework, the source and target data from SEED dataset at different training stages are visualized in a 2-dimension way by t-SNE [11] as shown in Fig. 4. To illustrate the influence of adversarial-training on marginal and conditional distributions more intuitively, samples from different subjects and emotion categories are visualized with different markers. However, in both pre-training and adversarial-training procedures, target labels are unknown to the framework.

Figure 4(a) depicts the distributions of the original data from the source subjects and the target subjects, which have diverse distributions due to inter-subject variability. From Fig. 4(a), we see that there is no any overlapping between the source subjects samples (small circles) and the target subjects samples (small triangles). This means that the original data from the source subjects and target subjects have diverse distributions due to inter-subject variability. Figure 4(b) depicts the distributions of $X'_s$ and $X'_t$ before adversarial-training procedure, which are the mappings of the original data from source and target subjects after target generator initialization. Note that although the samples from three emotion categories have been successfully clustered, the marginal distributions are not the same. Under this circumstance, the pre-trained classifier can only recognize the three emotions of the source subject. Figure 4(c) depicts the distributions of $X'_s$ and $X'_t$ after the adversarial training procedure. Now the marginal distributions of the source mappings are approximate to the target mappings, while the conditional distributions are similar as well. Thus the pre-trained classifier can recognize different emotions on target subject correctly.

## 5  Conclusion

In this paper, we have proposed a novel Wasserstein GAN domain adaptation framework for building cross-subject EEG-based emotion recognition models. The framework adopts adversarial strategy by using Wasserstein GAN gradient penalty version. The performance of our framework has been evaluated by conducting a leave-one-subject-out cross validation on two public EEG datasets for emotion recognition. By narrowing down the gap between probability distribution of different subjects, this adversarial domain adaptation method successfully handles inter-subject variability and domain shift problems of cross-subject EEG-based emotion recognition. By taking advantages of adversarial training, the proposed framework significantly outperforms the state-of-the-art methods with a mean accuracy of 87.07% on SEED dataset, and reaches 66.85% and 67.99% on DEAP dataset for arousal and valence classifications, respectively.

**Acknowledgments.** This work was supported in part by the grants from the National Key Research and Development Program of China (Grant No. 2017YFB1002501), the National Natural Science Foundation of China (Grant No. 61673266), and the Fundamental Research Funds for the Central Universities.

## References

1. Alarcao, S.M., Fonseca, M.J.: Emotions recognition using EEG signals: a survey. IEEE Trans. Affect. Comput. (2017)
2. Arjovsky, M., Chintala, S., Bottou, L.: Wasserstein GAN. arXiv preprint arXiv:1701.07875 (2017)

3. Duan, R.N., Zhu, J.Y., Lu, B.L.: Differential entropy feature for EEG-based emotion classification. In: IEEE EMBS NER 2013, pp. 81–84. IEEE (2013)
4. Goodfellow, I., et al.: Generative adversarial nets. In: NIPS 2014, pp. 2672–2680 (2014)
5. Gretton, A., Borgwardt, K.M., Rasch, M., Schölkopf, B., Smola, A.J.: A kernel method for the two-sample-problem. In: NIPS 2007, pp. 513–520 (2007)
6. Gulrajani, I., Ahmed, F., Arjovsky, M., Dumoulin, V., Courville, A.C.: Improved training of Wasserstein GANs. In: NIPS 2017, pp. 5769–5779 (2017)
7. Koelstra, S., et al.: DEAP: a database for emotion analysis; using physiological signals. IEEE Trans. Affect. Comput. **3**(1), 18–31 (2012)
8. Lin, Y.P., Yang, Y.H., Jung, T.P.: Fusion of electroencephalographic dynamics and musical contents for estimating emotional responses in music listening. Front. Neurosci. **8**, 94 (2014)
9. Long, M., Cao, Y., Wang, J., Jordan, M.: Learning transferable features with deep adaptation networks. In: ICML 2015, pp. 97–105 (2015)
10. Lu, Y., Zheng, W.L., Li, B., Lu, B.L.: Combining eye movements and EEG to enhance emotion recognition. In: IJCAI 2015, pp. 1170–1176 (2015)
11. Maaten, L.V.D., Hinton, G.: Visualizing data using t-SNE. J. Mach. Learn. Res. **9**(Nov), 2579–2605 (2008)
12. Morioka, H., et al.: Learning a common dictionary for subject-transfer decoding with resting calibration. NeuroImage **111**, 167–178 (2015)
13. Mühl, C., Allison, B., Nijholt, A., Chanel, G.: A survey of affective brain computer interfaces: principles, state-of-the-art, and challenges. Brain-Comput. Interfaces **1**(2), 66–84 (2014)
14. Pan, S.J., Tsang, I.W., Kwok, J.T., Yang, Q.: Domain adaptation via transfer component analysis. IEEE Trans. Neural Netw. **22**(2), 199–210 (2011)
15. Samek, W., Meinecke, F.C., Müller, K.R.: Transferring subspaces between subjects in brain-computer interfacing. IEEE Trans. Biomed. Eng. **60**(8), 2289–2298 (2013)
16. Sangineto, E., Zen, G., Ricci, E., Sebe, N.: We are not all equal: personalizing models for facial expression analysis with transductive parameter transfer. In: ACM Multimedia 2014, pp. 357–366. ACM (2014)
17. Schölkopf, B., Smola, A., Müller, K.-R.: Kernel principal component analysis. In: Gerstner, W., Germond, A., Hasler, M., Nicoud, J.-D. (eds.) ICANN 1997. LNCS, vol. 1327, pp. 583–588. Springer, Heidelberg (1997). https://doi.org/10.1007/BFb0020217
18. Sugiyama, M., Krauledat, M., MÃžller, K.R.: Covariate shift adaptation by importance weighted cross validation. J. Mach. Learn. Res. **8**, 985–1005 (2007)
19. Tzeng, E., Hoffman, J., Saenko, K., Darrell, T.: Adversarial discriminative domain adaptation. In: CVPR 2017, vol. 1, p. 4 (2017)
20. Villani, C.: Optimal Transport: Old and New, vol. 338. Springer, Heidelberg (2008). https://doi.org/10.1007/978-3-540-71050-9
21. Wang, X.W., Nie, D., Lu, B.L.: Emotional state classification from EEG data using machine learning approach. Neurocomputing **129**(4), 94–106 (2014)
22. Zheng, W.L., Lu, B.L.: Investigating critical frequency bands and channels for EEG-based emotion recognition with deep neural networks. IEEE Trans. Auton. Ment. Dev. **7**(3), 162–175 (2015)
23. Zheng, W.L., Lu, B.L.: Personalizing EEG-based affective models with transfer learning. In: IJCAI 2016, pp. 2732–2738. AAAI Press (2016)

# Improving Target Discriminability for Unsupervised Domain Adaptation

Fengmao Lv[1], Hao Chen[1], Jinzhao Wu[2], Linfeng Zhong[1], Xiaoyu Li[3], and Guowu Yang[1,2]($\boxtimes$)

[1] Big Data Center, School of Computer Science and Engineering, University of Electronic Science and Technology of China, Chengdu 611731, Sichuan, People's Republic of China
fengmaolv@126.com, chenhao0503@126.com, googlezlf@163.com, guowu@uestc.edu.cn
[2] Guangxi Key Laboratory of Hybrid Computation and IC Design Analysis, Guangxi University for Nationalities, Nanning 530006, Guangxi, People's Republic of China
gxmdwjzh@aliyun.com
[3] School of Information and Software Engineering, University of Electronic Science and Technology of China, Chengdu 611731, Sichuan, People's Republic of China
xiaoyu33521@163.com

**Abstract.** In the recent years, unsupervised domain adaptation has become increasingly attractive, since it can effectively relieve the annotation burden of deep learning through transferring knowledge from a different but related source domain. Domain shift is the major problem in domain adaptation. Although the recently proposed feature alignment methods, which reduce the domain shifts through maximum mean discrepancy or adversarial training at intermediate layers of deep neural network, can obtain domain-invariant representations, these deep features are not necessarily discriminative for the target domain as no mechanism is explicitly enforced to achieve such a goal. In this paper, we propose to improve the classifier's discriminative ability on the target domain through regularizing the entropies of the softmax predictions on the target data. We conduct our experiments on several standard adaptation benchmarks. The experiments demonstrate that our proposal can lead to significant performance improvement for unsupervised domain adaptation.

**Keywords:** Unsupervised domain adaptation · Transfer learning · Deep learning

L. Cheng et al. (Eds.): ICONIP 2018, LNCS 11305, pp. 287–298, 2018.
https://doi.org/10.1007/978-3-030-04221-9_26

# 1   Introduction

Although deep learning has propelled great advances in diverse machine learning tasks ranging from computer vision to natural language processing, its training heavily relies on huge amounts of labelled training data [13]. Transferring knowledges from a different but related source domain, domain adaptation can be a promising way to relieve the annotation burden of deep learning on a certain task at hand [20]. Specifically, domain adaptation aims at adapting the classifier trained with fully labeled source data to make it suitable for the target task when the labeled target data are scarce (semi-supervised domain adaptation), or even unavailable (unsupervised domain adaptation). In this work, we will focus on unsupervised domain adaptation, which is a much more difficult task than the semi-supervised setting.

Domain shift is the major problem in unsupervised domain adaptation [1]. Due to the discrepancy in the distribution of the source and target data, the classifier trained on the source domain cannot directly generalize to the target domain [20]. In order to reduce the data shift between the source and target domain, diverse unsupervised domain adaptation methods have been proposed over the past years [5,6,9,16,18,19,26]. In general, their main idea was to learn an isomorphic latent feature space, in which a distance metric of domain discrepancy is minimized. Although the previous shallow models produced promising adaptation results [5,9,19], the recent studies have demonstrated that the "deep" features obtained by deep neural networks are more transferable [4,8] since deep learning can effectively disentangle exploratory factors of variations behind data distributions and learn increasingly abstract representations that are invariant to the low-level details [2]. In the recent years, various methods, ranging from maximum mean discrepancy (MMD) [16,26] to adversarial training [6,25], have been proposed to learn domain-invariant representations through reducing the domain discrepancy at intermediate layers of convolutional neural networks. However, these deep features are not necessarily discriminative for the target domain as no mechanism is explicitly enforced to achieve such a goal.

In this paper, we propose an effective unsupervised domain adaptation method, which can effectively improve the classifier's discriminative ability in the target domain. Our method achieves this through regularizing the entropies of the softmax predictions on the target data. In particular, the entropies of the softmax probabilities of the target data are minimized to push the target samples away from the classification boundaries and increase the confidence of the classifier's prediction on the target data. The entropy minimization has also been used in unsupervised learning [12] and semi-supervised learning [23]. Though high prediction confidence does not necessarily imply correctness, it pushes the network to discover discriminative clusters underlying the target data. In addition, to avoid naively assigning all the target samples to the same class, we further balance the relative size of different classes through maximizing the entropy of the marginal distribution of class labels. We conduct experiments on several standard adaptation benchmarks. The experimental results demonstrate that our method can significantly improve the classifier's target discriminative ability on the target domain and achieve better adaptation performance than the compared methods.

To sum up, our contribution is mainly three-fold:

- We propose to minimize the entropies of the softmax predictions on the target data, in order to discover discriminative clusters underlying the target data.
- We propose to maximize the entropy of the marginal distribution of class labels, in order to enforce the class labels to be uniformly assigned across the target domain.
- Our method can significantly improve the classifier's target discriminability on several standard domain adaptation benchmarks.

This paper is organized as follows. Section 2 reviews the related works on unsupervised domain adaptation. Section 3 proposes to improve the classifier's discriminative ability on the target domain trough regularizing the target samples' prediction entropies. Section 4 demonstrates our method on several standard domain adaptation benchmarks. Section 5 summarizes this paper.

## 2   Related Works

Domain shift is the major problem in unsupervised domain adaptation. Before the deep learning era, multiple methods had been proposed to reduce the discrepancy in the distribution of the source and target data through learning a shallow feature space [5,9,17,19]. Specifically, Fernando et al. [5] and Gong et al. [9] proposed to align the subspaces described by eigenvectors, while Long et al. [17] and Pan et al. [19] tried to match the distribution means in the kernel-reproducing Hilbert space.

Recent studies have demonstrated that deep neural networks can obtain more transferable features than the shallow models [4,8]. However, in [27], it is revealed that deep features can only reduce, but not completely remove the domain discrepancy. Therefore, the current domain adaptation methods focus on improving the transferability of deep learning. In particular, Long et al. [16] and Tzeng et al. [26] proposed to learn "deep" domain-invariant features through minimizing MMD at intermediate layers of deep neural networks. On the other hand, Ganin et al. [6] adopted adversarial training, which was originally proposed in generative adversarial networks (GAN) [10], to align the "deep" features. In [7], Ghifary et al. proposed to encourage domain invariance through alternately learning source label prediction and target data reconstruction using a shared encoding representation.

Furthermore, Bousmalis et al. [3] explicitly modeled both private and shared components of the domain representations to improve a model's ability to extract domain-invariant features. In [22], transductive mechanism was incorporated into the deep adaptation framework to enhance the feature transferability. Different from the methods mentioned above, in which the weights of the network architecture were shared by both domains, Rozantsev et al. [21] and Tzeng et al. [25] proposed to extract "deep" features with two separate networks and perform domain confusion loss on their outputs, which led to significant performance improvement. Similarly, the coupled GAN (CoGAN) proposed in [15] trained two GANs to generate the source and target images respectively and achieved domain-invariant features by tying only the high-level layer parameters of the two GANs.

However, as indicated in the previous section, these adapted deep features does not necessarily imply good discriminative ability on the target domain. In this work, we aim to improve the classifier's target discriminability through regularizing the entropies of the softmax predictions on the target data. In the following section, we will present our method in detail.

## 3    Our Method

In unsupervised domain adaptation problem, we are given a fully labeled source dataset $\mathcal{D}_s = \{(\boldsymbol{x}_s^i, \boldsymbol{y}_s^i)\}_{i=1}^{N_s}$ ($\boldsymbol{y}_s^i$ is represented by a $K$-dimensional one-hot vector) and an unlabeled target dataset $\mathcal{D}_t = \{\boldsymbol{x}_t^i\}_{i=1}^{N_t}$. It should be noted that the source data and the target data are respectively sampled from two different but related distributions, $p_S(\boldsymbol{x}_s)$ and $p_T(\boldsymbol{x}_t)$. Unsupervised domain adaptation techniques aim to learn a classifier which can correctly predict the class labels of the target data.

### 3.1    Baseline: Domain Alignment

Due to the domain shift between the source and target data, the classifier trained on the labeled source domain cannot directly generalize to the target domain. Therefore, a good domain adaptation method need to consider both the prediction on the labeled source data and the data shift across different domains. In general, the cost function is formulated as

$$\min_{\boldsymbol{\theta}} \mathcal{L} = \mathcal{L}_S + \lambda_A \mathcal{L}_A,$$

where

$$\mathcal{L}_S = \frac{1}{N_s} J(p(\boldsymbol{y}|\boldsymbol{x}_s^i; \boldsymbol{\theta}), \boldsymbol{y}_s^i).$$

Specifically, $J$ is the cross-entropy loss function, $\boldsymbol{\theta}$ is the parameters of the classifier network and $p(\boldsymbol{y}|\boldsymbol{x}_s; \boldsymbol{\theta})$ is the softmax prediction on $\boldsymbol{x}_s$. $L_S$ and $L_A$ are respectively the source supervision term and the domain alignment term that encourages statistically similar representations. Domain alignment can be achieved through either performing MMD loss [26] or adversarial training [6] at intermediate layers of deep neural networks. In this work, we adopt the deep domain adaptation (DDC) model as our baseline, in which MMD is used to implement $\mathcal{L}_A$:

$$\mathcal{L}_A = \left\| \sum_{\boldsymbol{x}_s \in X_S} \phi(\boldsymbol{x}_s; \boldsymbol{\theta}) - \sum_{\boldsymbol{x}_t \in X_T} \phi(\boldsymbol{x}_t; \boldsymbol{\theta}) \right\|^2,$$

where $\phi()$ is evaluated at an intermediate layer of the classifier network. Through learning a domain-invariant feature, the classifier's predictive ability on the source domain can be transferred to the target domain. However, domain invariance does not necessarily imply capable discriminative ability on the target domain as the domain alignment term does not directly penalize the target prediction. As a result, the label assignments of the target samples may be unambiguous.

## 3.2 Entropy Regularization

To improve the classifier's discriminative ability on the target domain, we assume that the target samples are clustered according to their unknown class labels and the classification boundaries should pass through low-density regions. We propose to encourage the assumption through minimizing the entropies of the softmax predictions on the target data, which is denoted as $H(p(\boldsymbol{y}|\boldsymbol{x}_t^i;\boldsymbol{\theta}))$. As a result, the target samples are pushed away from the decision boundaries and classified with large margins. In other words, the softmax predictions are treated as the soft labels of the target data and entropy minimization works as a proxy of target supervision term. Though high prediction confidence does not necessarily imply correctness, it pushes the network to discover discriminative clusters underlying the target data.

However, simply minimizing the target entropy will lead to degenerate solutions, in which all the target samples are assigned to the same class. To circumvent the degenerate solutions, we further balance the relative size of each class through maximizing the entropy of the marginal distribution of class labels, which is denoted as $H(p(\boldsymbol{y};\boldsymbol{\theta}))$. In particular, the marginal distribution $p(\boldsymbol{y};\boldsymbol{\theta})$ is estimated by the empirical distribution on the target data:

$$p(\boldsymbol{y};\boldsymbol{\theta}) = \int p(\boldsymbol{x}_t)p(\boldsymbol{y}|\boldsymbol{x}_t;\boldsymbol{\theta})d\boldsymbol{x}_t$$

$$\approx \frac{1}{N_t}\sum_{i=1}^{N_t} p(\boldsymbol{y}|\boldsymbol{x}_t^i;\boldsymbol{\theta}).$$

The maximization of $H(p(\boldsymbol{y};\boldsymbol{\theta}))$ enforces the target samples to be evenly assigned to each class. In conclusion, our discriminability regularization term is formulated as

$$\mathcal{L}_T = \frac{1}{N_t}\sum_{i=1}^{N_t} H(p(\boldsymbol{y}|\boldsymbol{x}_t^i;\boldsymbol{\theta})) - H(p(\boldsymbol{y};\boldsymbol{\theta}))$$

$$= -\frac{1}{N_t}\sum_{i=1}^{N_t}\sum_{k=1}^{K} p(y_k=1|\boldsymbol{x}_t^i;\boldsymbol{\theta})\log p(y_k=1|\boldsymbol{x}_t^i;\boldsymbol{\theta})$$

$$+ \sum_{k=1}^{K} p(y_k=1;\boldsymbol{\theta})\log p(y_k=1;\boldsymbol{\theta}).$$

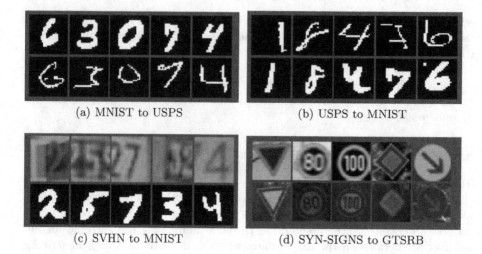

(a) MNIST to USPS                      (b) USPS to MNIST

(c) SVHN to MNIST                      (d) SYN-SIGNS to GTSRB

**Fig. 1.** Samples of the source (top row) and target images (bottom row) in each domain adaptation task.

Combining $\mathcal{L}_T$ with the source supervision term $\mathcal{L}_S$ and the domain alignment term $\mathcal{L}_A$, we define our final objective function as

$$\min_{\theta} \mathcal{L} = \mathcal{L}_S + \lambda_A \mathcal{L}_A + \lambda_T \mathcal{L}_T,$$

where $\lambda_A$ and $\lambda_T$ are the hyper-parameters that balance the importance of each term. The settings of $\lambda_A$ and $\lambda_T$ can be determined via a validation dataset. Through $\mathcal{L}_A$ and $\mathcal{L}_T$, we can obtain "deep" features that are both domain-invariant and discriminative for the target data.

## 4    Experimental Results

To validate our method, we conduct experiments on several adaptation benchmarks, including "MNIST to USPS", "USPS to MNIST", "Street View House Numbers (SVHN) to MNIST" and "Synthetic Signs (SYN-SIGNS) to German Traffic Signs Recognition Benchmark (GTSRB)". Figure 1 displays the samples from each domain adaptation task. In the experiments, our method is compared with the existing state-of-the-art unsupervised domain adaptation methods, including deep domain confusion (DDC) [26], domain-adversarial neural network (DANN) [6], deep reconstruction-classification network (DRCN) [7], domain transfer network (DTN) [24], domain separation network (DSN) [3], CoGAN [15] and Unsupervised Image-to-Image Translation (UNIT) [14].

### 4.1    Experimental Setup

We use the PyTorch deep learning framework to implement our method. In all the experiments, the images are preprocessed by the mean subtraction. To minimize our objective function $\mathcal{L}$, we use the Adam algorithm [11] as the optimizer.

The batch size and the learning rate are set to 64 and 0.0002, respectively. The MMD loss is performed at the first fully connected layer of the classifier network. Following the same strategy in [6], we gradually change the value of $\lambda_T$ from 0 to 1 instead of using a fixed value, in order to suppress incorrect inferences at the early stages of the learning procedure. As done in the recent state-of-the-art works, such as CoGAN [15] and DSN [3], we use a validation set that contains labeled target data to choose the hyper-parameters.

In each adaptation benchmark, we also report the source only result (obtained by training with only labeled source data) and the target only result (obtained by training with fully labeled target data), which are respectively considered as the lower bound and the upper bound of the unsupervised domain adaptation performance. For a fair comparison, we report the average accuracy over 5 trails with different random splits of each dataset.

## 4.2  "MNIST to USPS" and "USPS to MNIST"

The images in both MNIST and USPS are grayscale handwritten digits from 10 classes. However, the USPS and MNIST images are different in the resolution, shape, stroke style *etc*. To make them share the same network architecture, we uniformly resize the USPS images to the resolution of the MNIST digits. For a fair comparison, we follow the experiment protocol ("MNIST to USPS": 2,000 labeled MNIST samples and 1,800 unlabeled USPS samples for training, 1,000 labeled USPS samples for validation, the remaining USPS samples for testing; "USPS to MNIST": 1,800 labeled USPS samples and 2,000 unlabeled MNIST samples for training, 1,000 labeled MNIST samples for validation, the remaining MNIST samples for testing) in [15]. In both adaptation tasks, the classifier architecture is implemented with 2 convolutional layers[1] (CONV1: 20 $5 \times 5$ filters; CONV2: 50 $5 \times 5$ filters) and 2 fully connected layers (FC3: 500 activations; FC4: 10 activations), which is identical to that of [15].

Table 1 displays the adaptation results. As we can see, our method clearly surpasses the other adaptation methods. In particular, compared with DDC[2], our method is highly advantageous, which indicates that our discriminability regularization term significantly improves the classifier's discriminative ability in the target domain.

## 4.3  "SVHN to MNIST"

The SVHN dataset contains the images of street view house numbers. Different from the MNIST digits, the SVHN images contain significant variations, such as background clutter, rotation, slanting, embossing *etc*. As we can see in Fig. 1, the SVHN and MNIST samples are quite different in appearance, which makes

---

[1] The convolutional layers are followed by pooling layers, which is a default throughout the paper.
[2] Our method is equivalent to DDC when the discriminability regularization term $\mathcal{L}_T$ is removed.

**Table 1.** The classification accuracy (%) on the target domain obtained by our method and the recent state-of-the-art methods.

| Method | MNIST to USPS | USPS to MNIST | SVHN to MNIST | SYN-SIGNS to GTSRB |
|---|---|---|---|---|
| Source only | 80.2 | 65.8 | 54.9 | 70.2 |
| DDC [26] | 84.8 | 72.3 | 71.1 | 80.3 |
| DANN [6] | 82.1 | 89.7 | 73.9 | 78.9 |
| DRCN [7] | 91.8 | 73.7 | 82.0 | - |
| DTN [24] | - | - | 84.4 | - |
| kNN-Ad [22] | - | - | 78.8 | - |
| DSN [3] | - | - | 82.7 | 93.1 |
| CoGAN [15] | 91.2 | 89.1 | - | - |
| UNIT [14] | - | - | 90.5 | - |
| ADDA [25] | 89.4 | 90.1 | 76.0 | - |
| Ours | **92.3** | **91.9** | **93.1** | **93.7** |
| Target only | 96.1 | 98.7 | 99.4 | 99.8 |

the adaptation task of "SVHN to MNIST" fairly challenging. Also, its source only result in Table 1 clearly states the difficulty. Following [6], we use 73,257 labeled SVHN images and 59,000 unlabeled MNIST images for training, 1,000 MNIST samples for validation, and 10,000 MNIST samples for testing. In this experiment, the classifier network follows the architecture in [6], which contains 3 convolutional layers (CONV1: 64 $5 \times 5$ filters; CONV2: 64 $5 \times 5$ filters; CONV3: 128 $5 \times 5$ filters) and 3 fully connected layers (FC4: 3072 activations; FC5: 2048 activations; FC6: 10 activations). To make the network architectures suitable for both SVHN and MNIST, we convert the MNIST images to the RGB format and resize them to the resolution of the SVHN images.

As we can see in Table 1, our method obtains quite a strong advantage compared to the other methods, which clearly demonstrates that our method can really help to increase the classifier's discriminative ability on the target domain.

### 4.4 "SYN-SIGNS to GTSRB"

This task aims to demonstrate that our method is suitable for adapting from synthetic images to real images. Specifically, the samples in SYN-SIGNS are images of synthetic signs, simulating various imaging conditions. On the other hand, the samples in GTSRB are images of real traffic signs. Both datasets contain images from 43 classes. To conduct a fair comparison, we use the same experimental setting to that of [6]: 10,000 labeled SYN-SIGNS images and 31,367 unlabeled GTSRB images for training, 3,000 labeled GTSRB images for validation and the remaining GTSRB images for testing. Also, the classifier network follows

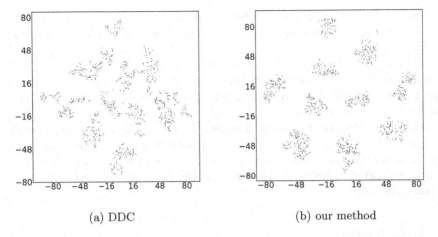

(a) DDC                                              (b) our method

**Fig. 2.** The visualization of the first fully connected layer on "MNIST to USPS", with source samples denoted by red points and target samples denoted by blue points. (a) the t-SNE embedding visualization for DDC; (b) the t-SNE embedding visualization for our method. (Color figure online)

the same structure of [6], which contains 3 convolutional layers (CONV1: 96 $5 \times 5$ filters; CONV2: 144 $3 \times 3$ filters; CONV3: 256 $5 \times 5$ filters) and 2 fully connected layers (FC4: 512 activations; FC5: 10 activations). In our preprocessing step, both the SYN-SIGNS and GTSRB images are resized to the resolution of $32 \times 32$.

The results displayed in Table 1 demonstrate that our method can effectively work in synthetic-to-real adaptation.

## 4.5  Discussion

In our ablation studies, we find that the domain alignment mechanism is important for our final performance. Only the source supervision loss and the target entropy regularization cannot result in desirable adaptation results. This can be attributed that simply minimizing the target samples' Softmax entropies may increase the risk of making incorrect decisions when the domain discrepancy is large. Therefore, the domain alignment loss that focuses on reducing the domain shift can help to lower this risk. Overall, domain alignment and the target entropy regularization work in a complementary way to improve the classifier's discriminative ability in the target domain.

## 4.6  Visualization

As indicated above, our method can significantly improve the target accuracy of unsupervised domain adaptation by regularizing the softmax probabilities of the target data. This indicates that the entropy minimization can really help to discover the discriminative clusters underlying the target data. In Fig. 2, we visualize the t-SNE embedding of the first fully connected layer on transfer task "MNIST to USPS".

As we expect, the target samples cluster well and keep accordance with the source data. Also, It is clear that the target samples are classified with large margins. Compared with the DDC model, which only considers domain alignment, our method can obtain "deep" feature that is both domain-invariant and discriminative for the target domain. Surprisingly, our method aligns the domains discrepancy much better than DDC, which qualitatively implies that the domain alignment term and the discriminability regularization term work complementarily. The entropy regularization mechanism provides an extra force to reduce the domain shift.

## 5    Conclusion

In this paper, we propose an effective unsupervised domain adaptation method. The main idea of our method is to improve the classifier's discriminative ability on the target domain through regularizing the classifier's prediction on the target samples. Specifically, the entropy minimization is performed on the softmax probabilities of the target data, which pushes the target samples away from the decision boundaries. In addiction, to prevent from simply assigning all the target samples to the same class, we further balance the relative size of each class through maximizing the entropy of the marginal distribution of class labels. Although high prediction confidence does not necessarily imply correctness, it pushes the network to discover discriminative clusters underlying the target samples.

Through combining the discriminability regularization term with the domain alignment term, we can obtain "deep" feature that is both domain-invariant and discriminative for the target domain. The experiments on several standard domain adaptation benchmarks clearly demonstrate that our method can significantly improve the classifier's discriminative ability on the target domain, and obtain better adaptation performance than the compared state-of-the-art unsupervised domain adaptation methods.

In our future work, we will consider how to leverage deep generative models like GANs or variational auto-encoders to generate target data with given class labels, in order to further improve the classifier's predictive accuracy on the target domain.

**Acknowledgments.** This paper is supported by the National Natural Science Foundation of China under grant No. 61572109, No. 11461006 and No. 61502082, and also the China Scholarship Council. Additionally, the authors would like to appreciate the anonymous reviewers for both the helpful and constructive comments.

# References

1. Ben-David, S., Blitzer, J., Crammer, K., Kulesza, A., Pereira, F., Vaughan, J.W.: A theory of learning from different domains. Mach. Learn. **79**(1), 151–175 (2010)
2. Bengio, Y., Courville, A., Vincent, P.: Representation learning: a review and new perspectives. IEEE Trans. Pattern Anal. Mach. Intell. **35**(8), 1798–1828 (2013)
3. Bousmalis, K., Trigeorgis, G., Silberman, N., Krishnan, D., Erhan, D.: Domain separation networks. In: Advances in Neural Information Processing Systems, pp. 343–351 (2016)
4. Donahue, J., et al.: DeCAF: a deep convolutional activation feature for generic visual recognition. In: International Conference on Machine Learning, pp. 647–655 (2014)
5. Fernando, B., Habrard, A., Sebban, M., Tuytelaars, T.: Unsupervised visual domain adaptation using subspace alignment. In: International Conference on Computer Vision, pp. 2960–2967 (2013)
6. Ganin, Y., et al.: Domain-adversarial training of neural networks. J. Mach. Learn. Res. **17**(59), 1–35 (2016)
7. Ghifary, M., Kleijn, W.B., Zhang, M., Balduzzi, D., Li, W.: Deep reconstruction-classification networks for unsupervised domain adaptation. In: Leibe, B., Matas, J., Sebe, N., Welling, M. (eds.) ECCV 2016. LNCS, vol. 9908, pp. 597–613. Springer, Cham (2016). https://doi.org/10.1007/978-3-319-46493-0_36
8. Glorot, X., Bordes, A., Bengio, Y.: Domain adaptation for large-scale sentiment classification: a deep learning approach. In: International Conference on Machine Learning, pp. 513–520 (2011)
9. Gong, B., Shi, Y., Sha, F., Grauman, K.: Geodesic flow kernel for unsupervised domain adaptation. In: 2012 IEEE Conference on Computer Vision and Pattern Recognition (CVPR), pp. 2066–2073. IEEE (2012)
10. Goodfellow, I., et al.: Generative adversarial nets. In: Advances in Neural Information Processing Systems, pp. 2672–2680 (2014)
11. Kingma, D.P., Ba, J.: Adam: a method for stochastic optimization. arXiv preprint arXiv:1412.6980 (2014)
12. Krause, A., Perona, P., Gomes, R.G.: Discriminative clustering by regularized information maximization. In: Advances in Neural Information Processing Systems, pp. 775–783 (2010)
13. Krizhevsky, A., Sutskever, I., Hinton, G.E.: Imagenet classification with deep convolutional neural networks. In: Advances in Neural Information Processing Systems, pp. 1097–1105 (2012)
14. Liu, M.Y., Breuel, T., Kautz, J.: Unsupervised image-to-image translation networks. arXiv preprint arXiv:1703.00848 (2017)
15. Liu, M.Y., Tuzel, O.: Coupled generative adversarial networks. In: Advances in Neural Information Processing Systems, pp. 469–477 (2016)
16. Long, M., Cao, Y., Wang, J., Jordan, M.I.: Learning transferable features with deep adaptation networks. arXiv preprint arXiv:1502.02791 (2015)
17. Long, M., Wang, J., Ding, G., Sun, J., Yu, P.S.: Transfer joint matching for unsupervised domain adaptation. In: Proceedings of the IEEE Conference on Computer Vision and Pattern Recognition, pp. 1410–1417 (2014)
18. Long, M., Zhu, H., Wang, J., Jordan, M.I.: Deep transfer learning with joint adaptation networks. arXiv preprint arXiv:1605.06636 (2016)
19. Pan, S.J., Tsang, I.W., Kwok, J.T., Yang, Q.: Domain adaptation via transfer component analysis. IEEE Trans. Neural Netw. **22**(2), 199–210 (2011)

20. Pan, S.J., Yang, Q.: A survey on transfer learning. IEEE Trans. Knowl. Data Eng. **22**(10), 1345–1359 (2010)
21. Rozantsev, A., Salzmann, M., Fua, P.: Beyond sharing weights for deep domain adaptation. IEEE Trans. Pattern Anal. Mach. Intell. (2018)
22. Sener, O., Song, H.O., Saxena, A., Savarese, S.: Learning transferrable representations for unsupervised domain adaptation. In: Advances in Neural Information Processing Systems, pp. 2110–2118 (2016)
23. Springenberg, J.T.: Unsupervised and semi-supervised learning with categorical generative adversarial networks. arXiv preprint arXiv:1511.06390 (2015)
24. Taigman, Y., Polyak, A., Wolf, L.: Unsupervised cross-domain image generation. arXiv preprint arXiv:1611.02200 (2016)
25. Tzeng, E., Hoffman, J., Saenko, K., Darrell, T.: Adversarial discriminative domain adaptation. arXiv preprint arXiv:1702.05464 (2017)
26. Tzeng, E., Hoffman, J., Zhang, N., Saenko, K., Darrell, T.: Deep domain confusion: maximizing for domain invariance. arXiv preprint arXiv:1412.3474 (2014)
27. Yosinski, J., Clune, J., Bengio, Y., Lipson, H.: How transferable are features in deep neural networks? In: Advances in Neural Information Processing Systems, pp. 3320–3328 (2014)

# Artificial Neural Networks Can Distinguish Genuine and Acted Anger by Synthesizing Pupillary Dilation Signals from Different Participants

Zhenyue Qin, Tom Gedeon$^{(\boxtimes)}$, Lu Chen, Xuanying Zhu, and Md. Zakir Hossain

Research School of Computer Science,
Australian National University, Canberra, Australia
{zhenyue.qin,lu.chen,xuanying.zhu,zakir.hossain}@anu.edu.au,
tom@cs.anu.edu.au

**Abstract.** Previous research has revealed that people are generally poor at distinguishing genuine and acted anger facial expressions, with a mere 65% accuracy of verbal answers. We aim to investigate whether a group of feedforward neural networks can perform better using raw pupillary dilation signals from individuals. Our results show that a single neural network cannot accurately discern the veracity of an emotion based on raw physiological signals, with an accuracy of 50.5%. Nonetheless, distinct neural networks using pupillary dilation signals from different individuals display a variety of genuineness for discerning the anger emotion, from 27.8% to 83.3%. By leveraging these differences, our novel Misaka neural networks can compose predictions using different individuals' pupillary dilation signals to give a more accurate overall prediction than even from the highest performing single individual, reaching an accuracy of 88.9%. Further research will involve the investigation of the correlation between two groups of high-performing predictors using verbal answers and pupillary dilation signals.

**Keywords:** Emotion veracity · Neural networks · Pupillary dilation

## 1 Introduction

Dilation of the pupil reflects a range of cognitive processes including interest [10], motivation [23], and emotionality [28]. Research has revealed that pupil dilation reflects the mechanisms of creating and retrieving memories [11]. Specifically, pupil dilation is positively correlated with people's confidence in their memories [21]. Hence, pupillary reflex has persisted as an index of cognitive demand [13]. Later research suggests pupillary reactions *summed index* of brain processes during cognitive activities [16]. In particular, pupillary dilation reflects the inner activity of the autonomic nervous system, which is vital for maintaining the equilibrium of the body [17], and is hence not under conscious control.

© Springer Nature Switzerland AG 2018
L. Cheng et al. (Eds.): ICONIP 2018, LNCS 11305, pp. 299–310, 2018.
https://doi.org/10.1007/978-3-030-04221-9_27

The autonomic nervous system consists of two sub-systems, namely the sympathetic nervous system and the parasympathetic nervous system [28]. The former encourages whereas the latter inhibits pupil dilation. That is, these two sub-components operate in an antagonistic fashion to support the processes of the autonomic nervous system [17], the effects of which are visible via pupil dilation. Studies on the relationship between cognitive activities and pupillary dilation often utilize discrete stimuli, since emotional stimuli can cause significant effects on the autonomic nervous system [1]. These stimuli can also result in distinctive waveforms corresponding to different mental activities [28]. For example, the pupil dilates when people observe beneficial as well as adverse pictures [27].

Emotions can be characterized as a combination of two dimensions, arousal and valence [18]. Arousal corresponds to how strong the emotion is, and valence shows how positive the emotion is, as Fig. 1 indicates. Despite the simplicity of this model for representing emotions, researchers have successfully utilized it in a range of tasks, such as emotion recognition and memory studies [15, 22, 29].

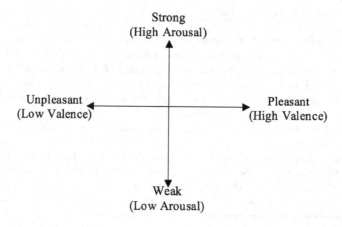

**Fig. 1.** The model of emotions [4].

Facial expressions are effective means of communication that can convey emotional information more promptly than languages, with which humans can rapidly recognize affective states of others [3]. Unlike the physiological signals mentioned above that are directly governed by the autonomic nervous system, people may perform acted facial expressions that contradict the expresser's true affective state [5]. For example, a salesperson may present acted smiles to pretend friendly attitudes. Research has indicated six basic facial expressions that are readily recognizable across dissimilar cultural backgrounds, namely anger, happiness, fear, surprise, disgust, and sadness [3]. In this paper, we consider anger due to its fundamental importance as a basic emotion.

Due to the divergent emotional strengths of genuine and acted emotions on the autonomic nervous system, there may exist differences between the physiological signals corresponding with genuine and acted emotions. Previous work done by Qin [25] revealed that peoples' capabilities for discerning the veracity of emotions vary. Neural networks can leverage this variety by assigning positive and negative weights to high and low accuracy discerners in order to aggregate peoples' responses and give a final higher precision prediction [25]. Preliminary statistical tests also indicate that people's physiological signals differ on distinguishing genuine and acted anger videos. That is, some participants' physiological signals corresponding to the two different kinds of videos vary more than others. Thus, we hypothesize that we can also aggregate physiological signals from different participants in order to give higher accuracy prediction of the source video label. That is, whether the stimulus is genuine or acted anger.

In this paper, we propose novel Misaka networks, which can predict the veracity of a person's expressed emotions by aggregating various observers' pupillary dilation signals. While previous related work focuses on using psychological signals from single participant, we novelly utilized physical signals from multiple individuals and showed better results. This research can be potentially applied to identify the true emotion of a person from his or her observers. Compared with collecting verbal answers from participants, utilizing physiological signals does not require a further process of interviewing and can possibly predict others' emotions ad-hoc and in-time. That is, if one can obtain observers' dynamic pupillary dilation signals in a timely fashion, one may predict the observed person's current emotion in real time.

This paper is organized as follows: We will introduce the structure of our Misaka neural networks using crowdsourcing techniques. Then, we will report our results and present discussion on the results obtained. We conclude this paper with a discussion of the limitations in our work and future work to tackle those limitations.

## 2   Method

### 2.1   Stimuli

This paper utilizes the pupillary dilation signals used by Chen [5]. The elicitation stimuli are videos sourced from YouTube, with 10 each corresponding with genuine and acted anger, respectively. We choose anger as the emotion to study because anger is one of the six basic emotions that are identifiable independent from cultural backgrounds [7], so we hypothesize that the results of this paper on anger can be generalized to other primary emotions.

Genuine emotion expressions were collected from live news reports and documentaries and acted ones were sourced from movies containing similar scenes. These videos were picked to balance ethnicity, gender and background context as far as possible. Further, they have been processed with greyscale normalization to reduce the differences between videos other than the veracity of emotions.

Avezier indicated that the contextual backgrounds for demonstrating emotional expressions are vital in order for humans to effectively discern different emotions [2]. Therefore, the stimuli adopted in this paper retain some contextual backgrounds to better simulate scenes from daily life.

Due to the different number of frames of different stimuli, we remove the two shortest videos due to their significantly lower number of frames, namely 60 and 89. Then, we truncate all the stimuli to only include the beginning 105 frames, which is the number of frames for the shortest remaining stimulus videos. Moreover, although there were 20 participants who took part in our experiment in total, due to the fact that some people were absent from some experiments, we only have 12 participants' complete pupillary dilation signals with all the videos. Thus, we only utilize data from these 12 participants in this study. This degree of loss of data is within the normal range with the mobile sensors use, trading non-intrusiveness and hence more natural behavior for occasional data loss.

Figure 2 demonstrated how the experiments were conducted. Participants were provided with oral instructions by the experimenters prior to the experiments. As Fig. 2 indicates, pupillary dilation was tracked with a remote Eye Tribe tracker at 60 Hz [6]. Furthermore, we also collected the subjects' skin conductance, blood volume pressure, and heart rate during the same experiments. They have not been utilized in the analysis of this paper and may be used for future work.

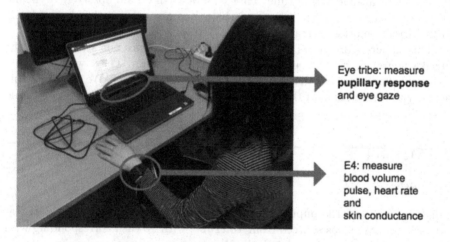

Eye tribe: measure **pupillary response** and eye gaze

E4: measure blood volume pulse, heart rate and skin conductance

**Fig. 2.** The experimental setup [5].

## 2.2   Neural Networks

Artificial neural networks are simulations of animal brain information processing. Theoretically, a multiple layer feedforward neural network can approximate any measurable functions [14]. Due to the recent increase of computational power and the vast amount of data, neural networks now have dramatically improved in applications, such as object detection and speech recognition [26].

## 2.3   Crowd Prediction

Social science researchers have demonstrated promising results of utilizing the crowd to predict future outcomes [30]. For example, the prediction results of five US presidential elections using crowd prediction were more accurate than the traditional polls, where the latter was basically random guessing [19]. Crowd prediction has also been extended to a range of industrial applications, including in healthcare companies and technology corporations [24].

Further research on crowd prediction has also acknowledged that people vary in capabilities. That is, instead of assembling every predictor's opinion into equal consideration, top-performing predictions will be extracted first and their answers synthesized as the representation responses of the whole crowd [20]. This elite-based method showed a 50% greater accuracy than composing crowd fore-casting teams [8].

## 2.4   Misaka Networks

In this paper, we combine both neural network and crowd prediction techniques in order to predict a person's true emotion from observer reactions to their video performance. A Misaka network includes a collection of feedforward neural networks to predict whether a pupillary dilation physiological signal corresponds to a genuine or acted anger. The name Misaka network? was inspired by a Japanese anime where clones of Misaka can demonstrate much more powerful capabilities when working as a cohesive group than as individuals [12]. Each of these neural networks is trained to predict whether a pupillary dilation signal belongs to genuine or acted anger, using one participant's data, trained on that participant's reactions to 18 videos. We call these neural networks discerners. In our case, we have trained 12 discerners corresponding to the 12 participants. Specifically, a discerner has an input layer with 105 nodes (corresponding to 105 frames), a hidden layer with 100 nodes and an output layer with 2 nodes. Each discerner is trained using leave-one-out cross-validation.

A Misaka network also contains another feedforward neural network to combine all the discerners' responses for one video, based on the discerners' previous accuracies. We call this neural network an aggregator. For example, if the accuracy of discerner $i$ was higher previously than discerner $j$, then the aggregator should assign more weight to a future prediction from discerner $i$. Moreover, the aggregator should also be able to reverse answers from the poor-performing discerners, and in general, learn to best aggregate the information from the discerners. An aggregator has an input layer with 12 nodes, a hidden layer with 120 nodes and an output layer with 2 nodes.

As Fig. 3 indicates, we collected 18 pupillary dilation time-series signals from each participant by letting him or her watch 18 videos, in an order balanced fashion to eliminate the effects of presentation order. Among these 18 videos, 10 correspond to genuine and the other 8 correspond to acted anger, respectively. Then, we train a discerner to predict the source of a signal. That is, whether a signal comes from watching a genuine anger video, or an acted one. We conduct leave-one-video-out cross-validation for each discerner to predict the label of every video. Thus, we utilize 17 signals to train and let the discerner predict the source of the remaining one signal. As a result, for each of the 12 discerners, we will have 18 predicted results for each video.

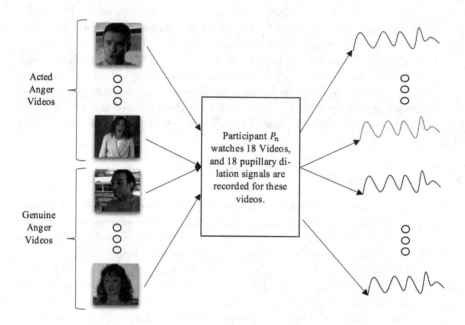

Acted Anger Videos

Participant $P_n$ watches 18 Videos, and 18 pupillary dilation signals are recorded for these videos.

Genuine Anger Videos

**Fig. 3.** An illustration of recording pupillary dilation signals from a participant.

Subsequently, we utilize the aggregator to learn the reliability of discerners in order to best combine their answers to give more accurate predictions. In our case, the input data is an $18 \times 12$ matrix, where each feature row represents a discerner's prediction for a given anger video. For example, a vector $<1, 0, ..., 1, 1>$ means that the predictions from the first two, ..., and the last two regarding an anger video are genuine, acted, ..., genuine and genuine, respectively. That is, the aggregator will learn from the historical answers from discerners in order to determine their reliabilities. Based on these reliabilities, the aggregator will eventually combine all the discerners' opinions and ideally, give a more precise prediction. We also conducted leave-one-video-out cross-validation for the aggregator. Nevertheless, due to our limited amount of participants, we test on *the same* videos. However, the aggregator will not know the correct labels, it only learns the credibilities from the discerners. Therefore, using the same data again will not result in unreliable issues.

Our previous work on crowdsourcing verbal responses has indicated that a minimal number of participants being 20 may be necessary for an accurate aggregator to work properly [25]. Here, we only have 12 participants. In order to reduce the unexpected effects caused by insufficient participants, we only utilize discerners that are strongly accurate or inaccurate (if significantly worse than chance) by only considering those outside $50\% \pm 10\%$ accuracy. This is because we wish to minimize the learning pressure on the aggregator by removing inputs that may confuse it, as discerners close to 50% (the chance value) provide noisy signals.

# 3   Results and Discussion

## 3.1   Discerning Reliabilities of Detecting Anger Differ with People's Varying Pu0pillary Dilation Signals

When observing genuine and acted anger facial expressions, people's pupillary dilation signals corresponding to those two kinds of emotional expressions can vary. Our preliminary analysis indicated this from Student's t-tests. For each participant, we averaged each signal so that we would collect 10 + 8 means, which correspond to genuine and acted stimuli, separately. The calculated statistical significances among the first 8 pairs of signal means differ among different participants, as Table 1 indicates.

**Table 1.** Statistical t-test significance calculated from the means of every individual's genuine and acted pupillary dilation signals.

| Participant | P1 | P2 | P3 | P4 | P5 | P6 |
|---|---|---|---|---|---|---|
| T-test significance | 0.8856 | 0.3494 | 0.6720 | 0.5765 | 0.8530 | 0.3363 |
| Participant | P7 | P8 | P9 | P10 | P11 | P12 |
| T-test significance | 0.0036 | 0.7592 | 0.2997 | 0.4486 | 0.5623 | 0.5959 |

In more detail, we may interpret the t-test significances as a reflection of the deviance between signals corresponding to genuine and acted anger emotions. These different values imply that some people's pupillary dilation signals may be more distinguishable for the veracity of the source anger videos than others.

Table 2 shows the accuracy of discerners, each corresponding to a participant's pupillary dilation signals. For instance, discerner D1 is trained with participant P1's physiological signals. The accuracy is calculated by averaging the cross-validation results of predicting the signal source with a discerner. Specifically, taking discerner D1 as an example, we train it with 17 signals from participant P1. Then we let D1 predict whether the remaining signal from P1 is sourced from a genuine anger video, or an acted one. The accuracy of D1 is defined as the proportion of correctly predicted signal sources over the total number of signals. In our data, discerner D1 demonstrated the highest accuracy, whilst discerner D5 showed a poor accuracy with merely 27.8%, which is noticeably worse than chance. We could speculate that this participant has had an unusual emotional background, such as only encountering anger in videos, and so judges genuine anger incorrectly, consistently.

**Table 2.** Different discerners have distinct prediction accuracies, with 6 close to chance.

| Discerner | D1 | D2 | D3 | D4 | D5 | D6 |
|---|---|---|---|---|---|---|
| Accuracy | 83.3% | 44.4% | 50.0% | 50.0% | 27.8% | 50.0% |
| Discerner | D7 | D8 | D9 | D10 | D11 | D12 |
| Accuracy | 50.0% | 38.9% | 44.4% | 61.1% | 66.7% | 38.9% |

## 3.2 Aggregating the Prediction Results from Various People's Pupillary Dilation Signals Can Increase the Accuracy of Prediction

As discussed previously, we used **the same** data for aggregation due to the limited amount of collected signals. The aggregator result is 88.9% accurate, showing that an aggregator, by combining multiple discerners' predictions based on their prior response accuracies, can outcompete the highest-accuracy discerner. The aggregator, as illustrated in Fig. 4, successfully learned the pattern of their reliabilities and that it should assign more accurate discerners like D1 mostly positive weights. Conversely, it gave poor-performing ones like D5 mostly negative weights. Eventually, this aggregator demonstrated 88.9% accuracy with cross-validation.

## 3.3 A Mathematical Explanation for the Feasibility of Misaka Networks

The problem faced by Misaka networks can be formally abstracted as given a collection of N discerners $D_1, D_2, ..., D_N$, each with an accuracy $A_1, A_2, ..., A_N$. If these $N$ discerners have reached a consensus on predicting a binary result as 1, what is the probability of the result actually being 1?

We apply Bayes' theorem. The probability of the result $R$ being 1 can be expressed as

$$P(R = 1|D_1 = 1, D_2 = 1, ..., D_N = 1)$$
$$= \frac{P(D_1 = 1, D_2 = 1, ..., D_N = 1|R = 1) \times P(R = 1)}{P(D_1 = 1, D_2 = 1, ..., D_N = 1)} \quad (1)$$

With a further assumption of $P(R = 1) = P(R = 0) = 0.5$ and $D_1, D_2, ..., D_N$ are conditionally independent given $R$, Eq. 1 can be further expressed as

$$\frac{\prod_{i=1}^{N} P(D_i = 1|R = 1) \times P(R = 1)}{\prod_{i=1}^{N} P(D_i = 1|R = 1) \times P(R = 1) + \prod_{i=1}^{N} P(D_i = 1|R = 1) \times P(R = 0)} \quad (2)$$

For example, given we have two discerners $D_1$ and $D_2$ with accuracies $A_1$ being 0.8 and $A_2$ being 0.7 and they have reached a consensus predicting the result being 1. Formula 2 tells us that the overall probability of the result being 1 is approximately 0.9032, which is higher than 0.8.

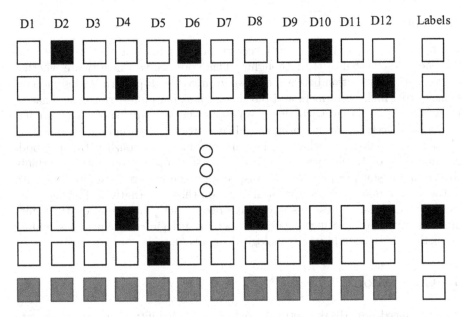

**Fig. 4.** An illustration of an aggregator. White and black represent genuine and acted anger respectively. Each row represents predictions from different discerners regarding the same video, whose correct label is given in the last "labels" column. The aggregator learns from the first 17 to predict the label for the last green row (genuine in the example shown). (Color figure online)

## 4    Limitations and Future Work

### 4.1    Limitations

The number of stimuli is limited, specifically, only 18 videos, also there are a limited number of participants. Thus, it is infeasible to split these videos into two groups, where the first half is to train discerners including cross-validation within that group, and then use the second half to train and test the synthesizer by predicting the genuineness of each video, based on the previous learnt accuracies of each discerner, again using cross-validation on that group. This is because it requires about 20 videos to consistently infer the reliability of each discerner. Instead, cross-validation was undertaken with the same videos for both discerners and aggregators. Nonetheless, our reported results are still relatively reliable since discerners will not provide correct labels to aggregators. In the future, we will collect more videos containing anger and other expressions as stimuli in order to more thoroughly test a Misaka network by splitting videos into two groups, one for training all the discerners and the other for training and testing the aggregator.

## 4.2  Future Work

Further, people's capabilities on discerning the veracity of emotions also differ when they verbalize their thoughts. Therefore, we will investigate whether there exists a correlation between the two groups of high-performing individuals, being verbal high-performers and physiological signal high-performers. That is, we would like to discover whether the people who are capable of giving quality emotional discerning results from verbalizing produce even more discriminating physiological signals, or whether they are just in better touch with their bodies/emotions, or to discover whether it is possible to be more correct verbally than from physiological signals. We suspect that the degree of emotional valence in the stimuli may have a substantial effect on these alternatives. That is, more exaggerated acted expressions may make people put more belief on their genuineness. We may also investigate whether fuzzy logic can help in assembling predictions from individual psychological signals [9].

## 5  Conclusion

We introduced our Misaka networks, which use reliability signals to aggregate outputs of discerner networks trained on participants' raw physiological signal data. We achieved state-of-the-art results on the (small) sizes of datasets common in close-to-real-world recording of emotional and physiological data.

In summary, we discovered that people's pupillary dilation signals vary in their ability to discern the veracity of anger facial expressions. This variety ranges from 27.8% to 83.3%. We can leverage these differences by training another neural network to learn these patterns of these different reliabilities in order to assign appropriate weights for each participant. After aggregating answers from different participants' physiological signals, the prediction accuracy can be boosted to 88.9%. This combination of discerning networks trained on individuals and an aggregator trained on their reliability compose our Misaka network.

**Acknowledgments.** The authors acknowledge Dongyang Li, Liang Zhang and Zihan Wang for the suggestion of applying Bayes' theorem in the probability calculation.

## References

1. Andreassi, J.L.: Psychophysiology: Human Behavior & Physiological Response, 5th edn. Lawrence Erlbaum Associates Publishers, Mahwah (2007)
2. Aviezer, H., Hassin, R., Bentin, S., Trope, Y.: Putting facial expressions back in context. In: Ambady, N., Skowronsky, J.J. (eds.) First Impressions, chap. 11, pp. 255–286. Guilford Press, New York (2008)
3. Batty, M., Taylor, M.J.: Early processing of the six basic facial emotional expressions. Cogn. Brain Res. **17**(3), 613–620 (2003)
4. Chanel, G., Ansari-Asl, K., Pun, T.: Valence-arousal evaluation using physiological signals in an emotion recall paradigm. In: 2007 IEEE International Conference on Systems, Man and Cybernetics, pp. 2662–2667 (2007)

5. Chen, L., Gedeon, T., Hossain, M.Z., Caldwell, S.: Are you really angry?: detecting emotion veracity as a proposed tool for interaction. In: Proceedings of the 29th Australian Conference on Computer-Human Interaction, Brisbane, Queensland, Australia, pp. 412–416. ACM (2017)

6. Dalmaijer, E.: Is the low-cost eyetribe eye tracker any good for research? Technical report, PeerJ PrePrints (2014)

7. Ekman, P.: An argument for basic emotions. Cogn. Emot. **6**(3–4), 169–200 (1992)

8. Frood, A.: Work the crowd. New Sci. **237**(3166), 32–35 (2018)

9. Gao, Y., Xiao, F., Liu, J., Wang, R.: Distributed soft fault detection for interval type-2 fuzzy-model-based stochastic systems with wireless sensor networks. IEEE Trans. Ind. Inf. (2018, early access version)

10. de Gee, J.W., Knapen, T., Donner, T.H.: Decision-related pupil dilation reflects upcoming choice and individual bias. Proc. Nat. Acad. Sci. **111**(5), E618–E625 (2014)

11. Goldinger, S.D., Papesh, M.H.: Pupil dilation reflects the creation and retrieval of memories. Curr. Dir. Psychol. Sci. **21**(2), 90–95 (2012)

12. Haimura, M.: A Certain Magical Index. ASCII Media Works, Tokyo (2013)

13. Hess, E.H., Polt, J.M.: Pupil size in relation to mental activity during simple problem-solving. Science **143**(3611), 1190–1192 (1964)

14. Hornik, K., Stinchcombe, M., White, H.: Multilayer feedforward networks are universal approximators. Neural Netw. **2**(5), 359–366 (1989)

15. Jirayucharoensak, S., Pan-Ngum, S., Israsena, P.: EEG-based emotion recognition using deep learning network with principal component based covariate shift adaptation. Sci. World J. (2014)

16. Kahneman, D., Beatty, J.: Pupil diameter and load on memory. Science **154**(3756), 1583–1585 (1966)

17. Kim, K.H., Bang, S.W., Kim, S.R.: Emotion recognition system using short-term monitoring of physiological signals. Med. Biol. Eng. Comput. **42**(3), 419–427 (2004)

18. Lang, P.J.: The emotion probe: studies of motivation and attention. Am. Psychol. **50**(5), 372 (1995)

19. Manski, C.F.: Interpreting the predictions of prediction markets. Econ. Lett. **91**(3), 425–429 (2006)

20. Mellers, B., et al.: Identifying and cultivating superforecasters as a method of improving probabilistic predictions. Perspect. Psychol. Sci. **10**(3), 267–281 (2015)

21. Papesh, M.H., Goldinger, S.D., Hout, M.C.: Memory strength and specificity revealed by pupillometry. Int. J. Psychophysiol. **83**(1), 56–64 (2012)

22. Partala, T., Jokiniemi, M., Surakka, V.: Pupillary responses to emotionally provocative stimuli. In: Proceedings of the 2000 Symposium on Eye Tracking Research & Applications, pp. 123–129. ACM (2000)

23. Pletti, C., Scheel, A., Paulus, M.: Intrinsic altruism or social motivationwhat does pupil dilation tell us about children's helping behavior? Front. Psychol. **8**, 2089 (2017)

24. Polgreen, P.M., Nelson, F.D., Neumann, G.R., Weinstein, R.A.: Use of prediction markets to forecast infectious disease activity. Clin. Infect. Dis. **44**(2), 272–279 (2007)

25. Qin, Z., Gedeon, T., Caldwell, S.: Neural networks assist crowd predictions in discerning the veracity of emotional expressions. arXiv Preprint arXiv:1808.05359 (2018)

26. Schmidhuber, J.: Deep learning in neural networks: an overview. Neural Netw. **61**, 85–117 (2015)

27. Steinhauer, S.: Pupillary dilation to emotional visual stimuli revisited. Psychophysiology **20**, S472 (1983)
28. Steinhauer, S.R., Siegle, G.J., Condray, R., Pless, M.: Sympathetic and parasympathetic innervation of pupillary dilation during sustained processing. Int. J. Psychophysiol. **52**(1), 77–86 (2004)
29. Wagner, J., Kim, J., André, E.: From physiological signals to emotions: implementing and comparing selected methods for feature extraction and classification. In: IEEE International Conference on Multimedia and Expo, ICME 2005, Amsterdam, Netherlands, pp. 940–943. IEEE (2005)
30. Wolfers, J., Zitzewitz, E.: Prediction markets. J. Econ. Perspect. **18**(2), 107–126 (2004)

# Weakly-Supervised Man-Made Object Recognition in Underwater Optimal Image Through Deep Domain Adaptation

Chaoqi Chen, Weiping Xie, Yue Huang[✉], Xian Yu, and Xinghao Ding

Fujian Key Laboratory of Sensing and Computing for Smart City,
School of Information Science and Engineering, Xiamen University, Xiamen 361005,
Fujian, China
yhuang2010@xmu.edu.cn

**Abstract.** Underwater man-made object recognition in optical images plays important roles in both image processing and oceanic engineering. Deep learning methods have received impressive performances in many recognition tasks in in-air images, however, they will be limited in the proposed task since it is tough to collect and annotate sufficient data to train the networks. Considered that large-scale in-air images of man-made objects are much easier to acquire in the applications, one can train a network on in-air images and directly applying it on underwater images. However, the distribution mismatch between in-air and underwater images will lead to a significant performance drop. In this work, we propose an end-to-end weakly-supervised framework to recognize underwater man-made objects with large-scale labeled in-air images and sparsely labeled underwater images. And a novel two-level feature alignment approach, is introduced to a typical deep domain adaptation network, in order to tackle the domain shift between data generated from two modalities. We test our methods on our newly simulated datasets containing two image domains, and achieve an improvement of approximately 10 to 20 % points in average accuracy compared to the best-performing baselines.

**Keywords:** Man-made object recognition
Underwater optical images · Weakly-supervised
Deep domain adaptation

The work is supported in part by National Natural Science Foundation of China under grants of 81671766, 61571382, 61571005, 81301278, 61172179 and 61103121, in part by Natural Science Foundation of Guangdong Province under grant 2015A030313007, in part by the Fundamental Research Funds for the Central Universities under Grants 20720180059 20720160075, in part of the Natural Science Foundation of Fujian Province of China (No. 2017J01126).

L. Cheng et al. (Eds.): ICONIP 2018, LNCS 11305, pp. 311–322, 2018.
https://doi.org/10.1007/978-3-030-04221-9_28

# 1    Introduction

Optical imaging is recommended as an important tool in underwater object detection. With the development of underwater optical image sensors, the underwater man-made target recognition in optical images has attracted more attention in both oceanic engineering and image processing. As shown in Fig. 1, the major challenge in underwater optical image analysis is poor image quality due to the strong turbidity caused by many factors, such as impurities and large water density in deep ocean [1,11,19]. Moreover, underwater images are relatively blue or green compared to the in-air images because the light is exponentially attenuated in water [2].

**Fig. 1.** Examples of underwater optical images with poor image quality.

Deep learning approaches, when trained on massive amounts of labeled data, can learn representations which have demonstrated impressive performances in object recognition task from natural images [6]. However, collecting and annotating large-scale underwater images is an extremely expensive and time-consuming process in real-world application. Meanwhile, it is much easier to acquire sufficient number of labeled in-air images of man-made objects (with a similar but different data distribution). It is nature to think that one can train a deep classifier on in-air images and directly applying it on underwater images. However, the distribution mismatch between in-air and underwater images leads to a significant performance drop [7], e.g. the average classification accuracy drops from 79.3% (if train on underwater images) to 21.7% (if train on in-air images, see Table 2). Motivated by this, we introduce a Weakly-Supervised Domain Adaptation Network (WSDAN) for solving this problem. Due to the scarce labeled underwater samples, we leverage the massive amount of fully labeled in-air images and sparsely labeled underwater images to train a network, after that, testing on underwater images. There is no overlap between training and testing data.

In the proposed work, we aims to transfer knowledge from a label-rich source domain (in-air images) to a sparsely labeled target domain (underwater images). This setting is closely to the practical application that we have numerous accesses to the labeled in-air images, and is often prohibitive to the labeled underwater images but sparsely labeled is feasible. We will introduce low-level and high-level feature alignment by adversarial learning and joint distribution adaptation to simultaneously reduce the domain discrepancy and learn target discriminative representations. The contributions of the proposed work can be concluded in three-folds: (1) we formally formulate the underwater man-made object recognition problem in a weakly-supervised domain adaptation setting; (2) we propose a novel end-to-end architecture, which simultaneously takes two-level feature alignment into consideration which allows two important and complementary aspects of knowledge to be transferred: low-level relatively generic features and high-level class-specific semantic information. (3) compared with the best-performing baselines, the proposed approach improves the recognition performances significantly even when very few (one or two in each category) labeled underwater samples is available.

## 2   Related Works

### 2.1   Underwater Man-Made Object Recognition

Recently, few studies on image analyses in underwater man-made objects recognition from underwater optical images have been reported. [3] proposed an underwater man-made object recognition framework by integrating different image processing techniques, including equalization as a preprocessing technique, line and edge detection, and Euclidean shape prediction. [4] proposed a detection method for underwater man-made objects based on color and shape features equipped with color correction. In [5], a system for detection of unconstrained man-made objects in unconstrained subsea videos was reported, where object contours were first extracted as stable features, and then Bayesian classifier was employed to predict if there is a man-made object in the image. [20] proposed to recognize underwater man-made objects in photo images by an exhaustive search of key points or small pieces of borders.

### 2.2   Domain Adaptation

Domain adaptation is an important topic in transfer learning which aims to transfer knowledge from source to target domains to substantially reduce the domain discrepancy. Many works are based on distance metric to measure the domain discrepancy, e.g. maximum mean discrepancy (MMD) [9,14], KL-divergence [8] and the mean and covariance of the two distributions [22], or using generative adversarial networks (GAN) [12,16], using adversarial losses to learn domain-invariant representations to reduce the domain discrepancy.

Our approach is different from previous works, because we simultaneously consider the two-level feature alignments and effectively leverage the weak supervision information to learn target-discriminative representations instead of only domain-invariant representations.

# 3    Methods

## 3.1    Simulation of Underwater Optical Images

Underwater images are mostly generated based on simulation of underwater environments. From Fig. 1, we can see that color is the most dominant feature of underwater images. Nguyen et al. [10] proposed a color transfer method based on illumination awareness and 3D gamut to manipulate the color value of reference images to generate images with the same appearance.

However, using only a color transfer we cannot realistically simulate the underwater environment. Therefore, we apply a turbidity simulation on top of color transfer to obtain a better representation. Therefore, the resultant signal is composed of two components, of which the first is the direct transmission component defined by:

$$D = I_{color}e^{-\eta z}, \tag{1}$$

where $I_{color}$ is the image obtained by color transfer, $\eta$ is the coefficient of diffusion attenuation obtained from a given real underwater patch, and $z$ represents the adjustable distance between $I_{color}$ and the reference underwater image; the higher the value of $z$ is, the higher turbidity will be.

The second term in the resultant signal is obtained by backscattering:

$$B = B_{\infty}(1 - e^{-\eta z}), \tag{2}$$

where $B_{\infty}$ is the backscatter in the line of sight which tends to infinity in water.

The resultant underwater image is generated by combining these two terms as follows:

$$I_{underwater} = D + B - D{\cdot}B, \tag{3}$$

where $\cdot$ denotes the element-wise multiplication.

## 3.2    Notations for WSDAN

In our WSDAN model, we are given a source domain $D_s = \{(x_i{}^s, y_i{}^s)\}_{i=1}^{n_s}$ of $n_s$ labeled in-air images, $x_i{}^s \in X^s$, with associated labels, $y_i{}^s \in Y^s$. Similarly, given a target domain $D_t = (\tilde{x}_i^t)\}_{i=1}^{n_t} + \{(x_i{}^t, y_i{}^t)\}_{i=1}^{m_t}$ of $n_t$ unlabeled underwater images, $\tilde{x}_i^t \in X^t$, as well as $m_t$ labeled underwater images, $x_i^t \in X^t$, with associated labels, $y_i{}^t \in Y^t$. And the number of unlabeled underwater images is assumed to be much larger than the number of labeled underwater images, $n_t \gg m_t$. We assumed that there is domain shift between $D_s$ and $D_t$, e.g. differences in the noise, resolution, illumination, color. And mathematically, we assume that there is a difference between the joint distributions $P(X^s, Y^s)$ and $Q(X^t, Y^t)$. The goal being to learn a function: $Y = f(X)$ that during testing is going to perform well on data from underwater images.

The proposed model consists of two parts, the architecture of our overall model is presented in Fig. 2. The main idea of our model is to exploit two-level feature alignment to learn a domain-invariant space that simultaneously maximizes the confusion between two domains while encourages the learning of target-discriminative representations.

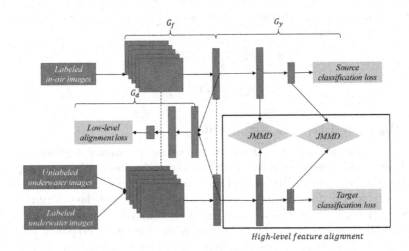

**Fig. 2.** Our overall architecture for weakly-supervised domain adaptation network. The dotted lines denote weight-sharing between source and target domains. We separate the network into three modules: feature extractor $G_f$, source label predictor $G_y$, domain discriminator $G_d$ and with associated parameters $\theta_f$, $\theta_y$, $\theta_d$.

### 3.3 Low-Level Feature Alignment

The low-level features usually contain relatively generic features between source and target domains. However, due to the large domain discrepancy, we still need to align them to be domain-invariant. Adversarial learning has been successfully introduced to domain adaptation by extracting domain-invariant features which can reduce the domain discrepancy [13,16]. The adversarial learning method utilizes two players to align distributions in an adversarial manner, where the first player is the domain discriminator $G_d$ trained to distinguish the source domain from the target domain, and the second player is the feature extractor $G_f$ fine-tuned simultaneously to confuse the domain discriminator.

The input $x$ is first embedded by $G_f$ to a $D$-dimensional feature vector $\mathbf{f} \in R^D$, i.e. $\mathbf{f} = G_f(x; \theta_f)$. In order to make $\mathbf{f}$ domain-invariant, the parameters $\theta_f$ of feature extractor $G_f$ are optimized by maximizing the loss of the domain classifier $G_d$, while simultaneously the parameters $\theta_d$ of domain discriminator $G_d$ are optimized by minimizing the loss of the domain discriminator. In addition, we also aim to minimize the loss of the label predictor $G_y$ for source data. And the objective of the network defined as

$$F_1(\theta_f, \theta_d, \theta_y) = \frac{1}{n_s} \sum_{x_i \in D_s} L_y(G_y(G_f(x_i; \theta_f); \theta_y), y_i)$$
$$-\frac{\lambda}{n} \sum_{x_i \in (D_s \cup D_t)} L_d(G_d(G_f(x_i; \theta_f); \theta_d), d_i), \tag{4}$$

where $n = n_s + n_t$ and $\lambda$ is a trade-off parameter between the two objectives. After training convergence, the parameters $\hat{\theta}_f, \hat{\theta}_y, \hat{\theta}_d$ will deliver a saddle point of the objective function (4):

$$(\hat{\theta}_f, \hat{\theta}_y) = arg \min_{\theta_f, \theta_y} F(\theta_f, \theta_d, \theta_y), \tag{5}$$

$$(\hat{\theta}_d) = arg \min_{\theta_d} F(\theta_f, \theta_d, \theta_y), \tag{6}$$

### 3.4 High-Level Feature Alignment

In the previous section, we introduced a method to enforce alignment of the global low-level features between source and target domains with no class specific transfer. The high-level features contain class-specific features. Here, we aim to not only reduce the domain discrepancy, but also learn target class-discriminative representations, i.e., we need to align the same class between source and target domains. We propose a variant of joint maximum mean discrepancy (JMMD) [17] as a regularization term to handle the problem that we mentioned above, and it is computed as the squared distance between the empirical kernel mean and a classification loss:

$$D_L(P, Q) = \frac{1}{n_s{}^2} \sum_{i=1}^{n_s} \sum_{j=1}^{n_s} \prod_{l \in L} K^l(z_i^{sl}, z_j^{sl}) + \frac{1}{n_t{}^2} \sum_{i=1}^{n_t} \sum_{j=1}^{n_t} \prod_{l \in L} K^l(z_i^{tl}, z_j^{tl})$$
$$-\frac{2}{n_s n_t} \sum_{i=1}^{n_s} \sum_{j=1}^{n_t} \prod_{l \in L} K^l(z_i^{sl}, z_j^{tl}) + \frac{1}{m_t} \sum_{i=1}^{m_t} \ell(f(x_i), y_i), \tag{7}$$

where $L = \{fc7, fc8\}$, $K$ is the Gaussian kernel function defined by $K(x_i, x_j) = e^{-\|x_i - x_j\|^2/\gamma}$ and $\ell$ is the standard cross-entropy loss. We use the joint distribution of activations in $L$ layers, namely $P(Z^{s1}, \cdots, Z^{s|L|})$ and $Q(Z^{t1}, \cdots, Z^{t|L|})$ to replace the original joint distributions $P(X^s, Y^s)$ and $Q(X^t, Y^t)$, respectively. Note that the first tree terms aim to reduce the joint distribution mismatch and the last term aims to learn class-discriminative representations in target domain.

### 3.5 End-to-End Training Procedure

We propose an end-to-end architecture that we can embed the proposed two-level feature alignment into deep neural networks (e.g. AlexNet [23]). Therefore, the overall objective function can be defined as:

$$\min_{\Theta} \mathcal{L}_{all} = \gamma F_1 + (1 - \gamma) D_L(P, Q), \tag{8}$$

where $\gamma$ denotes the trade-off parameters, $\Theta$ is the training parameters of the proposed model.

## 4 Experiments

### 4.1 Datasets, Baselines, and Protocol

**Datasets.** We use the public dataset Office-31 [18] as original man-made objects (source domain) which are downloaded and collected from *amazon.com*. There are 2817 images of man-made objects from 31 categories. We used the low resolution images acquired by web camera to simulate underwater optical images of man-made objects (target domain) with various turbidity. There are 800 images from the same 31 categories. The source and target domains share the same categories. Additionally, the objects in the target images have different background, views, sizes, colors and shapes, thus the recognition is more challenging compared to previous works [15]. Results of the simulation of underwater optical images are shown in Fig. 3, we generate three simulated underwater datasets by adjusting the turbidity factor $z$ in (3) denoting three different turbidity values.

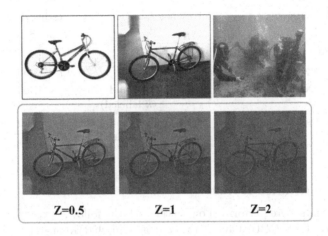

**Fig. 3.** Results of the simulation of underwater optical images. First row from left to right: an original image from the Amazon dataset; an image acquired by web camera which had various types of background; a real underwater optical image downloaded from the Internet, and it was used as reference image. Simulated underwater optical images were presented in the second row: the larger the value of $z$ is, the higher the turbidity is.

**Baseline Approaches.** We compare the proposed **WSDAN** with state of the art deep domain adaptation methods: We compare our approach with state of the art deep domain adaptation methods: Source Only, Target Only, Deep Adaptation Network (**DAN**), Reverse Gradient (**RevGrad**) and Residual Transfer Network (**RTN**). Source Only denotes training on in-air images and testing on underwater images, Target Only denotes training on underwater images and testing on underwater images, DAN [9] learns transferable features by embedding deep features of multiple task-specific layers to reproducing kernel Hilbert spaces (RKHSs) and matching different distributions optimally using multi-kernel MMD. RevGrad [16] improves domain adaptation by encouraging mistakes in domain prediction using adversarial training. RTN [14] jointly learns domain-invariant features and adapts different source and target classifiers by deep residual learning.

**Experimental Protocol.** For fair comparison, the proposed method and the other compared methods are all based on AlexNet. We follow standard evaluation protocols for domain adaptation [13,14]. At training time, for the underwater optical images, $k$ images were randomly labeled in each category, $k = \{1, 2, 3, 4, 5, 6\}$. We repeated each experiment 10 times, and calculated the average accuracy. There was no overlap between the images in the training and test datasets. The details of training and test data are provided in Table 1.

**Table 1.** Experimental settings: I stands for in-air images, U stands for underwater images; the $k$ labeled denotes there has $k$ labeled underwater images in each category and the rest is fully unlabeled; the 800 underwater images were randomly split into training and testing data.

| Experiments | Training data | | Testing data | |
|---|---|---|---|---|
| | Type | Size | Type | Size |
| Source only | I | 3100 | U | 400 |
| Target Only | U | 400 | U | 400 |
| DAN&RevGrad&RTN | I and U | 3100 and 400 (k labeled) | U | 400 |
| Ours | I and U | 3100 and 400 (k labeled) | U | 400 |

## 4.2 Implementation Details

The experiments were performed on server with the following components: NIVDIA GeForce 1080 GPU, 64G RAM and a56 Intel(R)Xeon(R) CPU E5-2683 V3@ 2.00 GHz. All the images were resized to the same size of $227 \times 227$. We implement all methods based on the **TensorFlow** and the pre-trained AlexNet on ImageNet [21], then fine-tuned using the proposed data. For the layers which are trained from scratch, we set its learning rate to be 10 times that of the other layers. The trade-off parameter $\gamma$ was fixed to 0.3. In order to suppress noisy signal from the domain discriminator at the early stages of the training, $\lambda$ was gradually changed from 0 to 1 by the following formula: $\lambda_x = \frac{2}{1+exp(-10x)} - 1$. We use mini-batch stochastic gradient descent (SGD) with momentum of 0.9 and the learning rate annealing strategy in [13]: the learning rate is not selected by a grid search due to high computational cost, it is adjusted during SGD using the following formula: the learning rate is not selected by a grid search due to high computational cost, it is adjusted during SGD using $\eta_0 = \frac{\eta_0}{(1+\alpha p)^\beta}$, where $p$ is the training progress linearly changing from 0 to 1, $\eta_0 = 0.01$, $\alpha = 10$ and $\beta = 0.75$ which is optimized to promote convergence and low error on source domain.

**Table 2.** Classification accuracy on the simulated datasets for different approaches (%).

| Turbidity | Source only | Target only | DAN | RevGrad | RTN | WSDAN ($k$ range from 1 to 6) | | | | | |
|---|---|---|---|---|---|---|---|---|---|---|---|
| | | | | | | 1 | 2 | 3 | 4 | 5 | 6 |
| 0.5 | 26.5 | 79.8 | 53.4 | 54.6 | 61.5 | 62.5 | 67.1 | 68.6 | 71.4 | 73.1 | 75.3 |
| 1 | 21.6 | 79.3 | 48.2 | 49.4 | 57.8 | 61.7 | 65.7 | 66.4 | 70.1 | 72.7 | 74.1 |
| 2 | 15.5 | 77.3 | 40.7 | 41.7 | 47.5 | 56.3 | 58.3 | 61.2 | 65.7 | 68.7 | 72.9 |

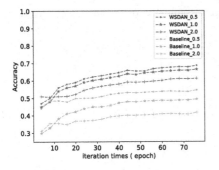

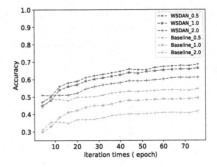

**Fig. 4.** Comparisons in three different turbidity.

**Fig. 5.** Ablation study of the proposed method.

### 4.3 Experimental Results

As shown in Table 2, we compare the accuracies of different experiments for datasets with different turbidities. We observe that our model outperforms compared methods by large margin even when very few labeled samples are available. Meanwhile, with the increase of $k$, the accuracy substantially goes up. It is noteworthy that our model *outperforms* the state-of-the-art approach using only one labeled underwater sample per category ($k = 1$). The encouraging results also highlight the importance of weakly-supervised domain adaptation in deep neural networks, and suggest that WSDAN is able to learn more transferable and class-specific representations for effective domain adaptation. In addition, we see that the accuracy of our model is just slightly worse than the Target Only (denotes the upper bound of our experiments) when only 6 labeled samples per category ($k = 6$) were used.

As shown in Fig. 4, we demonstrated comparisons between our method and the baseline (RevGrad) in the three different turbidity datasets. We empirically observe that with the increasing of turbidity, the improvements of our method is increased, revealing that the proposed method is robust for different turbidity and can better apply to the intricate underwater environment.

### 4.4 Ablation Study

We perform ablation study on task $z = 1$ by evaluating several variants of proposed WSDAN: (1) **only low-level alignment** which denotes only using

**Fig. 6.** Examples of diver in the real-world in-air and underwater images, respectively.

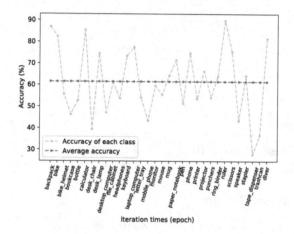

**Fig. 7.** The classification accuracy of each class.

low-level alignment of WSDAN; (2) **only high-level alignment** which denotes only using high-level alignment of WSDAN; (3) **unsupervised-WSADAN** which denotes compared with the proposed WSDAN we will not use the labeled underwater samples and the rest is the same. The comparisons with the proposed WSDAN and the baseline are provided in Fig. 5. The results show that (1) Both **only low-level alignment** and **only high-level alignment** outperform the baseline, revealing the effectiveness of the proposed two-level feature alignment respectively. (2) **WSDAN** outperforms **unsupervised-WSADAN**, highlighting the importance of the supervision information from target domain. (3) High-level alignment contributes more than low-level alignment.

### 4.5 Evaluation on Real-World Application

In order to evaluate the effectiveness of the proposed method on real-world application, we add one more class of *diver* in the real-world images to the simulated dataset ($z = 1$, $k = 1$). We collected 50 real-world diver images in the in-air scene and 50 in the underwater scene, the examples were shown in Fig. 6. The diver can be seen as class 32. The classification accuracy of each class were shown in Fig. 7. We can observe that the classification accuracy of diver outperforms most other classes and exceeds the average accuracy by large margin, revealing that the proposed method is effective in real-world application.

## 5    Conclusions

This paper has proposed a framework in weakly-supervised setting to address recognition of man-made objects in underwater optical images. We first simulated three underwater datasets with different turbidity. By introducing low-level and high-level feature alignments to a deep neural network, the proposed method shows superior performance than state-of-the-art domain adaptation approaches. Our method semantically align the distributions of different domains even when labeled underwater samples are very few. The experimental results also reveal that both alignments independently introduce a significant gain in performance.

## References

1. Jaffe, J.S.: Underwater optical imaging: the past, the present, and the prospects. IEEE J. Oceanic Eng. **40**(3), 683–700 (2015)
2. Ancuti, C.O., Ancuti, C., De Vleeschouwer, C., Bekaert, P.: Color balance and fusion for underwater image enhancement. IEEE Trans. Image Process. **27**(1), 379–393 (2018)
3. Hussain, S.S., Zaidi, S.S.H.: Underwater man-made object prediction using line detection technique. In: 2014 6th International Conference on Electronics, Computers and Artificial Intelligence (ECAI), pp. 1–6. IEEE (2014)
4. Hou, G.-J., Luan, X., Song, D.-L., Ma, X.-Y.: Underwater man-made object recognition on the basis of color and shape features. J. Coast. Res. **32**(5), 1135–1141 (2015)
5. Olmos, A., Trucco, E.: Detecting man-made objects in unconstrained subsea videos. In: BMVC, pp. 1–10 (2002)
6. Hinton, G.E., Osindero, S., Teh, Y.-W.: A fast learning algorithm for deep belief nets. Neural Comput. **18**(7), 1527–1554 (2006)
7. Pan, S.J., Yang, Q.: A survey on transfer learning. IEEE Trans. Knowl. Data Eng. **22**(10), 1345–1359 (2010)
8. Gong, B., Shi, Y., Sha, F., Grauman, K.: Geodesic flow kernel for unsupervised domain adaptation. In: 2012 IEEE Conference on Computer Vision and Pattern Recognition, pp. 2066–2073. IEEE (2012)
9. Long, M., Cao, Y., Wang, J., Jordan, M.: Learning transferable features with deep adaptation networks. In: International Conference on Machine Learning, pp. 97–105 (2015)
10. Nguyen, R.M.H., Kim, S.J., Brown, M.S.: Illuminant aware gamut-based color transfer. In: Computer Graphics Forum. Wiley Online Library, vol. 33, pp. 319–328 (2014)
11. Schechner, Y.Y., Karpel, N.: Clear underwater vision. In: 2004 IEEE Computer Society Conference on Computer Vision and Pattern Recognition, vol. 1, p. I. IEEE (2003)
12. Goodfellow, I., et al.: Generative adversarial nets. In: Advances in Neural Information Processing Systems, pp. 2672–2680 (2014)

13. Ganin, Y., Lempitsky, V.: Unsupervised domain adaptation by backpropagation. In: International Conference on Machine Learning, pp. 1180–1189 (2015)
14. Long, M., Zhu, H., Wang, J., Jordan, M.I.: Unsupervised domain adaptation with residual transfer networks. In: Advances in Neural Information Processing Systems, pp. 136–144 (2016)
15. Li, Y., Lu, H., Li, J., Li, X., Li, Y., Serikawa, S.: Underwater image de-scattering and classification by deep neural network. Comput. Electr. Eng. **54**, 68–77 (2016)
16. Ganin, Y., et al.: Domain-adversarial training of neural networks. J. Mach. Learn. Res. **17**(1), 1–35 (2016)
17. Long, M., Zhu, H., Wang, J., Jordan, M.I.: Deep transfer learning with joint adaptation networks. In: International Conference on Machine Learning, pp. 2208–2217 (2017)
18. Saenko, K., Kulis, B., Fritz, M., Darrell, T.: Adapting visual category models to new domains. In: Daniilidis, K., Maragos, P., Paragios, N. (eds.) ECCV 2010. LNCS, vol. 6314, pp. 213–226. Springer, Heidelberg (2010). https://doi.org/10.1007/978-3-642-15561-1_16
19. Srividhya, K., Ramya, M.M.: Accurate object recognition in the underwater images using learning algorithms and texture features. Multimedia Tools Appl. **76**(24), 25679–25695 (2017)
20. Pavin, A.: Underwater object recognition in photo images. In: OCEANS 2015, MTS/IEEE Washington, pp. 1–6. IEEE (2015)
21. Russakovsky, O., et al.: Imagenet large scale visual recognition challenge. Int. J. Comput. Vis. **115**(3), 211–252 (2015)
22. Sun, B., Saenko, K.: Deep CORAL: correlation alignment for deep domain adaptation. In: Hua, G., Jégou, H. (eds.) ECCV 2016. LNCS, vol. 9915, pp. 443–450. Springer, Cham (2016). https://doi.org/10.1007/978-3-319-49409-8_35
23. Krizhevsky, A., Sutskever, I., Hinton, G.E.: ImageNet classification with deep convolutional neural networks. In: Advances in Neural Information Processing Systems, pp. 1097–1105 (2012)

# Interactive Sketch Recognition Framework for Geometric Shapes

Abdelrahman Fahmy$^{(\boxtimes)}$, Wael Abdelhamid, and Amir Atiya

Computer Engineering Department, Cairo University, Cairo, Egypt
abdelrahman.h.fahmy@gmail.com, w.n.abdelhamid@gmail.com,
amir@alumni.caltech.com
http://eng.cu.edu.eg/en/

**Abstract.** With the recent advances in tablet devices industry, sketch recognition has become a potential replacement for existing systems' traditional user interfaces. Structured diagrams (flow charts, Markov chains, module dependency diagrams, state diagrams, block diagrams, UML, graphs, etc.) are very common in many science fields. Usually, such diagrams are created using structured graphics editors like Microsoft Visio. Structured graphics editors are extremely powerful and expressive, but they can be cumbersome to use. This paper presents an interactive sketch recognition system that converts user's sketch into structured geometric shapes in usable electronic format with minimal effort.

**Keywords:** Sketch recognition · Pattern recognition
Support Vector Machines (SVMs)
Human-Computer Interaction (HCI)

## 1 Introduction

With the rise of tablet devices era, developing a system that can convert hand-drawn sketches into structured (electronic) format became highly required. It will facilitate sketches creation and modification using Computer Aided Design (CAD) tools. The existing tools mainly use drag and drop feature for diagrams creation. The user constructs the diagram by incrementally selecting from the list of supported shapes. Sketching will provide a more intuitive replacement for this process. In this work, we propose an interactive sketch recognition framework, where user sketches using pen tablet or mouse, and the framework will recognize and convert the scene into well-formed structured diagram. The framework is capable of recognizing: lines, triangles, rectangles, squares, circles, ellipses, diamonds, arrows, arcs, and zigzag lines. User has the freedom of sketching primitives over multiple strokes. We have implemented a grouping algorithm which is capable of aggregating strokes that belong to the same primitive together. After that, the grouped strokes will be introduced for recognition by our trained classifiers. As can be noticed from the set of supported primitives, the proposed framework is a domain-independent sketch recognition system. We support a

© Springer Nature Switzerland AG 2018
L. Cheng et al. (Eds.): ICONIP 2018, LNCS 11305, pp. 323–334, 2018.
https://doi.org/10.1007/978-3-030-04221-9_29

set of primitives which are used in a wide range of diagrams and flow charts in various domains. Integrating this framework into a domain-dependent diagram recognition system shall be seamless and straight forward. A higher-level recognition system can be used to interpret the meaning of the recognized primitives (or group of primitives) into domain specific shapes (e.g., zigzag line is a resistor in the domain of electrical circuits). Therefore, the proposed framework output primitives can be used as input to domain-dependent frameworks.

This paper is organized as follows: Sect. 2 gives an overview of the related work. Section 3 explains the proposed sketch recognition framework architecture. The experimental results are presented in Sect. 4 followed by conclusion in Sect. 5.

## 2   Related Work

Sketch recognition has been an active area of research. Main focus was in the area of online sketch recognition systems and there were minimal research efforts in offline systems. However, there have been some remarkable contributions in offline research [15,19]. For online, we will highlight a set of remarkable sketch recognizers. PaleoSketch [11,16] used Neural Networks (NN) classifier and grouping technique based on spatial properties. CALI [9,10] used fuzzy logic classifier and grouping technique based on timeout. HHReco [12] utilizes Zernike moments as features in SVM and multi-stroke support using segmentation. Rubine [20] is a motion-based recognizer. Sezgin et al. [21] built Hidden Markov Models (HMM) sketch recognizer. $N [1] is template matching recognizer based on $1 uni-stroke recognizer [22]. $N supports multi-stroke sketch by representing a multi-stroke as a set of uni-strokes representing all possible stroke orders and directions. It is automatically generalizing from one multi-stroke template to all possible multi-stroke with alternative stroke orderings and directions. El Meseery et al. [7] proposed a system that attempts to find the optimal decomposition of the input stroke using Particle Swarm Optimization (PSO) algorithm and classification using SVM classifier.

There were also research efforts done in using sketch recognition in applications; Jayawardhana et al. [13] introduced a novel sketch based query language for database querying. Bresler et al. [5] introduced an online, stroke-based recognition system for hand-drawn arrow-connected diagrams. Dixon et al. [6] used sketch recognition to assist the user in creating a rendition of a human face with the intent of improving that person's ability to draw. Wu et al. [23] presented SmartVisio system which is a real-time sketch recognition system based on Visio. It recognizes hand-drawn flowchart/diagram with flexible interactions. Xu et al. [24] introduced Voltique Designer which is an Android-based circuits creation tool. Bohari et al. [2] presented a novel approach to recognize the users intention to draw or not to draw in a mid-air sketching task without the use of postures or controllers.

The sketch recognition framework proposed in this paper uses model-based classifiers and a set of effective geometric features. The grouping technique is based on spatial properties which overcomes timeout constraints and long computing time of template matching.

# 3   System Architecture

A key strength of this work is providing a simple yet robust system architecture. It is introducing guidelines for building domain-independent sketch recognition framework. The framework architecture is divided into five main stages: grouping, classification preprocessing, features extraction, classification, and system output.

## 3.1   Input Stage

The framework is currently supporting sketch recognition in interactive mode. We built a simple editor which captures user's sketch using pen tablet or mouse. Sketch's points are stored in the same order they were drawn. They are sent to the core engine on pen/mouse up event. They are represented in a form of array of points.

## 3.2   Grouping

Grouping stage is mandatory for multi-stroke primitives recognizers. In this stage, all the related sketched strokes will be grouped and sent to the primitives classifier. The focus is to apply minimal sketching constraints on the user. The user shall have the flexibility to draw an incomplete primitive, draw another primitive, and then continue the drawing of the first primitive. This flexibility requires a robust strokes grouping algorithm. We propose a grouping algorithm that uses quite simple grouping criteria, but proved to have very good results.

Only lines and paths[1] are considered as candidates for grouping. The user can edit a line to form a triangle, square, diamond, etc. Fig. 1.

**Fig. 1.** Multi-stroke sketched triangle by drawing line then edit it to triangle

---

[1] Path: the stroke that was not detected as a primitive. It is kept untouched, as we assume that it is part of multi-stroke primitive that the user is sketching.

*Euclidean Distance Grouping.* We used the Euclidean distance between strokes endpoints as the criteria for the grouping algorithm. The distance between two strokes endpoints divided by the average stroke length of the two strokes should be within some threshold (our experiments proved that 0.1 is a recommended value for this threshold). Thus, the gap between two strokes should be less than 10% of the average length of the two strokes. Any two strokes meet this criteria will be merged and sent to the classification preprocessing stage.

The proposed grouping technique uses the graph theory to ease the implementation [11]. The grouping is done over two steps: graph building, where we check if any strokes satisfy grouping criteria, and graph traversing, where we connect strokes satisfying the grouping criteria into one stroke. We will describe each of those two steps on the example in Fig. 2.

Each stroke index represents the order in which it was drawn. The user drew stroke1, so we built a graph with one vertex for stroke1. After the user drew stroke2, the grouping criteria is checked and found that it is applied on stroke1 and stroke2. Accordingly, another vertex is added for stroke2, and an edge is added between the two vertices to indicate that they shall be connected to form one stroke. After that, stroke3 is drawn, as it did not satisfy grouping criteria with neither stroke1 nor stroke2, we created another graph having one vertex for stroke3. Then stroke4 is drawn, and it is satisfying grouping criteria with stroke3, so we added vertex for stroke4 and added an edge in the second graph between stroke3 and stroke4 vertices. The algorithm continues similarly with the remaining drawn strokes (stroke5, stroke6, and stroke7), and builds the two graphs shown in Fig. 2. By the end of graph building step, each edge of the graph represents two strokes that shall be grouped. In graph traversing step, we traverse through the built graphs and group the strokes that their corresponding graph vertices are connected.

Using graphs for grouping technique implementation has many advantages: more structured and well organized implementation, prevents adding duplicate stroke as any vertex is added once, saves processing time as we avoid any duplicate processing for strokes, improves grouping technique debugging, readability, and ease of understanding.

### 3.3    Classification Preprocessing

Our experiments have shown that using one classifier to distinguish between the set of supported primitives is not an effective approach. It may obstruct achieving minimum classification errors and maximizing the recognition accuracy. The experiments have shown that having a preprocessing stage can significantly enhance the classification accuracy. In classification preprocessing, we calculate the shape closure ratio[2], which is the distance between the stroke's endpoints to the length of the stroke. Based on this ratio, we categorize user's stroke into one of three categories: closed shape (circle, ellipse, rectangle, square,

---

[2] Our experiments have shown that this ratio is less than 0.1 for closed shapes, 0.1-0.6 for paths, and greater than 0.6 for open shapes.

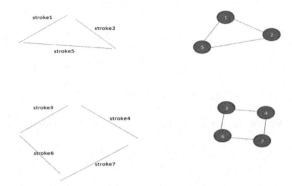

**Fig. 2.** Example explains grouping technique implementation using graphs

triangle, and diamond), open shape (line, arrow, arc, and zigzag line), or incomplete shape (path). According to the stroke category, we redirect the stroke to the appropriate classifier (will discuss that in details in classification stage). This approach empirically proved to improve shapes recognition accuracy and resolved many shapes classification confusions. It minimizes the number of shapes each classifier deals with, and sends each stroke to the designated classifier which is reflecting positively on the system overall classification accuracy.

### 3.4 Features Extraction

At this stage, we calculate a set of effective geometric features for the user's input stroke (grouped stroke in case of multi-stroke sketched primitive) to be used during classification stage. All geometric features used are size and orientation independent. The majority of the features rely on essential geometric traits of the input shapes, e.g., the area and the perimeter of the minimum area enclosing rectangle, the area of the maximum area inscribed k-gon that fits inside the convex hull of the shape, the shape thinness, and the shape straightness. Following are some of the geometric features used in our framework.

**Circle Thinness Value.** To distinguish circles from the other closed shapes, we use the thinness ratio $(P_{ch}^2/A_{ch})$, where $A_{ch}$ is the area of the convex hull, and $P_{ch}$ is its perimeter. The thinness of a circle is minimal, since it is the planar figure with the smallest perimeter enclosing a given area, yielding a value near $4\pi$.

**Intersections with Regression Line.** We calculate the regression line for the input sketch points. Then, we calculate the number of intersections between input sketch and this regression line. For zigzag lines we expect to have relatively big number of intersections compared to the other open shapes Fig. 3a.

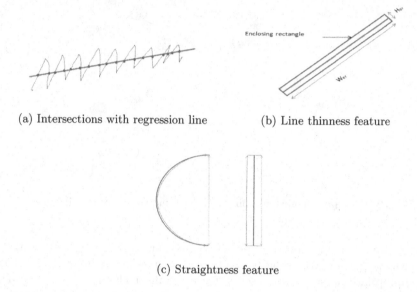

(a) Intersections with regression line        (b) Line thinness feature

(c) Straightness feature

**Fig. 3.** Geometric features

**Line Thinness Value.** We identify lines using another thinness ratio which compares the height of the (non-aligned) minimum area enclosing rectangle ($H_{er}$) with its width ($W_{er}$). The $H_{er}/W_{er}$ ratio will have values near zero for lines and bigger values for the other shapes Fig. 3b.

**Straightness Value.** This feature measures how straight the shape is by calculating the ratio between the direct distance between the stroke's two end points and the stroke length. For lines, this value shall be ~1.0. For zigzag lines, it shall be very small value, and for arcs, it will be ~0.5–0.7 Fig. 3c.

**Zigzag Value.** This feature used mainly to distinguish zigzag lines from the other open shapes. It measures the ratio between the convex hull perimeter and the stroke length. For zigzag lines, this value shall be very small compared to the other shapes.

### 3.5   Classification

One of the main challenges in any recognition problem is choosing a classifier that will result in the highest recognition accuracy. In this work, we are going to compare three classifiers: Support Vector Machines (SVMs) with Radial Basis Function (RBF) kernel, Random Forest (RF), and K-Nearest Neighbor (K-NN).

*Multi-stage Classification.* As we discussed in the previous section, during classification preprocessing, we categorize user's stroke into open shape, closed shape, or path. At classification stage, we introduce the stroke to the appropriate classifier based on the category provided by the preprocessing stage. In our current implementation, we have two classifiers, one for the closed shapes and another classifier for the open shapes. Paths are kept as is, not classified, because they represent an incomplete primitive (primitive that will be drawn on multi-stroke). Using multi-stage classification approach has minimized classification confusion, and facilitated the selection of effective features. For example, a geometric feature used to distinguish two closed shapes might introduce confusion to open shapes and vice-versa. The proposed two classifier approach has simplified the evaluation of the impact of the newly introduced feature to the framework overall accuracy.

## 3.6   Output Stage

The primitives' locations, orientations, and sizes need to be maintained as they were sketched by the user. Figuring out shape orientation from its points is not an easy task for some shapes (e.g., triangles). Therefore, we defined this problem as the problem of finding the minimum number of vertices that can be used to draw the sketched shape. The Ramer-Douglas-Peucker algorithm [18] is used for curves and polygons approximation. We used OpenCV library's algorithm implementation [3]. The implementation minimizes the distance between the original and the approximated shapes to the provided precision. We implemented a set of algorithms that use the vertices returned by OpenCV and extract the minimum number of vertices needed to draw each shape.

**Arcs Drawing.** For arcs, we use Casteljau's algorithm [8] which is a recursive method to evaluate Bezier curves given arc start and end points and the two control points.

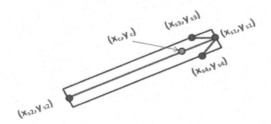

**Fig. 4.** Points needed to draw arrow

**Arrows Drawing.** For arrows, we calculate the minimum area enclosing rectangle for the input points. The arrow shaft points $(x_{s1}, y_{s1})$ and $(x_{s2}, y_{s2})$, Fig. 4, are the two points in the middle of the smaller enclosing rectangle sides. Then we average all the sketch points to get the centroid point $(x_c, y_c)$. After that, the distance between $(x_c, y_c)$ and each of $(x_{s1}, y_{s1})$ and $(x_{s2}, y_{s2})$ is calculated. As the density of the arrow points will be higher at the arrow head side, the $(x_c, y_c)$ will be closer to the arrow head than the other side. The arrow head points $(x_{s3}, y_{s3})$ and $(x_{s4}, y_{s4})$ to be located on the enclosing rectangle longer sides and to be on a distance of $\sim 0.25$ of the length of the longer enclosing rectangle side from $(x_{s1}, y_{s1})$. Then, we draw lines between $(x_{s1}, y_{s1})$ and each of the other points to draw the arrow.

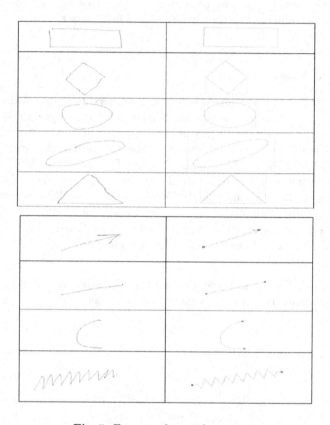

**Fig. 5.** Framework samples output

## 4    Experimental Results

We collected a training set of 1400 patterns of closed shapes and 1040 patterns of open shapes drawn by six users from different backgrounds. A test set of 615 patterns of closed shapes and 452 patterns of open shapes were drawn by other four users from different backgrounds as well. The framework achieved average accuracy of 96.05% with SVM classifier, 95.09% with RF classifier, and 85.78% with K-NN classifier. Table 1 shows a comparison between SVM, RF, and K-NN testing results and the accuracy per each supported shape.

As stated before, SVM is used with RBF kernel. We used 10-fold cross validation and grid search technique proposed in [14] to fine tune SVM parameters, C and $\gamma$. C and $\gamma$ values were incrementally increased exponentially (for example, $C = 2^{-5}, 2^{-4}, \ldots, 2^{10}$ and $\gamma = 2^{-10}, 2^{-5}, \ldots, 2^{5}$). Each time, 10-fold cross validation was executed while observing the effect of the new values on the output accuracy. The best accuracy was achieved with $C = 2^9$ and $\gamma = 2^{-6}$ for both the closed shapes and the open shapes. For RF, best results for the open shapes were with maximum depth set to 3 and maximum number of trees set to 1000. For the closed shapes, it was achieved with maximum depth set to 3 and maximum number of trees set to 5000. For K-NN, we ran 10-fold with three different values for K, K = 5, 11, and 51. K = 5 gave us the best cross validation results for both the closed shapes and the open shapes.

**Table 1.** Testing results: SVM with RBF kernel, $C = 2^9$ and $\gamma = 2^{-6}$, RF with maximum depth = 3, and K-NN with K = 5

| Classifier | SVM | RF | K-NN |
|---|---|---|---|
| Shape | Accuracy | Accuracy | Accuracy |
| *Closed shapes* | | | |
| Triangle | 100% | 100% | 94.31% |
| Circle | 97.56% | 97.56% | 95.93% |
| Ellipse | 96.75% | 96.75% | 92.68% |
| Diamond | 91.06% | 87.80% | 87% |
| Rectangle | 95.93% | 96.75% | 86.18% |
| Avg | 96.26% | 95.77% | 91.22% |
| *Open shapes* | | | |
| Line | 100% | 100% | 100% |
| Arc | 100% | 100% | 93.81% |
| Zigzag line | 92.92% | 83.19% | 53.10% |
| Arrow | 90.27% | 93.81% | 69.03% |
| Avg | 95.80% | 94.25% | 78.99% |
| Total Avg | 96.05% | 95.09% | 85.78% |

We compared the framework accuracy to set of widely known recognizers to give better understanding of our framework accuracy. We used the recognition accuracies data collected by Paulson [17] as shown in Table 2. The label "polyline" includes zigzag lines, rectangles, diamonds, and triangles. As can be noticed, the proposed framework has very competitive accuracy results. They have the advantage of recognizing spiral, helix, and complex shapes. On the

**Table 2.** Benchmarking: Recognition accuracy comparison of widely used sketch recognition systems

| | CALI | HHReco | Rubine | Sezgin | $1 | PaleoSketch 2011 | ProposedSVM |
|---|---|---|---|---|---|---|---|
| Arc | 97% | 96% | 51% | - | 98% | 99% | 100% |
| Circle | 97% | 89% | 37% | 83% | 97% | 90% | 97.56% |
| Complex | - | 79% | 19% | 99% | 93% | 84% | - |
| Curve | 61% | 77% | 47% | - | 85% | 94% | - |
| Ellipse | 100% | 44% | 61% | 56% | 95% | 99% | 96.75% |
| Helix | - | 92% | 75% | - | 96% | 100% | - |
| Line | 100% | 98% | 74% | 99% | 69% | 100% | 100% |
| Polyline | - | 67% | 44% | 92% | 56% | 97% | 98.23% |
| Spiral | - | 99% | 80% | - | 100% | 100% | - |
| Arrow | - | - | - | - | - | 96.3% | 90.27% |
| Average | 91% | 82.3% | 54.22% | 85.8% | 87.7% | 95.93% | 97.14% |

other hand, this work have the advantage of supporting arrows[3] and labeling rectangles, triangles, diamonds, and zigzag lines separately with very high accuracy. For domain-independent recognizer, it does not make much sense to recognize complex shapes. They can be more accurately recognized if domain-specific information is present. Accordingly, the proposed framework can be seamlessly integrated with different high-level domain-dependent recognizers.

## 5    Conclusion

In this paper, we had proposed a framework for domain-independent interactive sketch recognition. The proposed grouping technique allows the user to draw multi-stroke primitives. The grouping technique is based on spatial properties which overcomes timeout constraints and long computing time of template matching. The classification preprocessing stage minimized the number of primitives each classifier is dealing with, which reflected positively on the classification accuracy. The simple and robust grouping technique and the novel multi-stage classification aided in building constraints-free sketch recognition system having a competitive recognition accuracy. This work also introduced a set of novel geometric features which can be used to distinguish between wide set of geometric shapes. These shapes are considered as building blocks for a wide range of high-level diagrams domains. The framework have been evaluated by running a comparison between a set of famous sketch recognizers that are commonly used for sketch recognition benchmarking. The framework proved to have a very competitive accuracy.

---

[3] Arrow detection is one of the major challenges in any sketch recognition framework [4,5]. This work introduced novel geometric features which recognize arrows with very high accuracy, as well as novel arrow drawing technique which maintains the sketched arrows properties (size, orientation, etc.).

# References

1. Anthony, L., Wobbrock, J.O.: A lightweight multistroke recognizer for user interface prototypes. In: Proceedings of Graphics Interface 2010, pp. 245–252. Canadian Information Processing Society (2010)
2. Bohari, U., Chen, T.J., et al.: To draw or not to draw: recognizing stroke-hover intent in non-instrumented gesture-free mid-air sketching. In: 23rd International Conference on Intelligent User Interfaces, pp. 177–188. ACM (2018)
3. Bradski, G.: Dr. Dobb's Journal of Software Tools (2000)
4. Bresler, M., Prusa, D., Hlavac, V.: Detection of arrows in on-line sketched diagrams using relative stroke positioning. In: WACV, vol. 15, pp. 610–617 (2015)
5. Bresler, M., Prusa, D., Hlavac, V.: Online recognition of sketched arrow-connected diagrams. Int. J. Doc. Anal. Recogn. (IJDAR) 19(3), 253–267 (2016)
6. Dixon, D., Prasad, M., Hammond, T.: iCanDraw: using sketch recognition and corrective feedback to assist a user in drawing human faces. In: Proceedings of the SIGCHI Conference on Human Factors in Computing Systems, pp. 897–906. ACM (2010)
7. El Meseery, M., El Din, M.F., Mashali, S., Fayek, M., Darwish, N.: Sketch recognition using particle swarm algorithms. In: 2009 16th IEEE International Conference on Image Processing, pp. 2017–2020. IEEE (2009)
8. de Faget de Casteljau, P., Gardan, Y.: Mathematics and CAD: Shape Mathematics and CAD. Hermes (1985)
9. Fonseca, M.J., Jorge, J.A.: Using fuzzy logic to recognize geometric shapes interactively. In: The Ninth IEEE International Conference on Fuzzy Systems, vol. 1, pp. 291–296. IEEE (2000)
10. Fonseca, M.J., Pimentel, C., Jorge, J.A.: Cali: an online scribble recognizer for calligraphic interfaces. In: AAAI Spring Symposium on Sketch Understanding, pp. 51–58 (2002)
11. Hammond, T., Paulson, B.: Recognizing sketched multistroke primitives. ACM Trans. Interact. Intell. Syst. 1(1), 4 (2011)
12. Hse, H., Newton, A.R.: Graphic Symbol Recognition Toolkit (HHreco) Tutorial. Electronics Research Laboratory, College of Engineering, University of California (2003)
13. Jayawardhana, A., Ranathunga, L., Ahangama, S.: Sketch based database querying system. In: 2017 IEEE International Conference on Industrial and Information Systems (ICIIS), pp. 1–6. IEEE (2017)
14. Lin, C.J., Hsu, C., Chang, C.: A practical guide to support vector classification. National Taiwan U (2003). www.csie.ntu.edu.tw/cjlin/papers/guide/guide.pdf
15. Notowidigdo, M., Miller, R.C.: Off-line sketch interpretation. In: AAAI Fall Symposium, pp. 120–126 (2004)
16. Paulson, B., Hammond, T.: PaleoSketch: accurate primitive sketch recognition and beautification. In: Proceedings of the 13th International Conference on Intelligent User Interfaces, pp. 1–10. ACM (2008)
17. Paulson, B.C., Hammond, T.: Rethinking Pen Input Interaction: Enabling Freehand Sketching Through Improved Primitive Recognition. Texas A&M University (2011)
18. Ramer, U.: An iterative procedure for the polygonal approximation of plane curves. Comput. Graph. Image Process. 1(3), 244–256 (1972)

19. Refaat, K.S., Helmy, W.N., Ali, A., AbdelGhany, M.S., Atiya, A.F.: A new approach for context-independent handwritten offline diagram recognition using support vector machines. In: IEEE International Joint Conference on Neural Networks, pp. 177–182. IEEE (2008)
20. Rubine, D.: Specifying gestures by example. Comput. Graph. **25**(4), 329–337 (1991). https://dl.acm.org/citation.cfm?id=122753
21. Sezgin, T.M., Davis, R.: HMM-based efficient sketch recognition. In: Proceedings of the 10th International Conference on Intelligent User Interfaces, pp. 281–283. ACM (2005)
22. Wobbrock, J.O., Wilson, A.D., Li, Y.: Gestures without libraries, toolkits or training: a $1 recognizer for user interface prototypes. In: Proceedings of the 20th Annual ACM Symposium on User Interface Software and Technology, pp. 159–168. ACM (2007)
23. Wu, J., Wang, C., Zhang, L., Rui, Y.: SmartVisio: interactive sketch recognition with natural correction and editing. In: Proceedings of the ACM International Conference on Multimedia, pp. 735–736. ACM (2014)
24. Xu, D.D., Hy, M., Kalra, S., Yan, D., Giacaman, N., Sinnen, O.: Electrical circuit creation on Android (2014)

# Event Factuality Identification via Hybrid Neural Networks

Zhong Qian[✉], Peifeng Li[✉], Guodong Zhou[✉], and Qiaoming Zhu[✉]

Natural Language Processing Lab, School of Computer Science and Technology,
Soochow University, Suzhou, China
qianzhongqz@163.com, {pfli,gdzhou,qmzhu}@suda.edu.cn
http://nlp.suda.edu.cn/

**Abstract.** Event factuality identification aims at determining the factual nature of events, and plays an important role in NLP. This paper proposes a two-step framework for identifying the factuality of events in raw texts. Firstly, it extracts various basic factors related with factuality of events. Then, it utilizes a hybrid neural network model which combines Bidirectional Long Short-Term Memory (BiLSTM) networks and Convolutional Neural Networks (CNN) for event factuality identification, and considers lexical and syntactic features. The experimental results on FactBank show that the proposed neural network model significantly outperforms several state-of-the-art baselines, particularly on speculative and negative events.

**Keywords:** Event factuality · FactBank · LSTM · CNN

## 1 Introduction

Event factuality is defined as the information expressing the commitment of relevant sources towards the factual nature of events, conveying whether an event is evaluated as a fact, a possibility, or an impossible situation. In principle, event factuality is related with some basic factors, e.g., predicates, speculative and negative cues. For examples:

(S1) *This <u>scientist</u> **speculated** that liquid water **existed** on the planet.*

(S2) *The rescue <u>team</u> was not able to **contact** the climbers.*

In this paper, events are in **bold** and sources are <u>underlined</u> in example sentences. In S1, the event **existed** is a possibility according to <u>scientist</u> due to the predicate **speculated**, while in S2 the event **contact** is impossible according to the negative cue *not*. Table 1 shows that factuality is characterized by modality and polarity. Modality conveys the certainty degree of events, such as *certain* (CT), *probable* (PR), and *possible* (PS), while polarity expresses whether the event has happened, including *positive* (+) and *negative* (−). In addition, U/u means *underspecified*. Some combined values are not applicable (NA) grammatically (e.g., PRu, PSu, and U+/−), and are not considered [1,2].

© Springer Nature Switzerland AG 2018
L. Cheng et al. (Eds.): ICONIP 2018, LNCS 11305, pp. 335–347, 2018.
https://doi.org/10.1007/978-3-030-04221-9_30

**Table 1.** Various values of event factuality.

|  | + (positive) | − (negative) | u (underspecified) |
|---|---|---|---|
| CT (certain) | CT+ | CT− | CTu |
| PR (probable) | PR+ | PR− | (NA) |
| PS (possible) | PS+ | PS− | (NA) |
| U (underspecified) | (NA) | (NA) | Uu |

Previous studies usually employed rule-based approaches [1,2], machine learning models [3,4], or a combination of them [5,6]. Although the rule-based models can obtain high performance ([2] achieved the micro- and macro-averaged F1 of 85.45 and 73.59 on the Aquaint TimeML subcorpus in FactBank [7]), these rules were very complicated and relied on large-scale tables. Similarly, the machine learning methods depended on various kinds of features, e.g., 19 features in [2] and 15 features in [5]. Furthermore, previous researches relied on annotation information, such as source introducing predicates, relevant sources, and cues. Although [1,2,5] can obtain satisfactory results, these studies neglected the performance on raw texts.

Recently, neural networks have been widely-used in various NLP applications. Previous research [8,9] utilized neural networks to identify factuality of sentences. However, none of them addressed event factuality identification.

In order to address the issues described above, we propose a two-step framework with neural networks to identify event factuality in raw texts. First, those basic factors widely used to identify event factuality, i.e., events, Source Introducing Predicates (SIPs), relevant sources, and cues, are extracted from texts. Then, a hybrid neural network model which combines BiLSTM and CNN with attention is employed to identify the factuality of events according to these basic factors. **Shortest Dependency Paths (SDP)** are considered as the main syntactic features. Experimental results on FactBank show that our model outperforms the baselines significantly, especially on speculative and negative events. The code of this paper is released at https://github.com/qz011/event_factuality.

## 2    Basic Factor Extraction

This section presents the basic factors related with factuality, namely events, SIPs, sources and cues, and demonstrates the methods to identify them.

**Events** in FactBank are defined by TimeML [10]. We utilize the maximum entropy classification model developed by [11] for event detection.

**Source Introducing Predicates (SIPs)** are events that can not only introduce additional sources, but also influence the factuality of the embedded events. For example, in S1 the SIP *speculated* introduces _scientist_ as a new source, and _scientist_ evaluates the embedded event **existed** as PS+ according to **speculated**. To detect SIPs, we consider both **lexical level features** and **sentence level**

**features**. Similar to event detection, We employ *token, part-of-speech (POS)* and *hypernym* of the token as **lexical level features**.

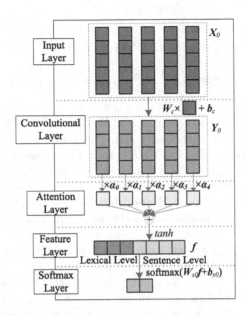

**Fig. 1.** Convolutional neural network for SIP detection.

An SIP has at least one embedded event [1]. Hence, we propose **Pruned Sentence (PSen)** as the **sentence level feature**: If a clause of the candidate event token contains events, the clause is replaced by the tag ⟨*event*⟩; Nouns, pronouns and the current candidate event are unchanged, while other tokens are replaced by the tag ⟨*O*⟩ to achieve more generality. S4 is the PSen of the event *said* in S3.

(S3) *Thomson, who was in India to **talk** to leaders, **said** the flights would **provide** extra **support** to the growing tourism market.*

(S4) *Thomson* ⟨O⟩ *who* ⟨O⟩ ⟨O⟩ *India* ⟨O⟩ ⟨O⟩ ⟨O⟩ *leaders* ⟨O⟩ **said** ⟨*event*⟩

PSen can clearly characterize whether there are events in the clause of the token. Furthermore, PSen is a simplified and effective structure, because only the candidate token and the tokens that are possible new sources are reserved, and the effects of other tokens are omitted.

We utilize the CNN shown in Fig. 1 to identify SIPs. Considering that PSen is a simplified structure, we learn the sentence level features $c$ through an attention-based CNN instead of an alternative RNN. We transfer the PSen to $X_0 \in \mathbb{R}^{d_0 \times n}$ according to pre-trained word embeddings and compute our CNN as follows:

$$Y_1 = W_c X_0 + b_c. \tag{1}$$

$$\alpha = \text{softmax}(v_c^T \tanh(Y_1)). \tag{2}$$

$$c = \tanh(Y_1 \alpha^T). \tag{3}$$

$$f = l \oplus c. \tag{4}$$

$$o = \text{softmax}(W_{s0} f_0 + b_{s0}). \tag{5}$$

$$J(\theta) = -\frac{1}{m} \sum_{i=0}^{m-1} \log p(y^{(i)} | x^{(i)}, \theta). \tag{6}$$

where $\oplus$ is the concatenation operator, and $W_c, b_c, v_c, W_{s0}, b_{s0}$ are parameters.

A **Relevant Source** is the participant of an event holding a stance with regard to the event factuality. *AUTHOR* is always the source of events by default. Further sources (e.g. *scientist* in S1) are represented in chain form [1,2]: *scientist_AUTHOR*, which means that we know about *scientist* perspective according to *AUTHOR* and *scientist* is an **Embedded Source** in *AUTHOR*. The grammatical subjects of SIPs are chosen as the **new sources**. Since we have identified events and SIPs, we can identify relevant sources $RS$, which is initialized as $RS = \{AUTHOR\}$ for each event. When traversing from the root of the dependency parse tree of the sentence to the current event, $RS$ is updated as $RS_n = RS_{n-1} \cup \{ns\_s\}$, where $ns$ is a new source introduced by the corresponding SIP and $s \in RS_{n-1}$. This is a recursive algorithm defined by [1].

**Cues** are words that have speculative or negative meanings. PR/PS events are governed by PR/PS cues, while events can be negated by negative (NEG) cues. Previous studies [12,13] concluded that lexical features can achieve excellent performances on cue detection task. Hence, we employ the lexical features developed by [13] to classify each token into PR/PS/NEG cue, or not cue.

## 3 Neural Networks for Event Factuality Identification

This section describes our neural network model considering both BiLSTM and CNN with attention for event factuality identification shown in Fig. 2. We design two outputs in the model: one indicates whether the event is Uu, Non-Uu or OTHER, and the other determines whether the event is governed by a speculative or negative cue and further classifies Non-Uu events as CT+/−, PR+/−, and PS+/−. We have the following main reasons for the architecture of two outputs:

(1) Speculative and negative factuality values can be identified more precisely with the help of corresponding cues;
(2) this design can address the problem of data imbalance, because the speculative and negative factuality values usually occupy the minority.

Finally, the factuality values of events are determined by the two outputs directly.

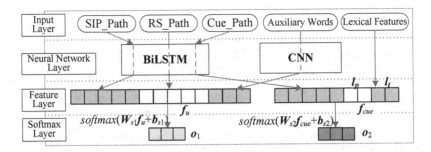

**Fig. 2.** Neural network architecture for event factuality identification.

### 3.1 Features

Previous studies [1–3] have proven the success of the dependency parse trees in event factuality identification. Therefore, we develop the following syntactic features based on the dependency parse trees and the basic factors:

**SIP_Path:** The SDP is from the ancestor SIPs to the event.

**RS_Path:** The SDP starts from the root of dependency tree, passes by all the relevant sources of the event, and ends with the current event.

**Cue_Path**: The SDP is from a cue to the event.

SIP_Path and RS_Path are used to determine whether an event is Uu or Non-Uu. In addition to Cue_Path, we also consider the following cue-related lexical features to judge whether the event is modified by a cue:

**Relative Position** is the surface distance from the cue to the event, and is mapped into vector $l_p$ with dimensions $d_p$;

**Type of Cue** includes PR, PS and NEG, is mapped into vector $l_t$ with dimensions $d_t$.

If there is more than one cue in the sentence, we consider whether the current event is governed by each cue separately. Besides, if a Non-Uu event is governed by both PR and PS cues, we adopt the cue with the highest confidence score to decide whether the modality of the event is PR or PS.

**Auxiliary Words (Aux_Words)** share the dependency relations *aux* or *mark* with the event, and can convey syntactic constructions of sentences. For each Aux_Word, both the word itself and the dependency relation are considered as the input features.

An example sentence and its features for our model are shown in Fig. 3.

### 3.2 Neural Network Modeling for Event Factuality Identification

Many sequence modeling tasks are beneficial from the access to the future as well as past context. To model the representations of syntactic paths, we utilize the bidirectional LSTM network that processes the syntactic path in both directions. It produces the forward hidden sequence $\overrightarrow{H}$, the backward hidden

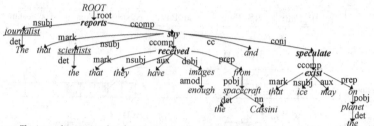

Sentence: *The journalist **reports** that the scientists say that they have **received** enough images from the Cassini spacecraft and **speculate** that ice may **exist** on the planet .* (Current Event: ***received***)

Current Relevant Source:   *scientists_journalist_AUTHOR*

RS_Path:*ROOT* root ***reports*** nsubj *journalist* nsubj ***reports*** ccomp *say* nsubj *scientists* nsubj *say* ccomp ***received***

Current SIPs: ***reports say***          SIP_Path: ***reports*** ccomp ***say*** ccomp ***received***

Current cue: *may*          Cue_Path: *may* aux ***exist*** ccomp ***speculate*** conj *say* ccomp ***received***

Relative Position from cue to event: 11     Aux_Words: (*that*, mark) (*have*, aux)

**Fig. 3.** An example sentence and features of an event for our neural network model.

sequence $\overleftarrow{H}$ and the output sequence $H_p = \overrightarrow{H} + \overleftarrow{H}$. To extract the most useful representations from sequences, we utilize the attention mechanism and produce the output vector $h_p$:

$$\alpha = \text{softmax}(v^T \tanh(H_p)). \tag{7}$$

$$h_p = \tanh(H_p \alpha^T). \tag{8}$$

where $p = sp(\text{SIP\_Path})$, $rp(\text{RS\_Path})$, $cp(\text{Cue\_Path})$. Noticing that the auxiliary words are a collection of tokens instead of a sequence with specific meanings, we employ the CNN in Sect. 2 to learn representations of auxiliary words and their dependency relations:

$$f_w = \text{CNN}(X_{aux\_words}). \tag{9}$$

We concatenate $h_{sp}$, $h_{rp}$ and $f_w$ into $f_u$ to judge whether the event is Uu, Non-Uu or other. To judge whether a Non-Uu event is governed by a speculative or negative cue, we consider not only Cue_Path $h_{cp}$ but the lexical features $l_p$ (Relative Position) and $l_t$ (Type of Cue):

$$f_u = h_{sp} \oplus h_{rp} \oplus f_w. \tag{10}$$

$$f_{cue} = l_p \oplus l_t \oplus h_{cp}. \tag{11}$$

Finally, the representations $f_u$ and $f_{cue}$ are fed into the softmax layer:

$$o_1 = \text{softmax}(W_{s1} f_u + b_{s1}). \tag{12}$$

$$o_2 = \text{softmax}(W_{s2} f_{cue} + b_{s2}). \tag{13}$$

where $W_{s1}, b_{s1}, W_{s2}, b_{s2}$ are parameters. $o_1$ represents whether the event is Uu, Non-Uu or other (label $y_1$), while $o_2$ determines whether the event is governed by the cue (label $y_2$), and Non-Uu events are classified as CT+/−, PR+/−, or

PS+/− according to $o_2$. The design of the two outputs can address imbalance among instances because speculative and negative values are in the minority. The objective function is designed as:

$$J(\theta) = \epsilon[-\frac{1}{m}\sum_{i=0}^{m-1}\log p(y_1^{(i)}|x^{(i)},\theta)] + (1-\epsilon)[-\frac{1}{m}\sum_{i=0}^{m-1}\log p(y_2^{(i)}|x^{(i)},\theta)]. \quad (14)$$

where given the training instances $(x, y_1)$ and $(x, y_2)$, $p(y_1^{(i)}|x^{(i)}, \theta)$ and $p(y_2^{(i)}|x^{(i)}, \theta)$ are the confidence scores of the golden label $y_1$ and $y_2$ in $o_1$ and $o_2$, respectively. $\epsilon$ is the trade-off.

## 4    Experimentation

This section introduces the experimental settings and then presents the detailed results and analysis of event factuality identification (Table 2).

### 4.1    Experimental Settings

We evaluate our models on FactBank [7]. Following previous studies [1–3], we focus on the five main categories of values, i.e., CT+, CT−, PR+, PS+, and Uu, which make up 99.05% of all the instances. We perform 10-fold cross-validation and employ Precision, Recall, F1-measure, micro- and macro-averaging to report the performance of factuality values. For SIP detection, we set the dimensions of *POS* and *hypernyms* embeddings as 50 and $n_c = 150$. For event factuality identification, we set $d_t = d_p = 10$, the hidden units in CNN $n_c = 50$, the hidden units in LSTM $n_{lstm} = 100$, and $\epsilon = 0.75$. We initialize word embeddings via Word2Vec [14], setting the dimensions as $d_0 = 100$. Other parameters are initialized randomly, and SGD with momentum is used to optimize our model.

**Table 2.** Distribution of factuality values in FactBank.

| Sources | CT+ | CT− | PR+ | PS+ | Uu | Other |
|---------|-----|-----|-----|-----|-----|-------|
| All | 7749/57.37% | 433/3.21% | 363/2.69% | 226/1.67% | 4607/34.11% | 128/0.95% |
| Author | 5412/57.05% | 206/2.17% | 108/1.14% | 89/0.94% | 3643/38.40% | 29/0.31% |
| Embed | 2337/58.15% | 227/5.65% | 255/6.34% | 137/3.41% | 964/23.99% | 99/2.46% |

### 4.2    Results and Analysis: Basic Factor Extraction

Table 3 displays the results of basic factor extraction tasks. For SIP detection task, an SIP is correctly identified means both the SIP and the new source introduced by it are correctly identified. We also utilize the maximum entropy classification model [11] and obtain the F1 = 72.56. Our CNN can achieve F1 = 73.66. We should note that one SIP can determine ALL the relevant sources of the events embedded in it. Therefore, the improvement of F1 is significant, which can prove the effectiveness of our CNN based on the pruned sentence structure.

**Table 3.** Performance of the basic factor extraction tasks.

|                  | P(%)  | R(%)  | F     |
|------------------|-------|-------|-------|
| Events           | 86.67 | 82.86 | 84.68 |
| SIPs             | 74.58 | 72.91 | 73.66 |
| Relevant sources | 80.70 | 77.44 | 78.99 |
| Cues             | 64.78 | 70.13 | 67.05 |

### 4.3 Results and Analysis: Event Factuality Identification

For the fair comparison with our neural network model, we adopt the following baselines whose input features are also developed according to the output of basic factor extraction task:

**SRules** is a rule-based model working on dependency parse trees [1,2].

**SVM** uses the features developed by [2]. Some studies [3,4] also utilized traditional machine learning models that only considered *AUTHOR* as the source. We re-implement these systems and obtain lower performance (macro-averaged F1 are 46.29 and 48.42, respectively) than [2] on *AUTHOR*.

**ME_QRules** is a two-step model with a combination of a maximum entropy classification model and a simple rule-based model [5].

**CNN** is a variant of our model in Sect. 3 whose LSTM is replaced by CNN defined in Sect. 2. The parameter of the convolutional layer is set as $n_c = 50$.

**Hybrid_NN_1** is a variant of our hybrid neural network model, and considers only ONE vector as the output of our model, i.e., vector representations $f_u$ and $f_{cue}$ in Eqs. (10) and (11) are concatenated into ONE vector and are fed into ONE softmax layer. Our hybrid model described in Sect. 3 has TWO outputs, and is denoted as **Hybrid_NN_2**.

Table 4 shows the performance of various models on the event factuality identification task. SRules gets low results on Uu, mainly due to the upstream error propagation from basic factor extraction tasks. SVM obtains much lower performance on CT−, PR+ and PS+, because they occupy the majority. ME_QRules achieves the micro- and macro-averaged F1 that are between SRules and SVM.

Our Hybrid_NN_2 model achieves the best performance not only on CT+ and Uu but on CT−, PR+, and PS+ with All sources. All the improvements are significant with $p < 0.05$ applying *two-sample two-tailed t-test* (the same below). The performance gaps in different factuality values illustrate that it is challenging to identify CT−, PR+ and PS+, which only cover 7.57% of all the factuality values. The improvement on CT−, PR+, and PS+ can show the effectiveness of features produced by speculative and negative cues. For example, compared with SVM, our model improves the F1 of CT−, PR+, and PS+ by 11.47, 19.28, and 18.20, respectively. In addition, Hybrid_NN_2 can also obtain satisfactory performance on CT+ and Uu, which can prove that our model can discriminate Non-Uu events from Uu ones effectively. It is critical to identify Non-Uu events because it is the previous step to identify CT+/−, PR+/−, and PS+/−.

**Table 4.** F1-measures of various systems on event factuality identification.

| Systems | Sources | CT+ | CT− | PR+ | PS+ | Uu | Micro-A | Macro-A |
|---------|---------|-----|-----|-----|-----|-----|---------|---------|
| SRules [1,2] | All | 61.83 | 54.52 | 20.75 | 39.89 | 26.08 | 50.71 | 40.62 |
| | Author | 64.83 | 48.83 | 13.35 | 31.93 | 26.17 | 53.49 | 37.02 |
| | Embed | **53.19** | **61.94** | 25.11 | **47.01** | 25.95 | 44.20 | 42.64 |
| SVM [2] | All | 64.94 | 44.80 | 26.54 | 25.90 | 57.68 | 60.78 | 43.97 |
| | Author | 71.39 | 42.61 | 34.67 | 35.00 | 65.64 | 68.14 | 52.59 |
| | Embed | 50.78 | 45.60 | 28.58 | 27.66 | 25.45 | 43.33 | 35.40 |
| ME_QRules [5] | All | 61.55 | 43.52 | 17.65 | 41.49 | 53.58 | 56.89 | 43.56 |
| | Author | 67.75 | 44.21 | 11.55 | 40.99 | 60.95 | 63.80 | 43.75 |
| | Embed | 46.39 | 40.40 | 22.22 | 40.78 | 30.46 | 40.73 | 36.05 |
| CNN | All | 62.35 | 52.11 | 39.60 | 40.30 | 55.65 | 58.92 | 50.00 |
| | Author | 68.28 | 52.26 | 52.03 | 45.92 | 62.93 | 65.54 | 56.50 |
| | Embed | 49.61 | 52.28 | 33.58 | 38.44 | 22.66 | 43.25 | 38.98 |
| Hybrid_NN_1 | All | 66.20 | 52.99 | 36.62 | 37.97 | 61.39 | 62.95 | 51.03 |
| | Author | 72.18 | 54.50 | 36.94 | 37.55 | 67.99 | 69.65 | 53.83 |
| | Embed | 52.73 | 51.39 | 37.56 | 36.79 | **33.97** | 47.07 | 41.79 |
| Hybrid_NN_2 | All | **66.41** | **56.27** | **45.82** | **44.10** | **61.80** | **63.81** | **54.88** |
| | Author | **72.73** | **58.11** | **54.14** | **48.85** | **68.29** | **70.41** | **59.90** |
| | Embed | 52.80 | 54.23 | **42.23** | 43.72 | 32.67 | **48.18** | **44.84** |

For events with *AUTHOR* as the sources, our Hybrid_NN_2 model can achieve the highest F1 on all the five main categories. For events with embedded sources, Hybrid_NN_2 model obtains lower performance due to their syntactic structures. The F1 of Uu on embedded sources is quite low (32.67), which is similar to other models. These results show that it is difficult to discriminate Uu from Non-Uu events with embedded sources.

To further verify the advantages of our Hybrid_NN_2 model, we implement other two neural network models as baselines, i.e., CNN and Hybrid_NN_1. The experimental results show that our model outperforms the two baselines on both micro- and macro-averaged F1-measures. The BiLSTM neural networks in our model can learn more information from both future and past context in syntactic paths, while CNN ignores the context of tokens. Compared with Hybrid_NN_1, the performance of CT−, PR+, and PS+ is significantly improved by our Hybrid_NN_2 model, which can prove that the design of the two outputs can identify speculative and negative factuality values more effectively.

To explore the upper bound of the performance of our model on event factuality identification, we present Table 5 that shows the performance of SRules and Hybrid_NN_2 models using annotated information. Comparing Tables 4 and 5, the micro- and macro-averaged F1 of SRules are improved by 30.81 and 29.38, and those of Hybrid_NN_2 by 17.87 and 16.75, respectively, indicating that the performance of SRules relies more heavily on annotated information.

**Table 5.** F1-measures of SRules and Hybrid_NN_2 models on event factuality identification with annotated information.

| Systems | Sources | CT+ | CT− | PR+ | PS+ | Uu | Micro-A | Macro-A |
|---------|---------|-----|-----|-----|-----|-----|---------|---------|
| SRules [1,2] | All | **86.69** | 73.72 | 57.83 | 55.64 | **76.13** | 81.52 | 70.00 |
| | Author | **88.32** | 68.59 | 53.93 | 56.18 | **81.18** | 84.56 | 69.29 |
| | Embed | **82.62** | **77.64** | **60.44** | 54.16 | **58.47** | 74.19 | 66.67 |
| Hybrid_NN_2 | All | 86.25 | **75.90** | **61.00** | **59.28** | 75.70 | **81.68** | **71.63** |
| | Author | 88.09 | **74.16** | **66.59** | 56.67 | 80.36 | **84.62** | **73.05** |
| | Embed | 82.13 | 76.62 | 58.54 | **63.50** | 53.81 | **74.67** | **66.92** |

For our Hybrid_NN_2 model, the performance of CT+ and Uu is lower than that of SRules. However, the F1 of CT−, PR+, and PS+ in our Hybrid_NN_2 model are higher, indicating that our model is better at identifying speculative and negative values. Furthermore, compared with SRules, Hybrid_NN_2 improves the micro- and macro-averaged F1 by 0.16 and 1.63, respectively. The performance obtained by Hybrid_NN_2 in Tables 4 and 5 means that the input features of our model are developed according to the basic factors, and do not rely on other intermediate ones.

### 4.4    Error Analysis

The errors produced by our model mainly include the following types:

**Incorrect Relevant Sources for Events (70.61%).** Our model fails to identify correct relevant sources for events due to the error propagation from the basic factor extraction tasks, which can prove the significance of the SIP and relevant source detection task. For example:

(S5) *The agency quoted witnesses as saying tanks are patrolling the streets.*

(Event: ***patrolling***, Source: *witnesses_agency_AUTHOR*, CT+)

In S5, our model fails to identify *witnesses* as the source introduced by the SIP ***saying***, and we cannot evaluate the factuality of the event ***patrolling*** from the perspective of *witnesses*.

**Incorrect Uu and Non-Uu (24.21%).** CT−, PR+/−, and PS+/− events cannot be correctly identified if the events are evaluated as Uu mistakenly.

(S6) *Officials said DNA test results showed a likelihood that a strand of hair discovered behind Slepian's home came from Kopp.*

(Event: ***came***, Source: *Officials_AUTHOR*, PR+)

In S6, the event ***came*** is assessed as Uu mistakenly, and it cannot be furtherly assigned as PR+ even if our Hybrid_NN_2 model correctly determine that ***came*** is governed by *likelihood*.

**Incorrect Modality or Polarity (4.42%).** Our model fails to determine whether the event is affected by a corresponding cue correctly.

(S7) _StatesWest_ **said** it may **purchase** more Mesa stock or **make** a tender offer directly to Mesa shareholders.

(Event: **make**, Source: _StatesWest_AUTHOR_, PS+)

(S8) _"Our company has not been able to **cope** very effectively with **changes** in the marketplace"_, **said** Ryosuke _Ito_, Sansui's president.

(Event: **changes**, Source: _Ito_AUTHOR_, CT+)

The event **make** in S7 is PS+ due to the PS cue _may_. Hybrid_NN_1 fails to judge that **make** is governed by _may_, while Hybrid_NN_2 gave the correct result. In S8, the event **changes** is CT+ according to _Ito_. But our Hybrid_NN_2 model determined that **changes** is negated by _not_ mistakenly.

# 5  Related Work

Early studies on event factuality identification limited the sources to _AUTHOR_. [15,16] classified predicates into Committed Belief (CB), Non-CB or NA under a supervised framework. Researchers also developed new corpus [17], scalable annotation schemes [4], and deep linguistic features [6] to predict factuality.

FactBank [7] considers both _AUTHOR_ and embedded sources. [1,2] proposed a rule-based top-down algorithm traversing dependency trees to identify event factuality on FactBank. [3] proposed a new annotation framework and identified the factuality of events in some sentences of FactBank. [5] utilized a two-step model combining machine learning and simple rule-based approaches. These studies utilized annotated information in FactBank directly.

Previous research showed that neural network methods can learn useful features from sentences [18] and syntactic paths [19]. Attention is an effective mechanism to capture important information in sequences and has achieved state-of-the-art performance in various NLP tasks [18,20]. In this paper, we employ a neural network model that considers both LSTM and CNN with attention to extract features from sequences and collections of words, respectively.

# 6  Conclusions

This paper focuses on event factuality identification. We propose a two-step framework with neural networks to identify the factuality of events with ALL sources (both author and embedded sources) in raw texts. For event factuality identification task, we utilize a hybrid neural network model with the combination of LSTM and CNN based on attention mechanism. The experimental results show that our model can identify speculative and negative factuality values effectively, and can outperform the state-of-the-art systems.

**Acknowledgments.** This research was supported by the National Natural Science Foundation of China under Grant Nos. 61751206, 61772354 and 61773276, and was also supported by the Strategic Pioneer Research Projects of Defense Science and Technology under Grant No. 17-ZLXD-XX-02-06-02-04.

# References

1. Saurí, R.: A factuality profiler for eventualities in text. Ph.D. thesis, Waltham, MA (2008)
2. Saurí, R., Pustejovsky, J.: Are you sure that this happened? Assessing the factuality degree of events in text. Comput. Linguist. **38**(2), 1–39 (2012)
3. de Marneffe, M.C., Manning, C.D., Potts, C.: Did it happen? The pragmatic complexity of veridicality assessment. Comput. Linguist. **38**(2), 301–333 (2012)
4. Lee, K., Artzi, Y., Choi, Y., Zettlemoyer, L.: Event detection and factuality assessment with non-expert supervision. In: EMNLP 2015, pp. 1643–1648. ACL, Stroudsburg (2015)
5. Qian, Z., Li, P.F., Zhu, Q.M.: A two-step approach for event factuality identification. In: IALP 2015, pp. 103–106. IEEE Press, New York (2015)
6. Stanovsky, G., Eckle-Kohler, J., Puzikov, Y., Dagan, I., Gurevych, I.: Integrating deep linguistic features in factuality prediction over unified datasets. In: ACL 2017, pp. 352–357. ACL, Stroudsburg (2017)
7. Saurí, R., Pustejovsky, J.: Factbank: a corpus annotated with event factuality. Lang. Resour. Eval. **43**(3), 227–268 (2009)
8. Marasović, A., Frank, A.: Multilingual modal sense classification using a convolutional neural network. In: Rep4NLP@ACL 2016, pp. 111–120. ACL, Stroudsburg (2016)
9. Schütze, H., Adel, H.: Exploring different dimensions of attention for uncertainty detection. In: EACL 2017, pp. 22–34. ACL, Stroudsburg (2017)
10. Pustejovsky, J., et al.: TimeML: robust specification of event and temporal expressions in text. In: New Directions in Question Answering, Papers from 2003 AAAI Spring Symposium, pp. 28–34. Stanford University, Stanford (2003)
11. Chambers, N.: NavyTime: event and time ordering from raw text. In: 7th International Workshop on Semantic Evaluation, SemEval@NAACL-HLT 2013, pp. 73–77. ACL, Stroudsburg (2013)
12. Øvrelid, L., Velldal, E., Oepen, S.: Syntactic scope resolution in uncertainty analysis. In: COLING 2010, pp. 1379–1387. Tsinghua University Press, China (2010)
13. Velldal, E., Øvrelid, L., Read, J., Oepen, S.: Speculation and negation: rules, rankers, and the role of syntax. Comput. Linguist. **38**(2), 369–410 (2012)
14. Mikolov, T., Sutskever, I., Chen, K., Corrado, G.S., Dean, J.: Distributed representations of words and phrases and their compositionality. In: NIPS 2013, pp. 3111–3119. Curran Associates, Inc. (2013)
15. Diab, M.T., Levin, L.S., Mitamura, T., Rambow, O., Prabhakaran, V., Guo, W.W.: Committed belief annotation and tagging. In: the Third Linguistic Annotation Workshop, pp. 68–73. ACL, Stroudsburg (2009)
16. Prabhakaran, V., Rambow, O., Diab, M.T.: Automatic committed belief tagging. In: 23rd International Conference on Computational Linguistics (COLING 2010), pp. 1014–1022. Chinese Information Processing Society of China, China (2010)
17. Prabhakaran, V., et al.: A new dataset and evaluation for belief/factuality. In: Fourth Joint Conference on Lexical and Computational Semantics, pp. 82–91. The *SEM 2015 Organizing Committee (2015)

18. Zhou, P., et al.: Attention-based bidirectional long short-term memory networks for relation classification. In: ACL 2016, pp. 207–212. ACL, Stroudsburg (2016)
19. Roth, M., Lapata, M.: Neural semantic role labeling with dependency path embeddings. In: ACL 2016, pp. 1092–1202. ACL, Stroudsburg (2016)
20. Wang, Y.Q., Huang, M.L., Zhu, X.Y., Zhao, L.: Attention-based LSTM for aspect-level sentiment classification. In: EMNLP 2016, pp. 606–615. ACL, Stroudsburg (2016)

# A Least Squares Approach
# to Region Selection

Liantao Wang[1,2(✉)], Yan Liu[1], and Jianfeng Lu[3]

[1] College of IoT Engineering, Hohai University, Changzhou, China
ltwang@hhu.edu.cn
[2] State Key Laboratory for Novel Software Technology, Nanjing University,
Nanjing, China
[3] School of Computer Science, Nanjing University of Science and Technology,
Nanjing, China

**Abstract.** Region selection is able to boost the recognition performance for images with background clutter by discovering the object regions. In this paper, we propose a region selection method under the least squares framework. With the assumption that an object is a combination of several over-segmented regions, we impose a selection variable on each region, and employ a linear model to perform classification. The model parameter and the selection parameter are alternatively updated to minimize a sum-of-squares error function. During the iteration, the selection parameter can automatically pick the discriminant regions accounting for the object category, then fine tunes the linear model with the objects, independently of the background. As a result, the learnt model is able to distinguish object regions and non-object regions, which actually generates irregular-shape object localization. Our method performs significantly better than the baselines on two datasets, and the performance can be further improved when combining deep CNN features. Moreover, the algorithm is easy to implement and computationally efficient because of the merits inherited from the least squares.

**Keywords:** Least squares · Region selection · Object localization

## 1 Introduction

Image classification is one of the fundamental problems in computer vision. The task aims to recognize the categories of the objects present in images. However, an object does not always cover the whole image. Severe background clutter in the images may disturb the recognition. Classification by detection [3,5,10] is tailored to this kind of situations. These methods annotate objects from the background by finding the bounding-box with maximum score, so as to provide natural comprehensive context for performance boosting.

Moreover, many other seemingly distinct computer vision tasks such as object detection [11,20] and segmentation [7,8] can be reduced to image classification. However, the classification models in these methods need carefully pruned

© Springer Nature Switzerland AG 2018
L. Cheng et al. (Eds.): ICONIP 2018, LNCS 11305, pp. 348–358, 2018.
https://doi.org/10.1007/978-3-030-04221-9_31

objects for training. Multiple-instance learning (MIL) has been exploited to automatically annotate the objects in images to train a classifier being capable of localizing objects [15,23,26]. In a MIL paradigm, data examples and labels are provided at bag-level. A negative bag contains only negative instances, and a positive bag contains at least one positive instance. This framework applies very naturally to image classification and localization, where an image is considered as a bag, and bounding-boxes in the image are considered as instances.

An alternative to the bounding-box annotation is region selection. Li et al. [12] propose KI-SVM to locate the regions of interest for content-based image retrieval. Based on convex relaxation of the MI-SVM [1], they maximize the margin via generating the most violated key instance step by step, and then combines them via efficient multiple kernel learning. Yakhnenko et al. [25] segment an image into a fixed grid, and assign a binary latent variable to indicate either selected or de-selected state for each segment, then incorporate the variable learning into a linear SVM model. Zhao et al. [26] consider an object as a combination of several over-segmented regions that can be obtained in unsupervised manners [2,24]. They use non-linear kernel SVM to discover discriminative regions in images and videos. Compared to bounding-box selection, region selection has the potential to be applied to irregular-shape object detection [19–22].

In this paper, we propose a region selection method under the least squares framework. Our algorithm aims to find the discriminant regions accounting for the object category. With unsupervised image segmentation methods, an image can be converted into a set of regions (or super-pixels). Instead of training on whole images, we impose the regions with selection variables to realize object discovery. We minimize a sum-of-squares error function to train a linear model. The model parameter and the selection parameter are learned in an alternate way. When the selection parameter is fixed, the model parameter is updated as a standard least squares problem with a closed-form solution. When the model parameter is fixed, we show the optimization with respect to the selection parameter is a standard quadratic programming. Compared to other methods, ours are computationally efficient and simple to implement because of the advantages inherited from least squares. The experimental results demonstrate the effectiveness for region selection and the potential for applications in weakly supervised irregular-shape object localization.[1]

The rest of the paper is organized as follows. The method is detailed in Sect. 2: Least squares for image classification is revisited in Sect. 2.1; The objective function for region selection in least squares is proposed in Sect. 2.2; Then the solving of the model parameter and selection parameter are described in Sects. 2.3 and Sect. 2.4, respectively; Sect. 3 presents the experimental setup and results analysis. Finally, we conclude with Sect. 4.

---

[1] The code will be published with the article.

## 2   Methods

### 2.1   Least Squares for Image Classification

Given a collection of image data $\{\mathcal{I}_1, \mathcal{I}_2, \cdots, \mathcal{I}_N\}$ with labels $y_i \in \{0,1\}$, traditional classification methods first extract feature $x_i \in \mathbb{R}^{D \times 1}$ for each image, comprising training data $\{x_1, x_2, \cdots, x_N\}$. A linear discriminant function $y = w^\top x$ can be used to predict the label. The conventional least squares approach to training the linear model is to minimize the sum-of-squares error function

$$E(w) = \frac{1}{2} \sum_{i=1}^{N} (y_i - w^\top x_i)^2. \tag{1}$$

The gradient of the error function (1) takes the form:

$$\nabla E(w) = - \sum_{i=1}^{N} (y_i - w^\top x_i) x_i^\top. \tag{2}$$

Setting the gradient to zero and solving for w gives,

$$w = \left( \sum_{i=1}^{N} x_i x_i^\top \right)^{-1} \sum_{i=1}^{N} y_i x_i. \tag{3}$$

By introducing a design matrix $\Phi \in \mathbb{R}^{N \times D}$, whose elements are given by $\Phi = [x_1, x_2, \cdots, x_N]^\top$, the solution (3) can be rewritten as:

$$w = (\Phi^\top \Phi)^{-1} \Phi^\top Y, \tag{4}$$

where $Y = [y_1, y_2, \cdots, y_N]^\top \in \mathbb{R}^{N \times 1}$.

### 2.2   Region Selection in Least Squares

With unsupervised over-segmentation methods, an image $\mathcal{I}_i$ is decomposed into a set of regions $\{x_{i1}, x_{i2}, \cdots, x_{im_i}\}$, $x_{ij} \in \mathbb{R}^{D \times 1}$. We impose a set of selection variables $P_i = [p_{i1}, p_{i2} \cdots, p_{im_i}]^\top$, $p_{ij} \in \{0,1\}$ on each region, and assume some of them comprising an object. Different assignments of $P_i$ correspond to selecting different regions to represent the image. With the bag-of-words histogram feature representation, the region selection intuition can be expressed as $X_i P_i$, where we have described an image by the matrix $X_i \in \mathbb{R}^{D \times m_i}$ whose columns are region vectors: $X_i = [x_{i1}, x_{i2}, \cdots, x_{im_i}]$. Note that the mathematical result of the selection $X_i P_i$ is still a $D \times 1$ column vector.

In order to make the optimization easier, we relax the constraint of $P_i$ and require only $|P_i| = 1, P_i \geq \mathbf{0}$, where $\mathbf{0}$ denotes a vector with the same size as $P_i$ and elements all zeros, and the inequality constraint needs to hold in element-wise manner.

We integrate this region selection idea into the least squares framework, and define the new sum-of-squares error function for least squares region selection as:

$$E_{rs}(\mathbf{w}, \mathbf{P}_1, \cdots, \mathbf{P}_N) = \frac{1}{2} \sum_{i=1}^{N} \left( y_i - \mathbf{w}^\top (\mathbf{X}_i \mathbf{P}_i) \right)^2. \tag{5}$$

The objective function consequently becomes:

$$\min_{\mathbf{w}, \mathbf{P}_1, \cdots, \mathbf{P}_N} \frac{1}{2} \sum_{i=1}^{N} \left( y_i - \mathbf{w}^\top (\mathbf{X}_i \mathbf{P}_i) \right)^2$$

$$\text{s.t. } |\mathbf{P}_i| = 1, \mathbf{P}_i \geq \mathbf{0}, i = 1, \cdots, N. \tag{6}$$

A regularization term (e.g., $L_2$ norm) of $\mathbf{w}$ can be added to the objective function to avoid over-fitting without increasing the optimization difficulty. Note that there are two coupled parameters, model parameter $\mathbf{w}$ and selection parameter $\mathbf{P}_i, i = 1, 2, \cdots, N$. We employ an alternate strategy to optimize the objective function.

### 2.3   Update Model Parameter

When the selection parameter $\mathbf{P}_i, i = 1, 2, \cdots, N$ is fixed, the optimization of (6) is equivalent to

$$\min_{\mathbf{w}} \frac{1}{2} \sum_{i=1}^{N} \left( y_i - \mathbf{w}^\top (\mathbf{X}_i \mathbf{P}_i) \right)^2. \tag{7}$$

As in the conventional least squares, we take the gradient of the region selection error function (5) with respect to $\mathbf{w}$:

$$\nabla E_{rs}(\mathbf{w}) = - \sum_{i=1}^{N} \left( y_i - \mathbf{w}^\top (\mathbf{X}_i \mathbf{P}_i) \right) (\mathbf{X}_i \mathbf{P}_i)^\top. \tag{8}$$

Setting the gradient to zero and solving for $\mathbf{w}$ we obtain,

$$\mathbf{w} = \left( \sum_{i=1}^{N} (\mathbf{X}_i \mathbf{P}_i)(\mathbf{X}_i \mathbf{P}_i)^\top \right)^{-1} \sum_{i=1}^{N} y_i (\mathbf{X}_i \mathbf{P}_i), \tag{9}$$

Similarly, if we define a new matrix $\Phi_{rs} \in \mathbb{R}^{N \times D}$ as $[\mathbf{X}_1 \mathbf{P}_1, \mathbf{X}_2 \mathbf{P}_2, \cdots, \mathbf{X}_N \mathbf{P}_N]^\top$, the solution (9) can be rewritten as:

$$\mathbf{w} = (\Phi_{rs}^\top \Phi_{rs})^{-1} \Phi_{rs}^\top \mathbf{Y}, \tag{10}$$

where $\mathbf{Y} = [y_1, y_2, \cdots, y_N]^\top$.

$\Phi_{rs}$ is a matrix with the same size as the design matrix $\Phi$, each row corresponding to the description of an image. Each row in the design matrix $\Phi$ is the feature description of the whole image, while that in $\Phi_{rs}$ is the description of the selected regions in the image. As a result, we call $\Phi_{rs}$ selected design matrix. The solution of the model parameter has the same form as the conventional least squares, and adding a $L_2$ regularization term for $\mathbf{w}$ is straightforward.

---

**Algorithm 1.** Least squares region selection (LSRS).

---

**Input**: A set of images $\{\mathcal{I}_1, \mathcal{I}_2, \cdots, \mathcal{I}_N\}$ with labels $y_i \in \{1, 0\}$.
**Output**: Selected region for each image.
1  Over-segment each image into regions;
2  Initialize $P_i$ ;
3  **while** *not converged* **do**
4  |    Update linear model parameter w using (10);
5  |    Update $\{P_1, P_2, \cdots, P_N\}$ by solving (13);
6  **end**
7  Return regions with 1 labels according to (14).

---

### 2.4  Update Selection Parameter

When the model parameter w is fixed, the objective function (6) becomes:

$$\min_{P_1,\cdots,P_N} \frac{1}{2} \sum_{i=1}^{N} \left( y_i - w^\top (X_i P_i) \right)^2$$
$$\text{s.t.} \quad |P_i| = 1, P_i \geq \mathbf{0}, i = 1, \cdots, N, \tag{11}$$

and this optimization can be equivalently decomposed into the following sub-optimization problems with $i = 1, 2, \cdots, N$:

$$\min_{P_i} \frac{1}{2} \left( y_i - w^\top (X_i P_i) \right)^2$$
$$\text{s.t.} \quad |P_i| = 1, P_i \geq \mathbf{0}. \tag{12}$$

We rearrange (12) and omit the constant term to have

$$\min_{P_i} \frac{1}{2} P_i^\top (X_i^\top w)(X_i^\top w)^\top P_i + (-y_i X_i^\top w)^\top P_i$$
$$\text{s.t.} \quad |P_i| = 1, P_i \geq \mathbf{0}, \tag{13}$$

which turns out to be a standard convex quadratic programming problem and can be solved by a typical optimization package.

The algorithm is shown in Algorithm 1. Given a set of images with image-level labels, we first over-segment each image using unsupervised methods to obtain regions. We impose a selection variable on each region, then alternately update the model parameter and selection parameter. During the iteration, the selection parameter can select the objects in the images, and the linear model is fine tuned with the objects independently of the background clutter. Therefore the discriminant function

$$y = sgn(w^\top x - 0.5) \tag{14}$$

is able to distinguish object regions and non-object regions in images.

# 3   Experimental Results

## 3.1   Dataset

We compare the LSRS with methods that obtain irregular-shape object localization through region selection: OBoW [22], CRANE [19], MILBoost [21], on Pittsburgh Car (PittCar) [15] dataset and YouTube-Objects (YTO) manually annotated in [19].

PittCar dataset contains 400 images of street scenes, where 200 images contain cars. The dataset is challenging for localization because the appearance of the cars in the images varies in shape, size, grayscale intensity and location. Furthermore, the cars occupy only a small portion of the images and may be partially occluded by other objects. Some examples are shown in the first row of Fig. 1.

YTO dataset contains videos of ten classes of objects collected from YouTube. The objects are moving, and the background is complicated. See the last two rows of Fig. 1 for some examples. Tang et al. [19] generated a groundtruthed set by manually annotating the segments for 151 selected shots. The segment-level ground truth is well suitable for the evaluation of the irregular-shape object localization.

**Fig. 1.** Some examples from the datasets: PittCar (row 1), YTO (rows 2 and 3).

## 3.2   Implementation

We extract local SIFT [13] descriptors densely for each image/frame, and randomly select 100,000 descriptors and obtain 1000 visual words by $k$-means clustering for each dataset. We use unsupervised methods to get over-segmentations for each image [2] and video [24]. Therefore each segment is represented by a 1000-dimensional histogram of visual word, and $L_2$ normalization is executed before feeding to the classifiers.

For each positive image/video, we impose selection variables on each segment. The variables are initialized with identical values $\frac{1}{m_i}$, where $m_i$ is the number of segments in the image/video. Then they are optimized iteratively. While for each negative image/video, we just use an average combination, i.e., the variables are set to $\frac{1}{m_i}$ constantly. The algorithm is terminated when the change rate of the objective function is less than $10^{-3}$.

### 3.3 Analysis

In order to give a quantitative comparison, we follow the popular evaluation metric for object localization [6, 16, 17]: the localization is considered correct when the overlapping of the detected region and the ground-truth is larger than 0.5 for images and 0.125 for videos, then the average precision (the fraction of the correctly detected images) is calculated. As shown in Fig. 2, our method outperforms the competitors on most classes in the PittCar and YTO datasets. This results from the automatic region selection when training the classifier. When minimizing the error function, the algorithm can adaptively select the regions, which are most different from the negative images, for each positive image/video. The selected regions usually comprises objects, and the classifier is actually trained with localized objects.

In order to further understand the algorithm, we plot the iteration process in Fig. 3. Four iterations are chosen whose APs are 0.07, 0.3, 0.5, 0.63 respectively, corresponding to the four rows in the figure. We selected three images from the PittCar dataset whose visualizations are indexed by (a), (b) and (c). Each set of visualization contains two columns: P-column is the visualization of the selection variables, where the warmer color indicates the larger variable value; Y-column is the visualization of predicted labels, where each region classified as positive is stained with red. As can be seen from the figure, since we initialize the selection parameter for each region with equal values $\frac{1}{m_i}$, the selection variables tend to be uniform in the first selected iteration. As the iteration continues, the selection variables show a trend of concentrating on cars, and the predicted labels yield satisfying localization result. It is interesting to see that the selection variables of the wheels are usually the largest, followed by those of the doors. It indicates that the wheels are the most discriminant parts for the positive images in the recognition of the PittCar dataset.

As for the time consumption, our algorithm consists of two steps: One is a least squares with closed-form solution; The other is a standard quadratic programming. It is (about 6 times) faster than CRANE that needs much nearest neighbour search on the same experimental platform.

### 3.4 Combining Deep Features

Recently deep convolutional neural networks (CNN) has achieved impressive results on visual recognition tasks compared to conventional feature representations [4]. CNN-based recognition methods typically use the output of the last layer as a feature representation, which has rich semantic information but little spatial information. Some researchers have investigated using multiple CNN layers to construct a deep feature representation for a pixel [9, 14].

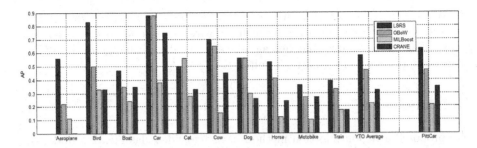

**Fig. 2.** Average precision on PittCar and YTO datasets.

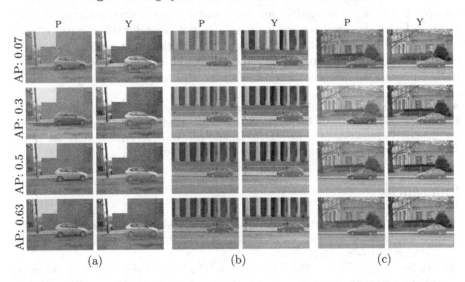

**Fig. 3.** Visualization of the LSRS iteration process. Four iterations with localization APs 0.07, 0.3, 0.5 and 0.63 are picked and plotted in each row. Three car images are picked and their visualization is indexed by (a), (b) and (c). P- and Y-columns indicate the visualization of the selection variables and the predicted labels for each region, respectively.

We test the deep feature from the VGG-NET [18] pre-trained on the ImageNet in this section. For each image, we first resize it to $224 \times 224$ and feed it into the pre-trained VGG model, then extract feature maps of Conv5-4, Conv4-4, and Conv3-4 layers. Since the convolution and pooling operations in CNN reduces the spatial resolution, we up-sample each feature map to the scale of the original image, then concatenate them into a 1280-dimensional hyper-column for each pixel. Then $k$-means is applied to the training points to obtain 1000 words, and therefore each segment can be represented by a 1000-bin histogram. $L_2$ normalization is used before feeding to the classifier.

We use the deep features in our LSRS, the localization average precision achieves 0.89 on PittCar dataset, which improves 41% compared to SIFT features. This demonstrates that our algorithm can take advantage of deep CNN features and obtain much better results.

## 4    Conclusion

We proposed a region selection method in image/video classification under the least squares framework. For each over-segmented region, a variable is imposed to indicate whether it is discriminant to the object category. The selection is integrated into a linear model, and the parameters are alternately optimized. It outperforms the baselines on two datasets, and can be further improved significantly when using deep CNN features. In addition, our method is easy to implement and computationally efficient, due to the advantages of least squares. In future work, we will study region selection with non-linear models.

**Acknowledgment.** This work was supported in part by the National Natural Science Foundation of China (NSFC) (No. 61703139), the Fundamental Research Funds for the Central Universities (No. 2016B12914), and the State Key Laboratory for Novel Software Technology (Nanjing University) (No. KFKT2017B09).

## References

1. Andrews, S., Tsochantaridis, I., Hofmann, T.: Support vector machines for multiple-instance learning. In: Advances in Neural Information Processing Systems, pp. 561–568 (2002)
2. Arbelaez, P., Maire, M., Fowlkes, C., Malik, J.: Contour detection and hierarchical image segmentation. IEEE Trans. Pattern Anal. Mach. Intell. **33**(5), 898–916 (2011)
3. Behmo, R., Marcombes, P., Dalalyan, A., Prinet, V.: Towards optimal naive bayes nearest neighbor. In: Daniilidis, K., Maragos, P., Paragios, N. (eds.) ECCV 2010. LNCS, vol. 6314, pp. 171–184. Springer, Heidelberg (2010). https://doi.org/10.1007/978-3-642-15561-1_13
4. Chatfield, K., Simonyan, K., Vedaldi, A., Zisserman, A.: Return of the devil in the details: delving deep into convolutional nets. In: British Machine Vision Conference, BMVC 2014, Nottingham, UK, 1–5 September 2014 (2014)
5. Chen, Q., Song, Z., Dong, J., Huang, Z., Hua, Y., Yan, S.: Contextualizing object detection and classification. IEEE Trans. Pattern Anal. Mach. Intell. **37**(1), 13–27 (2015)
6. Deselaers, T., Alexe, B., Ferrari, V.: Localizing objects while learning their appearance. In: Daniilidis, K., Maragos, P., Paragios, N. (eds.) ECCV 2010. LNCS, vol. 6314, pp. 452–466. Springer, Heidelberg (2010). https://doi.org/10.1007/978-3-642-15561-1_33
7. Farabet, C., Couprie, C., Najman, L., LeCun, Y.: Learning hierarchical features for scene labeling. IEEE Trans. Pattern Anal. Mach. Intell. **35**(8), 1915–1929 (2013)

8. Hariharan, B., Arbeláez, P., Girshick, R., Malik, J.: Simultaneous detection and segmentation. In: Fleet, D., Pajdla, T., Schiele, B., Tuytelaars, T. (eds.) ECCV 2014. LNCS, vol. 8695, pp. 297–312. Springer, Cham (2014). https://doi.org/10.1007/978-3-319-10584-0_20

9. Hariharan, B., Arbeláez, P.A., Girshick, R.B., Malik, J.: Hypercolumns for object segmentation and fine-grained localization. In: IEEE Conference on Computer Vision and Pattern Recognition, CVPR 2015, Boston, MA, USA, 7–12 June 2015, pp. 447–456 (2015)

10. Harzallah, H., Jurie, F., Schmid, C.: Combining efficient object localization and image classification. In: IEEE 12th International Conference on Computer Vision, ICCV 2009, Kyoto, Japan, 27 September–4 October 2009, pp. 237–244 (2009)

11. Lampert, C.H., Blaschko, M.B., Hofmann, T.: Beyond sliding windows: object localization by efficient subwindow search. In: 2008 IEEE Conference on Computer Vision and Pattern Recognition, pp. 1–8 (2008)

12. Li, Y.-F., Kwok, J.T., Tsang, I.W., Zhou, Z.-H.: A convex method for locating regions of interest with multi-instance learning. In: Buntine, W., Grobelnik, M., Mladenić, D., Shawe-Taylor, J. (eds.) ECML PKDD 2009. LNCS (LNAI), vol. 5782, pp. 15–30. Springer, Heidelberg (2009). https://doi.org/10.1007/978-3-642-04174-7_2

13. Lowe, D.G.: Distinctive image features from scale-invariant keypoints. Int. J. Comput. Vis. **60**(2), 91–110 (2004)

14. Ma, C., Huang, J., Yang, X., Yang, M.: Hierarchical convolutional features for visual tracking. In: 2015 IEEE International Conference on Computer Vision, ICCV 2015, Santiago, Chile, 7–13 December 2015, pp. 3074–3082 (2015)

15. Nguyen, M.H., Torresani, L., de la Torre, L., Rother, C.: Weakly supervised discriminative localization and classification: a joint learning process. In: IEEE International Conference on Computer Vision, pp. 1925–1932 (2009)

16. Pandey, M., Lazebnik, S.: Scene recognition and weakly supervised object localization with deformable part-based models. In: IEEE International Conference on Computer Vision, pp. 1307–1314 (2011)

17. Russakovsky, O., Lin, Y., Yu, K., Fei-Fei, L.: Object-centric spatial pooling for image classification. In: Fitzgibbon, A., Lazebnik, S., Perona, P., Sato, Y., Schmid, C. (eds.) ECCV 2012. LNCS, pp. 1–15. Springer, Heidelberg (2012). https://doi.org/10.1007/978-3-642-33709-3_1

18. Simonyan, K., Zisserman, A.: Very deep convolutional networks for large-scale image recognition. CoRR abs/1409.1556 (2014)

19. Tang, K.D., Sukthankar, R., Yagnik, J., Li, F.F.: Discriminative segment annotation in weakly labeled video. In: IEEE Conference on Computer Vision and Pattern Recognition, pp. 2483–2490 (2013)

20. Vijayanarasimhan, S., Grauman, K.: Efficient region search for object detection. In: 2011 IEEE Conference on Computer Vision and Pattern Recognition, pp. 1401–1408 (2011)

21. Viola, P.A., Platt, J.C., Zhang, C.: Multiple instance boosting for object detection. In: Advances in Neural Information Processing Systems, pp. 1–8 (2005)

22. Wang, L., Meng, D., Hu, X., Lu, J., Zhao, J.: Instance annotation via optimal bow for weakly supervised object localization. IEEE Trans. Cybern. **47**, 1313–1324 (2017)

23. Wu, J., Yu, Y., Huang, C., Yu, K.: Deep multiple instance learning for image classification and auto-annotation. In: IEEE Conference on Computer Vision and Pattern Recognition, CVPR 2015, Boston, MA, USA, 7–12 June 2015, pp. 3460–3469 (2015)

24. Xu, C., Xiong, C., Corso, J.J.: Streaming hierarchical video segmentation. In: Fitzgibbon, A., Lazebnik, S., Perona, P., Sato, Y., Schmid, C. (eds.) ECCV 2012. LNCS, vol. 7577, pp. 626–639. Springer, Heidelberg (2012). https://doi.org/10.1007/978-3-642-33783-3_45
25. Yakhnenko, O., Verbeek, J., Schmid, C.: Region-based image classification with a latent SVM model. INRIA Technical report, pp. 1–13 (2011)
26. Zhao, J., Wang, L., Cabral, R., la Torre, F.D.: Feature and region selection for visual learning. IEEE Trans. Image Process. **25**, 1084–1094 (2016)

# Adaptive Intrusion Recognition for Ultraweak FBG Signals of Perimeter Monitoring Based on Convolutional Neural Networks

Fang Liu[1,3], Sihan Li[2], Zhenhao Yu[2], Xiaoxiong Ju[3], Honghai Wang[1], and Quan Qi[4(✉)]

[1] National Engineering Laboratory of Fiber Optic Sensing Technology, Wuhan University of Technology, Wuhan 430070, China
[2] School of Computer Science and Technology, Wuhan University of Technology, Wuhan 430070, China
[3] Hubei Key Laboratory of Transportation Internet of Things, Wuhan University of Technology, Wuhan 430070, China
[4] No. 161 Hospital of PLA, Wuhan 430014, China
83258405@qq.com

**Abstract.** Intrusion recognition based on the fiber-optic sensing perimeter security system is a significant method in security technology. Nevertheless, it is of great challenge to distinguish among multitudinous intrusion signals. Many studies have been conducted to solve this problem, which are absolutely dependent on the handcrafted features, and the process of feature extraction is time-consuming and unreliable. In this paper, we present an adaptive intrusion recognition method for ultra-weak FBG signals of perimeter monitoring based on convolutional neural networks. The advantage of the proposed method is its ability to extract optimal vibration features automatically from the raw sensing vibration signals. A fiber-optic sensing perimeter security system was developed to evaluate the computational efficiency of the proposed recognition method. The experiment results demonstrated that the proposed method could recognize the intrusion in the perimeter security system effectively with the best recognition accuracy among all of the comparative methods.

**Keywords:** Convolutional neural networks · Intrusion recognition Feature engineering · Fiber-optic sensing · Perimeter security

## 1 Introduction

Among the many security technologies, fiber-optic sensing perimeter security has become a widely used method because of the outstanding advantages of light weight, flexible length, signal transmission security, easy installation, and immunity to electromagnetic interference [1, 2]. The fiber-optic sensing perimeter

© Springer Nature Switzerland AG 2018
L. Cheng et al. (Eds.): ICONIP 2018, LNCS 11305, pp. 359–369, 2018.
https://doi.org/10.1007/978-3-030-04221-9_32

security system monitors the deployed area in real-time by sensing fibers. With the open-air construction, the system is often affected by natural factors such as strong winds, heavy rain and hail, and it has the characteristics of high suddenness and strong randomness. The intrusion recognition of fiber-optic sensing perimeter security system could be regarded as signal recognition, and a lot of achievements has been made for the past decades.

The vast majority of existing signal recognition methods in the literature involve two processes, namely feature extraction and feature classification, which are also called feature engineering [3]. Common signal processing methods include time-domain statistical analysis, empirical mode decomposition, Fourier spectral analysis, and wavelet transformation [4–6]. The important step of signal recognition in feature engineering, which is well configured and trained to discriminate the extracted features, includes some parametric classifiers such as support vector machines (SVM) [7], neural networks (NN) [8] and k-nearest neighbors (KNN) [9]. The effectiveness of feature engineering in signal recognition is completely dependent on handcrafted extracted features, which is a time-consuming and challenging task.

The objective of this study is to address the laborious feature engineering of intrusion signal recognition by using convolutional neural networks (CNNs). CNNs have recently become the de-facto standard for deep learning tasks such as object recognition in image processing and speech recognition as they achieved state-of-the-art performance [10]. CNNs have successfully been applied for the classification of electrocardiogram (ECG) beats achieving excellent effectiveness in aspect of both speed and accuracy [11]. Additionally, CNNs have rapidly accomplished the diagnosis fault of rolling element bearings with high accuracy [12, 13]. The reason that CNNs achieves good results is that CNNs can fuse input raw data and extract basic information from it in its lower layers, fuse the basic information into higher representation information and decisions in its middle layers, and further fuse these decisions and information in its higher layers to form the final classification result [14–16].

In this paper, we propose a highly accurate and adaptive intrusion recognition approach for fiber sensing perimeter security based on CNNs. The main objective of the proposed method is to identify different intrusion behaviors by processing the raw intrusion signals collected by the installed fiber-optic sensors. This convenient signal processing of the proposed method considerably speeds up the intrusion recognition efficiency of the fiber-optic sensor perimeter security system, which makes the real-time signal data analysis possible. The advantages of the proposed CNNs model will be comparatively proved with time-consuming feature engineering and evaluated experimentally using a simulative system of perimeter security. Additionally, the better recognition results compared to those of conventional methods make the engineering applications possible.

## 2    Methodology

### 2.1    Brief Introduction to CNNs

As shown in Fig. 1, convolutional neural networks are multi-stage neural networks composed of some filter stages and one classification stage [14]. The filter stage is commonly considered to include the convolutional layer and the pooling layer, which is separately designed to have additional features from the inputs and reduce the dimensions of features. There are usually one or more fully-connected layers in the classification stage, which is regard as multi-layer perceptron and gives the decision (classification) vector.

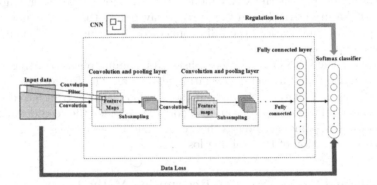

**Fig. 1.** Typical architecture of convolutional neural networks.

The convolutional layer convolutes the input with filters and obtains the output features by the activation function. To reduce the operation time and complexity of the model, each filter shares the same weighted parameters in different patches of the input [13]. The input of the convolutional layer is recorded as $I$, which size is $R^{H \times Z}$. Then the convolution process can be calculated as follows:

$$O_k^l = f(conv(W_k^l, I^l) + b^l) \tag{1}$$

where $O_k^l$ is the output of the $k$-th filter kernel in the $l$-th convolutional layer, and $l$ is between one and the number of layers. The weights and bias of the $k$-th filter kernel are denoted as $W_k^l$ and $b^l$ in the $l$-th layer, and each convolutional filter shares the same bias in the model; $f$ is an activation function, typically hyperbolic tangent or sigmoid function.

The pooling layer is also called as sub-sampling layer, which decimates the revolution of the features propagated from the previous layer [15]. The convolutional layer generally alternates with a pooling layer, which includes different kinds of operator. The max-pooling and average-pooling are the most commonly used layer, and the transformations are described as follows:

$$Max - pooling : P_{i,j}^l = \max_{0 \le m \le k-1, 0 \le n \le k-1} \{o_{i*d+m,j*d+n}^l\} \tag{2}$$

$$Average - pooling : P_{i,j}^l = \frac{1}{k^2} \sum_{m=1}^{k-1} \sum_{n=1}^{k-1} o_{i*d+m,j*d+n}^l \tag{3}$$

where $o^l$ denotes the input value of the pooling layer operation, $l$ is the number of layers, and the subscripted variable is the plane coordinates in the two-dimensional space. $P_{i,j}^l$ represents the output of the pooling operation in layer $l$, in which the abscissa is $i$ and the ordinate is $j$; $i$ and $j$ must comply with the following conditions: $0 \le i \le \frac{H-m}{d}$ and $0 \le j \le \frac{Z-n}{d}$. Further, $k$ is the size of pooling kernel, and $d$ is the stride of the pooling kernel.

As the final layer of the CNN model, the fully-connected layer maps the distributed feature learned from the previous layers into the sample mark space [12]. In practical application, the fully connected layer can be implemented by convolution operations.

In general, the CNNs model accomplishes its training phase by using a feedforward algorithm and a back-propagation algorithm. The feedforward process in the training is that the output from the previous layer is transferred to the next layer [15]. The back-propagation algorithm reverses the feedforward process, and the CNNs model updates the weights and biases of each layer through backward propagation of the training loss [16].

## 2.2 Proposed CNNs Intrusion Recognition Method

The CNNs model of proposed method is adjusted to satisfy the characteristics of intrusion behaviors, which is made up of a one-dimensional (1D) convolutional structure with a 1D filter bank. As shown in Fig. 2, the input of proposed model is the raw signal normalized data, and the output of proposed model is the classification result of the signal. There are some similarities between the architecture of the proposed model and the classical CNNs model. They are both made up of several convolutional pooling layers and one full connected layer. The main difference is that the first convolutional kernels are wide in the extracted-information layers, and the size of the following convolutional kernels is incremented at every step. The wide kernels in the first convolutional layer can obtain more effective information than the small kernels. The convolutional layers alternate with the pooling layers in the network help to obtain good representations of the input intrusion signals and improve the performance of the model. To speed up the training process in the training, batch normalization is designed before the activation function and after the convolutional layer and the fully-connected layer.

The major objective of the CNNs-based method is the effective recognition of different intrusion behaviors through the processing of corresponding raw intrusion signal. With the deep-layered architecture in the proposed model, the method has the adaptive ability to learn features automatically and makes good decisions about the classification of signals at the end of the model.

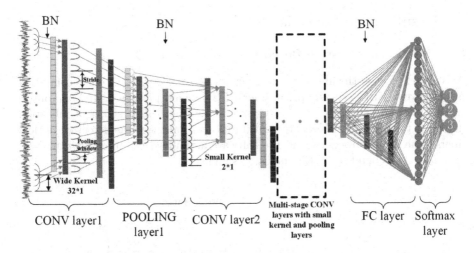

**Fig. 2.** Architecture of proposed CNNs model.

Figure 3 shows the flowchart of the proposed intrusion signal recognition method: (1) Simulation experiments are conducted to collect different intrusion signals; (2) Data pre-processing is applied to standardize each intrusion signal and divide it into segments; (3) The segments of the different signals are combined simply as one data sample to form the input data of the CNNs model; (4) CNNs is trained and tested with these fused input data of the intrusion signals, and their output is the classification result of the intrusion signals. The model parameters of the model initialized at the beginning of model training are adjusted to obtain higher accuracy in the training process. The trained model is directly tested with these random test samples. The effectiveness of the proposed intrusion signal recognition method is evaluated on the basis of the test accuracy of the output result.

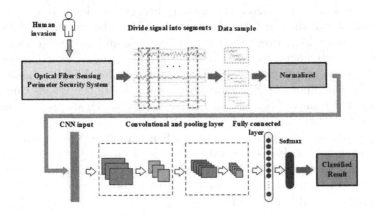

**Fig. 3.** Procedure of recognition method.

# 3    Results and Analysis

## 3.1    Monitoring System Based on Ultraweak FBG

As shown in Fig. 4, the fiber optic sensing perimeter security system is composed of ultra-weak FBG arrays, optical fibers, FBG signal processor, and computer. In this system, the ultra-weak FBG arrays were used as a string of vibration detectors to encapsulate the external vibration signals near the detection optical fibers. The signal processor and the software actualized the intelligent analysis and pattern recognition of the intrusion signals.

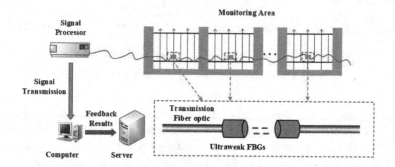

**Fig. 4.** Fiber-optic sensing perimeter security system

An experimental system of fiber-optic sensing perimeter security was set up to test the performance of various signal recognition methods. With the monitoring criterion of 5 m of singe area, the simulation system had 12 monitoring areas in all. As the steel railings were vulnerable to the erosion in the natural environment, the railings in this experiment were damaged to varying degrees. Different scenarios are shown in Fig. 5. The first is the damage of the upper horizontal rail joint; the second, the damage of the joint between the lower vertical rail and the horizontal rail; the third, the damage of the lower horizontal rail joint; and the fourth, the break between the lower vertical rail and the fixed cement.

Twelve targeted experiments were arranged to generate the undamaged and damaged signals, which were used as the input data for different classification approaches. The input of CNNs model was a total of 512 continuous time-series points selected from the normalized marked intrusion signals. Table 1 presents the parameters of the proposed CNNs model. In this experiment, the operating platform was based on PyCharm under Python, and the CNNs model was constructed using TensorFlow developed by Google. To minimize the loss function, the Adam stochastic optimization algorithm was applied to train the proposed CNNs model. Additionally, the action functions of the convolutional layers were all the functions of ReLu in the model.

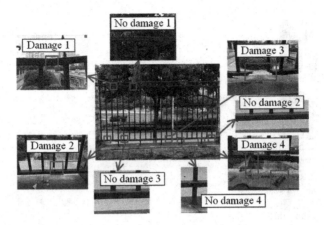

**Fig. 5.** Rail damage scenarios.

**Table 1.** Details of proposed CNN model used in experiments.

| Number | Layer type | Kernel size/stride | Kernel number | Output size (width × depth) | Padding |
|---|---|---|---|---|---|
| 1 | Convolution1 | 32 × 1/1 × 1 | 16 | 32 × 16 | No |
| 2 | Pooling1 | 2 × 1/2 × 1 | 16 | 16 × 16 | No |
| 3 | Convolution2 | 2 × 1/1 × 1 | 32 | 16 × 32 | No |
| 4 | Pooling2 | 2 × 1/2 × 1 | 32 | 8 × 32 | No |
| 5 | Convolution3 | 2 × 1/1 × 1 | 64 | 8 × 64 | No |
| 6 | Pooling3 | 2 × 1/2 × 1 | 64 | 4 × 64 | No |
| 7 | Convolution4 | 2 × 1/1 × 1 | 128 | 4 × 128 | No |
| 8 | Pooling4 | 2 × 1/2 × 1 | 128 | 2 × 128 | No |
| 9 | Fully-connected | 100 | 1 | 100 × 1 | - |
| 10 | SoftMax | 3 | 1 | 3 | - |

### 3.2   Evaluation of Proposed Recognition Method

The intrusion behaviors exhibited in the hanging-cable perimeter security system included tapping the railing, climbing the railing, and tapping the cable. Figure 6 displays those behaviors and the corresponding intrusion signals.

As a comparison of proposed method, the SVM and the BPNN were introduced as substitution of CNNs in the experiment. The particle swarm optimization algorithm was applied to look for the optimal solution in SVM, and the BPNN model was consisted of three layers with sigmoid activation functions. The damage scenarios and the testing results are displayed in Table 2, in which the result of the proposed method is marked in bold. As shown in Table 2, the signal samples in the different areas were trained with a single model trainer. Three domains of handcrafted features extracted from the raw signals were used as the dataset of the SVM and the BPNN, such as threshold coefficient, kurtosis factor, shape factor, frequency variances and wavelet energy, etc. Figure 7 presents the testing results of the 12 monitoring areas of the best three approaches.

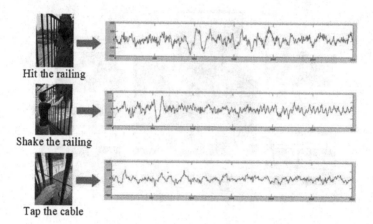

**Fig. 6.** Intrusion behaviors and signal waveforms in hanging-cable perimeter security.

**Table 2.** Average testing accuracy of different methods.

| Damage situation | Model | Raw signal | Manual feature extraction | | |
|---|---|---|---|---|---|
| | | | Time-domain features | Frequency-domain features | Wavelet features |
| NO | CNN1 | **96.56%** | – | – | – |
| | SVM1 | 35.18% | 90.42% | 75.84% | 82.32% |
| | BPNN1 | 38.23% | 89.94% | 76.04% | 82.18% |
| 1 | CNN2 | **96.16%** | – | – | – |
| | SVM2 | 45.87% | 92.15% | 82.51% | 85.46% |
| | BPNN2 | 46.62% | 91.42% | 80.38% | 83.93% |
| 2, 4 | CNN3 | **95.86%** | – | – | – |
| | SVM3 | 42.14% | 90.28% | 78.24% | 78.89% |
| | BPNN3 | 39.53% | 87.23% | 71.53% | 73.61% |
| 1, 3, 4 | CNN4 | **95.92%** | – | – | – |
| | SVM4 | 49.32% | 89.87% | 79.62% | 82.67% |
| | BPNN4 | 42.19% | 85.29% | 72.37% | 74.21% |

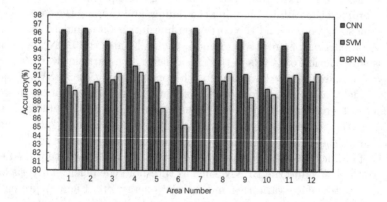

**Fig. 7.** Testing accuracy of each area for three methods.

The capability of the automatic feature learning of proposed CNNs model with raw signals was proven experimentally. The conclusion which can be obviously reached from Table 2 and Fig. 7 is that CNNs with feature learning achieves considerably better recognition accuracies than SVM and BPNN. The outcome proved that CNNs improved the performance of the intrusion signal classification method for perimeter security.

### 3.3    Analysis of Model Parameters

To evaluate the effectiveness of applying different numbers of stacked CNNs layers for the purpose of intrusion signal recognition, we chose four kinds of CNNs models with varying kernel sizes $m = [2, 8, 32, 64]$ and calculated the accuracy by $L = [1, 2, 3, 4, 5]$ numbers of CNNs layers. The number of kernels and the pooling dimensions were selected to ensure the same size in each layer. Because of the small feature maps extracted in the previous convolutional kernel, the pooling dimensions were defined as two in all of the layers. For example, the convolutional filter for the CNNs with kernel size $m = 8$ was set to $n = [8, 1, 32]$ in each layer and the pooling dimension was $p = [2, 1, 32]$. The accuracy as a function of the number of layers is shown in Fig. 8. In this study, the addition of multiple layers was not sure to increase the accuracy. However, when excessive layers were used, the accuracy was decreased in some case. Figure 8 displays the operation time as a function of different number of layers, and we can obviously acquire that the operation time is significantly increased with the addition of multiple layers.

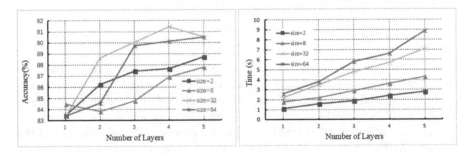

**Fig. 8.** Testing accuracy and simulation time of different numbers of layers and convolution kernel sizes.

Additionally, four convolution kernels with varying kernel sizes $m = [2, 8, 32, 64]$ were purposely chosen to evaluate the choice of kernels. The number of kernels and the pooling dimension were set to be the same. Figure 9 displays the classification results for some combinations of 1, 2, 3, or 4 of the convolutional kernels. For the example shown in the Fig. 9, 3 (8, 32, 64) represents that the number of model layers is 3 for different sizes of the convolutional kernel. The highest accuracy in various CNNs models was accomplished when the CNNs with the deepest layers, 4 (32, 2, 8, 64), was used. The larger kernels followed with a small kernel were particularly effective for extracting useful information from the input data, and the small kernels of multi-layered models in the first layer played a disappointing role in extracting sufficient and available information from the input data.

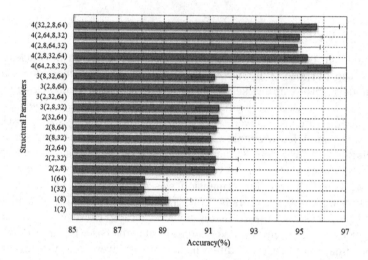

**Fig. 9.** Testing accuracy of different models in terms of intrusion signal recognition.

## 4   Conclusions

In this paper, we proposed an adaptive CNNs-based intrusion recognition method, which achieved astonishing performance on complex and uncorrelated signals. Compared with some of the existing methods of feature engineering, the proposed CNNs model worked directly on raw vibration signals without any time-consuming handcrafted feature extraction process. The performance of the proposed method was evaluated experimentally by using a simulated intrusion of fiber-optic sensing perimeter security. Our future work will focus on testing other approaches of deep learning on the recognition of the ultra-weak FBG signals.

**Acknowledgments.** This work is supported by National Natural Science Foundation of China under grant number 61735013 and 61402345.

# References

1. Kishore, P., Srimannarayana, K.: Vibration sensor using 2*2 fiber optic coupler. Opt. Eng. **52**(10), 107104 (2013)
2. Xin, L., Jin, B., Bai, Q., et al.: Distributed fiber-optic sensors for vibration detection. Sensors **16**(8), 1164 (2016)
3. Prieto, M.D., Cirrincione, G., Espinosa, A.G., et al.: Bearing fault detection by a novel condition-monitoring scheme based on statistical-time features and neural networks. Trans. Ind. Electron. **60**(8), 3398–3407 (2013)
4. Wang, X., Zheng, Y., Zhao, Z., et al.: Bearing fault diagnosis based on statistical locally linear embedding. Sensors **15**(7), 16225–16247 (2015)
5. Rai, V.K., Mohanty, A.R.: Bearing fault diagnosis using FFT of intrinsic mode functions in Hilbert-Huang transform. Mech. Syst. Signal Process. **21**(6), 2607–2615 (2007)
6. Lee, W., Chan, G.P.: Double fault detection of cone-shaped redundant IMUs using wavelet transformation and EPSA. Sensors **14**(2), 3428–3444 (2014)
7. Wang, Q., Li, Y., Liu, X.: Analysis of feature fatigue EEG signals based on wavelet entropy. Int. J. Pattern Recognit. Artif. Intell. **32**(8), 1854023 (2018)
8. Gharani, P., Suffoletto, B., Chung, T., et al.: An artificial neural network for movement pattern analysis to estimate blood alcohol content level. Sensors **17**(12), 2897 (2017)
9. Venkatesan, C., Karthigaikumar, P., Varatharajan, R.: A novel LMS algorithm for ECG signal preprocessing and KNN classifier based abnormality detection. Multimed. Tools Appl. **77**(8), 10365–10374 (2018)
10. Meier, U., Gambardella, L.M.: Deep, big, simple neural nets for handwritten digit recognition. Neural Comput. **22**(12), 3207–3220 (2010)
11. Kwon, Y.H., Shin, S.B., Kim, S.D.: Electroencephalography based fusion two-dimensional (2D)-convolution neural networks (CNN) model for emotion recognition system. Sensors **18**(5), 1383 (2018)
12. Zhang, W., Peng, G., Li, C., et al.: A new deep learning model for fault diagnosis with good anti-noise and domain adaptation ability on raw vibration signals. Sensors **17**(2), 425 (2017)
13. Ince, T., Kiranyaz, S., Eren, L., et al.: Real-time motor fault detection by 1-D convolutional neural networks. Trans. Ind. Electron. **63**(11), 7067–7075 (2016)
14. Lecun, Y., Bengio, Y., Hinton, G.: Deep learning. Nature **521**(7553), 436 (2015)
15. Schmidhuber, J.: Deep learning in neural networks: an overview. Neural Netw. **61**, 85–117 (2015)
16. Bengio, Y.: Learning deep architectures for AI. Found. Trends Mach. Learn. **2**(1), 1–127 (2009)

# Exploiting User and Item Attributes
# for Sequential Recommendation

Ke Sun and Tieyun Qian[✉]

School of Computer Science, Wuhan University, Hubei, China
{sunke1995,qty}@whu.edu.cn

**Abstract.** This paper exploits both the user and item attribute information for sequential recommendation. Attribute information has been explored in a number of traditional recommendation systems and proved to be effective to enhance the recommend performance. However, existing sequential recommendation methods model latent sequence patterns only and neglect the attribute information.

In this paper, we propose a novel deep neural framework which exploits the item and user attribute information in addition to the sequential effects. Our method has two key properties. The first one is to integrate the item attributes into the sequential modeling of purchased items. The second one is to combine the user attributes with his/her preference representation. We conduct extensive experiments on a widely used real-world dataset. Results prove that our model outperforms the state-of-the-art sequential recommendation approaches.

**Keywords:** Recommender systems · Sequential recommendation · Attribute information

## 1 Introduction

Recommender systems have become a core technology of many e-commerce or social networking websites. Traditional methods recommend the items only based on user's general preferences without considering the relationship between the items visited in a short period of time. Over the last decade, more and more researchers have paid attention to the sequential recommendation problem. Different from traditional recommendation system, sequential recommendation is highly dependent on current contexts. The users' next decision mainly depends on two factors: general preferences and recent decisions (also named as sequential patterns). For example, a user is more likely to buy a keyboard after purchasing a mouse while this user often buys books in his/her daily life. In this case, the last decision becomes the determining factor.

In general, there are two types of approaches for sequential recommendation problem. One is Markov chain based methods. FPMC [9] and Fossil [3] are in this line, and they work well in extracting local sequential patterns. However, they cannot model the global sequence. The other is the deep neural network based

© Springer Nature Switzerland AG 2018
L. Cheng et al. (Eds.): ICONIP 2018, LNCS 11305, pp. 370–380, 2018.
https://doi.org/10.1007/978-3-030-04221-9_33

Recurrent Neural Networks (RNN) methods [6,7,14,15] which aim to capture the global sequential patterns. More recently, other deep networks like memory network [1] and CNN [11] are also proposed for sequential recommendation. However, none of these methods takes the user or item attribute information into consideration.

Attribute information has been proved to be helpful in recommendation. For instance, females are more likely to watch romantic movies while males may prefer war movies. Many studies [2,10,16–18] have been proposed to integrate attribute information, and these methods perform well on recommendation tasks. Nevertheless, they are all designed for traditional rather than sequential recommendation problem.

In this paper, we investigate the problem of exploiting item and user attribute information for sequential recommendation. We propose a novel attribute enhanced neural framework to this end. Our model is inspired by the CNN based method [11] but moves one step further by integrating both user and item attributes into the network. More specifically, our framework has two distinct properties. The first one combines the item attributes with the sequential patterns, while the second one incorporates the user attributes into the user's preference representation. Extensive experiments on a real-world dataset demonstrate that our method achieves significant improvements over the state-of-the-art baselines.

## 2   Related Work

In the literature, a number of approaches have been proposed to model the sequential behaviors. By integrating Markov chain and matrix factorization, FPMC [9] builds a personal transition matrix for each user instead of a common matrix. Fossil [3] extends Markov chain with similarity-based method to tackle the data sparsity issue. Besides the above Markov chain based approaches, the RNN framework is introduced into this filed to model global sequential dependency. DREAM [15] puts user's transaction sequence into a RNN layer and treats the hidden states as the current preferences. CA-RNN [6] imports the adaptive context-specific input and transition matrices to model current contextual information. Some RNN-based methods are for session-based recommendation. For example, GRU4Rec [4] extracts the short-term preferences from previous behaviors. NARM [5] adopts an attention mechanism to explore the user's main purpose in the current session. In addition to RNN, a few other deep networks are employed for sequential recommendation. MANN [1] builds a memory network to model user's preference transformation. Caser [11] applies CNN to capture the union-level and point-level patterns, as well as the skip behaviors.

Attribute information is useful for recommendation. Recently, many meta-path based methods have been presented to integrate heterogeneous information for traditional recommendation. HeteRec [16] builds similarity matrices from different meta-paths while SemRec [10] predicts users' ratings based on weighted heterogeneous information network. Factorization machine (FM) is another basic strategy. DeepFM [2] combines deep neural network and FM to model both low and high order feature interactions. NSCR [13] creates a novel pooling layer to fuse user/item and their respective attributes.

In summary, all the sequential recommendation methods ignore the important attribute information, and the existing approaches considering attribute information cannot capture users' current preference. In this paper, we aim to combine attribute information with sequential patterns into a unified recommendation system.

## 3   The Proposed Model

This section we introduce our model. The basic idea is to incorporate the user and item attributes into the neural framework.

**Attribute Properties.** To make full use of attribute information, we consider the following three attribute properties.

1. An item or a user may have several attributes and these attributes may have different impacts on the item or user.
2. The representation of an item or a user should combine the user/item and the attributes together to get a comprehensive representation for the user or the item.
3. The attribute sequence generated from the transaction sequence should reflect the latent patterns. For example, a user may visit a hotel for a rest after he/she leaves an airport. This case reflects an underlying relationship between transportation and accommodation.

### 3.1   General Architecture

We first give basic definitions for our model. Let $U = \{u_1, u_2, u_3, ..., u_{|U|}\}$ and $I = \{i_1, i_2, i_3, ..., i_{|I|}\}$ be the user and item set, respectively. For each user $u$, a time-order transaction sequence is denoted by $S^u = \{S_1^u, S_2^u, S_3^u, ..., S_{|S^u|}^u\}$, where $S_i^u \in I$. The attribute set for an item $i$ and a user $j$ is denoted as $AI^i = \{AI_1^i, AI_2^i, ..., AI_{|AI^i|}^i\}$ and $AU^j = \{AU_1^j, AU_2^j, ..., AU_{|AU^j|}^j\}$, respectively. Our model aims to predict a list of new items to a user $u$ at time $t$ given his/her historical transaction sequence $Seq = \{S_{t-L}^u, S_{t-L+1}^u, ..., S_{t-1}^u\}$, where $L$ is the length of the sequence.

In order to predict a user's next decision, it is good to combine both sequential behaviors and user's general preferences. To this end, we designed three modules, i.e., the sequence, user, and prediction module, in our framework. Figure 1 depicts the general architecture of our model.

The left part in Fig. 1 is used to capture the sequence effect while the lower part in the middle is to model the user's preferences. Both of the sequence and user's preferences are fed into the right part to get the prediction score. Our

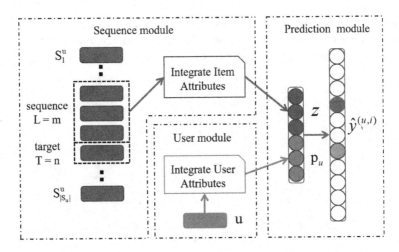

**Fig. 1.** The general architecture of our proposed model

main contributions are on the sequence and user modules. The prediction part is simply as same as that of Caser [11].

Specifically, the target of sequence module is to integrate item attribute with the sequence information. We fuse the latent features of item sequence, attribute sequence, and item-attribute-fusion sequence into a joint representation $z$, denoted as the model's sequential factor. The user module outputs a user-attribute integrated vector $p_u$ as the user's general preferences factor. We put the concatenation of $p_u$ and $z$ into a fully-connected layer which outputs a prediction score vector over all items denoted as $\hat{y}^{(u,t)}$. We adopt the widely used negative sampling strategy to train the model. Our *loss function* can be defined as:

$$loss = -log(\sigma(y_i^{(u,t)})) - \sum_{i \neq j} log(1 - \sigma(y_j^{(u,t)}))  \qquad (1)$$

where $i$ is the target item and $j$ is the negative one. The $\sigma$ is the sigmoid function.

## 3.2   Sequence Module

The objective of sequence module is to combine the item, the item attribute, and the item sequence in transactions together. We present our method based on three attribute properties as follows.

**Attribute Fusion:** We first consider the attribute property 1. Given an item $i$ and its attribute set $AI^i = \left\{ AI_1^i, AI_2^i, ..., AI_{|AI^i|}^i \right\}$ in an input sequence, different attributes should have different influences on the item $i$. Hence a high level attribute representation for $i$'s attribute set should be produced to reflect such difference. We choose max-pooling to aggregate all attributes for its low computational complexity. The max-pooling is a nonlinear operation [12] which can choose the most significant feature along one dimension of the input vectors. We adopt it to produce an overall attribute representation vector for item $i$'s attribute set. We illustrate this procedure in Fig. 2.

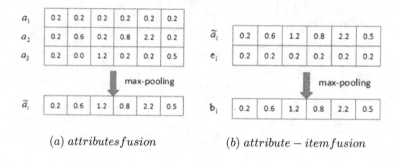

(a) $attributes\,fusion$                    (b) $attribute-item\,fusion$

**Fig. 2.** Examples for attributes fusion and attribute-item fusion

In Fig. 2, for a specific item $i$, we first get $i$'s attribute embeddings and stack them together, resulting in a matrix $EA^i$:

$$EA^i = \begin{bmatrix} a_1, \\ a_2, \\ ... \\ a_{|AI^i|} \end{bmatrix} \tag{2}$$

where $EA^i \in R^{|AI^i| \times d}$, $d$ is the latent dimension. After max-pooling operation, a high level representation can be produced as follows:

$$\widetilde{a}_i = [max(a_1[0], a_2[0], ...a_{|AI^i|}[0]), max(a_1[1], a_2[1], ...a_{|AI^i|}[1]), ...] \tag{3}$$

where $a_i[j]$ denotes the $j$th element in vector $a_i$.

**Attribute-Item Fusion:** We then consider the attribute property 2. Given an item $i$, a comprehensive representation can be produced by fusing the item $e_i$ and corresponding attribute representation $\widetilde{a}_i$. Following the attribute fusion method, we still adopt max-pooling to generate this vector. The integrated attribute-item fusion can be calculated by:

$$b_i = [max(e_i[0], \widetilde{a}_i[0]), max(e_i[1], \widetilde{a}_i[1]), ...]. \tag{4}$$

**Capture Attribute, Attribute-Item-Fusion Sequence Patterns:** We now consider the attribute property 3. We design the sequence module to integrate the item attribute information. The details are shown in Fig. 3.

The sequence module in our framework is the extension of Caser [11], which is shown as the red dashed rectangle in Fig. 3. Different from Caser, we add the attribute, attribute-item-fusion sequence into the framework, shown as the

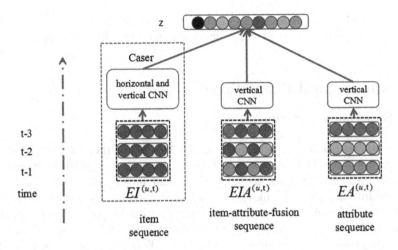

**Fig. 3.** Capture sequence patterns: we keep the red dashed rectangle part as same as Caser. The three matrices are made up of item vectors, item-attribute-fusion vectors, and attribute vectors in a time order, respectively. The sequential factor $z$ can be produced through the CNN layers. (Color figure online)

middle and right part in Fig. 3. Specifically, given a time-order sequence, we generate three kinds of stacked embedding matrices: $EI^{(u,t)}$ for item sequence, $EA^{(u,t)}$ for attribute sequence, and $EIA^{(u,t)}$ for item-attribute-fusion sequence:

$$EI^{(u,t)} = \begin{bmatrix} e_{S_{t-L}^u}, \\ e_{S_{t-L+1}^u}, \\ ... \\ e_{S_{t-1}^u} \end{bmatrix}, EA^{(u,t)} = \begin{bmatrix} \widetilde{a}_{S_{t-L}^u}, \\ \widetilde{a}_{S_{t-L+1}^u}, \\ ... \\ \widetilde{a}_{S_{t-1}^u} \end{bmatrix}, EIA^{(u,t)} = \begin{bmatrix} b_{S_{t-L}^u}, \\ b_{S_{t-L+1}^u}, \\ ... \\ b_{S_{t-1}^u} \end{bmatrix} \quad (5)$$

The time-order attribute sequence contains latent sequential patterns. Meanwhile, CNN performs well in capturing sequential patterns according to Caser [11]. Taking all these together, we adopt a vertical CNN layer to capture the latent attribute sequential patterns. In this CNN layer, there are $n$ vertical filters $F_k \in R^{L \times 1}, 1 \le k \le n$. Applying the $F_k$ filter to the matrix $EA^{(u,t)}$, a vector representing the attribute sequence features can be produced:

$$c_k^{(u,t)} = [c_k^{(u,t)}[0], c_k^{(u,t)}[1], ..., c_k^{(u,t)}[d-1]] \quad (6)$$

$$c_k^{(u,t)}[i] = r_i^T \cdot F_k \quad (7)$$

where $r_i$ is the $i$th column of $EA^{(u,t)}$. We also do the same calculation on $EIA^{(u,t)}$.

### 3.3   User Module

In the user module, we aim to produce a comprehensive representation for a user. As we discuss above, a user's general preference can be influenced by his/her attributes. To model a user's preference, a high level fusion will be produced. Once again, we resort to max-pooling to generate the user's representation vector $p_u$.

# 4    Experimental Evaluation

## 4.1    Experimental Setup

**Datasets.** We conduct experiments on the MovieLens dataset. This is a real-world dataset and has been widely used in the previous studies. It contains 1,000,209 ratings belonging to 6,040 users and 3,883 movies with abundant attribute information. The movies are labelled with several categories such as *Romance* and *Drama*. The users' profile information like gender, age and occupation information are also present in the dataset. We use the movies' category and users' profile information as the attributes. Following the previous studies, we filter out the items which have less than 5 records.

**Baselines:** We compare our method with the following state-of-the-art baselines: BPR [8], FPMC [9], Fossil [3], GRU4Rec [4] and Caser [11]. BPR method is based on matrix factorization for traditional recommendation problem while the others for sequential recommendation problem.

**Evaluation Metrics:** We evaluate the performance of recommendation methods using the Precision@N, Recall@N and MAP metrics.

**Settings:** For each user, we mask off his/her 20% most current feedbacks as testing set. The remaining 80% ratings are used as training data. We conduct two series of experiments on the MovieLens dataset. One is to compare the overall performance of our model with baselines. The other is to evaluate the effectiveness of different kinds of attribute information.

In the experiments, we set the latent dimension $d = 50$, the sequence's length $L = 5$ and the filters' number $n = 4$. The learning rate is set to $1e - 3$. We randomly generate 3 negative samples for each target.

## 4.2    Results

In this section, we report our experiments' results and give analyses on these results.

**Comparison with Baselines.** The comparison results between our method and the baselines are shown in Table 1.

We have the following three important observations from Table 1.

- Our model has the best performance in terms of all three metrics. For example, the Precision@1 of our method is 0.3278, while that of Caser is 0.3036, showing a 2.42 increase. This clearly proves that the attribute information is helpful for recommendation and our model can make good use of and benefit a lot from the attribute information.

**Table 1.** Comparison results on MovieLens in terms of MAP, Precision@N, and Recall@N Score

|  | Map | Pre@1 | Pre@5 | Pre@10 | Rec@1 | Rec@5 | Rec@10 |
|---|---|---|---|---|---|---|---|
| BPR | 0.0913 | 0.1478 | 0.1288 | 0.1193 | 0.0070 | 0.0312 | 0.0560 |
| FPMC | 0.1053 | 0.2022 | 0.1659 | 0.1460 | 0.0118 | 0.0468 | 0.0777 |
| Fossil | 0.1354 | 0.2306 | 0.2000 | 0.1806 | 0.0144 | 0.0602 | 0.1061 |
| GRU4Rec | 0.1440 | 0.2515 | 0.2146 | 0.1916 | 0.0153 | 0.0629 | 0.1093 |
| Caser | 0.1786 | 0.3036 | 0.2650 | 0.2356 | 0.0181 | 0.0765 | 0.1326 |
| Ours | **0.1861** | **0.3278** | **0.2773** | **0.2455** | **0.0201** | **0.0806** | **0.1382** |

- The Caser is the second best. It outperforms other sequential recommendation models. As discussed in the literature, FPMC and Fossil fail to model union-level sequential patterns and neglect the users' behaviors. Also, the RNN-based method GRU4Rec can't capture the relationships between actions which are not adjacent. Different from these methods, Caser adopts CNN layers to avoid these problems and thus achieves better performance.
- BPR has the worst performance. The reason is that it does not takes the sequences into account. This shows the effects of the latent sequential patterns.

**Effectiveness of Different Attribute Information.** We now show the results of different attribute information experiments in Table 2. Ours-MA denotes only using movies' attribute information. Ours-UA denotes using users' attribute information, and $-g$, $-a$, $-o$ denotes removing the gender, age, and occupation information, respectively. We also list the results of the best baseline Caser in Table 2 for a deep analysis.

**Table 2.** Effectiveness of different attribute information

|  | Map | Pre@1 | Pre@5 | Pre@10 | Rec@1 | Rec@5 | Rec@10 |
|---|---|---|---|---|---|---|---|
| Caser | 0.1786 | 0.3036 | 0.2650 | 0.2356 | 0.0181 | 0.0765 | 0.1326 |
| Ours-MA | 0.1829 | 0.3224 | 0.2721 | 0.2430 | 0.0189 | 0.0771 | 0.1344 |
| Ours-UA-g | 0.1827 | 0.3123 | 0.2717 | 0.2418 | 0.0185 | 0.0773 | 0.1335 |
| Ours-UA-a | 0.1823 | 0.3179 | 0.2726 | 0.2441 | 0.0189 | 0.0776 | 0.1343 |
| Ours-UA-o | 0.1784 | 0.3040 | 0.2677 | 0.2413 | 0.0173 | 0.0745 | 0.1315 |
| Ours | **0.1861** | **0.3278** | **0.2773** | **0.2455** | **0.0201** | **0.0806** | **0.1382** |

It is clear that all kinds of attribute information excepts occupation have the positive effects on our model. By taking only movie attribute information into consideration, our model can already perform much better than Caser, especially in terms of Map and Pre@1. It shows that the attributes also have the sequential patterns which are helpful for recommendation.

As for users' attribute information, gender and age play important roles in our model. However, the occupation attribute doesn't works well and it performs the worst in terms of Rec@N. This phenomenon reflects that a user's interest for a movie depends more on his/her gender and age.

In summary, we can get the best result when we integrate all kinds of attribute information despite that the single occupation attribute has negative effects.

### 4.3   Sensitivity

In this section, we investigate the impact of parameters, including the dimension $d$ and the iteration $r$. We conduct experiments on the same dataset and keep other parameters the same as we illustrate before. We report the Map score as a function of $d$, $r$ in Fig. 4.

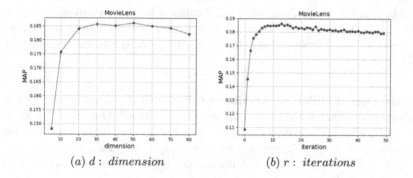

(a) $d$ : dimension          (b) $r$ : iterations

**Fig. 4.** Parameter sensitivity study

Figure 4(a) shows the impact of dimension $d$. We can see that the performance of our model becomes the best when $40 \leq d \leq 60$ and declines a little when $d > 60$. Figure 4(b) shows the impact of iteration $r$. The Map value increases rapidly when $r \leq 9$ and achieves the highest score when $r$ is around 14. Then it falls because of over-fitting. This can be due to the abundant sequential patterns in MovieLens which can easily lead the model to over-fitting.

## 5   Conclusion

In this paper, we propose a new model to incorporate the attribute information into the sequential recommendation system. By integrating the attribute information, our model can represent a user or an item more comprehensively. Furthermore, our model can capture the latent patterns in attribute sequence.

We conduct extensive experiments on a real-world dataset. The results demonstrate that our model takes advantages of attribute information and outperforms all baselines by a large margin.

**Acknowledgment.** The work described in this paper has been supported in part by the NSFC project (61572376).

# References

1. Chen, X., et al.: Sequential recommendation with user memory networks. In: Proceedings of WSDM, pp. 108–116. ACM (2018)
2. Guo, H., Tang, R., Ye, Y., Li, Z., He, X.: DeepFM: a factorization-machine based neural network for CTR prediction (2017). arXiv preprint arXiv:1703.04247
3. He, R., McAuley, J.: Fusing similarity models with Markov chains for sparse sequential recommendation. In: Proceedings of ICDM, pp. 191–200. IEEE (2016)
4. Hidasi, B., Karatzoglou, A., Baltrunas, L., Tikk, D.: Session-based recommendations with recurrent neural networks (2015). arXiv preprint arXiv:1511.06939
5. Li, J., Ren, P., Chen, Z., Ren, Z., Lian, T., Ma, J.: Neural attentive session-based recommendation. In: Proceedings of CIKM, pp. 1419–1428. ACM (2017)
6. Liu, Q., Wu, S., Wang, D., Li, Z., Wang, L.: Context-aware sequential recommendation. In: Proceedings of ICDM, pp. 1053–1058. IEEE (2016)
7. Quadrana, M., Karatzoglou, A., Hidasi, B., Cremonesi, P.: Personalizing session-based recommendations with hierarchical recurrent neural networks. In: Proceedings of RecSys, pp. 130–137. ACM (2017)
8. Rendle, S., Freudenthaler, C., Gantner, Z., Schmidt-Thieme, L.: BPR: Bayesian personalized ranking from implicit feedback. In: Proceedings of UAI, pp. 452–461 (2009)
9. Rendle, S., Freudenthaler, C., Schmidt-Thieme, L.: Factorizing personalized Markov chains for next-basket recommendation. In: Proceedings of WWW, pp. 811–820. ACM (2010)
10. Shi, C., Zhang, Z., Luo, P., Yu, P.S., Yue, Y., Wu, B.: Semantic path based personalized recommendation on weighted heterogeneous information networks. In: Proceedings of CIKM, pp. 453–462. ACM (2015)
11. Tang, J., Wang, K.: Personalized top-n sequential recommendation via convolutional sequence embedding. In: Proceedings of WSDM, pp. 565–573 (2018)
12. Wang, P., Guo, J., Lan, Y., Xu, J., Wan, S., Cheng, X.: Learning hierarchical representation model for nextbasket recommendation. In: Proceedings of SIGIR, pp. 403–412. ACM (2015)
13. Wang, X., He, X., Nie, L., Chua, T.: Item silk road: recommending items from information domains to social users. In: Proceedings of SIGIR, pp. 185–194 (2017)
14. Wu, C.Y., Ahmed, A., Beutel, A., Smola, A.J., Jing, H.: Recurrent recommender networks. In: Proceedings of WSDM, pp. 495–503. ACM (2017)
15. Yu, F., Liu, Q., Wu, S., Wang, L., Tan, T.: A dynamic recurrent model for next basket recommendation. In: Proceedings of SIGIR, pp. 729–732. ACM (2016)
16. Yu, X., et al.: Personalized entity recommendation: a heterogeneous information network approach. In: Proceedings of WSDM, pp. 283–292. ACM (2014)

17. Zhang, Y., Ai, Q., Chen, X., Croft, W.B.: Joint representation learning for top-n recommendation with heterogeneous information sources. In: Proceedings of CIKM, pp. 1449–1458. ACM (2017)
18. Zhao, H., Yao, Q., Li, J., Song, Y., Lee, D.L.: Meta-graph based recommendation fusion over heterogeneous information networks. In: Proceedings of SIGKDD, pp. 635–644. ACM (2017)

# Leveraging Similar Reviews to Discover What Users Want

Zongze Jin[1,2], Yun Zhang[1,2(✉)], Weimin Mu[1,2], and Weiping Wang[2,3]

[1] School of Cyber Security, University of Chinese Academy of Sciences,
Beijing, China
[2] Institute of Information Engineering, Chinese Academy of Sciences, Beijing, China
{jinzongze,zhangyun,muweimin,wangweiping}@iie.ac.cn
[3] National Engineering Research Center Information Security Common Technology,
NERCIS, Beijing, China

**Abstract.** With the development of deep learning techniques, recommender systems leverage deep neural networks to extract both the features of users and items, which have achieved great success. Most existing approaches leverage both the descriptions and reviews to represent the features of an item. However, for some items, such as newly released products, they lack users' reviews. In this case, only the descriptions of these items can be used to represent their features, which may result in bad representations of these items and further influence the performance of recommendations. In this paper, we present a deep learning based framework, which can use the reviews of the items that are similar to the target items to complement the descriptions. At last, we do experiments on three real world datasets and the results demonstrate that our model outperforms the state-of-the-art methods.

**Keywords:** Recommendation system · Deep learning
Knowledge representation

## 1 Introduction

With the increasing of users and the number of products, more and more websites have leveraged recommender systems as an effective technique to find what the users need and resist information overload [1,15]. Collaborative filtering (CF) [9,14], as the typical traditional recommender system, mainly utilizes the users' historical behavior to discover the users' preferences. However, CF has some limitations: (1) The CF techniques cannot solve the sparsity problem very well. (2) The interpretability of the CF techniques is poor. Recently, some researches [4,20,21] have adopted both descriptions and reviews to enhance the performance of recommendations in order to solve these problems.

McAuley et al. [11] has adopted HFT to develop statistical models that combine latent dimensions in rating data with topics in review text. With the development of deep learning, researchers have more effective ways to make full use

© Springer Nature Switzerland AG 2018
L. Cheng et al. (Eds.): ICONIP 2018, LNCS 11305, pp. 381–391, 2018.
https://doi.org/10.1007/978-3-030-04221-9_34

of the rich information of reviews. Tang et al. [17] has proposed UWCVM to introduce a novel neural network method for review rating prediction by taking user information into account. Zheng et al. [22] has used convolutional neural network (CNN) to deal with reviews in order to get the users' features and the items' features. Song et al. [16] not only captures the static features of reviews, but also considers the temporal changes in users' reviews. It leverages long-short term memory (LSTM) networks to capture the dynamic features.

However, the above approaches mainly have the following problems: (1) The descriptions of items are ignored by them. (2) Some items, such as some newly released products, lack users' reviews. For example, a newly released wristband only has descriptions like "plastic", "colorful" and "waterproof" and their users' reviews are not enough. In this case, users usually cannot decide whether the wristband is worth buying or not simply based on these descriptions. The users' reviews like "the plastic material is fragile" and "the white one is easy to be soiled" should be combined with the descriptions to help users make decisions.

In this paper, we propose a novel deep model, named Similar Reviews Discovering Networks (SRDN), which can use the reviews of the items that are similar to the target items to complement the descriptions to generate the features of items. SRDN including two key parts is illustrated in Fig. 1. Firstly, we use the triplet network proposed in [8] to discover the items with similar descriptions. After that, we use two CNN networks to generate the features for users and items, respectively. In general, our contributions can be summarized as follows:

- We present a deep learning model, SRDN, which can use the reviews of the items that are similar to the target items to complement the descriptions. To the best of our knowledge, we are the first to leverage the users' reviews of the items that are similar to the target items.
- We do some experiments to demonstrate the effectiveness of our model through comparing with the state-of-the-art methods on several datasets.

## 2   Approach

### 2.1   Problem Definition

Given an item $I$, we adopt $D_I = \{D_1^I, D_2^I, \cdots, D_k^I\}$ to represent the descriptions of items ($k$ denotes the $k$-th item), and use $R_I = \{R_1^I, R_2^I, \cdots, R_j^I\}$ to denote the reviews of items ($j$ denotes the $j$-th item). But in real scenarios, the number of items $k$ will be much larger than the number of reviews $j$. Therefore, only the descriptions of these items $D_I$ can be used to represent the features, which may result in bad representations and further influence the performance. To recommend suitable items to each user, we use the reviews $R_I$ that are similar to the target items to complement the descriptions $D_I$.

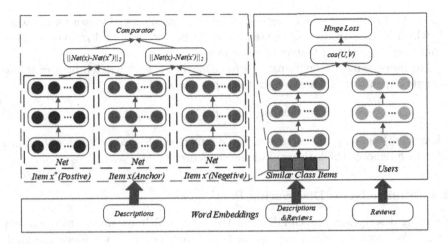

**Fig. 1.** The architecture of SRDN

## 2.2 Overview

As shown in Fig. 1, SRDN has three key parts: (1) Word Embeddings Part (2) Similar Items Discovering Part (3) Recommendation Part. Firstly, we adopt word embeddings to deal with descriptions and reviews. Secondly, we adopt a triplet network to find the items which are similar to the target items based on descriptions. Finally, we utilize two CNN networks to generate the features for users and items, respectively. For the user part, we utilize reviews as the input. For the item part, we utilize reviews and descriptions as the input.

## 2.3 Word Embeddings Part

We leverage the word embedding, which is better than the bag-of-words techniques, to deal with the semantics of reviews and descriptions. We use the embedding layers to get the dense vector representation. With the development of the semantic analysis, many researches [3,6] have demonstrated word embeddings have achieved great success. In SRDN, we use word2vec[1] to pre-train the dictionary of words. Reviews contain both user information and item information. Descriptions contain the properties of items. The document matrix $D$ in our approach is:

$$D = \begin{pmatrix} & | & | & | & \\ \cdots & w_{i-1} & w_i & w_{i+1} & \cdots \\ & | & | & | & \end{pmatrix} \tag{1}$$

For items, we merge all the reviews about the item $i$ into a single sequence $S_n^i$. In this look-up layer, we find the corresponding vectors and concatenate them. The input word embedding matrix $V_i^R$ of reviews is:

$$V_i^R = concat(v(s_1^i), v(s_2^i), \cdots, v(s_k^i)) \tag{2}$$

---

[1] https://code.google.com/archive/p/word2vec/.

where $v(s_k^i)$ represents the corresponding $c$-dimensional word vector for the word $s_k^i$, which is the k-th word of the sequence $S_n^i$. Besides, we use $V_I^D$ to denote the input word embedding matrix of descriptions.

For users, same as the items, we merge all the reviews written by user $u$ into a single sequence $S_n^u$. Then we get the input matrix $V_u$:

$$V_u = concat(v(s_1^u), v(s_2^u), \cdots, v(s_k^u)) \tag{3}$$

where $v(s_k^u)$ denotes the corresponding $c$-dimensional word vector for the word $s_k^u$, which denotes the k-th word of the sequence $S_n^u$.

### 2.4  Similar Items Discovering Part

By using descriptions, we can discover the items which are similar to the target items. As shown in Fig. 1, we use a triplet network [8] to achieve the goal. In our model, we use a convolutional network, consisting of 3 convolutional and $2 \times 2$ max-pooling layers, followed by a fourth convolutional layer. Same as [8], we utilize the $L_2$ distances to represent the distance amongs the embedded representations. In our model, we use Anchor ($x$) to present the anchor item, utilize Positive ($x^+$) to present the similar item and adopt Negative ($x^-$) to present the unsimilar item, respectively. In addition, we use $Net(x)$, $Net(x^+)$ and $Net(x^-)$ to denote the output of the networks, respectively. In this subsection, we first randomly select the items as the anchor items. Then we select the similar items as the positive items and we adopt the unrelated items as negative ones. We use the following equation to measure the distances between each of $x^+$ and $x^-$ against the anchor $x$:

$$TripletNet(x, x^+, x^-) = \begin{bmatrix} ||Net(x) - Net(x^-)||_2 \\ ||Net(x) - Net(x^+)||_2 \end{bmatrix} \tag{4}$$

### 2.5  Recommendation Part

When we discover the items which are similar to the target items through similar items discovering part, we should combine reviews to help users make decisions. As shown in Fig. 1, we adopt two same CNN networks to find whether the items should be recommended to the users or not in recommendation part.

***CNN Layers.*** Convolutional neural network extracts the features in our approach, which includes convolutional layers, max pooling and fully connected layers. We use the CNN model proposed in [10]. For each convolutional kernel $K_j$, we can get the output from the convolutional layer. For each user, we can get the output from the convolutional layer. In our model, we adopt ReLUs (Rectified Linear Units) [13] as our activation function to avoid the problem of vanishing gradient. In [10], they demonstrate that ReLU can get better performance than other non-linear activation functions. Then, the convolutional feature $c^j$ of a sequence is constructed, and $i$ presents the $i$-$th$ convolutional kernel. The pooling layers extract representative features from convolutional layers. Our task follows [6], and we can use max pooling operation to get the reduced fixed size vector. The results from the max-pooling layer are passed to a fully connected layer with a weight matrix $W_u$. Last we can get the output $X_u$. The details are as follows:

$$c_i^j = ReLU(V_u * K_j + b_j) \tag{5}$$

$$c^j = [c_1^j, c_2^j, \cdots, c_i^j, \cdots, c_k^j] \tag{6}$$

$$m_u = [max(c^1), max(c^2), max(c^3), \ldots, max(c^n)] \tag{7}$$

$$X_u = f(W_u * m_u + b_u) \tag{8}$$

where $K_j$ is convolutional kernel, $*$ is the convolutional operator, $b_j$ is a bias for $V_u$ and $b_u$ is the bias term for $X_u$. Same as the user part, we can obtain the output of item part $X_i$:

$$X_i = f(W_i * m_i + b_i) \tag{9}$$

where $W_i$ denotes the fully connected weight, $m_i$ denotes the output of the maxpooling operations and $b_i$ denotes the bias term.

**The Output Layer.** In this part, we adopt the cosine distance between $X_u$ and $X_i$ as the similarity between users and items.

$$cos(X_u, X_i) = \frac{\sum_{i=1}^n (X_u * X_i)}{\sqrt{\sum_1^n (X_u)^2} * \sqrt{\sum_1^n (X_i)^2}} \tag{10}$$

At last, we adopt hinge loss [19] as the loss function to train SRDN.

$$L = max(0, 1 - t_h \cdot cos(X_u, X_i)) \tag{11}$$

where $t_h = \pm1$ represents whether the users are interested in the items ($t_h = 1$) or not ($t_h = -1$).

## 2.6  Training

In this subsection, we firstly train the similar items discovering part and then we train the recommendation part. In similar items discovering part, we regard the task to be a 2-classification problem. We adopt the same loss function as [8]:

$$Loss(d_+, d_-) = ||(d_+, d_- - 1)||_2^2 = const \cdot d_+^2 \tag{12}$$

where

$$d_+ = \frac{e^{||Net(x) - Net(x^+)||_2}}{e^{||Net(x) - Net(x^+)||_2} + e^{||Net(x) - Net(x^-)||_2}} \tag{13}$$

and

$$d_- = \frac{e^{||Net(x) - Net(x^-)||_2}}{e^{||Net(x) - Net(x^+)||_2} + e^{||Net(x) - Net(x^-)||_2}} \tag{14}$$

where $Net(x)$ denotes the output of the networks. In addition, we adopt backpropagation (BP) to take the derivative of the loss with respect to the whole set of parameters. Besides, stochastic gradient descent (SGD) should be updated the parameters.

In recommendation part, by using the same shared parameters network, we also use BP algorithm to deal with the derivative of the loss, meanwhile, we leverage SGD with mini-batch to update the parameters. To avoid over-fitting, we adopt dropout to deal with it. We set the word vector dimension as 200 in word embeddings.

## 3    Experiment

### 3.1    Datasets

In our experiments, we adopt three real-world datasets as follows.

*Clothing.* This dataset, which not only contains 278677 reveiws from 39387 users, but also contains the metadata spanning from 1996 to 2014. We download the dataset by McAuley's [7,12] website[2].

*Yelp16.* It is a large-scale dataset consisting of restaurant reviews and descriptions from Yelp. It is released by the seventh round of the Yelp Dataset Challenge in 2017. There are more than 1M ratings and reviews in Yelp2016[3].

*Trip.* It is one of the world's top travel sites[4]. We collect the dataset from 2015 to 2016. This dataset contains users' reviews and the whole descriptions of items.

### 3.2    Metrics

In our experiments, we use 2 popular metrics for evaluation: precision at position 1 ($P@1$) and 5 ($P@5$) and Mean Reciprocal Rank (MRR).

$$P@N = \frac{\sum_{i=1}^{|N|} hasTrue(N_i)}{|N|} \tag{15}$$

where $hasTrue(N_i) = 1$ means the $i$-th item $N_i$ should be recommended, and vice versa.

$$MRR = \frac{1}{|Q|} \sum_{i=1}^{|Q|} \frac{1}{rank_i} \tag{16}$$

where $|Q|$ is the number of query and $rank_i$ refers to the rank position of the first relevant document for the $i$-th query.

---

[2] http://jmcauley.ucsd.edu/data/amazon/.
[3] http://www.yelp.com.
[4] http://www.trip.com.

## 3.3  Comparison Methods

We compare SRDN with the following methods.

LDA [2]: Latent Dirichlet Allocation is a well-known topic modeling algorithm presented in [2].

HFT [11]: Hidden Factors and Hidden Topics is proposed to employ LDA to learn a topic distribution from a set of reviews for each item. By treating the learned topic distributions as latent features for each item, latent features for each user is estimated by optimizing rating prediction accuracy with gradient descent.

CNN [3,5]: Convolution Neural Network is a state-of-the-art performer on sentence-level sentiment analysis tasks.

CDL [18]: Collaborative Deep Learning tightly couples a Bayesian formulation of the stacked denoising autoencoders and PMF.

DeepCoNN [22]: Deep Cooperative Neural Networks (DeepCoNN) uses two coupled neural networks to deal with reviews to find the users' interest.

We split the original corpus into train, val and test sets with a 80:10:10 split. We adopt SGD and set batch size as 128. Meanwhile, we train each model for 500 epochs. We set word embedding size as 200 in our experiments. We set $K = 10$ for LDA and we set $\alpha = 0.1$, $\lambda_u = 0.02$ and $\lambda_v = 10$ for HFT-10. We use the recommended parameters to deal with the others. For SRDN, we set the dropout as 0.2. Our models are implemented in Keras[5]. All models are trained and tested on an NVIDIA GeForce GTX1080 GPU.

## 3.4  Results

*Performance Evaluation.* In Table 1, we list the performance of SRDN and that of the baselines. Table 1 shows the experimental results on the three datasets and we select the best performance through 10-fold cross validation for comparisons. In addition, the averages are reported with the best performance shown in bold. We can get the following conclusions through the experimental results: (1) Traditional methods have not achieved good effects, of which LDA is the worst among the whole approaches. The reason is likely to be that traditional algorithms cannot mine the deep latent knowledge. (2) Although the typical deep learning algorithms (CNN and CDL) are better than the traditional algorithms, the effects are still not ideal. CNN can capture the complex latent features, but it has not considered the user information and item information, respectively. As CDL uses unsupervised learning, its effectiveness is worse than DeepCoNN and SRDN. (3) DeepCoNN and SRDN are better than the above approaches, which consider the users' reviews and items' reviews, respectively. Because DeepCoNN ignores the impact of the items which are similar to the target items, the performance of DeepCoNN is not as good as SRDN. (4) SRDN achieves the best

---

[5] http://keras.io.

performances, which demonstrates that using the reviews of the items which are similar to the target items to complement the descriptions is effective.

**Table 1.** P@1, P@5 and MRR comparision with baselines

| Method | Clothing | | | Yelp16 | | | Trip | | |
|---|---|---|---|---|---|---|---|---|---|
| | P@1 | P@5 | MRR | P@1 | P@5 | MRR | P@1 | P@5 | MRR |
| LDA | 0.154 | 0.071 | 0.092 | 0.176 | 0.091 | 0.105 | 0.152 | 0.081 | 0.094 |
| HFT-10 | 0.157 | 0.073 | 0.112 | 0.179 | 0.092 | 0.127 | 0.161 | 0.085 | 0.119 |
| CNN | 0.202 | 0.113 | 0.131 | 0.212 | 0.121 | 0.142 | 0.198 | 0.101 | 0.136 |
| CDL | 0.213 | 0.127 | 0.133 | 0.231 | 0.126 | 0.148 | 0.218 | 0.119 | 0.143 |
| DeepCoNN | 0.226 | 0.132 | 0.143 | 0.242 | 0.139 | 0.150 | 0.226 | 0.125 | 0.149 |
| **SRDN** | **0.235** | **0.143** | **0.154** | **0.255** | **0.155** | **0.163** | **0.238** | **0.135** | **0.157** |

(a) Clothing          (b) Yelp16          (c) Trip

**Fig. 2.** The similar items on three datasets. (Color figure online)

**Impact of Other Parameters.** To measure the impact of similar items, we classify 3 datasets through leveraging descriptions to categorize 10 classes and show the results of Clothing, Yelp16 and Trip, respectively, in Fig. 2. As shown in Fig. 2, we can observe the data distributions through data visualization. Besides, different colors represent different classes.

Our model focuses on leveraging the reviews of items which are similar to the target items to complement for the descriptions. We measure the impact of descriptions and reviews, as shown in Fig. 3(a). We use $C$, $Y$ and $T$ to denote Clothing, Yelp16 and Trip, respectively. Besides, $D$ denotes descriptions and $R$ denotes reviews. In this experiment, we use recommedation part to measure the impact of descriptions and reviews. As shown in Fig. 3(a), on the experimental datasets, we observe the phenomenon which the performance of using reviews and descriptions is significantly higher than only using descriptions. This experimental results demonstrate that using reviews and descriptions can improve the performance of recommendations.

Since SRDN contains a lot of text processing, we show the word embedding experiment in Fig. 3(b). Firstly, we need to fix other factors to avoid affecting the

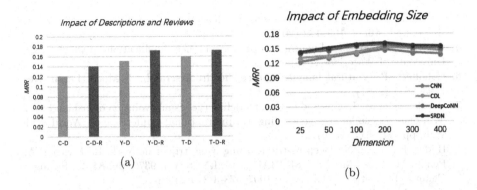

**Fig. 3.** The impact of other parameters

results of the experiment. Then we set the embedding size from 25 to 400. With the increasing of embedding size, we find that the impact of the word embedding size tends to be stable, as shown in Fig. 3(b). When we set the size as 200, our model can present the best performance. When we tune the size from 150 to 300, the experiment shows that the others can achieve the best perfomance.

## 4    Conclusions

We propose a novel deep model, named Similar Reviews Discovering Networks (SRDN) in this paper. The SRDN model contains two key parts: Similar Items Discovering Part and Recommendation Part. In Similar Item Discovering Part, we use a triplet network to find the items which are simliar to the target items based on descriptions. In Recommendation Part, we utilize two CNN networks to generate the features for users and items, respectively. The experimental results show that our proposed SRDN model can improve the performance of recommendations through leveraging the reviews of items which are similar to the target items. In addition, SRDN outperforms the above algorithms.

**Acknowledgment.** This task was supported by National Key Research and Development Plan (2016QY02D0402).

## References

1. Adomavicius, G., Tuzhilin, A.: Toward the next generation of recommender systems: a survey of the state-of-the-art and possible extensions. IEEE Trans. Knowl. Data Eng. **17**(6), 734–749 (2005)
2. Blei, D.M., Ng, A.Y., Jordan, M.I.: Latent dirichlet allocation. J. Mach. Learn. Res. **3**(Jan), 993–1022 (2003)
3. Blunsom, P., Grefenstette, E., Kalchbrenner, N.: A convolutional neural network for modelling sentences. In: Proceedings of the 52nd Annual Meeting of the Association for Computational Linguistics, pp. 499–509 (2014)

4. Bobadilla, J., Ortega, F., Hernando, A.: Recommender systems survey. Knowl.-Based Syst. **46**(1), 109–132 (2013)
5. Chen, Y.: Convolutional neural network for sentence classification. Master's thesis, University of Waterloo (2015)
6. Collobert, R.: Natural language processing from scratch. J. Mach. Learn. Res. 2493–2537 (2011)
7. He, R., Mcauley, J.: Ups and downs: modeling the visual evolution of fashion trends with one-class collaborative filtering. In: International Conference on World Wide Web, pp. 507–517 (2016)
8. Hoffer, E., Ailon, N.: Deep metric learning using triplet network. In: Feragen, A., Pelillo, M., Loog, M. (eds.) SIMBAD 2015. LNCS, vol. 9370, pp. 84–92. Springer, Cham (2015). https://doi.org/10.1007/978-3-319-24261-3_7
9. Koren, Y., Bell, R., Volinsky, C.: Matrix factorization techniques for recommender systems. Computer **42**(8), 30–37 (2009)
10. Krizhevsky, A., Sutskever, I., Hinton, G.E.: ImageNet classification with deep convolutional neural networks. In: International Conference on Neural Information Processing Systems, pp. 1097–1105 (2012)
11. Mcauley, J., Leskovec, J.: Hidden factors and hidden topics: understanding rating dimensions with review text. In: ACM Conference on Recommender Systems, pp. 165–172 (2013)
12. Mcauley, J., Targett, C., Shi, Q., Van Den Hengel, A.: Image-based recommendations on styles and substitutes. In: International ACM SIGIR Conference on Research and Development in Information Retrieval, pp. 43–52 (2015)
13. Nair, V., Hinton, G.E.: Rectified linear units improve restricted Boltzmann machines. In: International Conference on International Conference on Machine Learning, pp. 807–814 (2010)
14. Sarwar, B., Karypis, G., Konstan, J., Riedl, J.: Item-based collaborative filtering recommendation algorithms. In: International Conference on World Wide Web, pp. 285–295 (2001)
15. Shen, H.W., Wang, D., Song, C., Barabsi, A.: Modeling and predicting popularity dynamics via reinforced poisson processes. In: Twenty-Eighth AAAI Conference on Artificial Intelligence, pp. 291–297 (2014)
16. Song, Y., Elkahky, A.M., He, X.: Multi-rate deep learning for temporal recommendation. In: International ACM SIGIR Conference on Research and Development in Information Retrieval, pp. 909–912 (2016)
17. Tang, D., Qin, B., Yang, Y., Yang, Y.: User modeling with neural network for review rating prediction. In: International Conference on Artificial Intelligence, pp. 1340–1346 (2015)
18. Wang, H., Wang, N., Yeung, D.Y.: Collaborative deep learning for recommender systems. In: Proceedings of the 21th ACM SIGKDD International Conference on Knowledge Discovery and Data Mining, pp. 1235–1244 (2015)
19. Wu, Y., Liu, Y.: Robust truncated hinge loss support vector machines. J. Am. Stat. Assoc. **102**(479), 974–983 (2007)
20. Zhang, F., Yuan, N.J., Lian, D., Xie, X., Ma, W.Y.: Collaborative knowledge base embedding for recommender systems. In: ACM SIGKDD International Conference on Knowledge Discovery and Data Mining, pp. 353–362 (2016)

21. Zhang, Y., Ai, Q., Chen, X., Croft, W.B.: Joint representation learning for top-N recommendation with heterogenous information sources. In: ACM International Conference on Information and Knowledge Management, pp. 1449–1458 (2017)
22. Zheng, L., Noroozi, V., Yu, P.S.: Joint deep modeling of users and items using reviews for recommendation. In: Tenth ACM International Conference on Web Search and Data Mining, pp. 425–434 (2017)

# Deep Learning for Real Time Facial Expression Recognition in Social Robots

Ariel Ruiz-Garcia[(⊠)], Nicola Webb, Vasile Palade, Mark Eastwood, and Mark Elshaw

School of Computing, Electronics and Mathematics, Faculty of Engineering, Environment and Computing, Coventry University, Coventry, UK
{ariel.ruiz-garcia,webbn4,vasile.palade,
mark.eastwood,mark.elshaw}@coventry.ac.uk

**Abstract.** Human robot interaction is a rapidly growing topic of interest in today's society. The development of real time emotion recognition will further improve the relationship between humans and social robots. However, contemporary real time emotion recognition in unconstrained environments has yet to reach the accuracy levels achieved on controlled static datasets. In this work, we propose a Deep Convolutional Neural Network (CNN), pre-trained as a Stacked Convolutional Autoencoder (SCAE) in a greedy layer-wise unsupervised manner, for emotion recognition from facial expression images taken by a NAO robot. The SCAE model is trained to learn an illumination invariant downsampled feature vector. The weights of the encoder element are then used to initialize the CNN model, which is fine-tuned for classification. We train the model on a corpus composed of gamma corrected versions of the CK+ , JAFFE, FEEDTUM and KDEF datasets. The emotion recognition model produces a state-of-the-art accuracy rate of 99.14% on this corpus. We also show that the proposed training approach significantly improves the CNN's generalisation ability by over 30% on nonuniform data collected with the NAO robot in unconstrained environments.

**Keywords:** Deep convolutional neural networks · Emotion recognition
Greedy layer-wise training · Social robots · Stacked convolutional autoencoders

## 1 Introduction

Robots have been incorporated into the lives of humans for years, particularly in industry. However, the field of social robotics has been growing in recent years. What defines a social robot outside of a typical robot is its ability to interact with humans in a supportive and assistive manner, as part of a larger society. In this work, we aim to contribute to the creation of an empathic robot by furthering their ability to recognise emotions in real time. By providing social robots with the ability to recognise emotions in real time,

A. Ruiz-Garcia and N. Webb are joint first authors and contributed equally. Names ordered in alphabetical order.

© Springer Nature Switzerland AG 2018
L. Cheng et al. (Eds.): ICONIP 2018, LNCS 11305, pp. 392–402, 2018.
https://doi.org/10.1007/978-3-030-04221-9_35

human-robot interaction (HRI) can become more immersive and social robots will be able to mimic natural conversations with the addition of nonverbal communication [1]. Contemporary work into facial emotion recognition using Deep Learning (DL) models is usually conducted using facial expression images obtained in controlled environments, and, as a result, high accuracy rates have been achieved. However, DL models often fail to generalise on data with nonuniform conditions, making real time emotion recognition in unconstrained environments a challenge difficult to overcome.

In this work, we propose a deep CNN model for emotion recognition from facial expression images obtained in uncontrolled environments. This model is pre-trained in a greedy layer-wise (GLW) unsupervised fashion [2] as a deep SCAE. The SCAE produces illumination invariant hidden representations of the input images by learning to reconstruct gamma corrected versions of an input image as the input image before these transformations. Once the SCAE is trained, we use the encoder weights to initialize our CNN and fine-tune it for classification. The proposed training approach shows that by using an autoencoder to reconstruct different versions of an image, i.e., the same image with different luminance levels, as the same image, the emotion recognition model improves its ability to generalise on data with nonuniform conditions. We show that our training approach improves classification performance on data with nonuniform conditions by over 30%.

The following section introduces related work on social robots and existing DL approaches for real time emotion recognition. Section 3 introduces our experimental setup. Section 4 presents our results and a discussion of these results. Finally, Sect. 5 presents our conclusions and ideas for future work.

## 2 Related Work

Social robots are beginning to be introduced into many real-world applications, such as elderly care [3], which means an increase in everyday HRI. In order for these interactions to become more natural, it is imperative that robots are able to understand the emotions being expressed by the user. The addition of emotional awareness will provide social robots with the ability to interpret the full intentions behind a conversation and will be able to provide appropriate responses.

### 2.1 Social Robots

A social robot is defined as a robot that interacts with humans and can understand them, in order to support them [4]. The distinction from a typical autonomous robot and a social robot lies in its communicative abilities [5] and in some case in its anthropomorphic features. In a survey of socially interactive robots, Fong et al. [6] define a social robot as having the ability to 'express and perceive emotions', an important part of regular human interaction. With the ability to perceive emotions, the robot can imitate human empathy to an abstract level and can then go on to adjust its following actions accordingly. According to Tapus et al. [7], for a robot to emulate empathy, the robot should be able to recognise the human's emotional state and convey the ability of taking perspective.

An example of a socially assistive robot comes from Fasola et al. [3] who presented a socially assistive robot used for encouraging elderly participants to exercise. Participants were asked to follow the arm movements of the robot and copy them themselves, whilst the robot gave vocal feedback and encouragement. Feedback was given throughout the exercise in real time, based on the performance of the user. The addition of vocal feedback is beneficial as it was helpful in keeping the participant engaged. However, the addition of emotion recognition would have been beneficial, as the patient's attitude during the exercises could be monitored to see whether any action needed to be taken.

Social robots are also making progress in the areas of education and social care. A study into the effectiveness of socially assistive robots from Shamsuddin et al. [8] demonstrated that by using a humanoid robot during simple activities, including speaking and moving, with children with Autism Spectrum Disorder (ASD) reduced the display of typical autistic behaviour. Although not conducted over a long period, this study demonstrates the use that social robots have in aiding people with ASD, by reducing autistic behaviour, and shows the potential of social robots in human care.

## 2.2   Emotion Recognition

Recognizing human emotions can be done by analysing a person's facial expressions, speech signals, or body language. This work explores more into emotion recognition using facial expression images as it is commonly easier to obtain facial expression images compared to audio or body language in unconstrained environments.

Regarding emotion recognition on static images, high levels of accuracy have been achieved on facial expression corpora collected in controlled environments. For example, Burkert et al. [9] proposed a new CNN architecture for emotion recognition, which includes two parallel feature extraction blocks. The datasets used were the Extended Cohn-Kanade (CK+) dataset [10] and MMI [11]. On the CK+ dataset, they obtained an accuracy of 99.6%, an increase from the previous benchmark of 99.2% using a conventional CNN. Most notably, they improve on the previous benchmark of 93.33%, to 98.63% for the MMI dataset. Real time emotion recognition has not reached these levels of classification accuracy.

Duncan et al. [12] used a CNN for classifying facial expressions in real time using transfer learning. They employed the VGG_S network architecture using both the CK+ and the Japanese Female Facial Expressions (JAFEE) [13] datasets for training. The result was a training accuracy of 90.7% and a test accuracy of 57.1% on the live stream images. On the live feeds, they included a feature of superimposing a corresponding emoji over the subject's face, which demonstrates a unique application for real time emotion recognition by providing a response of some sort to the person's emotion. Gilligan et al. [14] retrained LeNet and AlexNet networks for their own approach to real time emotion recognition. Included in their work are self-taken images to supplement the CK+ and JAFFE datasets. On their custom dataset, they achieved an accuracy of 98.5%. In addition to this, they took a series of live feed images. Out of the 50 live images taken, 28, or 56%, were classified correctly. Looking closer at the incorrectly classified images, either the second and third strongest classified emotion, out of seven emotions, was the actual correct class, demonstrating a potential for this methodology.

Ruiz-Garcia et al. [15] proposed a hybrid model for real time emotion recognition with the use of a NAO robot. The authors employed a deep CNN for feature extraction and a Support Vector Machine for classification of the KDEF and CK+ datasets. Additionally, they test their model in a real-life application on images taken by a NAO robot in an uncontrolled environment. The authors report an average accuracy rate of 68.75% on the self-taken image and show that most of the misclassifications happen on images with relatively low illumination.

Work on emotion recognition with variation in lighting has been done by Ma and Mohamed [16]. The authors altered the BU-4DFE dataset by changing the lighting conditions and observed whether their classification rate was improved. They found that when compared to no pre-processing, their pre-processing methods improved classification accuracy from 55% to 86%, showing an importance of considering illumination variation in facial emotion recognition. This work aims to build on the findings of [15] and [16] by proposing an illumination invariant emotion recognition architecture for a NAO robot. The following section introduces the experimental setup.

## 3  Methodology

### 3.1  Facial Expression Corpora

#### CK+ , KDEF, JAFFE and FEEDTUM Datasets

The facial expression corpora used in training our emotion recognition model is composed of four different datasets: CK+ , the Karolinska Directed Emotional Faces (KDEF) dataset [17], JAFEE and the Facial Expressions and Emotions (FEEDTUM) database [18]. The CK+ dataset is a labelled set of 486 images from 97 people, including seven emotion categories, with a split of 65/35 female to male participants. The KDEF dataset is also a labelled set of 4900 images from 70 people with seven emotion categories taken from 5 different angles, only frontal facing images are considered in this work, with an even number of female and male participants. The JAFFE dataset consists of 213 images from 10 Japanese female subjects posing seven emotions. The FEEDTUM database contains video streams from 18 participants' reactions to stimuli videos, capturing 7 affective states, from neutral to the peak of the emotion.

The emotion categories include Ekman's six universal emotions: angry, disgust, fear, happy, neutral, sad, and surprise, plus neutral. Neutral has been added considering that all other emotions develop from a neutral state. All seven categories of images were used for training and testing. The CK+ , KDEF and JAFFE datasets were randomly split into 70% for training and 30% for testing. For the FEEDTUM database, we discarded the first 30% along with the last 10% of each sequence of images. Since each sequence starts with a neutral face and transitions to an emotion, we wanted to ensure that the images used contained the most emotion related information rather than neutral faces. The resulting images were also split into 70% training and 30% testing. All four corpora were then combined into one.

## Own Facial Expression Corpus

In addition to the combined dataset, we collected facial expression images in unconstrained environments. Examples are shown in Fig. 1. The additional images were taken to test the model on a dataset that demonstrated images in a more realistic environment, such as a classroom where a social robot could be used. Images were taken in three sessions and two different classrooms, as to make the images more realistic. All images were collected using a NAO robot, a 58-centimetres-tall humanoid robot with a 1.22-megapixel camera with an output of 30fps.

**Fig. 1.** Sample images from our dataset illustrating seven emotions.

In both sessions, participants were asked to sit at a table across from the NAO robot. No specific instructions were provided on how far to sit from the robot or at what angle. Similarly, participants were not asked to remove glasses, hats or scarves as long as their face was visible. Participants were asked to express one of seven emotions at a time and hold it for three seconds for the robot to capture it. A total of 196 images were collected from 21 male participants and 7 females. Participants are undergraduate and postgraduate students between ages 20-55 and from at least five different ethnic backgrounds. Participants were also asked to express the emotion as natural as possible. No other factors were controlled and the rooms had varying lighting conditions due to windows being present. The NAO was in a sitting position to better suit the height of the participants. Note that this meant varying face tilt for every participant, as opposed to zero tilt on the images from the combined corpus. This corpus also differs from the combined corpus datasets, as the extraneous variables were significantly less controlled, the lighting in the images was subject to variability due to the gradual change in the natural external lighting and faces are not centred. After collecting images, we asked three independent parties to label each image with one of seven emotions and tested our model on the images which were labelled with the same emotion by all parties, a total of 121 images. Note that none of these images are used for training.

## Data Pre-processing

For all images, the faces were extracted using a Histogram of Oriented Gradients face recognition model [19]. The resulting images were converted to grey-scale for dimensionality reduction. As the resulting cropped images were of different sizes, they were resized using bipolar interpolation and down sampled to $100 \times 100$ in order to speed up training and recognition times.

In an attempt to improve the CNN model's generalisation performance, we use gamma correction to alter image luminance on the training set. Gamma correction alters the luminance of an image with a non-linear alteration of the input values and the output values. Given an input image $i$, the gamma corrected image $x$ is defined by:

$$x = \left(\frac{i}{255}\right)^{\frac{1}{\gamma}} \times 255 \tag{1}$$

where $\gamma \in \{0.4, 0.6, 0.8, 1.0, \ldots, 3.4\}$. This approach inflates our training dataset over a magnitude of ten. Note that where $\gamma = 1.0$ the input image remains unchanged, thus in this case $x = i$. For this reason, when training the SCAE model, the gamma corrected image $x$ with $\gamma = 1.0$, referred to as $x$ hereafter, becomes the target reconstruction image for all the other gamma corrected images, referred to as $x_\gamma$, including itself.

### 3.2   Illumination Invariant Feature Learning - SCAE

Considering that CNNs often fail to generalise on data with nonuniform conditions, e.g. varying image luminance, in this work we propose pre-training a deep CNN, described in Sect. 3.3, as a deep SCAE to deal with illumination invariance. Autoencoders are neural networks composed of an encoder function $f(x)$ that maps an input vector $x$ to a hidden representation $h$ and a decoder function $g$ that maps $h$ to a reconstruction $y$ which in turn is an approximation of $x$. In effect, empirically, autoencoders are designed to learn an identity function $g(f(x)) = x$. However, since in this case we are interested in learning a function than can map $x_\gamma$ to $x$, learning an identity function is not suitable and thus the objective becomes $g(f(x_\gamma)) = x$. Note that, since the combined corpus is composed of images taken in controlled environments, we assume that $x$ has a good degree of luminance and therefore is a good reconstruction target.

We build a shallow autoencoder for each convolutional layer of the CNN model and treat it as the encoder element along with its activation, batch normalization and max pooling, if any, layers. For the decoder, we replicate the same layers and replace max pooling with nearest neighbour upsampling layers. Each shallow autoencoder is trained greedily using the inter-layer GLW approach described in [20] to improve training by reducing error accumulations. Training and fine-tuning of each shallow autoencoder is done using mini-batch stochastic gradient descent (SGD) and nesterov momentum for 100 epochs using a mean absolute value criterion. The mean absolute value $C$ of the element-wise difference between the reconstruction $y$ and target image $x$ is defined by:

$$c = \frac{\sum_{i=1}^{n} |x_i - y_i|}{n} \tag{2}$$

where $x$ and $y$ are both vectors with a total of $n$ elements. Learning rate was set to 0.1 and momentum to 0.75.

### 3.3   Feature Classification – CNN

Once the SCAE is trained, the decoder element is discarded and a fully connected layer with 100 hidden units is attached to the encoder. This CNN model is then fine-tuned for classification using a cross-entropy criterion. Cross-entropy loss is defined by:

$$L_i = -f_{y_i} + log \sum_j e^{f_j} \tag{3}$$

where $f_j$ equals the $j$ th element of the vector of the $f$ class scores and $y_i$ is the true label of example $i$. Our CNN model is composed of four convolutional layers, followed by rectifier linear unit (ReLU) activation functions, batch normalization (BN), and $2 \times 2$ max pooling layers, except for the last convolutional layer where there is no max pooling. The first two convolutional layers have 20 and 40 $5 \times 5$ filters, and the second two have 60 and 80 $3 \times 3$ filters. Refer to Fig. 2 for a pictorial description. Fine-tuning is done for 100 epochs using SGD, a learning rate of 0.001, and momentum of 0.4.

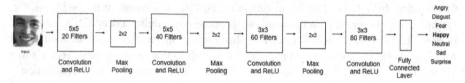

**Fig. 2.** CNN architecture and sample image from our dataset.

## 4   Results and Discussion

The deep CNN emotion recognition model proposed in this work achieves a state-of-the-art accuracy rate of 99.14% on the combined facial expression corpus composed of the CK+ , JAFEE, KDEF, and FEEDTUM datasets. Note that only the training subset contained gamma corrected images. When tested on our own dataset of images collected with the NAO robot in unconstrained environments, our model achieved a classification performance of 73.55%, as observed in Table 1. One of the main observations in Table 1 is the decrease in performance from 99.14% to 73.55% when tested on the dataset collected with the NAO robot, even though we trained the CNN with many image luminance variations. Nonetheless, we attribute this decrease in performance due to the fact that even though the CNN can deal with changes in image illumination, it was not trained to deal with other factors such as face pose or tilt, subjects wearing glasses or scarves, or from many different ethnic backgrounds. It can also be observed that our CNN model fails to generalise on neutral expressions. This can be explained by the fact that in the FEEDTUM database all sequences of emotions start, and most of the time finish, with a neutral face. Even though we discarded the first 30% and the last 10% of the sequences, many of the images with low emotion intensities remained in the dataset. Therefore, if our test subjects illustrated expressions with very low intensities instead of neutral faces, our model will misclassify them. This can also justify neutral being the most misclassified emotion on the testing subset of the combined corpus.

Another discrepancy observed in Table 1 is that most of the classes, except for happy which only had one image misclassified, have misclassified images with the class angry even though angry was one of the best performing classes. This is also true for fear, most of the classes in our dataset have images misclassified as fear, but it is not

**Table 1.** Confusion matrix for the CNN emotion recognition model for the combined corpus (top) and confusion matrix for our own dataset (bottom).

|     | A | D | F | H | N | S | SU |
|-----|-----|-----|-----|-----|-----|-----|-----|
| A  | 99.34 | 0.53 | 0.13 | 0 | 0 | 0 | 0 |
| D  | 0.18 | 99.18 | 0 | 0.18 | 0 | 0.36 | 0.09 |
| F  | 0.26 | 0 | 99.04 | 0.09 | 0.09 | 0.35 | 0.17 |
| H  | 0 | 0 | 0.11 | 99.31 | 0.11 | 0.34 | 0.11 |
| N  | 0.8 | 0 | 0 | 0.4 | 97.21 | 2.79 | 0 |
| S  | 0.32 | 0.06 | 0.25 | 0 | 0.25 | 99.05 | 0.06 |
| SU | 0 | 0 | 0.12 | 0.25 | 0 | 0 | 99.63 |
| A  | 85.14 | 7.14 | 7.14 | 0 | 0 | 0 | 0 |
| D  | 8.33 | 66.66 | 16.66 | 0 | 0 | 8.33 | 0 |
| F  | 9.09 | 0 | 72.72 | 0 | 0 | 0 | 18.18 |
| H  | 0 | 3.57 | 0 | 96.43 | 0 | 0 | 0 |
| N  | 3.85 | 3.85 | 7.69 | 19.23 | 42.3 | 15.38 | 7.69 |
| S  | 9.09 | 0 | 18.18 | 0 | 0 | 72.72 | 0 |
| SU | 5.26 | 0 | 15.80 | 0 | 0 | 0 | 78.95 |

the same case for the combined dataset. On the contrary, happy, surprise and angry are the three classes with the highest classification rates across both testing sets. This can be explained by the fact that these emotions have stronger facial expressions, for instance a standard image expressing happy involves a person with an upturned smiling mouth, pushed up cheeks and slightly squinted eyes. These characteristics are not largely shared with the other emotions, especially the upturned mouth, meaning images depicting happy are not often confused.

The CNN emotion recognition model proposed in this paper outperforms the hybrid emotion recognition model proposed by [15]. The authors employed a CNN for feature extraction, and a SVM for classification. When tested on data also collected by a NAO robot in unconstrained environments, the authors obtained a classification rate of 68.75%, compared to 73.55% achieved by our model. Moreover, [15] only tested with four subjects, whereas we evaluated our model on images from 28 participants.

When training our model without gamma corrected versions of the combined corpus, i.e., training the SCAE model to simply learn an identity function, the best performance we obtained on the testing set of the combined corpus was of 92%. Similarly, the best performance our own corpus was below 40%. With these observations we conclude that it is imperative to train emotion recognition models designed for real time emotion recognition in unconstrained environments with data that represents the environment settings where the application is to be used.

One issue highlighted was the variation in expressions of emotion in different cultures. The authors of [21] showed that between Eastern and Western cultures, there is a difference in the levels of emotional arousal. Western cultures were found to prefer and expressed more higher-arousal emotions, whereas Eastern cultures were found to prefer lower arousal emotions. When taking the images to add to the dataset, participants of different cultures expressed some emotions stronger than others. For example, in our dataset a Chinese participant expressed all emotions, excluding neutral, in a subtler way than participants from a Western cultural group, which resulted in some of their images being misclassified as neutral.

In addition, there was an issue with the ratio between the genders of participants. The participants who took part in the images consisted of 7 females and 21 males, which is not truly representative. Kring and Gordon [22] conducted a study where participants were asked to view emotional films and complete a self-report of how they viewed their expressivity. They found that the female participants were more expressive of emotions in their facial expressions than the male participants were. This shows the importance of collecting a representative set of images with an equal split in gender, as emotions are expressed at different levels between males and females.

One consideration that has not been made is variance in the pose of a face. All images used for training and testing were forward facing and showed the entirety of the face. For emotion recognition in real time to be effective, models should be trained to be able to classify images of faces that are not looking directly at a camera.

## 5 Conclusions and Future Work

In this work, we looked into emotion recognition in unconstrained environments with an application in social robotics. We proposed pre-training a deep CNN model as a SCAE model in a greedy layer-wise unsupervised manner. Instead of learning an identity function like empirical autoencoder models, our SCAE learns to reconstruct images from the same person, illustrating the same emotion and with varying luminance levels produced with gamma correction, as an image with fixed luminance. We demonstrated that this training approach improves the classifier's ability to generalise on unseen data with nonuniform conditions and improves classification rates by over 30%.

Images from 28 participants were collected in an uncontrolled environment to test our CNN emotion recognition model, resulting in a classification rate of 73.55%. Similarly, our model achieved a state-of-the-art classification performance on a combined corpus composed of the CK+ , JAFEE, KDEF and FEEDTUM databases with 99.14% accuracy. From the training results, we can determine that real time emotion recognition in uncontrolled environments with consideration to illumination variance is conceivable. Future work will look at improving this method to also deal with pose invariance, which is a common problem for emotion recognition models designed to work in real time in unconstrained environments.

In this work, we have made progress towards real-time emotion recognition in unconstrained environments for social robots, an essential step in the development of empathic robots. In future work we will look at improving this approach, to be able to deal with varying poses and face tilt, as well as with different emotion intensities.

# References

1. Kulic, D., Croft, E.: Affective state estimation for human–robot interaction. IEEE Trans. Robot. **23**, 991–1000 (2007)
2. Bengio, Y., Lamblin, P., Popovici, D., Larochelle, H.: Greedy layer-wise training of deep networks. In: Proceedings of the 19th International Conference on Neural Information Processing Systems, pp. 153–160 (2006)
3. Fasola, J., Mataric, M.: A socially assistive robot exercise coach for the elderly. J. Hum. Robot Interaction **2**, 3–32 (2013)
4. Breazeal, C.: Toward sociable robots. Robot. Auton. Syst. **42**, 167–175 (2003)
5. Bartneck, C., Forlizzi, J.: A design-centred framework for social human-robot interaction. In: 13th IEEE International Workshop on Robot and Human Interactive Communication, pp. 591–594 (2004)
6. Fong, T., Nourbakhsh, I., Dautenhahn, K.: A survey of socially interactive robots. Robot. Auton. Syst. **42**, 143–166 (2003)
7. Tapus, A., Mataric, M., Scassellati, B.: Socially assistive robotics [grand challenges of robotics]. IEEE Robot. Autom. Mag. **14**, 35–42 (2007)
8. Shamsuddin, S., Yussof, H., Ismail, L., Mohamed, S., Hanapiah, F., Zahari, N.: Initial response in HRI- a case study on evaluation of child with autism spectrum disorders interacting with a humanoid robot NAO. Proc. Eng. **41**, 1448–1455 (2012)
9. Burket, P., Afzal, M., Dengel, A., Liwicki, M.: Dexpression: deep convolutional neural network for expression recognition. Arxiv (2016)
10. Lucey, P., Cohn, J., Kanade, T., Saragih, J., Ambadar, Z., Matthews, I.: The extended Cohn-Kanade dataset (CK+): A complete dataset for action unit and emotion-specified expression. In: 2010 IEEE Computer Society Conference on Computer Vision And Pattern Recognition, pp. 94–101 (2010)
11. Valstar, M., Pantic, M.: Induced disgust, happiness and surprise: an addition to the mmi facial expression database. In: Proceedings of the Internation Conference on Language Resources and Evaluation, pp. 65–70 (2010)
12. Duncan, D., Shine, G., English, C.: Facial emotion recognition in real time (2017)
13. Lyons, M., Akamatsu, S., Kamachi, M., Gyoba, J.: Coding facial expressions with gabor wavelets. In: Third IEEE International Conference on Automatic Face and Gesture Recognition, pp. 200–205 (1998)
14. Gilligan, T., Akis, B.: Emotion AI, real-time emotion detection using CNN (2015)
15. Ruiz-Garcia, A., Elshaw, M., Altahhan, A., Palade, V.: A hybrid deep learning neural approach for emotion recognition from facial expressions for socially assistive robots. Neural Comput. Appl. **29**, 359–373 (2018). https://doi.org/10.1007/s00521-018-3358-8
16. Ma, R., Mohamed, A.: Image processing pipeline for facial expression recognition under variable lighting (2015)
17. Lundqvist, D., Flykt, A., Öhman, A.: The karolinska directed emotional faces CD ROM from department of clinical neuroscience, psychology (1998)
18. Wallhoff, F., Schuller, B., Hawellek, M., Rigoll, G.: Efficient recognition of authentic dynamic facial expressions on the feedtum database. In: IEEE International Conference on Multimedia and Expo, pp. 493–496 (2006)
19. Déniz, O., Bueno, G., Salido, J., De la Torre, F.: Face recognition using histograms of oriented gradients. Pattern Recogn. Lett. **32**, 1598–1603 (2011)

20. Ruiz-Garcia, A., Palade, V., Elshaw, M., Almakky, I.: Deep learning for illumination invariant facial expression recognition. In: Proceedings of the International Joint Conference on Neural Networks, Rio de Janeiro (2018). https://doi.org/10.1109/IJCNN.2018.8489123
21. Lim, N.: Cultural differences in emotion: differences in emotional arousal level between the east and the west. Integr. Med. Res. **5**, 105–109 (2016)
22. Kring, A., Gordon, A.: Sex differences in emotion: expression, experience, and physiology. J. Pers. Soc. Psychol. **74**, 686–703 (1998)

# Cross-Subject Emotion Recognition Using Deep Adaptation Networks

He Li[1], Yi-Ming Jin[1], Wei-Long Zheng[1], and Bao-Liang Lu[1,2,3(✉)]

[1] Center for Brain-Like Computing and Machine Intelligence,
Department of Computer Science and Engineering, Shanghai Jiao Tong University,
800 Dong Chuan Road, Shanghai 200240, China
{polarsky,jinyiming,weilong,bllu}@sjtu.com
[2] Key Laboratory of Shanghai Education Commission for Intelligent
Interaction and Cognitive Engineering, Shanghai Jiao Tong University,
800 Dong Chuan Road, Shanghai 200240, China
[3] Brain Science and Technology Research Center, Shanghai Jiao Tong University,
800 Dong Chuan Road, Shanghai 200240, China

**Abstract.** Affective models based on EEG signals have been proposed
in recent years. However, most of these models require subject-specific
training and generalize worse when they are applied to new subjects. This
is mainly caused by the individual differences across subjects. While, on
the other hand, it is time-consuming and high cost to collect subject-
specific training data for every new user. How to eliminate the individual differences in EEG signals for implementation of affective models is
one of the challenges. In this paper, we apply Deep adaptation network
(DAN) to solve this problem. The performance is evaluated on two publicly available EEG emotion recognition datasets, SEED and SEED-IV,
in comparison with two baseline methods without domain adaptation
and several other domain adaptation methods. The experimental results
indicate that the performance of DAN is significantly superior to the
existing methods.

**Keywords:** Affective brain-computer interface
Emotion recognition · EEG · Deep neural network · Domain adaptation

## 1 Introduction

Emotion plays a critical role in human lives, which affects our behavior and
thought almost anytime and anywhere. As a result, the technology of emotion
recognition has various applications in many fields, including assistance for people everyday life, improvement of working performance, and even implementation of emotional intelligence. On the other hand, EEG singals are considered
to reflect the internal temporal states of human brains and has been studied in
the field of Brain-computer interface (BCI). In recent years, BCIs have also seen

---

H. Li and Y.-M. Jin—The first two authors contributed equally to this work.

© Springer Nature Switzerland AG 2018
L. Cheng et al. (Eds.): ICONIP 2018, LNCS 11305, pp. 403–413, 2018.
https://doi.org/10.1007/978-3-030-04221-9_36

progressive growth in affective Brain-computer interface (aBCI) that aims at recognizing emotions from brain signals [13,18]. Many studies have been made in detecting human emotion states with EEG signals [4,8,12].

Though existing studies have achieved many successes in emotion recognition, most of them only focus on training specific models for particular subjects. These subject-specific models suffer from degraded performance when they are applied to new subjects. The phenomenon is caused by the large domain shift introduced by individual differences across subjects. The naive solution for the problem is to train subject-specific models for every subject, which takes a lot of effort to collect labeled dataset. Another path to solve the problem is to apply domain adaptation methods. As the matter of fact, domain adaptation methods have been applied in various fields to solve the domain shift problem.

There are already several studies on the application of domain adaptation methods to EEG-based emotion recognition. In our previous work [21], Zheng and Lu adopted Transfer component analysis (TCA) [14], Kernel principle analysis (KPCA) [17], and Transductive parameter transfer (TPT) [16] for emotion recognition. Lan *et al.* explored various domain adaptation methods applied on two EEG emition recognition datasets [9]. Jin *et al.* proposed to use Domain-adversarial neural network (DANN) [6] to eliminate the subject differences and achieved appreciable improvement in recognition performance [7]. Lin *et al.* proposed a conditional transfer learning method for emotion recognition task to avoid negative transfer [10].

In this paper, we introduce Deep adaptation network (DAN) to EEG-based emotion recognition and compare DAN with two baseline methods without domain adaptation and several other domain adaptation methods. As far as we know, this is the first work to apply DAN to deal with the subject transfer problem in EEG-based emotion recognition on two publicly available datasets: SEED and SEED-IV. From experimental results we find that DAN achieves the best performance and improves the accuracy of recognition significantly in comparison with the baseline methods.

## 2    Materials and Methods

### 2.1    Dataset Description

Two publicly available emotion recognition datasets, SEED [21] and SEED-IV[1] [20], are used in this paper to evaluate the proposed methods.

The SEED dataset contains EEG signals of 15 subjects recorded while they were watching Chinese film clips. A total number of 15 film clips were selected by a preliminary study and labeled as being negative, positive, or neutral. For each of the subjects, three experiments were performed at an interval of no less than one week. During the experiments, the 15 film clips were played in 15 trials and the subjects were required to watch the clips patiently. The EEG signals

---

[1] The SEED and SEED-IV datasets are available at http://bcmi.sjtu.edu.cn/~seed/index.html.

were recorded with a 62-channel cap according to the 10–20 system using ESI Neuroscan system.

The SEED-IV dataset also contains EEG signals of 15 subjects while they were watching Chinese film clips. The main difference between SEED-IV and SEED is there are film clips of four emotion categories: happy, sad, fear, and neutral. A total number 72 film clips were selected by a preliminary study (18 clips for each emotion category). Each experiment contains 24 trials so that the subjects watched all of the 72 film clips. The EEG signals were recorded with 62-channel cap according to the 10–20 system using ESI Neuroscan system.

## 2.2 Data Preprocessing

The EEG signals are firstly downsampled to 200 Hz and processed with a 1–75 Hz bandpass filter. The filtered signals are then segmented into 1-s and 4-s segments for SEED and SEED-IV datasets, respectively. The segments are attached with the label of the corresponding film clips. Differencial entropy (DE) features are extracted from the segment at the frequency band of delta (1–4 Hz), theta (4–8 Hz), alpha (8–14 Hz), beta (14–31 Hz), and gamma (31–50 Hz) [5,19]. The DE feature is a robust EEG feature that has been applied in our previous studies [7,21]. The definition of the DE feature on a one-dimensional signal $X$ drawn from a Gaussian distribution $N(\mu, \delta^2)$ is

$$h(X) = -\int_{-\infty}^{\infty} P(x) \log(P(x)) = \frac{1}{2} \log 2\pi e \delta^2. \tag{1}$$

For SEED, because there are three duplicate experiment for each subject, we select one of them to reduce the scale of the data. After the preprocessing, there are 3394 and 2505 data samples for each subject in the SEED and SEED-IV datasets, respectively. The feature dimension is 310 (62 EEG channels by 5 frequency bands).

## 2.3 Domain Adaptation Methods

According to Pan *et al.* [15], a domain $\mathcal{D} = \{\mathcal{X}, P(X)\}$ consists of a feature space $\mathcal{X}$ and the corresponding marginal probability distribution $P(X)$, where $X \in \mathcal{X}$. Given a domain $\mathcal{D}$, a task $\mathcal{T} = \{\mathcal{Y}, f(\cdot)\}$ consists of a label space $\mathcal{Y}$ and the corresponding objective predictive function $f(\cdot)$, where $y = f(x), x \in \mathcal{X}$, and $y \in \mathcal{Y}$.

Traditional machine learning approaches focus on solving the task with data samples and the corresponding labels from the same domain. However, in the field of transfer learning, the goal is to solve tasks in a domain when there is no or little observation of data sample while some data samples are available from related domains.

The objective domain where the task lies is called the target domain $\mathcal{D}_T = \{\mathcal{X}_T, P(X_T)\}$ and the related domain is called source domains $\mathcal{D}_S =$

$\{\mathcal{X}_S, P(X_S)\}$. Additionally, when source and target domains share the same feature space and task, the problem is a subset of transfer learning, and is called domain adaptation. In this paper, we study the domain adaptation problem with the target domains being the data from the target subjects, the source domains being the data from the other subjects.

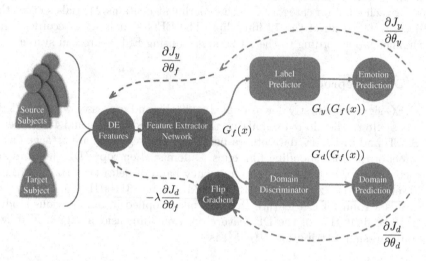

**Fig. 1.** Structure of DANN. The arrows in solid lines indicate the forward propagation path, while the arrows in dotted lines indicate the backpropagation path.

**Domain-Adversarial Neural Network.** Domain-adversarial neural network (DANN) is a domain adapation method based on deep adversarial network. It is composed of three sub-networks as shown in Fig. 1: a feature extractor, a label predictor, and a domain discriminator whose network functions are denoted by $G_f(\cdot)$, $G_y(\cdot)$, and $G_d(\cdot)$ parameterized by $\theta_f$, $\theta_y$, and $\theta_d$, respectively. The method aims to train a feature extractor that eliminates domain discrepancies as well as keep objective task related component of the input features.

In the forward propagation phase, the feature extractor projects the input features into a new feature space. The output is directed to the label predictor and the domain discriminator, simultaneously. The label predictor produces predictions of the labels according to the input, while the domain discrimintor produces predictions of the corresponging domain. The loss of the whole network is

$$\frac{1}{n}\sum_{i=1}^{n} J_y(G_y(G_f(x_i)), y_i) + \alpha J_d(G_d(G_f(x_i)), d_i), \tag{2}$$

where $J_y(G_y(G_f(x_i)), y_i)$ denotes the loss for the label prediction $G_y(G_f(x_i))$ when the true label is $y_i$, $J_d(G_d(G_f(x_i)), d_i)$ denotes the loss for the domain prediction $G_d(G_f(x_i))$ when the true domain is $d_i$, $\alpha$ is a tradeoff hyperparameter, and $n$ is the data sample number.

During the backpropagation phase, the label prediction and domain discrimination losses ($J_y$ and $J_d$) are propagated along the network as in ordinary networks. However, the derivatives are inverted when it is passed from the domain discriminator to the feature extractor: the feature extractor is updated in the direction of maximizing the domain discrimination losses (i.e., deceiving the domain discriminator). In this way, the feature extractor finally discards the domain-specific component of the input (i.e., eliminates the domain discrepancies) in order to keep the domain discrimination losses $J_d$ high. In the test phase, the prediction is made by the feature extractor and the label predictor.

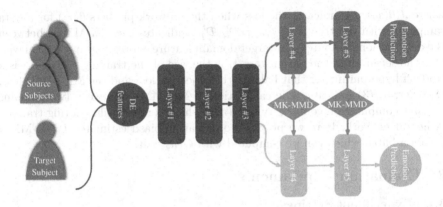

**Fig. 2.** Structure of DAN. The first 3 layers are ordinary layers. The MK-MMD values between source and target domains are calculated in the forth and fifth layers.

**Domain Adaptation Network.** According to the recent findings, neural networks extract features that transition from general to domain and task specific with the growth of their depths. Basing on this idea, Long *et al.* proposed to use multi-kernel Maximum mean discrepancies (MK-MMDs) as a measurement for domain discrepancies of hidden features extracted by deep layers in neural networks [11]. By jointly minimizing the MK-MMDs and the task related loss, the proposed Deep adaptation network (DAN) can eliminate domain discrepancies across domains as well as maintaining task related features.

The structure of DAN is shown in Fig. 2. The first several layers are ordinary ones that behave the same as in traditional networks in forward-propagation and back-propagation phases. Because the feature representation transition to be task and domain specific as the layers become deeper, the deep layers must be treated differently to eliminate domain discrpancies. MK-MMD is applied to achieve this goal in DAN. MK-MMD is multiple kernel variation of MMD that is used for distribution discrepancy measurement. The MK-MMD distance between two probability distributions $p$ and $q$ is defined as the distance between their mean embeddings in a reproducing kernel Hilbert space (RKHS) endowed with a characteristic kernel $k$:

$$d_k^2(p, q) \triangleq ||E_p[\phi_k(x)] - E_q[\phi_k(x)]||^2_{\mathcal{H}_k}, \tag{3}$$

where $\phi_k(\cdot)$ is the corresponding projection function associated with the kernel. If the probability distributions $p$ and $q$ are the ones of the source and target domains, respectively, the MK-MMD value can then measure the domain discrepany. In order to eliminate the domain discrepancies in the deep layers, the MK-MMDs between the distributions of source and target domain feature expressions in the deep layers should be minimized. As a result, the final objective for DAN is

$$\min_{\Theta} \frac{1}{n} \sum_{i=1}^{n} J(\theta(x_i), y_i) + \lambda \sum_{l=4}^{5} d_k^2(\mathcal{D}_S^l, \mathcal{D}_T^l), \tag{4}$$

where $J(\theta(x_i), y_i)$ indicates the loss when the network predicts $\theta(x_i)$ for a data sample $x_i$ with the true label $y_i$, $d_k^2(\mathcal{D}_S^l, \mathcal{D}_T^l)$ indicates the MK-MMDs between the distributions of source and target domain feature expressions in the $l$th layer, $\Theta$ is the set of all of the parameters, $n$ is the size of the training set, and $\lambda$ is a tradeoff hyperparameter that balance the objective loss and the MK-MMD loss. However, as (3) indicates, the calculation of MK-MMDs between two domains requires computation complexity of $O(n^2)$, which is not feasible during training of neural network. Here we propose to use an unbiasd estimate of MK-MMD within a batch which can be computed with $O(n)$ cost.

## 3    Evaluation Experiments

### 3.1    Experiment Settings

We applied leave-one-subject-out cross validation to evaluate the domain adaptation methods on SEED and SEED-IV datasets: for each subject, an emotion recognition model is trained with the subject as target domain, and other subjects as source domain. The deep learning based methods are compared with several traditional methods and two baseline methods to show their adavantages. Both DAN and DANN contain convolutional layers to extract features from images in there original papers [6,11]. In this paper, general features are used instead of images, so the network structures are modified to adapt our problem. For DANN, the feature extractor consists of two fully connected layers, and the label predictor and the domain discriminator consist of three fully connected layers. For DAN, there are three ordinary fully connected layers, two specialized fully connected layers attached with MK-MMD losses, and one output layer for the label prediction. The specific structure of the two networks are described in Table 1. Other methods are described as follows:

(1) KPCA projects the original features into a reproduce Hilbert kernel space (RHKS) with a projection function $\phi(\cdot)$ [17]. A low dimensional subspace of the RHKS is then found by maintaining the variance of the data distribution.

(2) TCA is similar to KPCA in projecting the original features into a RHKS and find a low dimensional subspace [14]. The difference lies in that the subspace is found by minimizing the MMD distance between the source and target domain distributions as well as preserving data properties that are useful for the target supervised learning task.

(3) TPT is a parameter based domain adaptation method on multiple source domains [16]. The method consists of three steps. First, domain-specific models are learnt on each domain. Then, a regression algorithm is applied to project the source domain distributions to the domain-specific model parameters. Finally, the domain-specific model for the target domain is constructed with the target domain distribution and the regression algorithm.

(4) Baseline methods consist of training Support vector machine (SVM) and Multi-layer perceptron (MLP) models on the source domain and applying the trained models directly on the target domain.

**Table 1.** Structure description of DANN and DAN

| Method | Description |
|--------|-------------|
| DANN | The feature extractor has 2 layers, both with node number of 128. The label predictor and domain discriminator have 3 layers with node numbers of 64, 64, and $C$, repectively. $C$ indicates the number of emotion classes to be recognized |
| DAN | There are 5 layers in total, each of them with node numbers of 128, 128, 64, 64, and $C$ from the input end to the output end, respectively. The last two layers are attached with MK-MMD losses |

### 3.2 Results and Discussion

The mean accuracies and standard deviations of each method for the two datasets are shown in Tables 2 and 3, respectively. The specific statistics when each subject is trained as target domain are shown in Figs. 3 and 4.

**Table 2.** Means and standard deviations of the accuracies for each method applied to the SEED dataset

| Method | SVM | MLP | TCA | KPCA | TPT | DANN | DAN |
|--------|-----|-----|-----|------|-----|------|-----|
| Mean | 0.5818 | 0.6101 | 0.6400 | 0.6902 | 0.7517 | 0.7919 | **0.8381** |
| SD | 0.1385 | 0.1238 | 0.1466 | 0.0925 | 0.1283 | 0.1314 | **0.0856** |

For SEED, we compare our previous results in [7] with the results of DAN. As Table 2 shows, DAN achieves the mean accuracy of 0.8381, which outperforms any other methods. DAN also achieves the smallest standard deviation value of 0.0856. It outperforms DANN, which was the best method in [7], by 4.62% in terms of mean recognition accuracy (but with no statistical significance with $p = 0.2645$ in ANOVA test). To show the advantages of DAN, we further compare it with results on the SEED dataset from other papers. Chai

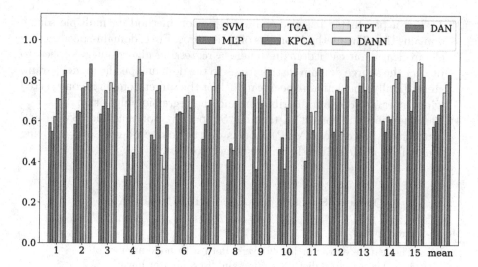

**Fig. 3.** The specific accuracies of each method for all the subjects and the averages in the SEED dataset.

and colleagues applied several novel domain adaptaiton methods to cross-subject emotion recognition from EEG data and evaluated those methods on the SEED dataset. They reported that the mean accuracies of 77.88%, 80.46%, and 79.61% were obtained in their studies [1–3]. Though the evaluation strategies are slightly different (mostly on the selection of the data), DAN outperforms all of the existing methods, which confirms it to be the state-of-the-art approach on the dataset for the cross-subject problem.

**Table 3.** Means and standard deviations of the accuracies for each method applied to the SEED-IV dataset

| Method | SVM | MLP | TCA | KPCA | TPT | DANN | DAN |
|--------|-----|-----|-----|------|-----|------|-----|
| Mean | 0.5178 | 0.4935 | 0.5397 | 0.5176 | 0.5243 | 0.5463 | **0.5887** |
| SD | 0.1285 | 0.0974 | 0.0805 | 0.1289 | 0.1443 | **0.0803** | 0.0813 |

As for SEED-IV, DAN still achieves the best performance, followed with DANN, TCA, TPT, SVM, KPCA, and MLP (in order of declining performance). The method outperforms the baseline SVM and DANN with 6.09% and 4.24% in terms of mean accuracy, respectively. The other deep learning based method, DANN, achieves the second best mean accuracy and the smallest standard deviation. TCA achieves the best performance among the three tranditional methods, but still falls behind DAN with 4.90% of the mean accuracy. In the original paper of SEED-IV [20], Zheng and colleagues achieved a mean accuracy of 70.58% in a within-subject and within-experiment evaluation experiment (training and test

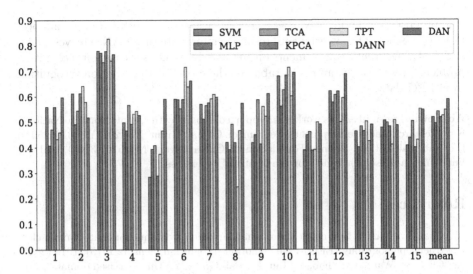

**Fig. 4.** The specific accuracies of each method for all the subjects and the averages in the SEED-IV dataset.

data are from the same subject, in the same experiment), compared with 58.87% in our results. There is an 11.72% gap of accuracy between the two mean accuracies. However, considering the great difference in the evaluation settings, our results should be a desirable one.

For both of the datasets, DAN outperforms DANN and achieves the best performance in terms of mean accuracy. It outperforms the baseline method significantly for SEED (with $p < 0.01$ in ANOVA test). For SEED-IV, it outperforms the baseline method with weaker statistical significance (with $p < 0.1$ in ANOVA test). Besides, it also has the smallest and the second smallest standard deviation of the reconition accuracies for the two dataset, respectively. These clues demonstrate that DAN is suitable for the EEG-based cross-subject emotion recognition, and can achieve more stable performance in comparison with the other domain adaptation methods.

By observing the accuracies on the two datasets, we find that the overall performance of the methods is worse in SEED-IV compared with those in SEED. There might be two reasons for this phenomenon. The first one is that SEED-IV contains four emotional states for recognition, which makes its task a harder one. The second one lies in that each subject has 2505 data samples in SEED-IV, compared with 3394 data samples in SEED, which adds to the difficulty for the methods to capture and eliminate the domain discrepancies.

## 4   Conclusion

In this paper, we have adopted Deep adaptation network (DAN) for dealing with the cross-subject problem in EEG-based emotion recognition. Two publicly available datasets SEED and SEED-IV have been used for performance evaluation.

The proposed method, DAN, was compared with several other domain adaptation approaches. The experimental results demonstrate that DAN achieves 4.62% and 4.24% accuracy improvements on three and four classes emotion recognition problems, respectively, and is suitable for the cross-subject emotion recognition from EEG data.

**Acknowledgement.** This work was supported in part by the grants from the National Key Research and Development Program of China (Grant No. 2017YF-B1002501), the National Natural Science Foundation of China (Grant No. 6167-3266), and the Fundamental Research Funds for the Central Universities.

# References

1. Chai, X., et al.: A fast, efficient domain adaptation technique for cross-domain electroencephalography (EEG)-based emotion recognition. Sensors **17**(5), 1014 (2017)
2. Chai, X., Wang, Q., Zhao, Y., Liu, X., Bai, O., Li, Y.: Unsupervised domain adaptation techniques based on auto-encoder for non-stationary EEG-based emotion recognition. Comput. Biol. Med. **79**, 205–214 (2016)
3. Chai, X., Wang, Q., Zhao, Y., Liu, X., Liu, D., Bai, O.: Multi-subject subspace alignment for non-stationary EEG-based emotion recognition. Technol. Health Care **26**, 1–9 (2018)
4. Daniela, S., Maren, G., Thomas, F., Stefan, K.: Music and emotion: electrophysiological correlates of the processing of pleasant and unpleasant music. Psychophysiology **44**(2), 293–304 (2007)
5. Duan, R., Zhu, J., Lu, B.: Differential entropy feature for EEG-based emotion classification. In: International IEEE/EMBS Conference on Neural Engineering, pp. 81–84. IEEE Press, San Diego (2013)
6. Ganin, Y., Lempitsky, V.: Unsupervised domain adaptation by backpropagation. In: International Conference on Machine Learning, vol. 37, pp. 1180–1189. PMLR, Lille (2015)
7. Jin, Y.M., Luo, Y.D., Zheng, W.L., Lu, B.L.: EEG-based emotion recognition using domain adaptation network. In: International Conference on Orange Technologies, Singapore, pp. 222–225 (2017)
8. Knyazev, G.G., Slobodskoj-Plusnin, J.Y., Bocharov, A.V.: Gender differences in implicit and explicit processing of emotional facial expressions as revealed by event-related theta synchronization. Emotion **10**(5), 678–687 (2010)
9. Lan, Z., Sourina, O., Wang, L., Scherer, R., Müller-Putz, G.R.: Domain adaptation techniques for EEG-based emotion recognition: a comparative study on two public datasets. IEEE Trans. Cogn. Dev. Syst. 1 (2018)
10. Lin, Y.P., Jung, T.P.: Improving EEG-based emotion classification using conditional transfer learning. Front. Hum. Neurosci. **11**, 334 (2017)
11. Long, M., Cao, Y., Wang, J., Jordan, M.: Learning transferable features with deep adaptation networks. In: International Conference on Machine Learning, vol. 37, pp. 97–105. PMLR, Lille (2015)
12. Mathersul, D., Williams, L.M., Hopkinson, P.J., Kemp, A.H.: Investigating models of affect: relationships among EEG alpha asymmetry, depression, and anxiety. Emotion **8**(4), 560–572 (2008)

13. Mühl, C., Allison, B., Nijholt, A., Chanel, G.: A survey of affective brain computer interfaces: principles, state-of-the-art, and challenges. Brain-Comput. Interfaces **1**(2), 66–84 (2014)
14. Pan, S.J., Tsang, I.W., Kwok, J.T., Yang, Q.: Domain adaptation via transfer component analysis. IEEE Trans. Neural Netw. **22**(2), 199–210 (2011)
15. Pan, S.J., Yang, Q.: A survey on transfer learning. IEEE Trans. Knowl. Data Eng. **22**(10), 1345–1359 (2010)
16. Sangineto, E., Zen, G., Ricci, E., Sebe, N.: We are not all equal: personalizing models for facial expression analysis with transductive parameter transfer. In: ACM International Conference on Multimedia, pp. 357–366. ACM Press, New York (2014)
17. Schölkopf, B., Smola, A., Müller, K.-R.: Kernel principal component analysis. In: Gerstner, W., Germond, A., Hasler, M., Nicoud, J.-D. (eds.) ICANN 1997. LNCS, vol. 1327, pp. 583–588. Springer, Heidelberg (1997). https://doi.org/10.1007/BFb0020217
18. Wang, X.W., Nie, D., Lu, B.L.: Emotional state classification from eeg data using machine learning approach. Neurocomputing **129**, 94–106 (2014)
19. Zheng, W., Lu, B.: Investigating critical frequency bands and channels for eeg-based emotion recognition with deep neural networks. IEEE Trans. Auton. Ment. Dev. **7**(3), 162–175 (2015)
20. Zheng, W.L., Liu, W., Lu, Y., Lu, B.L., Cichocki, A.: Emotionmeter: a multimodal framework for recognizing human emotions. IEEE Trans. Cybern. **99**, 1–13 (2018)
21. Zheng, W.L., Lu, B.L.: Personalizing EEG-based affective models with transfer learning. In: International Joint Conference on Artificial Intelligence, pp. 2732–2738. AAAI Press, New York (2016)

# Open Source Dataset and Machine Learning Techniques for Automatic Recognition of Historical Graffiti

Nikita Gordienko[1], Peng Gang[2], Yuri Gordienko[1(✉)], Wei Zeng[2], Oleg Alienin[1], Oleksandr Rokovyi[1], and Sergii Stirenko[1]

[1] National Technical University of Ukraine "Igor Sikorsky Kyiv Polytechnic Institute", Kyiv, Ukraine
yuri.gordiienko@gmail.com
[2] School of Information Science and Technology, Huizhou University, Huizhou, China

**Abstract.** Machine learning techniques are presented for automatic recognition of the historical letters (XI–XVIII centuries) carved on the stoned walls of St. Sophia cathedral in Kyiv (Ukraine). A new image dataset of these carved Glagolitic and Cyrillic letters (CGCL) was assembled and pre-processed for recognition and prediction by machine learning methods. The dataset consists of more than 4000 images for 34 types of letters. The explanatory data analysis of CGCL and notMNIST datasets shown that the carved letters can hardly be differentiated by dimensionality reduction methods, for example, by t-distributed stochastic neighbor embedding (tSNE) due to the worse letter representation by stone carving in comparison to hand writing. The multinomial logistic regression (MLR) and a 2D convolutional neural network (CNN) models were applied. The MLR model demonstrated the area under curve (AUC) values for receiver operating characteristic (ROC) are not lower than 0.92 and 0.60 for notMNIST and CGCL, respectively. The CNN model gave AUC values close to 0.99 for both notMNIST and CGCL (despite the much smaller size and quality of CGCL in comparison to notMNIST) under condition of the high lossy data augmentation. CGCL dataset was published to be available for the data science community as an open source resource.

**Keywords:** Machine learning · Explanatory data analysis
t-distributed stochastic neighbor embedding · Stone carving dataset
notMNIST · Multinomial logistic regression · Convolutional neural network
Deep learning · Data augmentation

## 1 Introduction

Various writing systems have been created by humankind, and they evolved based on the available writing tools and carriers. The term graffiti relates to any writing found on the walls of ancient buildings, and now the word includes any graphics applied to surfaces (usually in the context of vandalism) [1]. But graffiti are very powerful source of historical knowledge, for example, the only known source of the Safaitic language is

© Springer Nature Switzerland AG 2018
L. Cheng et al. (Eds.): ICONIP 2018, LNCS 11305, pp. 414–424, 2018.
https://doi.org/10.1007/978-3-030-04221-9_37

graffiti inscriptions on the surface of rocks in southern Syria, eastern Jordan and northern Saudi Arabia [2]. In addition to these well-known facts, the most interesting and original examples of the Eastern Slavic visual texts are represented by the medieval graffiti that can be found in St. Sophia Cathedral of Kyiv (Ukraine) (Fig. 1) [3]. They are written in two alphabets, Glagolitic and Cyrillic, and vary by the letter style, arrangement and layout [4, 5]. The various interpretations of these graffiti were suggested by scholars as to their date, language, authorship, genuineness, and meaning [6, 7]. Some of them were based on the various image processing techniques including pattern recognition, optical character recognition, etc.

**Fig. 1.** Example of the original image for graffito #1 (c. 1022) (a) and preprocessed glyphs (b) from the medieval graffiti in St. Sophia Cathedral of Kyiv (Ukraine) [3].

The main aim of this paper is to apply some machine learning techniques for automatic recognition of the historical graffiti, namely letters (XI–XVIII centuries) carved on the stoned walls of St. Sophia cathedral in Kyiv (Ukraine) and estimate their efficiency in the view of the complex geometry, barely discernible shape, and low statistical representativeness (small dataset problem). The Sect. 2 contains the short characterization of the basic terms and parameters of the methods used. The Sect. 3 includes description of the datasets used and methods applied for their characterization. The Sect. 4 gives results of the initial analysis of preprocessed graffiti images. The Sect. 5 contains results of machine learning approaches to the problem. The Sect. 6 is dedicated to discussion of the results obtained and lessons learned.

## 2  State of the Art

The current and previous works on recognition of handwriting were mainly targeted to pen, pencil, stilus, or finger writing. The high values of recognition accuracy (>99%) were demonstrated on the MNIST dataset [8] of handwritten digits by a convolutional neural networks [9]. But stone carved handwriting has usually much worse quality and shabby state to provide the similar values of accuracy. At the moment, most work on character recognition has concentrated on pen-on-paper like systems [10]. In all cases the methods were based on the significant preprocessing actions without which accuracy falls significantly. And this is especially important for analysis and recognition of the carved letters like historical graffiti. Usually, the preprocessing requires a priori knowledge about entire glyph, but the Glagolitic and Cyrillic glyph datasets are not available at the moment as open source databases except for some cases of their publications [3, 4]. The recent progress of computer vision and machine learning methods allows to apply some of them to improve the current recognition, identification, localization, semantic segmentation, and interpretation of such historical graffiti of various origin from different regions and cultures, including Europe (ancient Ukrainian graffiti from Kyivan Rus) [3], Middle East and Africa (Safaitic graffiti) [2], Asia (Chinese hieroglyphs) [11], etc. Moreover, the progress of diverse mediums in the recent decades determined the need for many more alphabets and methods of their recognition for different use cases, such as controlling computers using touchpads, mouse gestures or eye tracking cameras. It is especially important topic for elderly care applications [12] on the basis of the newly available information and communication technologies based on multimodal interaction through human-computer interfaces like wearable computing, brain-computing interfaces [13], etc.

## 3   Datasets and Models

Currently, more than 7000 graffiti of St. Sophia Cathedral of Kyiv are detected, studied, preprocessed, and classified (Fig. 1) [14–16]. The unique corpus of epigraphic monuments of St. Sophia of Kyiv belongs to the oldest inscriptions, which are the most valuable and reliable source to determine the time of construction of the main temple of Kyivan Rus. For example, they contain the cathedral inscriptions-graffiti dated back to 1018–1022, which reliably confirmed the foundation of the St. Sophia Cathedral in 1011. A new image dataset of these carved Glagolitic and Cyrillic letters (CGCL) from graffiti of St. Sophia Cathedral of Kyiv was assembled and pre-processed to provide glyphs (Fig. 1a) for recognition and prediction by multinomial logistic regression and deep neural network [17]. At the moment the whole dataset consists of more than 4000 images for 34 types of letters (classes), but it is permanently enlarged by the fresh contributions.

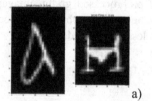

a)          b)

**Fig. 2.** Examples of glyphs obtained from: CGCL dataset from graffiti of St. Sophia Cathedral of Kyiv (a) and from notMNIST dataset (b).

The second dataset, notMNIST, contains some publicly available fonts for 10 classes (letters A–J) taken from different fonts and extracted glyphs from them to make a dataset similar to MNIST [8]. notMNIST dataset consists of small (cleaned) part, about 19k instances (Fig. 1b), and large (unclean) dataset, 500k instances [18]. It was used for comparison of the results obtained with GCCL dataset.

## 4  Explanatory Data Analysis

The well-known t-distributed stochastic neighbor embedding (t-SNE) technique was applied [19, 20]. It allowed us to embed high-dimensional glyph image data into a 3D space, which can then be visualized in a scatter plot (Fig. 3). The cluster of glyphs of the carved graffiti from CGCL dataset is more scattered (Fig. 3a) than the cluster of glyphs from notMNIST dataset (Fig. 3b) (note the difference of >30 times for scales of these plots).

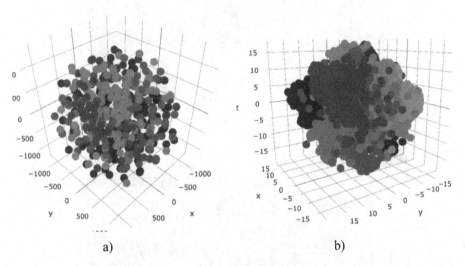

a)                                    b)

**Fig. 3.** Results of tSNE analysis for 10 letters from A (red) to H (blue) glyphs (Fig. 2) from: CGCL dataset (a) and notMNIST dataset (b). The similar letters are modeled by nearby points and dissimilar ones are mapped to distant points. The same color corresponds to the same class (type of letter from A to H). (Color figure online)

The explanatory data analysis of CGCL and notMNIST datasets shown that the carved letters can hardly be differentiated by dimensionality reduction methods, for example, by t-distributed stochastic neighbor embedding (tSNE) due to the worse letter representation by stone carving in comparison to hand writing.

For the better representation of the distances between different images the cluster analysis of differences by calculation of pairwise image distances was performed for subsets of the original CGCL dataset and notMNIST datasets that contained glyphs of A and H letters only. Then the clustered distance map was constructed for CGCL dataset (Fig. 4a) and notMNIST (Fig. 4b) datasets. The crucial difference of the visual quality of glyphs from CGCL and notMNIST datasets consists in the more pronounced

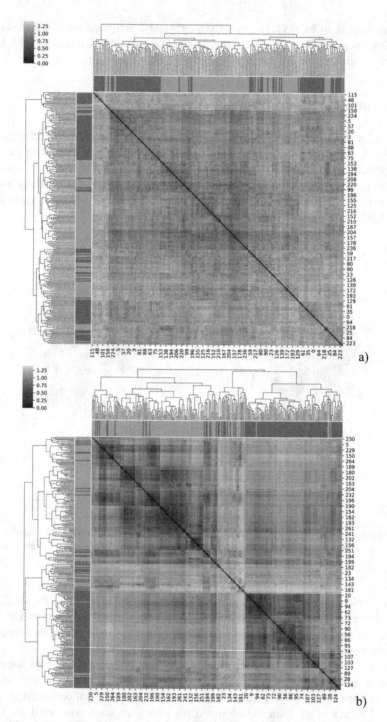

**Fig. 4.** The clustered distance maps for distances between A (red) and H (blue) letters in CGCL (a) and notMNIST (b) datasets. (Color figure online)

clustering in two distinctive sets (denoted by separate blue and red parts of the legend ribbons) and more darker regions inside map (the darker region means the lower distance between letters) for notMNIST dataset (Fig. 4b). Even from the first look these maps are very different and have the clear understanding about correlation among the glyphs in both datasets.

# 5 Machine Learning for Automatic Recognition of Graffiti

## 5.1 Multinomial Logistic Regression

To estimate the possibility to predict the letters by glyphs the multinomial logistic regression (MLR) was applied for subsets with 10 classes of letters (Fig. 5). The MLR model demonstrated that the area under curve (AUC) values for receiver operating characteristic (ROC) for separate letters were not lower than 0.92 and 0.60 for notMNIST and CGCL, respectively, and the averaged AUC values were 0.99 and 0.82 for notMNIST and CGCL, respectively.

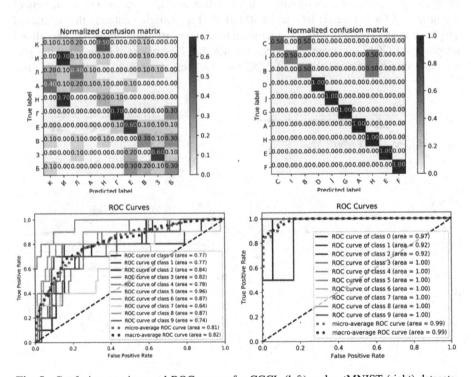

**Fig. 5.** Confusion matrixes and ROC-curves for CGCL (left) and notMNIST (right) datasets.

## 5.2 Convolutional Neural Network

2D convolutional neural network (CNN) was applied to check the feasibility of application of neural networks for the small dataset like CGCL in comparison to notMNIST dataset to recognize two letters A and H from their glyphs. The CNN had pyramid like architecture with 5 convolutional/max-pooling layers and 205 217 trainable parameters, rectified linear unit (ReLU) activation functions, a binary cross-entropy as a loss function, and RMSProp (Root Mean Square Propagation) as an optimizer with a learning rate of $10^{-4}$. In Fig. 6 accuracy and loss results of training and validation attempts are shown for subsets of the original CGCL (Fig. 6, left) and notMNIST (Fig. 6, right) datasets that contained glyphs of A and H letters only.

The model becomes overtrained very soon after 5 epochs for CGCL (Fig. 6, left) and after 10 epochs for notMNIST (Fig. 6, right) datasets. The prediction accuracy for test subset of 70 images and AOC for ROC-curve was very small ($\sim 0.5$) and it is explained by the small size of datasets in comparison to the complexity of the CNN model.

To avoid such overtraining the lossless data augmentation with addition of the random horizontal and vertical flips of the original images was applied. In Fig. 7 accuracy and loss results of training and validation attempts are shown for subsets of the original CGCL (Fig. 7, left) and notMNIST (Fig. 7, right) datasets that contained glyphs of A and H letters only. Again model became overtrained a little bit later: after 7 epochs for CGCL and after 15 epochs for notMNIST datasets. In this case the prediction accuracy for test subset and AOC for ROC-curve was very small ($\sim 0.5$) also.

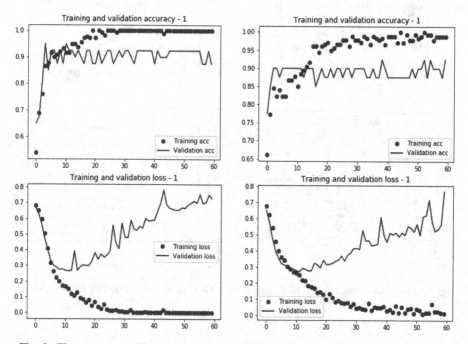

**Fig. 6.** The accuracy and loss for the original CGCL (left) and notMNIST (right) datasets.

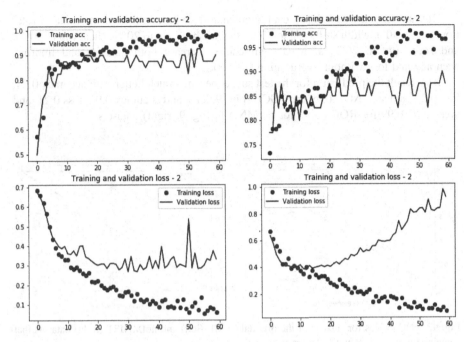

**Fig. 7.** The accuracy and loss for CGCL (left) and notMNIST (right) with lossless data augmentation.

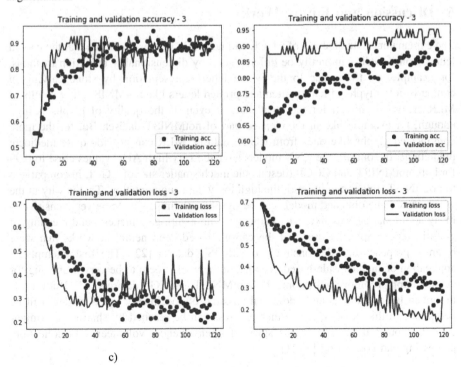

c)

**Fig. 8.** The accuracy and loss of training and validation attempts for the original CGCL (left) and notMNIST (right) datasets with lossy data augmentation.

But application of the lossy data augmentation with addition of the random rotations (up to 40°), width shifts (up to 20%), height shifts (up to 20%), shear (up to 20%), and zoom (up to 20%) allowed significantly increase both datasets and improve accuracy and loss without overtraining (Fig. 8).

As a result the prediction for the test subset became much better with accuracy 0.94, loss 0.21, and area AOC 0.99 for CGCL (Fig. 9, left), and accuracy 0.91, loss 0.21, and area AOC 0.99 for ROC-curve for notMNIST (Fig. 9, right) datasets.

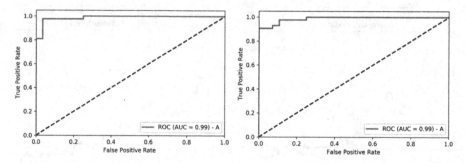

**Fig. 9.** ROC-curves for subsets of the original CGCL (left) and notMNIST (right) datasets that contained glyphs of A and H letters only.

## 6 Discussion and Future Work

The explanatory data analysis of CGCL and notMNIST datasets shown that the carved letters from CGCL can hardly be differentiated by dimensionality reduction methods, for example, by tSNE due to the worse letter representation by stone carving in comparison to glyphs of handwritten and printed letters like notMNIST. The results of MLR are good enough for the small dataset even, if the quality of glyphs is high enough, for example like in the cleaned part of notMNIST dataset. But for the more complicated glyphs like ones from CGCL dataset, MLR can provide quite mediocre predictions. In contrast, the CNN models gave the very high AUC values close to 0.99 for both notMNIST and CGCL (despite the much smaller size of CGCL in comparison to notMNIST) under condition of the high lossy data augmentation. That is why in the wider context the obtained models can be significantly improved to be very sensitive to many additional aspects like date, language, authorship, genuineness, and meaning of graffiti, for example, by usage of the capsule-based deep neural networks that were recently proposes and demonstrated on MNIST dataset [22]. The first attempts of application of such capsule-based deep neural networks seem to be very promising for classification problems in the context of notMNIST and CGL datasets [23]. But for this the much larger datasets and additional research of specifically tuned models will be necessary. In this context, the further progress can be reached by sharing the similar datasets around the world in the spirit of open science, volunteer data collection, processing and computing [2, 21].

In conclusion, the new image dataset of the carved Glagolitic and Cyrillic letters was prepared and tested by MLR and deep CNNs for the letter recognition. The dataset was published for the data science community as an open source resource.

**Acknowledgements.** The work was partially supported by Huizhou Science and Technology Bureau and Huizhou University (Huizhou, P.R. China) in the framework of Platform Construction for China-Ukraine Hi-Tech Park Project. The glyphs of letters from the graffiti [3] were prepared by students and teachers of National Technical University of Ukraine "Igor Sikorsky Kyiv Polytechnic Institute" and can be used as an open science dataset under CC BY-NC-SA 4.0 license (https://www.kaggle.com/yoctoman/graffiti-st-sophia-cathedral-kyiv).

# References

1. Ancelet, J.: The History of Graffiti. University of Central London, London (2006)
2. Burt, D.: The Online Corpus of the Inscriptions from Ancient North Arabia (OCIANA). http://krc2.orient.ox.ac.uk/ociana. Accessed 30 Aug 2018
3. Nikitenko, N., Kornienko, V.: Drevneishie Graffiti Sofiiskogosobora v Kieve i Vremya Ego Sozdaniya (Old Graffiti in the St. Sofia Cathedral in Kiev and Time of Its Creation). Mykhailo Hrushevsky Institute of Ukrainian Archeography and Source Studies, Kiev (2012). (in Russian)
4. Vysotskii, S.A.: Drevnerusskie Nadpisi Sofii Kievskoi XI–XIV vv. (Old Russian Inscriptions in the St. Sofia Cathedral in Kiev, 11th–14th Centuries). Naukova Dumka, Kiev (1966). (in Russian)
5. Nazarenko, T.: East Slavic visual writing: the inception of tradition. Can. Slavon. Pap. **43**(2–3), 209–225 (2001)
6. Drobysheva, M.: The difficulties of reading and interpretation of Old Rus Graffiti (the inscription Vys. 1 as example). Istoriya **6**(6(39)), 10–20 (2015)
7. Pritsak, O.: An eleventh-century Turkic Bilingual (Turko-Slavic) Graffito from the St. Sophia Cathedral in Kiev. Harv. Ukr. Stud. **6**(2), 152–166 (1982)
8. LeCun, Y., Cortes, C., Burges, C.J.: MNIST Handwritten Digit Database. AT&T Labs. http://yann.lecun.com/exdb/mnist. Accessed 30 Aug 2018
9. LeCun, Y., Bottou, L., Bengio, Y., Haffner, P.: Gradient-based learning applied to document recognition. Proc. IEEE **86**(11), 2278–2324 (1998)
10. Hafemann, L.G., Sabourin, R., Oliveira, L.S.: Offline handwritten signature verification—Literature review. In: 2017 Seventh International Conference on Image Processing Theory, Tools and Applications, pp. 1–8. IEEE (2017)
11. Winter, J.: Preliminary investigations on Chinese ink in far eastern paintings. In: Archaeological Chemistry, pp. 207–225 (1974)
12. Gang, P., et al.: User-driven intelligent interface on the basis of multimodal augmented reality and brain-computer interaction for people with functional disabilities. In: Proceedings of Future of Information and Communication Conference, 5–6 April 2018, pp. 322–331. IEEE, Singapore (2018)
13. Gordienko, Yu., et al.: Augmented coaching ecosystem for non-obtrusive adaptive personalized elderly care on the basis of Cloud-Fog-Dew computing paradigm. In: Proceedings of IEEE 40th International Convention on Information and Communication Technology, Electronics and Microelectronics, pp. 387–392. IEEE, Opatija, Croatia (2017)

14. Nikitenko, N., Kornienko, V.: Drevneishie Graffiti Sofiiskogosobora v Kieve i Ego Datirovka (The Ancient Graffiti of St. Sophia Cathedral in Kiev and Its Dating). Byzantinoslavica **68**(1), 205–240 (2010). (in Russian)
15. Kornienko, V.V.: Korpus Hrafiti Sofii Kyivskoi, XI - pochatok XVIII_st, chastyny I–III (The Collection of Graffiti of St. Sophia of Kyiv, 11th–17th centuries), Parts I–III. Mykhailo Hrushevsky Institute of Ukrainian Archeography and Source Studies, Kiev (2010–2011). (in Ukrainian)
16. Sinkevic, N., Kornienko, V.: Nowe Zrodla do Historii Kosciola unickiego w Kijowie: Graffiti w Absydzie Glownego Oltarza Katedry Sw. Zofii. Studia Zrodloznawcze **50**, 75–80 (2012). http://rcin.org.pl/Content/31104/WA303_44631_B88-SZ-R-50-2012_Sinkevic.pdf
17. Glyphs of Graffiti in St. Sophia Cathedral of Kyiv. https://www.kaggle.com/yoctoman/graffiti-st-sophia-cathedral-kyiv. Accessed 30 Aug 2018
18. Bulatov, Y.: notMNIST dataset. Google (Books/OCR), Technical report. http://yaroslavvb.blogspot.it/2011/09/notmnist-dataset.html. Accessed 30 Aug 2018
19. Maaten, L.V.D., Hinton, G.: Visualizing data using t-SNE. J. Mach. Learn. Res. **9**(1), 2579–2605 (2008)
20. Schmidt, P.: Cervix EDA and model selection. https://www.kaggle.com/philschmidt. Accessed 30 Aug 2018
21. Gordienko, N., Lodygensky, O., Fedak, G., Gordienko, Yu.: Synergy of volunteer measurements and volunteer computing for effective data collecting, processing, simulating and analyzing on a worldwide scale. In: Proceedings of 38th International Convention on Information and Communication Technology, Electronics and Microelectronics, pp. 193–198. IEEE, Opatija (2015)
22. Sabour, S., Frosst, N., Hinton, G.E.: Dynamic routing between capsules. In: Advances in Neural Information Processing Systems, pp. 3856–3866 (2017)
23. Gordienko, N., Kochura, Y., Taran, V., Gang, P., Gordienko, Y., Stirenko, S.: Capsule deep neural network for recognition of historical Graffiti handwriting. In: IEEE Ukraine Student, Young Professional and Women in Engineering Congress, Kyiv, Ukraine, 2–6 October 2018. IEEE (2018, Submitted)

# Supervised Two-Dimensional CCA
# for Multiview Data Representation

Yun-Hao Yuan[✉], Hui Zhang, Yun Li, Jipeng Qiang, and Wenyan Bao

School of Information Engineering, Yangzhou University, Yangzhou 225137, China
yhyuan@yzu.edu.cn

**Abstract.** Since standard canonical correlation analysis (CCA) works with vectorized representations of data, an limitation is that it may suffer small sample size problems. Moreover, two-dimensional CCA (2D-CCA) extracts unsupervised features and thus ignores the useful prior class information. This makes the extracted features by 2D-CCA hard to discriminate the data from different classes. To solve this issue, we simultaneously take the prior class information of intra-view and inter-view samples into account and propose a new 2D-CCA method referred to as supervised two-dimensional CCA (S2CCA), which can be used for multi-view feature extraction and classification. The method we propose is available to face recognition. To verify the effectiveness of the proposed method, we perform a number of experiments on the AR, AT&T, and CMU PIE face databases. The results show that the proposed method has better recognition accuracy than other existing multi-view feature extraction methods.

**Keywords:** Multi-view learning · Canonical correlation analysis
Two-dimensional analysis · Dimensionality reduction

## 1 Introduction

In recent years, data representation has become more and more diverse, especially in the domains of pattern recognition and computer vision. In practical applications, the same objects can be characterized by various feature vectors in different high-dimensional feature spaces. For instance, a person can be expressed by facial image feature and fingerprint feature; a speaker may be characterized by audio feature and image/video feature. In general, the data with multiple kinds of features are often called multiple-view data, each of which depicts a specific feature space. Because the multiple-view data can reveal different viewpoints or attributes of same objects, they provide a more comprehensive and more accurate description of the objects than single-view data. Therefore, learning from high-dimensional multiple-view data, often referred to as multi-view learning, has attracted an increasing attention.

To date, there have been lots of dedicated multi-view learning approaches, in which subspace learning is undoubtedly one of the most important multi-view

© Springer Nature Switzerland AG 2018
L. Cheng et al. (Eds.): ICONIP 2018, LNCS 11305, pp. 425–434, 2018.
https://doi.org/10.1007/978-3-030-04221-9_38

learning topics. Canonical Correlation Analysis (CCA) [5], proposed by Hotelling in 1936, is the most widely used multi-view subspace analysis algorithm, which linearly transforms two sets of random variables into a lower-dimensional subspace where they are maximally correlated. In real world, multiple view data usually have complex nonlinear dependency rather than only linear relations. But, CCA can not effectively deal with this nonlinear case owing to its linearity. To overcome this shortcoming, some nonlinear variants of CCA have been proposed. For example, kernel CCA [4] uses two nonlinear mappings to transform the multi-view original data into high-dimensional Hilbert spaces where standard CCA is adopted to analyze the correlations between two views. In addition, a famous nonlinear version of CCA is the deep CCA [2], which uses deep neural networks instead of kernel mappings to achieve a more flexible representation of high-dimensional data. More details about deep CCA can be found in [2].

It is well-known that CCA does not consider the class label information of training data in learning. Thus, it is an unsupervised algorithm. This makes the CCA-extracted features difficult to effectively differentiate unseen samples. In order to solve this issue, generalized CCA (GCCA) [10] and discriminant CCA (DCCA) [11] have been proposed to extract discriminative feature vectors for pattern classification tasks. However, GCCA merely utilizes intra-view-sample class label information and DCCA merely inter-view-sample class label information. To make full use of the class label information, Yuan et al. [12] proposed a new supervised CCA algorithm, which takes the class information of intra-view and inter-view samples into account at the same time.

However, in image recognition, CCA and related methods above need to first reshape two-dimensional (2D) image matrices into the vectors before they are applied. Such transformation leads to two issues. First, it loses the intrinsic spatial relationships among image pixels. Second, it results in the high dimensionality of image data due to the vectorized representation. Moreover, vector representation-based CCA always suffers from the singularity problem of within-view covariance matrices in small samples size cases [7] where the dimensionality of feature vectors is larger than the number of training samples. To exploit the image structure information, Lee et al. [6] proposed a two-dimensional CCA (2D-CCA) method, which directly measures the relations between two sets of two-dimensional image matrices rather than image vectors and thus reduces the computational cost. Later, An and Bhanu [1] employed 2D-CCA to present a new facial image high-resolution reconstruction algorithm. More recently, Gao et al. [3] proposed two new 2D canonical correlation approaches, i.e., 2D-LPCCA and 2D-SPCCA. Experimental results have shown the effectiveness of both methods in pattern classification.

The foregoing 2D-CCA-related methods are unsupervised. From the perspective of pattern classification, the extracted features by those 2D methods may not be optimal for recognition tasks. Thus, in this paper we propose a supervised two-dimensional CCA approach referred to as S2CCA for multi-view feature representation, which simultaneously considers the prior label information of intra-view and inter-view training samples. S2CCA can find more discriminative low-dimensional representation than 2D-CCA due to the use of supervised information. Numerous experiments show that the proposed S2CCA can obtain encouraging results.

## 2    Related Work

### 2.1    CCA

Consider $n$ pairwise centered samples $\{(x_i, y_i)\}_{i=1}^n$, where $x_i \in \Re^p$ and $y_i \in \Re^q$ with $p$ and $q$ as the sample dimension. Assume $X = (x_1, x_2, \cdots, x_n) \in \Re^{p \times n}$ and $Y = (y_1, y_2, \cdots, y_n) \in \Re^{q \times n}$. CCA aims to find a pair of base vectors, $w_x \in \Re^p$ and $w_y \in \Re^q$, which maximize the correlation coefficient of canonical projections $w_x^T X$ and $w_y^T Y$ defined as

$$\rho(w_x, w_y) = \frac{w_x^T XY w_y}{\sqrt{w_x^T XX^T w_x}\sqrt{w_y^T YY^T w_y}} = \frac{w_x^T C_{xy} w_y}{\sqrt{w_x^T C_{xx} w_x}\sqrt{w_y^T C_{yy} w_y}}, \quad (1)$$

where $\rho$ is the correlation coefficient, $C_{xx}$ and $C_{yy}$ are the intra-view covariance matrices of $X$ and $Y$, and $C_{xy}$ is the inter-view covariance matrix of $X$ and $Y$. As $\rho$ is scale-invariant w.r.t. $w_x$ and $w_y$, maximizing (1) can be reformulated as

$$\max_{w_x, w_y} w_x^T C_{xy} w_y$$
$$s.t. \ w_x^T C_{xx} w_x = 1, \ w_y^T C_{yy} w_y = 1. \quad (2)$$

The optimization problem (2) can be solved by the generalized eigenvalue problem. More details of CCA can be found in [4].

### 2.2    2D-CCA

Let $n$ pairs of training samples of two-dimensional random variables $X$ and $Y$ be given as $\{(X_i, Y_i)\}_{i=1}^n$, where $X_i \in \Re^{m_x \times n_x}$ and $Y_i \in \Re^{m_y \times n_y}$. 2D-CCA tries to simultaneously find left transformations, $l_x$ and $l_y$, and right transformations, $r_x$ and $r_y$, so that the correlation between $l_x^T X r_x$ and $l_y^T Y r_y$ is maximized. Concretely, the optimization problem of 2D-CCA can be formulated as

$$\max_{l_x, l_y, r_x, r_y} \text{cov}\left(l_x^T X r_x, l_y^T Y r_y\right)$$
$$s.t. \ \text{var}\left(l_x^T X r_x\right) = 1, \ \text{var}\left(l_y^T Y r_y\right) = 1, \quad (3)$$

where $\text{cov}(\cdot)$ and $\text{var}(\cdot)$ denote the covariance and variance operators, respectively. Note that, in practice, the covariance and variance are computed using $n$ pairwise samples $\{(X_i, Y_i)\}_{i=1}^n$.

2D-CCA needs to make use of an alternating iteration algorithm to solve left and right transforms. Existing study [6] has shown that iteration-based 2D-CCA can converge fast after a few iterations. The details of solving (3) can be found in [6].

## 3   Proposed S2CCA

The proposed S2CCA simultaneously considers the prior label information of intra-view and inter-view 2D training samples. Specifically, let two sets (views) of 2D training samples of $c$ classes be $\{X^{(i)}\}_{i=1}^c$ and $\{Y^{(i)}\}_{i=1}^c$, respectively, where $X^{(i)} = \{X_j^{(i)} \in \Re^{m_x \times n_x}\}_{j=1}^{n_i}$, $Y^{(i)} = \{Y_j^{(i)} \in \Re^{m_y \times n_y}\}_{j=1}^{n_i}$, $X_j^{(i)}$ and $Y_j^{(i)}$ are separately the $j$-th 2D samples in the $i$-th class, $n_i$ is the number of samples in $i$-th class, and $\sum_{i=1}^c n_i = n$. Assume all training samples of two views in class $i$ have been centered, i.e., $\sum_{j=1}^{n_i} X_j^{(i)} = 0$ and $\sum_{j=1}^{n_i} Y_j^{(i)} = 0$, $i = 1, 2, \cdots, c$. The goal of our S2CCA is to find left transforms, $l_x$ and $l_y$, and right transforms, $r_x$ and $r_y$, such that the intra-class correlation between $l_x^T X^{(i)} r_x$ and $l_y^T Y^{(i)} r_y$ is maximized. It can be formulated as

$$\max_{l_x,l_y,r_x,r_y} \rho^{(i)} = \frac{\text{cov}(l_x^T X^{(i)} r_x, l_y^T Y^{(i)} r_y)}{\sqrt{\text{var}(l_x^T X^{(i)} r_x)}\sqrt{\text{var}(l_y^T Y^{(i)} r_y)}}, \quad i = 1, 2, \cdots, c. \tag{4}$$

To obtain the left and right transformations, (4) shows that we need to simultaneously solve $c$ optimization problems, which is generally difficult to find $c$ global optimal values at the same time. In order to simplify our model, summing all $c$ objective functions in (4) leads to the resulting optimization model of S2CCA, as follows

$$\max_{l_x,r_x,l_y,r_y} \sum_{i=1}^c \frac{\text{cov}(l_x^T X^{(i)} r_x, l_y^T Y^{(i)} r_y)}{\sqrt{\text{var}\left(l_x^T X^{(i)} r_x\right)}\sqrt{\text{var}(l_y^T Y^{(i)} r_y)}}. \tag{5}$$

Due to the scaling of left and right transforms, we reformulate the optimization problem (5) as

$$\max_{l_x,r_x,l_y,r_y} \sum_{i=1}^c \text{cov}(l_x^T X^{(i)} r_x, l_y^T Y^{(i)} r_y)$$
$$\text{s.t.} \begin{cases} \text{var}(l_x^T X^{(i)} r_x) = 1, \\ \text{var}(l_y^T Y^{(i)} r_y) = 1, \\ i = 1, 2, \cdots, c. \end{cases} \tag{6}$$

Using the idea in [8], we further relax the optimization problem (6) by reducing the $2c$ constraints to 2 ones, which is the following

$$\max_{l_x,r_x,l_y,r_y} \sum_{i=1}^c \text{cov}(l_x^T X^{(i)} r_x, l_y^T Y^{(i)} r_y)$$
$$\text{s.t.} \begin{cases} \sum_{i=1}^c \text{var}(l_x^T X^{(i)} r_x) = 1, \\ \sum_{i=1}^c \text{var}(l_y^T Y^{(i)} r_y) = 1. \end{cases} \tag{7}$$

Now, let us define

$$C_{xy}^r = \sum_{i=1}^{c} \left\langle X^{(i)} r_x r_y^T Y^{(i)T} \right\rangle = \sum_{i=1}^{c} \frac{1}{n_i} \sum_{j=1}^{n_i} X_j^{(i)} r_x r_y^T Y_j^{(i)T}, \tag{8}$$

$$C_{xx}^r = \sum_{i=1}^{c} \left\langle X^{(i)} r_x r_x^T Y^{(i)T} \right\rangle = \sum_{i=1}^{c} \frac{1}{n_i} \sum_{j=1}^{n_i} X_j^{(i)} r_x r_x^T X_j^{(i)T}, \tag{9}$$

$$C_{yy}^r = \sum_{i=1}^{c} \left\langle Y^{(i)} r_y r_y^T Y^{(i)T} \right\rangle = \sum_{i=1}^{c} \frac{1}{n_i} \sum_{j=1}^{n_i} Y_j^{(i)} r_y r_y^T Y_j^{(i)T}. \tag{10}$$

Since

$$\sum_{i=1}^{c} \mathrm{cov}(l_x^T X^{(i)} r_x, l_y^T Y^{(i)} r_y) = \sum_{i=1}^{c} \left\langle l_x^T X^{(i)} r_x r_y^T Y^{(i)T} l_y \right\rangle = l_x^T C_{xy}^r l_y,$$

the optimization problem (7) can be rewritten as

$$\begin{aligned} \max \quad & l_x^T C_{xy}^r l_y \\ s.t. \quad & l_x^T C_{xx}^r l_x = l_y^T C_{yy}^r l_y = 1. \end{aligned} \tag{11}$$

In addition, we also notice that

$$\sum_{i=1}^{c} \mathrm{cov}(l_x^T X^{(i)} r_x, l_y^T Y^{(i)} r_y) = \sum_{i=1}^{c} \mathrm{cov}(r_x^T X^{(i)T} l_x, r_y^T Y^{(i)T} l_y).$$

Thus, the optimization problem (7) can be also rewritten as

$$\begin{aligned} \max \quad & r_x^T C_{xy}^l r_y \\ s.t. \quad & r_x^T C_{xx}^l r_x = r_y^T C_{yy}^l r_y = 1, \end{aligned} \tag{12}$$

where

$$C_{xy}^l = \sum_{i=1}^{c} \left\langle X^{(i)T} l_x l_y^T Y^{(i)} \right\rangle = \sum_{i=1}^{c} \frac{1}{n_i} \sum_{j=1}^{n_i} X_j^{(i)T} l_x l_y^T Y_j^{(i)}, \tag{13}$$

$$C_{xx}^l = \sum_{i=1}^{c} \left\langle X^{(i)T} l_x l_x^T X^{(i)} \right\rangle = \sum_{i=1}^{c} \frac{1}{n_i} \sum_{j=1}^{n_i} X_j^{(i)T} l_x l_x^T X_j^{(i)}, \tag{14}$$

$$C_{yy}^l = \sum_{i=1}^{c} \left\langle Y^{(i)T} l_y l_y^T Y^{(i)} \right\rangle = \sum_{i=1}^{c} \frac{1}{n_i} \sum_{j=1}^{n_i} Y_j^{(i)T} l_y l_y^T Y_j^{(i)}. \tag{15}$$

In S2CCA, we first fix right transforms $r_x$ and $r_y$, and then solve optimization problem (11) to obtain left transforms $l_x$ and $l_y$. By the Lagrange multipliers technique, the problem (11) can be converted into the following generalized eigenvalue problem:

$$\begin{bmatrix} 0 & C_{xy}^r \\ C_{yx}^r & 0 \end{bmatrix} \begin{bmatrix} l_x \\ l_y \end{bmatrix} = \lambda \begin{bmatrix} C_{xx}^r & 0 \\ 0 & C_{yy}^r \end{bmatrix} \begin{bmatrix} l_x \\ l_y \end{bmatrix} \tag{16}$$

Alternately, we fix left transforms $l_x$ and $l_y$, and then solve optimization problem (12) to obtain right transforms $r_x$ and $r_y$. Likewise, the problem (12) can be transformed into the following generalized eigenvalue problem:

$$\begin{bmatrix} 0 & C_{xy}^l \\ C_{yx}^l & 0 \end{bmatrix} \begin{bmatrix} r_x \\ r_y \end{bmatrix} = \lambda \begin{bmatrix} C_{xx}^l & 0 \\ 0 & C_{yy}^l \end{bmatrix} \begin{bmatrix} r_x \\ r_y \end{bmatrix}. \tag{17}$$

We select the top $e_1$ eigenvectors of the generalized eigenvalue problem in (16) to form left transformation matrices $L_x \in \Re^{m_x \times e_1}$ and $L_y \in \Re^{m_y \times e_1}$, and the top $e_2$ eigenvectors of the generalized eigenvalue problem in (17) to determine right transformation matrices $R_x \in \Re^{n_x \times e_2}$ and $R_y \in \Re^{n_y \times e_2}$.

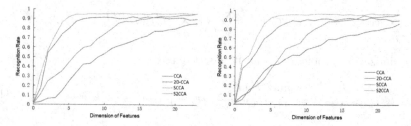

**Fig. 1.** Recognition rates of CCA, 2D-CCA, SCCA, and our S2CCA with 1NN classifier and different number of training samples on the AR database, where the left is the $l = 10$ case and the right is the $l = 12$ case.

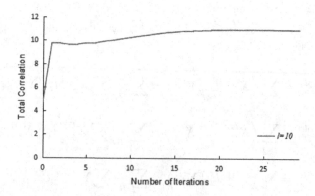

**Fig. 2.** Total correlation versus iteration number when $l = 10$ on AR database.

## 4    Experiment

In order to verify the effectiveness of the proposed S2CCA, we perform a number of experiments on the AR, AT&T, and CMU PIE face databases and compare it with existing related methods including CCA, supervised CCA (SCCA) [12], and 2D-CCA. To obtain multi-view data, we extract wavelet features [9] from original image data as the first set of features, and use the original image data to form the second set of features. For CCA and SCCA, we separately use PCA to reduce the dimensionality of the two-set features to 150. For 2D-CCA and S2CCA, we directly use the two sets of features. Finally, the 1-Nearest-Neighbor (1NN) classifier is applied for classification.

### 4.1  AR Face Database

The AR database consists of more than 4,000 color images of 126 people. All these image data are the frontal view of the face with different expressions, illumination conditions and characteristic changes. In our experiment, we select 120 persons and each of which has 14 unobstructed images. All images are grayscale and resized by $50 \times 40$ pixels.

On this database, we separately choose the first $l = 10$ and $l = 12$ images per person for training and the remaining images for testing. We use CCA, SCCA, 2D-CCA, and the proposed S2CCA for feature extraction. As seen in Fig. 1, our S2CCA method overall achieves better recognition rates than CCA, SCCA, and 2D-CCA, regardless of the variation of dimensions. In addition, Fig. 2 shows the convergence curve of our S2CCA. It is obvious that S2CCA only takes a few iterations to reach the convergence on the whole.

### 4.2  AT&T Face Database

The AT&T face database consists of 400 images from 40 people, and each person has 10 grayscale images taken under different lighting conditions, facial expressions, facial details and at different times. Each image of the resolution is $92 \times 112$ and the head in the image is slightly titled or rotated.

On this database, we separately select the first $l = 5$ and $l = 6$ images per class for training and the rest for testing. The 1NN classifier is used to evaluate recognition performances of CCA, SCCA, 2D-CCA, and our proposed S2CCA. As seen in Fig. 3, our S2CCA method overall outperforms again other methods on different training cases. The trend of recognition rates in 2D-CCA is not stable. Moreover, Fig. 4 shows the convergence curve of our S2CCA. Clearly, our S2CCA is of fast convergence after some iterations, which is in accordance with that obtained from the previous experiment in Sect. 4.1.

### 4.3  CMU PIE Database

The CMU PIE face database consists of 41,368 images of 68 individuals, each of whom has thirteen poses, forty-three different conditions, and four different expressions. We select the subset containing 3329 images of 68 individuals and each individual approximately has 49 images. Each image is grayscaled and cropped to $32 \times 28$ pixels.

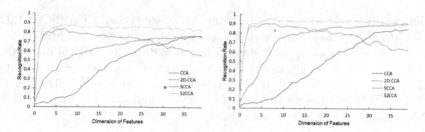

**Fig. 3.** Recognition rates of CCA, 2D-CCA, SCCA, and our S2CCA with 1NN classifier and different number of training samples on the AT&T database, where the left is the $l = 5$ case and the right is the $l = 6$ case.

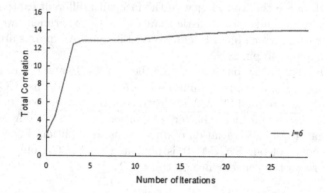

**Fig. 4.** Total correlation versus iteration number when $l = 6$ on AT&T database.

On this database, we separately select the first $l = 20$ and $l = 30$ images of each individual for training and the remaining for testing. We use CCA, SCCA, 2D-CCA, and our S2CCA to extract low-dimensional features. As seen in Fig. 5, our S2CCA method and 2D-CCA performs better than CCA and SCCA on the whole. 2D-CCA performs comparably with our S2CCA. Figure 6 shows that our S2CCA takes a few iterations for convergence.

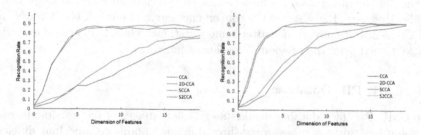

**Fig. 5.** Recognition rates of CCA, 2D-CCA, SCCA, and our S2CCA with 1NN classifier and different number of training samples on the CMU PIE database, where the left is the $l = 20$ case and the right is the $l = 30$ case.

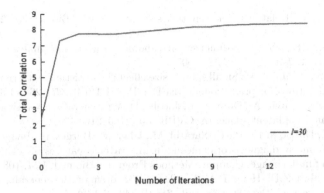

**Fig. 6.** Total correlation versus iteration number when $l = 30$ on CMU PIE database.

## 5  Conclusion

In this paper, a supervised 2D-CCA method has been proposed referred as S2CCA for feature extraction and classification, which simultaneously takes the prior class information of intra-view and inter-view samples into account and extracts more discriminative features. S2CCA has the following three advantages: (1) effectively reducing computational complexity; (2) minimizing the loss of intrinsic spatial structure information; (3) enhancing the discriminative power of low-dimensional features. The experiments on the AR, AT&T, and CMU PIE face databases show that our method has better recognition performance than existing multiview feature extraction methods.

**Acknowledgments.** This work is supported by the National Natural Science Foundation of China under Grant Nos. 61402203, 61472344, 61611540347, 61703362, Natural Science Fund of Jiangsu under Grant Nos. BK20161338, BK20170513, and Yangzhou Science Fund under Grant Nos. YZ2017292, YZ2016238. Moreover, it is also sponsored by the Excellent Young Backbone Teacher (Qing Lan) Fund and Scientific Innovation Research Fund of Yangzhou University under Grant No. 2017CXJ033.

## References

1. An, L., Bhanu, B.: Face image super-resolution using 2D CCA. Sig. Process. **103**, 184–194 (2014)
2. Andrew, G., Arora, R., Bilmes, J.A., Livescu, K.: Deep canonical correlation analysis. In: ICML, pp. 1247–1255 (2013)
3. Gao, X., Sun, Q., Xu, H., Li, Y.: 2D-LPCCA and 2D-SPCCA: two new canonical correlation methods for feature extraction, fusion and recognition. Neurocomputing **284**, 148–159 (2018)
4. Hardoon, D.R., Szedmak, S.R., Shawe-Taylor, J.R.: Canonical correlation analysis: an overview with application to learning methods. Neural Comput. **16**(12), 2639–2664 (2004)

5. Hotelling, H.: Relations between two sets of variates. Biometrika **28**, 321–377 (1936)
6. Lee, S.H., Choi, S.: Two-dimensional canonical correlation analysis. IEEE Sig. Process. Lett. **14**(10), 735–738 (2007)
7. Raudys, S.J., Jain, A.K.: Small sample size effects in statistical pattern recognition: recommendations for practitioners. IEEE T-PAMI **13**(3), 252–264 (1991)
8. Sharma, A., Kumar, A., Daume, H., Jacobs, D.W.: Generalized multiview analysis: a discriminative latent space. In: CVPR, pp. 2160–2167 (2012)
9. Singh, A., Dutta, M.K., ParthaSarathi, M., Uher, V., Burget, R.: Image processing based automatic diagnosis of glaucoma using wavelet features of segmented optic disc from fundus image. Comput. Methods Programs Biomed. **124**, 108–120 (2016)
10. Sun, Q.S., Liu, Z.D., Heng, P.A., Xia, D.S.: A theorem on the generalized canonical projective vectors. Pattern Recognit. **38**(3), 449–452 (2005)
11. Sun, T., Chen, S., Yang, J., Shi, P.: A novel method of combined feature extraction for recognition. In: ICDM, pp. 1043–1048 (2008)
12. Yuan, Y., Lu, P., Xiao, Z., Liu, J., Wu, X.: A novel supervised CCA algorithm for multiview data representation and recognition. In: CCBR, pp. 702–709 (2015)

# Word, Text and Document Processing

# Constructing Pseudo Documents with Semantic Similarity for Short Text Topic Discovery

Heng-yang Lu, Yun Li, Chi Tang, Chong-jun Wang[✉], and Jun-yuan Xie

National Key Laboratory for Novel Software Technology, Nanjing University,
Nanjing, China
{hylu,ctang}@smail.nju.edu.cn, liycser@gmail.com,
{chjwang,jyxie}@nju.edu.cn

**Abstract.** With the popularity of the Internet, short texts become common in our daily life. Data like tweets and online Q&A pairs are quite valuable in application domains such as question retrieval and personalized recommendation. However, the sparsity problem of short text brings huge challenges for learning topics with conventional topic models. Recently, models like Biterm Topic Model and Word Network Topic Model alleviate the sparsity problem by modeling topics on biterms or pseudo documents. They are encouraged to put words with higher semantic similarity into the same topic by using word co-occurrence. However, there exist many semantically similar words which rarely co-occur. To address this limitation, we propose a model named SEREIN which exploits word embeddings to find more comprehensive semantic representations. Compared with existing models, we improve the performance of topic discovery significantly. Experiments on two open-source and real-world short text datasets also show the effectiveness of involving word embeddings.

**Keywords:** Topic model · Word embeddings · Short text

## 1 Introduction

Short text has become quite popular with the explosive development of the Internet. Data examples include tweets, news headlines and online Q&A pairs. These massive data are of great value but we can hardly analyse them directly. Topic model, which can represent each document with a topic-level vector, is able to organize and summarize digital data automatically. probabilistic Latent Semantic Analysis (pLSA) [5] and Latent Dirichlet Allocation (LDA) [3] are two conventional topic models and are widely used for topic discovery. However, they are designed for long text. When it comes to short text, they will suffer from the sparsity problem brought by lack of words.

There exist several strategies to address this problem. It is intuitive to aggregate similar texts to form a longer one [7]. This solution over-relies on extra data

© Springer Nature Switzerland AG 2018
L. Cheng et al. (Eds.): ICONIP 2018, LNCS 11305, pp. 437–449, 2018.
https://doi.org/10.1007/978-3-030-04221-9_39

which may be unavailable in some cases. It is also common to add restrictions on model assumptions, such as each document only has one topic, known as Dirichlet Multinomial Mixture (DMM) [19]. But this kind of restriction is too strong because some short text documents can also contain several topics. Another one is to learn topic distributions over word pairs or word groups instead of original corpus. Biterm Topic Model (BTM) [18] and Word Network Topic Model (WNTM) [21] are two representative models. They only treat co-occurrent words as semantically similar ones. But semantic similarity is not only determined by the word co-occurrence. Examples from Online Question dataset are as Table 1 shows. They are all semantically similar but rarely co-occurrent words.

**Table 1.** Examples of semantically similar words and their co-occurrence

| Word-1 (Frequency) | Word-2 (Frequency, co-occurrence with Word-1) |
|---|---|
| Essay (162) | Dissertation (12, 2) |
| Muscle (204) | Dumbells (13, 3) |
| Trading (106) | Trader (14, 2) |

We believe that capturing these latent semantically similar words is necessary for the following two reasons: (1) involving latent semantically similar words can further increase the length of pseudo documents, which we will introduce later, and is beneficial for alleviating the sparsity problem; (2) these latent semantically similar words have better topic coherent representations than simply using co-occurrent relationship. Based on this fact, we propose a short text topic model with word semantic representations involved (SEREIN) to discover better topics. The main contributions include:

- We propose a short text topic model by alleviating the sparsity problem from the perspective of constructing pseudo documents with word embeddings.
- We involve semantic representations in pseudo documents construction procedure, including quantifying the relationship between co-occurrent words with similarity, discovering semantically similar but not co-occurrent words by arithmetic relationship of word vectors and involving words with high semantic similarity calculated by word vectors.
- We discover word embeddings' positive effects on SEREIN and show the utility of word embeddings.

This paper is organized as follows. Section 2 shows related researches. Section 3 presents our short text topic model named SEREIN. Section 4 contains the experiments as well as analysis and finally Sect. 5 concludes.

## 2   Related Work

Recently, short text topic model has attracted much attention. Early works are mainly based on aggregation. For example: Hong and associates aggregated tweets which shared the same keywords [6]; Jin and associates enlarged short text documents with auxiliary related texts searched from the Internet [7]. This may fail if auxiliary data are limited or difficult to achieve. Some other researchers develop their models with extra assumptions: Yin and associates added the restriction that each document should only contain one topic [19]; Lin and associates assumed that each document would contain the most related subset of topics [10]. These restrictions are too strong and the rationality of assumptions overly depends on the content of the corpus. A more popular strategy is to use word co-occurrence for learning, such as BTM [18] RIBS-TM [12] and WNTM [21]. WNTM constructs pseudo documents for each word in the corpus by using all words co-occurrent with it. This is based on the idea that if two words appear in the same context, they are more likely to share the same topic. Instead of learning topic distributions over the original short-text documents, WNTM learns topic distributions over pseudo documents for each word. Because each pseudo document is composed of co-occurrent words, the length is longer than that in the original document, thus the sparsity problem can be relieved to some extent. However, WNTM still ignores some semantically similar but rarely co-occurrent words.

Word embeddings, early introduced in [16], can involve useful syntactic and semantic properties in each dimension of learned word vectors. This kind of method usually uses deep neural network to learn vectors by predicting words with their context [2,13]. So it can offer dense, low-dimensional and real-valued word vectors with semantic information. There exist some methods [8,15] utilizing word embeddings for topic discovery. For example, GPU-DMM [8] extends DMM by involving word embeddings. These models have too strong restrictions on model assumptions brought by DMM. Different from existing models, we solve the sparsity problem from another perspective of constructing pseudo documents by utilizing word embeddings.

## 3   SEREIN Topic Model

The problem setting of topic discovery for short text is as Definition 1 shows.

**Definition 1.** *Given the corpus $D$ with $N_D$ documents whose vocabulary size is $N_W$, topic model aims to discover topics of each document and learn topic representations with words. If the corpus has $K$ topics, topic model should give an $N_D \times K$ matrix $\theta$ for topic distributions over documents and a $K \times N_W$ matrix $\phi$ for word distributions over topics by learning observed words. In short text scenarios, each document consists of only few words.*

### 3.1   Model Description

SEREIN learns topic distributions over pseudo documents, which are constructed according to semantic similarity between words. Following is its generative procedure:

1. For each word $w_i$, learn semantic vector $\overrightarrow{w_i}$ and construct pseudo document $pd_i$ for $w_i$ with semantically similar words, denoted as $z_i^{PD}$.
2. For each $z$, draw $\phi_z^{PD} \sim$ Dirichlet $(\beta)$.
3. Draw $\theta_i^{PD} \sim$ Dirichlet $(\alpha)$, treated as topic distributions over word $w_i$.
4. For each word $w_j \in pd_i$
   (a) draw $z_j^{PD} \sim \theta_i^{PD}$.
   (b) draw the semantically similar word $w_j \sim \phi_{z_j}^{PD}$.

where $pd_i$ contains all semantically similar words for $w_i$, $\alpha$ and $\beta$ are the symmetric Dirichlet priors for $\theta$ and $\phi$.

LDA generates a collection of documents by using topics and words under those topics. Different from that, SEREIN learns to generate pseudo documents, which are long enough, for each word by using semantically similar words.

## 3.2 Learn Word Semantic Similarity

Most existing short text topic models lack quantifiable description for the semantic similarity between words. Fortunately, word embeddings have shown effectiveness and are successfully applied in NLP tasks [4, 20]. Inspired by the recent work which uses word-embeddings procedure to further learn semantic similarity [8], we also choose Word2Vec [14] to learn word representations for two reasons:

- Context-predict models can obtain more semantic information than count-based models [1].
- The arithmetic ability of vectors learned by Word2Vec can help mine latent semantically similar words [17], which can enrich the pseudo documents.

In SEREIN, we calculate the semantic similarity with *cosine similarity* just as same as many related works [11,14]. Given word representations $\overrightarrow{w_i}, \overrightarrow{w_j}$ for word $w_i, w_j$, we can have the semantic similarity $sim(i,j) = \frac{\overrightarrow{w_i} \overrightarrow{w_j}}{\|\overrightarrow{w_i}\| \|\overrightarrow{w_j}\|}$. We will use this learned knowledge for pseudo document construction.

## 3.3 Construct Pseudo Documents

In SEREIN, pseudo documents are designed to describe each word with a collection of semantically similar words. So we can learn topic distributions over pseudo documents to infer original ones. Most existing works only use word co-occurrence to describe semantic similarity. But there exist many semantically similar words which don't co-occur. We incorporate this factor to construct pseudo documents for SEREIN. The pseudo documents are as Definition 2 shows. And some notations we will use are listed in Table 2.

**Definition 2.** *Given the corpus $D$ with $N_D$ documents whose vocabulary size is $N_W$. The goal is to construct pseudo documents $PD$ whose size $N_{PD}$ exactly equals to the vocabulary size of $D$. For each pseudo document $pd_i \in PD$, $pd_i$ refers to a word collection semantically describes word $w_i$. It is composed of a weighted co-occurrent word list $L_{cooccur}(i)$, a latent semantically similar word list $L_{latent}(i)$ and a high similarity word list $L_{similar}(i)$.*

**Table 2.** Notations used for pseudo document construction

| Variable | Description |
|---|---|
| $S_{cooccur}(i)$ | A set contains all words co-occur with $w_i$ |
| $Count(i, j)$ | The co-occurrent count of $w_i$ and $w_j$ |
| $Avr_i$ | The average co-occurrent count for $w_i$ |
| $\tau(\vec{a}, \delta)$ | Return the most similar word with vector $\vec{a}$ with similarity larger than $\delta$ |

We can divide the construction procedures into three phases. The first phase aims to utilize co-occurrent relationship. $L_{cooccur}(i)$ is a set of words which co-occur with $w_i$ in a certain frequency. The co-occurrent frequency of $w_i$ and $w_j$, denoted as $fr_{i,j}$, can be defined as Eq. (1) shows.

$$fr_{i,j} = \lceil \sigma(sim(i,j)) \times Count(i,j)/Avr_i \rceil. \tag{1}$$

where $Avr_i = \frac{\sum_{w_j \in S_{cooccur}(i)} Count(i,j)}{|S_{cooccur}(i)|}$ and $\sigma$ is a sigmoid function.

$fr_{i,j}$ is to reflect the strength of co-occurrence whose value means the frequency $w_j$ appears in $L_{cooccur}(i)$. A higher value indicates a stronger similarity. The $\sigma$ part of $fr_{i,j}$ is designed to involve semantic similarity knowledge, where a higher $sim(i,j)$ value will increase the value of $fr_{i,j}$. The rest part of $fr_{i,j}$ is based on co-occurrent count, we use $Avr_i$ to reduce the length of $L_{cooccur}(i)$ without losing co-occurrent information.

The second phase aims to find latent semantic similar words. $L_{latent}(i)$ is such a word list which contains all latent semantically similar words of $w_i$. We can capture various syntactical and semantic relationship by mathematical operations [17] with the arithmetic ability of Word2Vec. For example, given the learned word vectors, we can find arithmetic relationship like $vec(phone) + vec(song) \approx vec(ringtone)$. This example indicates that addictive operation of vectors can produce meaningful results. The learning procedure gives the theoretical explanation of addictive property. It's because skip-gram learns word vector representations by predicting words in a context and words in the analogous context may obtain similar vector representations. So the sum of two words is proportional to composite of the two corresponding context distributions, which may be semantically similar to these two words. Based on this idea, we propose an approach to find a latent word $w_{latent}$ for $w_i$, as Eq. (2) shows.

$$w_{latent} = \tau(\vec{w_i} + \vec{w_j}, \delta). \tag{2}$$

where $w_j \in S_{cooccur}(i)$ and $w_{latent} \notin S_{cooccur}(i)$. For $\tau(\vec{a})$, We use *cosine similarity* as a metric to find the most similar word representation with $\vec{a}$. $\delta$ is a threshold parameter to filter semantic-irrelevant words. After traversing $S_{cooccur}(i)$, we can find all latent words for $w_i$.

The third phase aims to ensure the words of each pseudo document are not too few. If $w_i$ has few co-occurrent words in the original corpus, $pd_i$ may still lack words after the above two phases. This situation may have bad influence on using topic model. So we involve $L_{similar}(i)$, which contains the top $M$ most similar words for each $w_i(i \in [1, N_W])$ according to $sim(i, j)$ we've learned.

Algorithm 1 shows the construction of pseudo document for SEREIN. By applying it, pseudo documents are long enough to discover topic distributions.

---

**Algorithm 1.** Pseudo document construction algorithm for SEREIN

---

**Input:**     Corpus $D$, word representations $\overrightarrow{W}$, context window size $l$, number of most similarly words $M$, threshold $\delta$.
**Output:**     Pseudo Documents $PD$.
  Traverse $D$ to get $S_{cooccur}$ and $Count$.
  **for** $i \leftarrow 1$ to $W$ **do**
    Calculate $Avr_i$.
    **for** every $w_j \in S_{cooccur}(i)$ **do**
      Calculate $fr_{i,j}$.
      Calculate $\overrightarrow{w_i} + \overrightarrow{w_j}$ to find $w_{latent}$.
      Add $w_j$ to $L_{cooccur}(i)$ with $fr_{i,j}$ times, Add $w_{latent}$ to $L_{latent}(i)$.
    **end for**
    Add the top $M$ most similar words to $L_{similar}(i)$.
    Construct $pd_i$ with $L_{cooccur}(i)$, $L_{latent}(i)$ and $L_{similar}(i)$.
    Add $pd_i$ to $PD$.
  **end for**

---

Figure 1 is a simple example for better understanding the construction procedure. Solid lines connect two co-occurrent words like $(A, B)$. There are also two kinds of dotted lines in this figure. One represents connecting latent semantic similar words through mathematical operations, like $D$ is captured by $A + B$. The other one which connects $A$ and $(G, K, M)$ represents connecting the most semantically similar words. Finally, we can bring more semantically similar words in pseudo document $pd_A$. Note that, all the words we used to construct pseudo documents are from $\mathbb{D}$, so we will not bring noise to SEREIN.

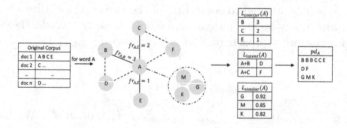

**Fig. 1.** An example for pseudo document construction for word $A$

### 3.4  Topic Inference

Because SEREIN doesn't model topics on the original documents, we have to infer the topic distributions with knowledge learned from pseudo documents. Equation (3) shows the derivation of topic $z_k$'s proportion of a document $d \in D$.

$$p(z|d) = \sum_{w_i \in W_d} p(z|w_i)p(w_i|d). \tag{3}$$

where $W_d$ is a word set for all words in $d$. $p(z|w_i)$ denotes the topic distributions over word $w_i$. Because the pseudo document $pd_i$ is the semantic representation of word $w_i$, we think the topic distributions over pseudo document $pd_i$ can stand for the topic distributions over word $w_i$, so we can get $p(z|w_i) = \theta_{i,z}^{PD}$. $\theta_{i,z}^{PD}$ is the topic proportion of topic $z$ learned from pseudo document $pd_i$.

$p(w_i|d)$ refers the word distributions over document. We can simply treat it as a counting problem, calculation is as Eq. (4) shows.

$$p(w_i|d) = \frac{n_d(w_i)}{|d|}. \tag{4}$$

where $n_d(w_i)$ is the frequency of word $w_i$ in document $d$ and $|d|$ is the number of words in it. So the final topic inference is calculated as Eq. (5) shows.

$$\theta_{d,z}^{D} = p(z|d) = \sum_{w_i \in W_d} \theta_{i,z}^{PD} \frac{n_d(w_i)}{\text{Size}(d)}. \tag{5}$$

## 4  Experiments and Analysis

We use two open-source and real-world datasets for experiment. Online Questions dataset[1] is offered by Yahoo! Research. Each question is attached with a label according to the forum it was posted. We have over 80,000 question contents for experiments. There are 24 categories in the corpus, the vocabulary size is 9696 and the average length of a single question is 4.950 words. Online News dataset[2] is offered by UCI Machine Learning Repository. We have over 170,000 news headlines for experiments. Each headline is annotated with a category by the provider. There are 4 categories in the corpus, the vocabulary size is 7247 and the average length of a single question is 6.876 words.

### 4.1  Experiment Settings

We compare SEREIN with four baseline topic models:

- LDA is a famous topic model which performs really well in long text scenarios. We use a standard open-source LDA implemented by Gibbs sampling.

---

[1] https://webscope.sandbox.yahoo.com/catalog.php?datatype=l&did=10.
[2] http://archive.ics.uci.edu/ml/datasets/News+Aggregator.

- BTM is a recently proposed topic model for short text. We do experiments with code provided by the authors[3].
- WNTM is a state-of-the-art short text topic model, which construct pseudo documents with word co-occurrence. We implement this model by ourselves.
- GPU-DMM is a state-of-the-art topic model for short text with auxiliary knowledge learned by word2vec. We use code provided by the authors[4].

To show the advantages of using semantic similarity, we remain parts of pseudo document construction for ablation studies:

- SEREIN-1 constructs pseudo documents only with weighted co-occurrent word list $L_{cooccur}$.
- SEREIN-2 constructs pseudo documents with both $L_{cooccur}$ and latent semantically similar word list $L_{latent}$.
- SEREIN-3 constructs pseudo documents with both $L_{cooccur}$ and word list $L_{similar}$ which contains top $M$ semantically similar words.

Parameter settings are as follows. For pseudo document construction, we set context window size $l = 4, \delta = 0.7$ for $L_{latent}$ and $M = 10$ for $L_{similar}$. For training documents, we set $\alpha = 50/K, \beta = 0.1$ and iteration $= 1000$ for all models. For learning word representations, we set layerSize $= 200$, window $= 8$, sampling threshold $= 1e-4$ and learning rate $= 0.025$.

## 4.2 Quality of Topic Discovery

The most important goal of topic model is to discover topics. This experiment is designed to show SEREIN has a better performance in topic discovery than baselines. A good topic should consist of words in cohesive semantic similarity. A simple but rational way is intuitive display, so we select two representative topics from two datasets respectively when $K = 10$ and display 10 words with the highest probability. Italic words are topic irrelevant comprehensively judged by volunteers, as Table 3 shows.

**Table 3.** Topic display of Online Questions dataset and Online News dataset

| QUESTIONS. TOPIC FINANCE | | | | | QUESTIONS. TOPIC COMMUNICATION | | | | | NEWS. TOPIC HEALTH | | | | | NEWS. TOPIC FILM & TV SERIES | | | | |
|---|---|---|---|---|---|---|---|---|---|---|---|---|---|---|---|---|---|---|---|
| LDA | BTM | WNTM | GPU-DMM | SEREIN | LDA | BTM | WNTM | GPU-DMM | SEREIN | LDA | BTM | WNTM | GPU-DMM | SEREIN | LDA | BTM | WNTM | GPU-DMM | SEREIN |
| money | money | estate | money | loan | phone | yahoo | yahoo | phone | email | health | health | disease | health | disease | season | season | *vii* | movie | episode |
| business | credit | county | credit | estate | *person* | real | phone | yahoo | yahoo | *prices* | *study* | alzheimers | *study* | mers | game | game | batman | *box* | *vii* |
| credit | business | tax | business | investment | *child* | baby | cell | music | messages | *study* | cancer | mers | cancer | salmonella | day | video | episode | *office* | season |
| online | job | airport | job | tax | internet | day | *game* | ipod | e-mail | *oil* | *change* | patients | *climate* | patients | thrones | thrones | potter | trailer | marvels |
| *bad* | buy | loan | card | capital | cell | *month* | email | computer | message | *gas* | risk | infections | risk | infections | *chris* | watch | featurette | film | trilogy |
| buy | company | taxes | buy | fund | address | *create* | send | *songs* | phone | cancer | *climate* | *breast* | report | alzheimers | *review* | *awards* | *dawn* | america | ninja |
| card | pay | mortgage | company | debt | email | *stay* | *play* | download | mail | risk | virus | measles | *change* | *breast* | *record* | trailer | ninja | xmen | captain |
| company | *online* | rent | car | trading | send | *search* | address | cell | text | *paul* | report | *protein* | virus | obesity | finale | finale | casting | captain | batman |
| *calculate* | real | credit | pay | income | care | web | *search* | dvd | search | *dvd* | *west* | cells | *west* | epidemic | *dead* | movie | apes | *review* | thrones |
| pay | card | *filed* | *online* | mortgage | *family* | site | *change* | video | messenger | life | *obama* | *mosquitoes* | drug | symptoms | *recap* | *recap* | thrones | star | superman |

[3] https://github.com/xiaohuiyan/BTM.
[4] https://github.com/NobodyWHU/GPUDMM.

SEREIN is quite good at discovering topics. The cohesive semantic similarity among each topic is very high. We believe this is attributed to semantic representations involved. For example, in 'topic communication', WNTM selects word *change* mistakenly. This may due to the co-occurrence with word *email*, which is up to 25 times. Instead, word *messenger* rarely co-occurs with *email*, but they share high similarity, which helps SEREIN to select them into the same topic.

## 4.3    Performance of Clustering and Classification

Clustering is to divide unlabeled documents into clusters. We use the same method as BTM and WNTM does: Take each topic $z$ as a cluster, then assign each document $d$ to the topic cluster $z$ according to the highest value of conditional probability $P(z|d)$. We set $K$ from 10 to 30 with step size as 5. We use purity and entropy as evaluation metrics. Results are as Fig. 2 shows.

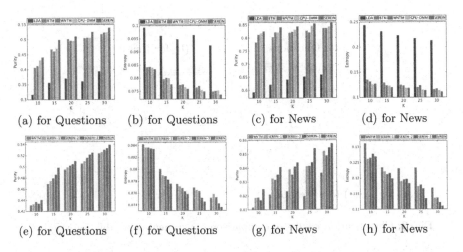

(a) for Questions    (b) for Questions    (c) for News    (d) for News

(e) for Questions    (f) for Questions    (g) for News    (h) for News

**Fig. 2.** Clustering results of comparison experiments and ablation studies

For purity, a larger value indicates a better performance. For entropy, the smaller the better. Figure 2(a)–(d) are comparisons between baselines and SEREIN. For both purity and entropy metrics, LDA performs much worse than other four short text topic models. This result indicates that the sparsity problem will affect the performance of conventional topic model significantly. Compared with other three short text topic models, no matter what value $K$ is, SEREIN can get best performance in most cases. This means SEREIN can extract higher-quality and stable topics to cluster documents. We think the better performance of SEREIN benefits from using semantic similarity. So we conduct ablation studies, as Fig. 2(e)–(h) shows. Take $K = 15$ in Question dataset as an example, SEREIN-1 uses semantic similarity to quantify co-occurrent relationship, which indeed improves the clustering performance to some extent compared with simply using co-occurrence like WNTM does. SEREIN-2 involves semantically similar words captured by arithmetic relationship and SEREIN-3 involves words with high similarity calculated by word vectors. The purity and entropy results show that both factors have a positive effect on improving clustering performance.

Classification aims to predict a label for each document by learning from labeled ones. We use topic distributions as features of documents and naïve Bayes as classification algorithm. Results are as Fig. 3 shows.

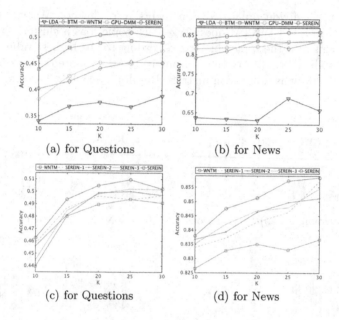

(a) for Questions

(b) for News

(c) for Questions

(d) for News

**Fig. 3.** Classification results of comparison experiments and ablation studies

Figure 3(a) and (b) compare accuracy for both datasets. We find the performance of BTM and GPU-DMM is sensitive to different datasets. They perform quite badly on Questions dataset but can achieve approximately the same results as WNTM does on News dataset. SEREIN can stably have advantages over other models. It shows the effectiveness of constructing pseudo documents and makes us believe that the topic distributions learned by SEREIN can describe documents much better. Figure 3(c) and (d) are ablation studies to show the effectiveness of involving semantic representations. Take 3(c) for example, when $K = 10$, the contribution ranking of improvement is SEREIN-1 > SEREIN-2 > SEREIN-3. When $K = 20$, it changes to SEREIN-2 > SEREIN-3 > SEREIN-1. And when $K = 30$, SEREIN-3 has the best improvement. This indicates that all $L_{cooccur}$, $L_{latent}$ and $L_{similar}$ can contribute to the pseudo document construction to some extent. It's essential to combine them together instead of using part of them.

## 4.4   Utility of Word Embeddings

Since SEREIN is naturally dependent on Word2Vec, this experiment aims to show how different word vectors learned by Word2Vec affect the performance of SEREIN. The quality of word vectors is closely related to the scale of training corpus, so we select documents whose amount ranges from about 300,000 to 1,500,000 for training word vectors respectively.

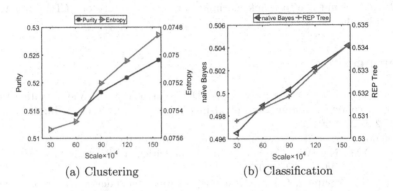

(a) Clustering                              (b) Classification

**Fig. 4.** Performance with the change of training data scale

Figure 4(a) is the clustering performance with the change of data scale. With the training corpus getting larger, the performance of both purity and entropy is getting better meanwhile. Figure 4(b) is the classification performance with the change of data scale. Changing trends of accuracy with both naïve Bayes and decision tree algorithms are consistent in most cases. Such a discovery is quite encouraging. It not only shows the utility of Word2Vec, but also indicates we can optimize SEREIN by using optimized word embeddings models. It offers a new idea to improve short text topic model and makes SEREIN more extensible.

## 5   Conclusion and Future Work

The Internet has totally changed our life style these years. People prefer to express their opinions through online social platforms. This may produce massive amount of valuable short text data. Unlike long text, short text data faces the sparsity challenge for lacking words. We propose a short text topic model called SEREIN, which not only takes word co-occurrence into consideration, but also utilizes word semantic similarity learned from Word2Vec. This model covers the semantic information ignored by most existing models, which makes SEREIN have better performance in both topic discovery and document characterization. In our experiments, we surprisingly find the performance of SEREIN is positively related to the Word2Vec. This may imply studies on short text topic model like optimizing our model can be undertaken from the perspective of improving word semantic representations. We think the follow-up work is promising because the dramatic development of neural networks may promote the research on word representations. We will also do some future researches based on this idea.

**Acknowledgments.** This paper is supported by the National Key Research and Development Program of China (Grant No. 2016YF- B1001102), the National Natural Science Foundation of China (Grant Nos. 61502227, 61876080), the Fundamental Research Funds for the Central Universities No.020214380040, the Collaborative Innovation Center of Novel Software Technology and Industrialization at Nanjing University. We also would like to thank machine learning repository of UCI [9] and Yahoo! Research for the datasets.

# References

1. Baroni, M., Dinu, G., Kruszewski, G.: Don't count, predict! A systematic comparison of context-counting vs. context-predicting semantic vectors. In: ACL, vol. 1, pp. 238–247 (2014)
2. Bengio, Y.: Neural net language models. Scholarpedia **3**(1), 3881 (2008)
3. Blei, D.M., Ng, A.Y., Jordan, M.I.: Latent Dirichlet allocation. J. Mach. Learn. Res. **3**(Jan), 993–1022 (2003)
4. Chen, D., Manning, C.D.: A fast and accurate dependency parser using neural networks. In: EMNLP, pp. 740–750 (2014)
5. Hofmann, T.: Probabilistic latent semantic indexing. In: Proceedings of the 22nd Annual International ACM SIGIR Conference on Research and Development in Information Retrieval, pp. 50–57. ACM (1999)
6. Hong, L., Davison, B.D.: Empirical study of topic modeling in Twitter. In: Proceedings of the First Workshop on Social Media Analytics, pp. 80–88. ACM (2010)
7. Jin, O., Liu, N.N., Zhao, K., Yu, Y., Yang, Q.: Transferring topical knowledge from auxiliary long texts for short text clustering. In: Proceedings of the 20th ACM International Conference on Information and Knowledge Management, pp. 775–784. ACM (2011)
8. Li, C., Duan, Y., Wang, H., Zhang, Z., Sun, A., Ma, Z.: Enhancing topic modeling for short texts with auxiliary word embeddings. ACM Trans. Inf. Syst. (TOIS) **36**(2), 11 (2017)
9. Lichman, M.: UCI machine learning repository (2013). http://archive.ics.uci.edu/ml
10. Lin, T., Tian, W., Mei, Q., Cheng, H.: The dual-sparse topic model: mining focused topics and focused terms in short text. In: Proceedings of the 23rd International Conference on World Wide Web, pp. 539–550. ACM (2014)
11. Liu, Y., Liu, Z., Chua, T.S., Sun, M.: Topical word embeddings. In: AAAI, pp. 2418–2424 (2015)
12. Lu, H., Xie, L.Y., Kang, N., Wang, C.J., Xie, J.Y.: Don't forget the quantifiable relationship between words: using recurrent neural network for short text topic discovery. In: AAAI, pp. 1192–1198 (2017)
13. Mikolov, T., Karafiát, M., Burget, L., Cernockỳ, J., Khudanpur, S.: Recurrent neural network based language model. In: Interspeech, vol. 2, p. 3 (2010)
14. Mikolov, T., Sutskever, I., Chen, K., Corrado, G.S., Dean, J.: Distributed representations of words and phrases and their compositionality. In: Advances in Neural Information Processing Systems, pp. 3111–3119 (2013)

15. Nguyen, D.Q., Billingsley, R., Du, L., Johnson, M.: Improving topic models with latent feature word representations. Trans. Assoc. Comput. Linguist. **3**, 299–313 (2015)
16. Rumelhart, D.E., Hinton, G.E., Williams, R.J., et al.: Learning representations by back-propagating errors. Cognit. Model. **5**(3), 1 (1988)
17. Wang, P., Xu, B., Xu, J., Tian, G., Liu, C.L., Hao, H.: Semantic expansion using word embedding clustering and convolutional neural network for improving short text classification. Neurocomputing **174**, 806–814 (2016)
18. Yan, X., Guo, J., Lan, Y., Cheng, X.: A biterm topic model for short texts. In: Proceedings of the 22nd International Conference on World Wide Web, pp. 1445–1456. ACM (2013)
19. Yin, J., Wang, J.: A Dirichlet multinomial mixture model-based approach for short text clustering. In: Proceedings of the 20th ACM SIGKDD International Conference on Knowledge Discovery and Data Mining, pp. 233–242. ACM (2014)
20. Zamani, H., Croft, W.B.: Relevance-based word embedding. In: Proceedings of the 40th International ACM SIGIR Conference on Research and Development in Information Retrieval, pp. 505–514. ACM (2017)
21. Zuo, Y., Zhao, J., Xu, K.: Word network topic model: a simple but general solution for short and imbalanced texts. Knowl. Inf. Syst. **48**(2), 379–398 (2016)

# Text Classification Based on Word2vec and Convolutional Neural Network

Lin Li, Linlong Xiao, Wenzhen Jin, Hong Zhu, and Guocai Yang[(⊠)]

School of Computer and Information Science, Southwest University,
Chongqing, China
cqkxxn@163.com, paul.g.yang@gmail.com

**Abstract.** Text representations in text classification usually have high dimensionality and are lack of semantics, resulting in poor classification effect. In this paper, TF-IDF is optimized by using optimization factors, then word2vec with semantic information is weighted, and the single-text representation model CD_STR is obtained. Based on the CD_STR model, the latent semantic index (LSI) and the TF-IDF weighted vector space model (T_VSM) are merged to obtain a fusion model, CD_MTR, which is more efficient. The text classification method MTR_MCNN of the fusion model CD_MTR combined with convolutional neural network is further proposed. This method first designs convolution kernels of different sizes and numbers, allowing them to extract text features from different aspects. Then the text vectors trained by the CD_MTR model are used as the input to the improved convolutional neural network. Tests on two datasets have verified that the performance of the two models, CD_STR and CD_MTR, is superior to other comparable textual representation models. The classification effect of MTR_MCNN method is better than that of other comparison methods, and the classification accuracy is higher than that of CD_MTR model.

**Keywords:** Text classification · Text representation · Word2vec
Convolutional neural network

## 1 Introduction

Most of the data generated by Internet are stored in a format of text, and text data occupy an important position. Manually organizing and managing textual information has been unable to adapt to the ever-expanding digital information of the Internet age. With such a large amount of data and a variety of data forms, finding the right method to effectively manage and use these text data is very important. Efficient feature extraction and text representation are challenges for text classification, and it is also the first problem that should be solved in text classification.

Most of the early text representation methods used were vector space models. Later, most researchers used the distributed representation of words [1], which was proposed by Hinton in 1986 and could overcome the shortcomings of the one-hot representation. Bengio proposed to use a three-layer neural network to train text representation model in 2003 [2]. Hinton used hierarchical ideas in 2008 to improve the training process from

© Springer Nature Switzerland AG 2018
L. Cheng et al. (Eds.): ICONIP 2018, LNCS 11305, pp. 450–460, 2018.
https://doi.org/10.1007/978-3-030-04221-9_40

the hidden layer to the output layer in the Bengio method, speeding up the training model [3]. Mikolov proposed a neural network model to train distributed word vectors in 2013. The training tool word2vec implemented by this model has been widely used [4, 5]. Hu used a convolutional neural network to extract semantic combination information from local words in a sentence through a neural network in 2014 [6].

We find traditional text representation methods, such as Boolean models and vector space models, have problems of data sparseness and dimensional disaster. With the rapid development of machine learning and deep learning technologies, researchers have begun to use various neural network to construct text representation models and map texts to low-dimensional continuous vectors through neural network, improving the model's representation ability. However, the existing neural network text representation model also has some problems. First of all, though the neural network obtains better semantic information for the text representation, its class distinction ability is lacking. Secondly, the existing method for extracting text features using convolutional neural network is based on the length of the longest text in the data set. Texts that are shorter than this length are filled with special characters. The introduction of too many non-semantic characters in this text affects the original information of the text and results in poor classification effect.

## 2   Related Work

Text representation is an important step in text classification. The original texts are unstructured data. You must find a suitable representation method to convert the text content into information that computer can recognize. The text representation mainly contains two aspects: representation and calculation, respectively referring to definition of feature selection and feature extraction, and the definition of computational weighting and semantic similarity [7]. The Vector Space Model (VSM) was proposed by Salton in the 1970 [8]. It simplifies the process of processing text content into vector operations in vector space, and expresses the similarity of text semantics by calculating spatial similarity. VSM is a common and very classic text representation. But the problem of dimension disaster exists in vector space model.

The Convolutional Neural Network (CNN) is a deep neural network that has made major breakthroughs in computer vision and speech recognition. It is widely used in image understanding [9, 10]. In recent years, continuous development of convolutional neural network has been used in natural language processing tasks such as text classification and element identification [11]. Wang proposed a semi-supervised convolutional neural network to enhance the semantic relevance of the context [12]. The c-lstm model proposed by Zhou, c-lstm uses the convolutional neural network to extract text sentence features and uses short-term memory recursive neural network to obtain sentence representations [13]. Lai proposed recursive convolutional neural network for text classification, which introduces less noise than traditional neural network [14].

## 3   Methods

### 3.1   Word2vec

Word2vec can train word vectors quickly and efficiently. There are two Word2vec models, CBOW model and Skip-gram model. The CBOW model uses the c words before and after the word w(t) to predict the current word; whereas the Skip-gram model does the opposite. It uses the word w(t) to predict the c words before and after it. The two model training methods are respectively shown in Fig. 1 left and right. This paper uses the CBOW model to train word vectors.

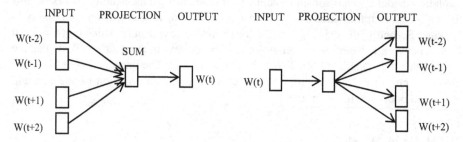

**Fig. 1.**  CBOW and Skip-gram

### 3.2   Convolutional Neural Network

**Convolution Layer**
The convolutional layer is also called feature extraction layer. This layer is the core part of the convolutional neural network and can describe the local characteristics of the input data. The convolution kernel $w \in Q^{hk}$ included in the convolution operation [11] will generate a new feature value each time it passes through a word sequence window with a height of $h$ and a width of $k$. For example, a feature point $c_i$ in a feature map is the result of the window $x_{i:i+h-1}$ after convolution operation, that is, each feature value can be obtained by formula 1:

$$c_i = f(w \cdot x_{i:i+h-1} + b) \tag{1}$$

Where, $x_i \in Q^k$, $w$ is the weight parameter of the convolution kernel; $b$ is the offset term of the convolution layer; and $f$ is a nonlinear activation function.

When training a convolutional neural network, it is necessary to establish a convolution kernel sliding stride, which can be set to be 1, 2 or more. The convolution kernel can convolve the input data to get a feature map, as shown in 2:

$$c = [c_1, c_2, \cdots, c_{s-h+1}] \tag{2}$$

**Pooling Layer**

The pooling layer is generally disposed between two consecutive convolution layers. The pooling layer down-samples the feature map of the convolutional layer output, aggregates the statistics of all the feature maps of the convolutional layer, simplifies the information output from the convolutional layer through the pooling layer, and reduces the features and network parameters.

### 3.3    The Idea of the CD_STR Model

The main idea of IDF in TF-IDF algorithm is: if there are fewer documents containing characteristic words $t$, larger IDF indicates that characteristic words $t$ have better category discrimination ability. The simple structure of IDF in TF-IDF cannot effectively reflect the importance of words and the distribution of feature words. The CD_STR model first considers that if a word appears in each text, and the frequency of occurrence in each text or in each type of text does not differ much, then the word contributes very little to the category distinction and should be filtered out or given a smaller weight. Conversely, the feature words should be given a higher weight value.

If there are three characteristic words $t_1$, $t_2$, $t_3$, in the three categories $c_1$, $c_2$, $c_3$, the distributions are (8, 8, 8), (5, 8, 5), (1, 8, 5). Then, the weights of these three feature words should be increased successively, because the frequency of $t_3$ in each category is relatively uneven compared to $t_1$, $t_2$. So, it will be better to distinguish categories.

CD_STR optimizes the TF-IDF algorithm mainly based on the distribution of feature words in various category, and specific improved algorithm is shown in formula 3:

$$G(t, d) = \frac{tf_{ij}(t,d) \times idf_i(t) \times \sum_k p(t|c_k)^2 p(c_k|t)^2}{\sqrt{\sum_{i=1}^{n} [tf_{ij}(t,d) \times idf_i(t)]^2}} \tag{3}$$

Among them, $m_{ij}$ is the number of occurrences of feature word $t_i$ appearing in the text $d_j$; $n_i$ is the total number of texts containing feature word $t_i$; $N$ is the total number of texts in the corpus; $F$ is a normalization factor; $\sum_k m_{kj}$ is the total number of feature word in the text $d_j$; and $p(t|c_k)$ is the probability that the feature word $t$ appears in the category $c_k \cdot p(c_k|t)$ is the conditional probability that the feature belongs to the category $c_k$ when the feature word appears.

Then use the optimized TF-IDF weighting word2vec to train the word vector, assign a weight to each feature word vector, and accumulate each weighted word vector according to the corresponding dimension to obtain the vector representation of each text, that is, updating each text vector according to the formula 4:

$$GW(d) = \sum_{t \in d} G(t,d) \times Word2vec(t) \tag{4}$$

### 3.4    The Idea of the CD_MTR Model

With the advantages of the TF-IDF weighted vector space model, the LSI model and the CD_STR, they express the text information in different ways. Therefore, the three single-text representation models are combined to allow the three models to complement each other and to better express the content of the text. Thus, a text representation model (CD_MTR) that integrates multiple models is proposed.

The main idea of the CD_MTR model is that each single-text representation model selects an appropriate dimension to vectorize the original text and obtains three different sets of text representation vectors. The union of the text vectors corresponding to these text vector sets is determined as the final text vector. The specific solution is shown in Eq. 5:

$$CD\_MTR(d_i) = LSI(d_i) \oplus T\_VSM(d_i) \oplus CD\_STR(d_i) \tag{5}$$

Among them, $d_i$ is the $i$th text in the data set $D$; $\oplus$ is a splice operator; $LSI(d_i)$ is the text vector representation obtained from the LSI model training text $d_i$; $T\_VSM(d_i)$ is vector representation obtained by the TF-IDF weighted vector space model; and $CD\_STR(d_i)$ is the text vector representation obtained by the CD_STR model training text $d_i$.

### 3.5    Structural Improvements for Convolutional Neural Network (MCNN)

In general, only one convolution kernel is included in each convolution layer, and the number of convolution kernels is set to a fixed value. For text data, in order to take the contextual information of each feature word in the text into consideration, a variety of convolution kernels of different sizes can be designed. In this paper, three convolution kernels with different sizes are designed as $3 \times 360$, $4 \times 360$ and $5 \times 360$. The respectively number of corresponding convolution kernels are 150, 100, and 50.

### 3.6    CD_MTR Combined with Convolutional Neural Network (MTR_MCNN)

For the input features of convolutional neural network, the references [11, 15–17] use the length of the longest text in all the texts of the data set as a benchmark, and the rest of the texts shorter than this length are filled with special characters. For example, if the text is insufficiently long, padding is used to fill it. This method is also a commonly used processing method for input data based on the convolutional neural network text classification method. Two disadvantages are shown in this method:

(1)  Taking the length of the longest text in a data set as a benchmark, texts that are shorter than this length are filled with special characters. There will be too many non-semantic characters in short texts, which affects the classification effect.
(2)  The text represented by single model is used as the input of the convolutional neural network. The feature representation of the text is relatively single, which is not conducive to text classification.

In order to correct the drawbacks of the above method, a method of combining the CD_MTR model with a convolutional neural network (MTR_MCNN) is proposed. The dimension of each text vector obtained by this method is the same, so it does not need to be filled with special characters. Then, the text retains the original semantic information. And the text vector trained by the CD_MTR model expresses each text in multiple ways as an input to the convolutional neural network, allowing it to extract deeper features and achieve better classification results.

## 4   Experimental Design

### 4.1   Experiment Data Set

In order to verify the performance of the CD_MTR model on text classification, two classification data sets were selected for experimentation. A total of 24,000 texts were selected from 6 categories of automotive, culture, economics, medicine, military, and sports of the NetEase News Corpus. There are 7691 texts in 8 categories of Fudan Text Classification Corpus: art, history, computer, environment, agronomy, economics, politics, and sports. The number of corpora categories and the proportionality of texts are not the same. This experiment uses a ten-fold cross validation method to evaluate the effectiveness of this method.

### 4.2   The Influence of Single-Text Representation Model Dimension on the Effect of CD_MTR Model

The CD_MTR model proposed in this paper combines three single models of T_VSM, LSI and CD_STR. In order to have a good text representation effect for the CD_MTR model, it is necessary to fuse three dimensions of the T_VSM, LSI, and CD_STR. The effect of testing the CD_MTR on two datasets for three single models with different dimensions were chosen (the number of topics for the LSI is 400). This paper tests the

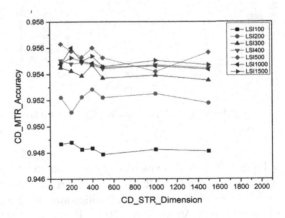

**Fig. 2.** The effect of LSI and CD_STR dimensions on CD_MTR classification effect when T_VSM dimension is 100

various combinations of the three single models of T_VSM, LSI and CD_STR when dimensions of [100, 200, 300, 400, 500, 1000, 1500] are selected. The combination is too much. This paper only takes part of them to explain. Respectively shown in Figs. 2, 3, and 4.

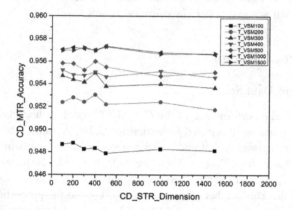

**Fig. 3.** The effect of T_VSM and CD_STR dimensions on the CD_MTR classification effect when the LSI dimension is 100

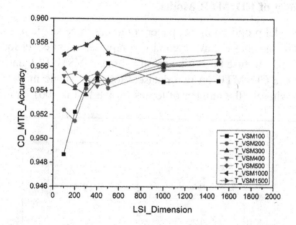

**Fig. 4.** The effect of T_VSM and LSI dimensions on CD_MTR classification effect when CD_STR dimension is 100

The results in Figs. 2, 3 and 4 show that changes in the three single-model dimensions of T_VSM, LSI, and CD_STR affect the text representation capability of the fusion model CD_MTR. Considering the classification effect and classification speed of the four models of T_VSM, LSI, CD_STR and CD_MTR in different dimensions, the dimensions of the three models T_VSM, LSI and CD_STR are selected to be 1000, 500 and 400 respectively. The number of LSI model topics was selected as 400, so the dimension of the CD_MTR model is 1800.

## 4.3    Text Representation Model Comparison and Analysis

In order to verify the validity of the CD_MTR model, we compared the effect of different models: A_word2vec (an average of each word vector per text), T_word2vec (TF-IDF+word2vec), CD_STR, LDA fusion word2vec (LDA+word2vec), T_VSM fusion LSI (T_VSM+LSI), LSI fusion CD_STR (LSI+CD_STR), and T_VSM fusion CD_STR (T_VSM+CD_STR). The classification effect of each model is shown in Table 1.

**Table 1.** Classification effect of each model on two datasets

| Methods | NetEase news text (%) | | Fudan text (%) | |
|---|---|---|---|---|
| | Micro-average $F_1$ | Macro-average $F_1$ | Micro-average $F_1$ | Macro-average $F_1$ |
| A_word2vec | 91.79 | 91.83 | 92.20 | 90.59 |
| T_word2vec | 93.24 | 93.25 | 91.92 | 90.18 |
| **CD_STR** | 94.24 | 94.25 | 93.08 | 92.39 |
| LDA+word2vec | 92.99 | 93.00 | 93.80 | 93.01 |
| T_VSM+LSI | 94.84 | 94.85 | 95.97 | 95.66 |
| LSI+CD_STR | 95.58 | 95.59 | 96.78 | 96.49 |
| T_VSM +CD_STR | 95.70 | 95.70 | 96.76 | 96.44 |
| **CD_MTR** | 95.85 | 95.86 | 96.93 | 96.56 |

The results from Table 1 show that:

(1)  The micro-average F1 value and the macro-average F1 value obtained by CD_STR on the two data sets are superior to the single models A_word2vec and T_word2vec. This result also verifies that the CD_STR model considers the influence of a single word on the entire document and has better class discrimination ability.

(2)  Compared with other combined models, the CD_MTR model presented in this paper improves both the micro-average F1 value and the macro-average F1 value.

## 4.4    Ten-Fold Cross Result

In order to test the effectiveness of the MTR_MCNN method proposed in this chapter, a ten-fold cross validation was used. The ten-fold cross-validation method divides the data set into 10 equal and disjoint sub-samples each time. In 10 sub-samples, one sub-sample is used as the data of the test model, and the other 9 samples are used for training. Verifying each sub-sample for one time and repeat cross validation for 10 times. Figure 5 show the classification effect of the NetEase news text and Fudan text under different sub-samples, respectively.

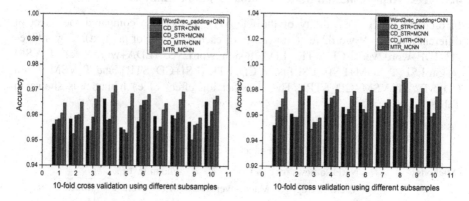

**Fig. 5.** Different methods of ten-fold cross-validation

The results in Fig. 5 show that:

(1) Compared with CD_STR+CNN, CD_STR+MCNN in the distribution of ten different training sets/test sets pairs is better in most cases. MTR_MCNN is superior to CD_MTR+CNN.
(2) The classification accuracy of CD_MTR+CNN is better than CD_STR+CNN, and the classification accuracy of MTR_MCNN is also superior to CD_STR+MCNN. Furthermore, under the same convolutional neural network structure, the fusion model CD_MTR presented in this paper can better represent text information, distinguish categories, and improve the accuracy of text classification.
(3) The model of word2vec_padding+CNN performs better under certain training set/test set pairs in normal conditions. But compared with the MTR_MCNN presented in this chapter, the classification accuracy is lower than that of MTR_MCNN. In the experiment, the word2vec_padding+CNN method needs to fill in most of the texts in the text sets with special characters, and there is a case where the supplemented text information and the original text information are deviated which affects the classification effect. In this method, the matrix dimension of the input convolutional neural network is the number of words per text multiplied by the dimension of each feature word. When the number of text feature words is too large, the text matrix dimension of the input convolutional neural network will be very large, thus affecting the training speed.

## 4.5   Method Comparison and Analysis

To further verify the validity of the MTR_MCNN method, this paper compares it with other classification methods. Table 2 shows the average classification accuracy of each method under the 10-fold crossover method.

**Table 2.** Classification accuracy of different text classification methods (%)

| Methods | NetEase news text | Fudan text |
|---|---|---|
| | Accuracy (%) | Accuracy (%) |
| CD_MTR | 95.85 | 96.93 |
| CD_STR+CNN | 95.55 | 96.21 |
| CD_STR+MCNN | 95.88 | 96.46 |
| Word2vec_padding+CNN | 95.91 | 96.99 |
| CD_MTR+CNN | 96.36 | 97.46 |
| MTR_MCNN | 96.70 | 97.87 |

From the results in Table 2, we can conclude:

(1) The text vector represented by the single model CD_STR is lower in classification accuracy than the CD_MTR proposed in Sect. 3.4 of this paper, whether it is an input as a non-optimized or optimized convolutional neural network. The classification accuracy of the method once again proves the performance of the CD_MTR model.

(2) The MTR_MCNN method proposed in this paper has the highest classification accuracy among all the methods. The accuracy values on the two data sets respectively are 96.70% and 97.87%, and its classification is also more effective than the common used word2vec_padding+CNN method.

The above results are mainly because the MTR_MCNN method proposed in this paper introduces a convolutional neural network to improve the feature extraction of the CD_MTR method which is superficial. The MTR_MCNN method designs different convolution kernels of different sizes and numbers, which extracts text features from different angles. The MTR_MCNN method uses the CD_MTR model to vectorize the text and convert the resulting text vectors into a matrix form as input to convolutional neural network. Additionally, the MTR_MCNN method do not need to fill shorter texts with special characters, which affects the expression of the original text information, so it improved the text classification accuracy.

## 5  Conclusion

This paper proposes a single model CD_STR and a fusion model CD_MTR, which using optimized TF-IDF weighting word2vec with semantic information combines with LSI and TF-IDF weighted vector space model to complement each other. Based on the CD_MTR model, the classification method MTR_MCNN combined with CD_MTR and convolutional neural network is proposed in this paper. In this method, the convolutional neural network structure is improved, and different sizes and numbers of convolution kernels are designed to extract text features from different angles. In addition, the text vectors obtained from the CD_MTR are converted into a matrix form as an input to the convolutional neural network. For MTR_MCNN method, it does not need special characters to fill the text, avoiding meaningless additional text information and reducing the dimension of the input matrix so as to improve the training speed of the convolutional neural network. The experiment results show that both the CD_STR model and CD_MTR model and the MTR_MCNN method in this paper have achieved good classification effect and are superior to other methods.

# References

1. Hinton, G.E.: Learning distributed representations of concepts. In: Eighth Conference of the Cognitive Science Society, pp. 1–12 (1986)
2. Bengio, Y., Ducharme, R., Vincent, P., et al.: A neural probabilistic language model. J. Mach. Learn. Res. **3**(2), 1137–1155 (2003)
3. Mnih, A., Hinton, G.: A scalable hierarchical distributed language model. In: International Conference on Neural Information Processing Systems, pp. 1081–1088. Curran Associates Inc., (2008)
4. Mikolov, T., Chen, K., Corrado, G., et al.: Efficient estimation of word representations in vector space. Comput. Sci. (2013)
5. Mikolov, T., Yih, W.T., Zweig, G.: Linguistic regularities in continuous space word representations. In: HLT-NAACL (2013)
6. Hu, B., Lu, Z., Li, H., et al.: Convolutional neural network architectures for matching natural language sentences. In: International Conference on Neural Information Processing Systems, pp. 2042–2050. MIT Press (2014)
7. Yan, Y.: Text representation and classification with deep learning. University of Science and Technology, Beijing (2016)
8. Salton, G.: A vector space model for automatic indexing. Commun. ACM **18**(11), 613–620 (1975)
9. Lecun, Y., Boser, B., Denker, J.S., et al.: Backpropagation applied to handwritten zip code recognition. Neural Comput. **1**(4), 541–551 (2014)
10. Bouvrie, J.: Notes on convolutional neural network. Neural Nets (2006)
11. Liu, X., Zhang, Y., Zheng, Q.: Sentiment classification of short texts on internet based on convolutional neural network model. Comput. Mod. **2017**(4), 73–77 (2017)
12. Wang, P., Xu, J., Xu, B., et al.: Semantic clustering and convolutional neural network for short text categorization. In: Proceedings of the 53rd Annual Meeting of the Association for Computational Linguistics and the 7th International Joint Conference on Natural Language Processing, vol. 2, pp. 352–357 (2015)
13. Zhou, C., Sun, C., Liu, Z., et al.: A C-LSTM neural network for text classification. Comput. Sci. **1**(4), 39–44 (2015)
14. Lai, S., Xu, L.H., Liu, K., et al.: Recurrent convolutional neural networks for text classification. In: Proceedings of the Twenty-Ninth AAAI Conference on Artificial Intelligence, pp. 2267–2273 (2015)
15. Cai, H.: Research of short-text classification method based on convolution neural network. Southwest University (2016)
16. Yin, Y., Yang, W., Yang, H., et al.: Research on short text classification algorithm based on convolutional neural network and KNN. Comput. Eng. (2017)
17. Kim, Y.: Convolutional Neural network for Sentence Classification. Eprint Arxiv (2014)

# CNN-Based Chinese Character Recognition with Skeleton Feature

Wei Tang[1,2,3], Yijun Su[1,2,3], Xiang Li[1,2,3], Daren Zha[3], Weiyu Jiang[3(✉)],
Neng Gao[3], and Ji Xiang[3]

[1] School of Cyber Security, University of Chinese Academy of Sciences,
Beijing, China
[2] State Key Laboratory of Information Security, Chinese Academy of Sciences,
Beijing, China
[3] Institute of Information Engineering, Chinese Academy of Sciences, Beijing, China
{tangwei,suyijun,lixiang9015,zhadaren,jiangweiyu,gaoneng,
xiangji}@iie.ac.cn

**Abstract.** Recently, the convolutional neural networks (CNNs) show the great power in dealing with various image classification tasks. However, in the task of Chinese character recognition, there is a significant problem in CNN-based classifiers: insufficient generalization ability to recognize the Chinese characters with unfamiliar font styles. We call this problem the Style Overfitting. In the process of a human recognizing Chinese characters with various font styles, the internal skeletons of these characters are important indicators. This paper proposes a novel tool named Skeleton Kernel to capture skeleton features of Chinese characters. And we use it to assist CNN-based classifiers to prevent the Style Overfitting problem. Experimental results prove that our method firmly enhances the generalization ability of CNN-based classifiers. And compared to previous works, our method requires a small training set to achieve relatively better performance.

**Keywords:** Chinese character recognition
Convolutional neural networks · Style Overfitting · Skeleton feature

## 1 Introduction

Chinese characters have been widely used (modified or extended) in many Asian countries such as China, Japan, Korea, and so on [22]. There are more than tens of thousands of different Chinese characters with variable font styles. Most of them can be well recognized by most people. However, in the field of artificial intelligence, Chinese character automatic recognition is considered as an extremely difficult task due to the very large number of categories, complicated structures, similarity between characters and the variability of font styles [3]. Because of its unique technical challenges and great social needs, during the last five decades there are intensive research in this field and a rapid increase

© Springer Nature Switzerland AG 2018
L. Cheng et al. (Eds.): ICONIP 2018, LNCS 11305, pp. 461–472, 2018.
https://doi.org/10.1007/978-3-030-04221-9_41

of successful applications [8,10,15,17]. However, higher recognition performance is continuously needed to improve the existing application and to exploit new applications.

Recently, the convolutional neural networks (CNNs) show their great power in dealing with multifarious image classification tasks [5,6,11,14,16,21]. CNN-based classifiers break the bottleneck of Chinese character recognition and achieve excellent performance even better than human on ICDAR'13 Chinese Character Recognition Competition [1,4,18,20,24]. But there is a significant problem in CNN-based classifiers: insufficient generalization ability to recognize Chinese characters with unfamiliar font styles (e.g. it perform poorly when test on the characters with the font style that the trained model have never seen). We call this problem the *Style Overfitting*. However, most people are able to deal with this problem easily. In the process of a human recognizing Chinese characters with unfamiliar font style, the internal skeletons of these characters are significant indicators. If a Chinese character printed in two or more different font styles, the inherent skeletons of them are usually the same. Figure 1 shows three pairs of same Chinese characters with different font styles that challenge CNN-based classifiers.

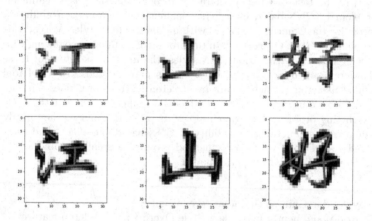

**Fig. 1.** Three pairs of same Chinese characters with different font styles that challenge CNN-based classifiers. It could be seen that the inherent skeleton (highlighted by the red lines) of the above character and the bottom character are essentially the same. (Color figure online)

In this paper, we propose a novel tool named *Skeleton Kernel* to capture skeleton features of Chinese characters to prevent the Style Overfitting problem. A Skeleton Kernel is designed as a long narrow rectangle window sliding along the horizontal or vertical axis of the input image. It calculates the cumulative distributions of pixel values to capture skeleton features of input images. Then the same pattern of the same Chinese character printed in different fount styles could be easily recognized as the same one. Figure 2 illustrates the same pattern extracted by Skeleton Kernel of the same Chinese character printed in different font styles.

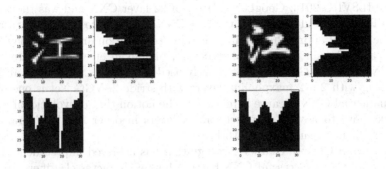

**Fig. 2.** The right histograms of each character depict the cumulative distributions of pixel values calculated by a long narrow rectangle window sliding along the vertical axis. And the bottom histograms do the same along the cross direction. The same inherent skeletons of each character are represented by the same patterns in these histograms.

Overall, our contributions are as follows:

1. We introduce the Style Overfitting problem of CNN-based classifiers in dealing with Chinese character recognition: insufficient generalization ability to recognize Chinese characters with unfamiliar font styles.
2. We propose a novel tool named Skeleton Kernel to extract skeleton features, which are important indicators to identify Chinese characters. And we use it to assist CNN-based classifiers to prevent aforementioned Style Overfitting problem.
3. We have done a series of experiments to prove that the Style Overfitting problem indeed exists, and our method alleviates this problem by firmly enhancing the generalization performance of CNN-based classifiers.

The rest of this paper is organized as follows. Section 2 summarizes the related works. Section 3 discusses the Style Overfitting problem from the practical view. Section 4 introduces the proposed Skeleton Kernel in details. Section 5 presents the experimental results. Finally in Sect. 6, we conclude our work and discuss the future work.

## 2   Related Work

Convolution neural networks has greatly promoted the development of image recognition technology. Lecun [7] first introduces the convolutional neural network specifically designed to deal with the variability of 2D shapes. Krizhevsky et al. [6] create a large, deep convolutional neural network (named Alex Net) that was used to win the ILSVRC 2012 (ImageNet Large-Scale Visual Recognition Challenge). Zeiler et al. [21] explain a lot of the intuition behind CNNs and showing how to visualize the filters and weights correctly, and their ZF Net won the ILSVRC 2013. GoogLeNet [16] is a 22 layer CNN and was the winner of ILSVRC 2014, and it is one of the first CNN architectures that really strayed from the general approach of simply stacking convolution and pooling layers on top of each other in a sequential structure. Simonyan et al. [14] created a 19 layer CNN (name VGG Net) that strictly used $3 \times 3$ filters with stride and pad of 1, along with $2 \times 2$ maxpooling layers with stride 2. VGG Net is one of the most influential CNN because it reinforced the notion that convolutional neural networks have to have a deep network of layers in order for this hierarchical representation of visual data to work.

CNN-based Chinese character recognition has achieved unprecedented success. But all these successful CNN-based Chinese character classifiers are test on the character set with the same font styles as the training set. Wu et al. [18] propose a handwriting Chinese character recognition method based on relaxation convolutional neural network, and took the 1st place in ICDAR'13 Chinese Handwriting Character Recognition Competition [20]. Meier et al. [4] create a multi-column deep neural network achieving first human-competitive performance on the famous MNIST handwritten digit recognition task. And this CNN-based classifier classifies the 3755 classes of handwritten Chinese characters in ICDAR'13 with almost human performance. Zhong et al. [24] design a streamlined version of GoogLeNet for handwritten Chinese character recognition outperforming previous best result with significant gap. Chen et al. [1] propose a CNN-based character recognition framework employ random distortion [13] and multi-model voting [12]. This classifier performed even better than human on MNIST and ICDAR'13.

Few works are devoted to reducing overfitting in CNN-based Chinese character recognition. Xu et al. [19] propose a artificial neural network architecture called cooperative block neural networks to address the variation in the shape of Chinese characters by considering only three different fonts. Lv [9] successfully applied the stochastic diagonal Levenberg-Marquardt method to a convolutional neural network to recognize a small set of multi-font characters used in Baidu CAPTCHA, which consists of the Arabic numerals and English letters without the Chinese characters. Zhong et al. [23] propose a CNN-based multi-font Chinese character recognizer using multi-pooling and data augmentation achieving acceptable result. But they use 240 fonts for training and 40 fonts for test. The size of the training set is more than 500% of the test set, and the huge training set containing a great deal of font styles especially reducing the overfitting. Different from training on a predetermined large training set, we use a relatively small training set with its size being 10% of the test set. And we gradually increase the number of font styles contained in the training set to dynamically verify the effectiveness of our method.

# 3   Style Overfitting

The font styles of Chinese characters are multifarious. In practice, to train a CNN-based Chinese character classifier, it is hard to provide a perfect training set with all font styles. And it is absolutely expensive to use such massive amount of images to train a deep neural network. In view of this situation, a well trained classifier should obtain excellent generalization ability to recognize characters with unfamiliar font styles.

However, CNN-based classifiers could not satisfy this demand. We have done a lot of experiments showing that the CNN-based classifiers perform poorly when it test on Chinese character sets with unfamiliar font styles. Even adding more amount of convolution layers, this problem still exists. We call this problem the Style Overfitting. It is a significant problem seriously restricting the generalization ability of CNN-based Chinese character classifiers.

We believe that the main cause of this problem is that CNNs excessively focus on the texture features that strongly indicate the font styles, and relatively ignore the inherent skeleton features. Inherent skeletons are important indicators to classify Chinese characters. One Chinese character could be printed in diverse font styles, and the inherent skeleton of each Chinese character are commonly fixed. Therefore, to solve Style Overfitting problem, the influence of skeleton features on CNN-based classifiers must be strengthened.

# 4   Method

To prevent aforementioned Style Overfitting problem, we propose a novel tool named Skeleton Kernel to extract skeleton features of Chinese characters. And we use Skeleton Kernel to enhance the generalization performance of CNN-based classifiers.

## 4.1   Skeleton Kernel

A Skeleton Kernel is a window with a specific shape sliding along a specific path (like the Convolution Kernel [6]). Considering the deformation and scaling of strokes in Chinese characters, we design the Skeleton Kernel as a long narrow rectangle window sliding along the horizontal or vertical axis of the input image. There are 3 main differences between Skeleton Kernel and Convolution Kernel: (1) the Skeleton Kernel appear long and narrow, while Convolution Kernel is always a relatively small square; (2) the weights in Skeleton Kernel are predetermined, which in Convolution Kernel are learning from training set; (3) the output of the Skeleton Kernel is a vector, and the Convolution Kernel produce a matrix. Figure 3 briefly illustrates how Skeleton Kernels work.

The Skeleton Kernel has two important hyperparameters: size and stride. It should be noted that the excessively thin or fat window is bad for capture skeleton features. It is because a excessively thin window has insufficient resilience to tolerate the deformation and scaling of a stroke (the main difference between diverse font styles), and a excessively fat window could be confused by too many strokes in it.

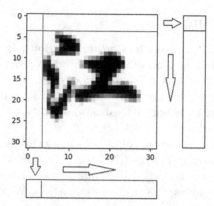

**Fig. 3.** A pair of Skeleton Kernels slide along the coordinate axis of the input image and output two skeleton feature vectors. The right and bottom vectors are separately produced by the blue and red window. These two vectors contain the skeleton information by calculate the cumulative distributions of pixel values in the input image. (Color figure online)

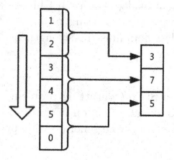

**Fig. 4.** This figure briefly show how Chunk-Sum Pooling works. Specifically, in this figure, the hyperparameters (size, stride) of the sliding window are (2, 2). The input vector is cut into many chunks by this sliding window. We add 0 to the end of the input vector if the sliding window exceeds the bound of it. CSP calculates the sum of all the elements in each chunk and then produce a new vector.

## 4.2    Chunk-Sum Pooling

To avoid the aforementioned problem, we could designed more than one Skeleton Kernel with different sizes and strides. More effectively, we use only one Skeleton Kernel with its size to be $1 \times$ the height of input image (or the width of input image $\times 1$) and stride to be 1. Then, we employ Chunk-Sum Pooling (noted as CSP) to process the output vector to achieve the same effect as using a lot of Skeleton Kernels. CSP is a variant of Chunk-Max Pooling [2]. It cuts a vector into a lot of chunks and calculates the sum of all the elements in each chunk to produce a new vector, just like a sliding window with certain hyperparameters: size and stride. Figure 4 shows how CSP works.

### 4.3   CNN-Based Framework with Skeleton Kernel

In order to prevent the Style Overfitting problem of CNN-based classifiers, we propose a new CNN-based framework employing Skeleton Kernel. The main difference between our new framework and the original framework is that the new framework contain a extra bypass consisting of several Skeleton Kernels and parallel CSP modules. A pair of Skeleton Kernels are used to extract skeleton features, and these CSP modules are used to process the feature vectors. The diagram of the new framework is illustrated in Fig. 5.

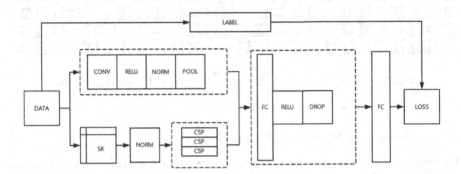

**Fig. 5.** Diagram of the new CNN-based Chinese character recognition framework employing Skeleton Kernel. The CONV–RELU–NORM–POOL and FC–RELU–DROP modules constitute the main structures of popular convolutional networks (e.g. Alex nets [6]). Skeleton Kernels (SK) in bypass produced several vectors that represent skeleton features of input image. There are several parallel CSP modules with different sizes and strides to process these skeleton features.

Using Skeleton Kernel does not affect the convergence and complexity of the original convolutional neural network. It is because our method is equivalent to the addition of extra feature extraction kernels, and each of these extra kernels obtain fixed weights and biases.

## 5   Experiments

We have done a series of experiments to evaluate the generalization ability of CNN-based framework and our new framework to recognize Chinese characters with unfamiliar font styles. Results prove that the Style Overfitting problem does exist, and our method indeed alleviate this problem by firmly enhancing the generalization performance of CNN-based classifiers.

### 5.1   Data

The data are extracted from True Type font (TTF) files. TTF is a font file format jointly launched by Apple and Microsoft. With the popularity of Microsoft

Windows operating systems, it has become the most commonly used format of font file. We extract 130 candidate sets from 130 TTF files with widely varying font styles. Each of them contains 3755 frequently-used Chinese characters (level-l set of GB2312-80). Each Chinese character is presented by a $32 \times 32$ PNG image. Figure 6 shows the example of Chinese character 'JIANG' in 130 different fonts.

**Fig. 6.** Example of Chinese character 'JIANG' in 130 different fonts.

## 5.2 Existence of Style Overfitting

We evaluate the generalization ability of the popular deep convolutional neural network VGG [14] to recognize Chinese characters with unfamiliar font styles. We chose two typically VGG nets: VGG-11 and VGG-19. VGG-11 is a 11 layers CNN consisting of 8 convolution layers and 3 full connection layers. VGG-19 is a 19 layers CNN consisting of 16 convolution layers and 3 full connection layers. These two CNNs are used to assess the impact of network depth on the Style Overfitting problem.

Tables 1 and 2 show the experimental results of VGG-11 and VGG-19. In this experiment, there are 10 character sets randomly selected from the 130 candidate sets. The 10 randomly selected character sets are tagged with the ID 1 to 10. We use 1 of the 10 selected sets for training and other 9 sets for test. The first rows of these tables list the IDs of the training sets (noted as tra-ID, e.g. tra-1); the first columns of these tables list the IDs of the test sets (noted as test-ID, e.g. test-1).

It could be seen that the train accuracy are very high (highlighted with bold font), but the test accuracy on other character sets are extremely low. It prove that CNN-based classifiers lack the generalization ability to recognize Chinese characters with unfamiliar font styles. Even adding more amount of convolution layers, the Style Overfitting problem still exists.

## 5.3 Performance Comparison

In order to prove that our method indeed alleviate the Style Overfitting problem and to compare the generalization performance between our method and CNN-based classifier using data augmentation and multi-pooling, we set up 3 controlled groups for the experiments: (1) VGG-19 [14]; (2) VGG-19 using data

**Table 1.** Results of VGG-11 on 10 randomly selected Chinese character sets.

| IDs | tra-1 | tra-2 | tra-3 | tra-4 | tra-5 | tra-6 | tra-7 | tra-8 | tra-9 | tra-10 |
|---|---|---|---|---|---|---|---|---|---|---|
| test-1 | **0.9827** | 0.0749 | 0.1209 | 0.0817 | 0.1368 | 0.1118 | 0.1172 | 0.1007 | 0.0945 | 0.1119 |
| test-2 | 0.0818 | **0.9752** | 0.1356 | 0.1393 | 0.0970 | 0.1241 | 0.1265 | 0.0821 | 0.1127 | 0.0868 |
| test-3 | 0.1100 | 0.1268 | **0.9776** | 0.1460 | 0.1030 | 0.1369 | 0.0937 | 0.1358 | 0.1239 | 0.1063 |
| test-4 | 0.0852 | 0.1286 | 0.1449 | **0.9550** | 0.1281 | 0.0919 | 0.1403 | 0.1382 | 0.1155 | 0.1368 |
| test-5 | 0.1427 | 0.0842 | 0.1017 | 0.1310 | **0.9830** | 0.1334 | 0.1076 | 0.0804 | 0.0943 | 0.1446 |
| test-6 | 0.1233 | 0.1358 | 0.1414 | 0.0823 | 0.1414 | **0.9744** | 0.1419 | 0.1142 | 0.1334 | 0.0999 |
| test-7 | 0.1305 | 0.1313 | 0.0860 | 0.1443 | 0.1193 | 0.1443 | **0.9795** | 0.0964 | 0.1382 | 0.1132 |
| test-8 | 0.1079 | 0.0951 | 0.1465 | 0.1385 | 0.0820 | 0.1198 | 0.1007 | **0.9664** | 0.1145 | 0.0765 |
| test-9 | 0.0897 | 0.1244 | 0.1356 | 0.1025 | 0.0908 | 0.1206 | 0.1278 | 0.1089 | **0.9664** | 0.1395 |
| test-10 | 0.1119 | 0.0815 | 0.0948 | 0.1387 | 0.1387 | 0.1039 | 0.1196 | 0.0810 | 0.1342 | **0.9704** |

**Table 2.** Results of VGG-19 on 10 randomly selected Chinese character sets.

| IDs | tra-1 | tra-2 | tra-3 | tra-4 | tra-5 | tra-6 | tra-7 | tra-8 | tra-9 | tra-10 |
|---|---|---|---|---|---|---|---|---|---|---|
| test-1 | **0.9947** | 0.0895 | 0.1156 | 0.0905 | 0.1491 | 0.1366 | 0.1412 | 0.1167 | 0.0998 | 0.1178 |
| test-2 | 0.0824 | **0.9861** | 0.1396 | 0.1390 | 0.0919 | 0.1411 | 0.1409 | 0.1012 | 0.1356 | 0.0874 |
| test-3 | 0.1262 | 0.1433 | **0.9859** | 0.1566 | 0.1076 | 0.1505 | 0.0919 | 0.1532 | 0.1460 | 0.1063 |
| test-4 | 0.0894 | 0.1497 | 0.1588 | **0.9606** | 0.1393 | 0.0924 | 0.1512 | 0.1505 | 0.1126 | 0.152 |
| test-5 | 0.1493 | 0.1079 | 0.1089 | 0.1401 | **0.9902** | 0.1502 | 0.1323 | 0.0897 | 0.1004 | 0.1451 |
| test-6 | 0.1233 | 0.1350 | 0.1494 | 0.1044 | 0.1446 | **0.9869** | 0.1552 | 0.1297 | 0.1318 | 0.1154 |
| test-7 | 0.1289 | 0.1382 | 0.1049 | 0.1462 | 0.1169 | 0.1518 | **0.9870** | 0.1127 | 0.1342 | 0.1316 |
| test-8 | 0.1119 | 0.0888 | 0.1441 | 0.1449 | 0.0932 | 0.1243 | 0.1057 | **0.9752** | 0.1148 | 0.0927 |
| test-9 | 0.1009 | 0.1196 | 0.1298 | 0.1278 | 0.1010 | 0.1411 | 0.1446 | 0.1262 | **0.9752** | 0.1472 |
| test-10 | 0.1247 | 0.0983 | 0.1132 | 0.1488 | 0.1523 | 0.1114 | 0.1255 | 0.0840 | 0.1504 | **0.9832** |

augmentation and multi-pooling (noted as VGG-19 with MP) [23]; (3) VGG-19 with Skeleton Kernel (our method, noted as VGG-19 with SK).

According to experience, the integrity of training data will affect the generalization ability of the model. Therefore, the controlled variable is the number of font styles in training set. We first randomly choose 1 candidate set for training and randomly choose other 10 candidate sets for test, and then we add the number of font styles for training until there are 9 font styles for training and other 90 for test. Each experiment is carried out 10 times by randomly change the training set and test set, and the size of the training set is constantly 10% of the test set. Finally, we calculate the mean accuracy. Figure 7 shows the results of these experiments.

In the experiments of VGG nets with Skeleton Kernel, there is a pair of Skeleton Kernels to extract skeleton features of the input image. The size and stride of the window sliding along the horizontal axis of the input image are $32 \times 1$ and 1; the size and stride of the window sliding along the vertical axis of the input image are $1 \times 32$ and 1. And the weights and bias of all Skeleton Kernels are

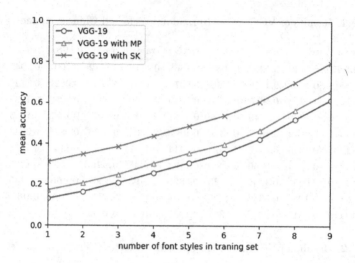

**Fig. 7.** Experimental results of VGG-19, VGG-19 with MP, and VGG-19 with SK.

predetermined to be constant 1 and 0. Each feature vector produce by Skeleton Kernels is normalized by the norm of it. We employ 5 CSPs to process these skeleton feature vectors. The hyperparameters (size, stride) of these CSPs are $(1, 1)$, $(2, 1)$, $(2, 2)$, $(4, 2)$ and $(4, 4)$.

The experimental results show that our method remarkably enhances the generalization ability of CNN-based classifier to recognize Chinese characters with unfamiliar font styles. And compared to previous works, our method only requires a small training set to achieve relatively better results on widely varying test sets. It prove that our method is effective for alleviating the Style Overfitting problem.

## 6    Conclusion and Future Work

CNN-based classifiers lack generalization ability to recognize Chinese characters with unfamiliar font styles (Style Overfitting problem). To solve this problem, we propose Skeleton Kernel to extract skeleton features of Chinese characters to assists CNN-based classifiers. Experimental results prove that our method alleviate the Style Overfitting problem by firmly enhancing the generalization performance of CNN-based Chinese character recognizers.

This paper design the Skeleton Kernel as a long narrow window with fixed weights and bias sliding along the axis of input images. In the future, it could be designed a train-able Skeleton Kernel with more kinds of shape and sliding path to achieve better performance.

**Acknowledgment.** We thank Lei Wang, Yujin zhou, Zeyi Liu, Jiahui Sheng and Yuanye He for their useful advice. This work is partially supported by National Key Research and Development Program of China.

# References

1. Chen, L., Wang, S., Fan, W., Sun, J.: Beyond human recognition: a CNN-based framework for handwritten character recognition. In: IAPR Asian Conference on Pattern Recognition, pp. 695–699 (2015)
2. Chen, Y., Xu, L., Liu, K., Zeng, D., Zhao, J.: Event extraction via dynamic multi-pooling convolutional neural networks. In: The Meeting of the Association for Computational Linguistics (2015)
3. Dai, R., Liu, C., Xiao, B.: Chinese character recognition: history, status and prospects. Front. Comput. Sci. China 1(2), 126–136 (2007)
4. Dan, C., Meier, U.: Multi-column deep neural networks for offline handwritten Chinese character classification. In: International Joint Conference on Neural Networks, pp. 1–6 (2015)
5. Girshick, R., Donahue, J., Darrell, T., Malik, J.: Rich feature hierarchies for accurate object detection and semantic segmentation. In: Computer Science, pp. 580–587 (2013)
6. Krizhevsky, A., Sutskever, I., Hinton, G.E.: ImageNet classification with deep convolutional neural networks. In: Pereira, F., Burges, C.J.C., Bottou, L., Weinberger, K.Q. (eds.) Advances in Neural Information Processing Systems, vol. 25, pp. 1097–1105. Curran Associates Inc, Red Hook (2012)
7. Lecun, Y., Bottou, L., Bengio, Y., Haffner, P.: Gradient-based learning applied to document recognition. Proc. IEEE 86(11), 2278–2324 (1998)
8. Long, T., Jin, L.: Building compact MQDF classifier for large character set recognition by subspace distribution sharing. Pattern Recogn. 41(9), 2916–2925 (2008)
9. Lv, G.: Recognition of multi-fontstyle characters based on convolutional neural network. In: Fourth International Symposium on Computational Intelligence and Design, pp. 223–225 (2011)
10. Mori, S., Yamamoto, K., Yasuda, M.: Research on machine recognition of handprinted characters. IEEE Trans. Pattern Anal. Mach. Intell. 6(4), 386–405 (1984)
11. Ren, S., He, K., Girshick, R., Sun, J.: Faster R-CNN: towards real-time object detection with region proposal networks. In: International Conference on Neural Information Processing Systems, pp. 91–99 (2015)
12. Schmidhuber, J., Meier, U., Ciresan, D.: Multi-column deep neural networks for image classification. In: Computer Vision and Pattern Recognition, pp. 3642–3649 (2012)
13. Simard, P.Y., Steinkraus, D., Platt, J.C.: Best practices for convolutional neural networks applied to visual document analysis. In: International Conference on Document Analysis and Recognition, p. 958 (2003)
14. Simonyan, K., Zisserman, A.: Very deep convolutional networks for large-scale image recognition. In: Computer Science (2014)
15. Stallings, W.: Approaches to Chinese character recognition. Pattern Recogn. 8(2), 87–98 (1976)
16. Szegedy, C., et al.: Going deeper with convolutions. In: IEEE Conference on Computer Vision and Pattern Recognition, pp. 1–9 (2015)
17. Umeda, M.: Advances in recognition methods for handwritten Kanji characters. IEICE Trans. Inf. Syst 79(5), 401–410 (1996)
18. Wu, C., Fan, W., He, Y., Sun, J., Naoi, S.: Handwritten character recognition by alternately trained relaxation convolutional neural network. In: International Conference on Frontiers in Handwriting Recognition, pp. 291–296 (2014)

19. Xu, N., Ding, X.: Printed Chinese character recognition via the cooperative block neural networks. In: IEEE International Symposium on Industrial Electronics, vol. 1, pp. 231–235 (1992)
20. Yin, F., Wang, Q.F., Zhang, X.Y., Liu, C.L.: ICDAR 2013 Chinese handwriting recognition competition (ICDAR), pp. 1464–1469 (2013)
21. Zeiler, M.D., Fergus, R.: Visualizing and understanding convolutional networks. In: Fleet, D., Pajdla, T., Schiele, B., Tuytelaars, T. (eds.) ECCV 2014. LNCS, vol. 8689, pp. 818–833. Springer, Cham (2014). https://doi.org/10.1007/978-3-319-10590-1_53
22. Zhang, X.Y., Yin, F., Zhang, Y.M., Liu, C.L., Bengio, Y.: Drawing and recognizing Chinese characters with recurrent neural network. IEEE Trans. Pattern Anal. Mach. Intell. (2016)
23. Zhong, Z., Jin, L., Feng, Z.: Multi-font printed Chinese character recognition using multi-pooling convolutional neural network. In: International Conference on Document Analysis and Recognition, pp. 96–100 (2015)
24. Zhong, Z., Jin, L., Xie, Z.: High performance offline handwritten Chinese character recognition using GoogleNet and directional feature maps. In: International Conference on Document Analysis and Recognition, pp. 846–850 (2015)

# Fuzzy Bag-of-Topics Model for Short Text Representation

Hao Jia and Qing Li[✉]

School of Computer Engineering and Science, Shanghai University, Shanghai, China
qli@shu.edu.cn

**Abstract.** Text representation is the keystone in many NLP tasks. For short text representation learning, the traditional Bag-of-Words model (BoW) is often criticized for sparseness and neglecting semantic information. Fuzzy Bag-of-Words (FBoW) and Fuzzy Bag-of-Words Cluster (FBoWC) model are the improved model of BoW, which can learn dense and meaningful document vectors. However, word clusters in FBoWC model are obtained by K-means cluster algorithm, which is unstable and may result in incoherent word clusters if not initialized properly. In this paper, we propose the Fuzzy Bag-of-Topics model (FBoT) to learn short text vector. In FBoT model, word communities, which are more coherent than word clusters in FBoWC, are used as basis terms in text vector. Experimental results of short text classification on two datasets show that FBoT achieves the highest classification accuracies.

**Keywords:** Short text · Representation learning
Word communities

## 1 Introduction

With the vigorous development of Web applications, massive amount of data are generated on the Internet every day. Among all forms of data, short text has become the prevalent format for the information on the Internet, including microblogs, product reviews, short messages, advertising messages, search snippets, and so on. A great deal of valuable information exists in these short texts, hence there is a growing need for effective text mining technology for short texts.

For most text mining tasks, the representation of texts is vital. As a traditional text representing method, Bag-of-Words model (BoW) has been widely used in various text mining and NLP tasks. In BoW model, a document $d$ is represented as a numerical vector $\mathbf{d} = [x_1, x_2, \ldots, x_l]$, where $x_i$ denotes tf-idf value of the $i$th word in basis terms and $l$ is the number of basis terms. Generally, basis terms are selected from the whole words in corpus by frequency [1].

However, for short text, the document vector in BoW model would be extremely sparse due to tf-idf is calculated based on the normalized number of occurrence of term and short text contains only a small number of words. For instance, consider the sentence: *"Trump speaks to the media in Illinois."*.

© Springer Nature Switzerland AG 2018
L. Cheng et al. (Eds.): ICONIP 2018, LNCS 11305, pp. 473–482, 2018.
https://doi.org/10.1007/978-3-030-04221-9_42

After the filtering of stop words, only four words left, which means the weights of all words but these four are zero in the text vector. This sparsity would result in some problems in NLP tasks. For instance, there are two sentences: *"Trump speaks to the media in Illinois."* and: *"The President greets the press in Chicago.".* These two sentences have no common words except stop words, however, their meanings are similar. In BoW model, however, their vectors are orthogonal, thus we cannot discover the similarity between them by similarity measure methods such as cosine. In addition, semantic information in texts is neglected in BoW model.

To improve BoW model, Zhao et al. proposed Fuzzy Bag-of-Words (FBoW) and Fuzzy Bag-of-Words Cluster (FBoWC) model based on BoW [2]. FBoW adopts a fuzzy mapping between words in text and basis terms instead of a hard mapping in BoW. In a document vector, the weight of a basis term $t_i$ depend on the sum of cosine similarities between $t_i$ and all the words in the document. FBoWC is an improved model of FBoW for the sake of redundancy reduction. In FBoWC, basis terms in document vector are replaced by word clusters, which were obtained by clustering of basis terms.

FBoW and FBoWC can vastly alleviate the sparsity in BoW model since words in document contribute to all the entries in text vector. Word embeddings [3] are applied to calculate the similarities in FBoW and FBoWC, thus semantic information is integrated in the document vector. In FBoWC, K-means clustering algorithm is applied to obtain word clusters. Therefore, the performance of FBoWC is influenced by K-means. However, K-means is an unstable algorithm. Different initialization would generate different clusters. Thus the performance of FBoWC may be unstable, i.e., the word clusters are incoherent and meaningless.

Some researchers enhanced the performance of short text mining tasks by using auxiliary external knowledge, such as Wikipedia [4], WordNet [5], HowNet [6], and Web search results [7], etc. Inspired by their works and FBoWC model, in this paper, we present the Fuzzy Bag-of-Topics (FBoT) model to represent short texts. Different from FBoWC, in our proposed FBoT model, pre-trained word communities are used as basis terms instead of word clusters. To obtain word communities, we first run LDA [8] on an external corpus and obtain K topics. Then, top high-probability words in each topic are selected to form word communities. Finally, document vectors are obtained based on similarities between words in topics and word communities. More details of the FBoT model are presented in Sect. 3. Compared with word clusters in FBoWC, word communities is more coherence, thus text vector learned by FBoT is more meaningful. The experiment results in Sect. 4 show the effectiveness of our proposed FBoT model.

The rest of this article is organized as follows. In the next section, we review some related work about text representation. Details of our proposed FBoT model are described in Sect. 3. Section 4 introduces our experiments about this work and the results are analyzed. In Sect. 5, we conclude this paper.

## 2   Related Work

In most traditional text mining tasks, the Bag-of-Words model (BoW) is used to represent normal texts, in which tf-idf is used to obtain the weights of terms. Although BoW can work well in many NLP tasks of long texts, however, it is not a proper model to represent short text due to the sparsity problem. In addition, BoW is often criticized for ignorance of contextual information and semantic knowledge.

Based on the intuition that words appearing in similar contexts tend to related to same topic, some models have been proposed to represent text instead of VSM including latent semantic analysis (LSA) and topic models [8–10]. In LSA [9], the corpus is represented as a $V \times D$ matrix where the rows denote the terms and the columns denote the weight values of terms in their corresponding documents. Singular Value decomposition (SVD) is utilized to the document-term matrix to obtain low-dimension representation of text. However, the SVD is computationally expensive when deal with a large corpus. Moreover, LSA lacks semantic interpretation.

Latent Dirichlet Allocation (LDA) [8] is a representative topic model, and has achieved great success in many text mining tasks. By applying gibbs sampling, topic distribution of documents can be obtained, which can be utilized as document vector. However, when applied directly to short texts, LDA will generate extremely terrible text representation, because short text cannot provide enough word co-occurrence information.

In 2014, Le and Mikolov proposed Paragraph Vector, a dense representation for document using a simplified neural language model [11]. PV-DM and PV-DBOW model were proposed in their work. In the PV-DM model, the paragraph vector were asked to contribute to the prediction task of a word, based on the given context words in the same paragraph. In the PV-DBOW, the model was asked to predict words sampled from the output.

The most related work with ours is the fuzzy bag-of-words model (FBoW), which can be considered as an improved model of BoW [2]. FBoW adopts a fuzzy mapping between words in documents and basis terms in document vector, i.e., each word in document contributes all the entries in document vector. The amount of contributions are based on similarities calculated by word embeddings. We will discuss more details of FBoW in Sect. 3.

## 3   Fuzzy Bag-of-Topics Model

In this section, our proposed FBoT model is presented. Since FBoT model is constructed based on word embedding and FBoW, we begin with a brief review of them.

### 3.1   Word Embedding

Word embeddings have been successfully utilized in many NLP tasks, such as named entity recognition and parsing [3]. This is because they can encode syntactic and semantic information of words into low-dimension continuous vectors, thus semantically similar words tend to close to each other in the vector space. Hence, the similarity between two words are often measured by the similarity of their corresponding word embeddings. To give an illustration, the top five similar words to two example words *book* and *computer* and their cosine similarity scores are given in Table 1. In our model, pre-trained word embeddings and cosine similarity measure are applied to obtain similarity between two words, and cosine similarity is computed as:

$$\cos(w_i, w_j) = \frac{\mathsf{w_i} \cdot \mathsf{w_j}}{||\mathsf{w_i}|| \, ||\mathsf{w_j}||} \tag{1}$$

where $\mathsf{w_i}$ and $\mathsf{w_j}$ denote word embeddings of two words $w_i$ and $w_j$, respectively.

**Table 1.** Top five similar words to words: *book* and *computer*.

| Inquiry | Similar words | Cosine similarity scores |
|---|---|---|
| Book | Books | 0.693 |
| | Novel | 0.655 |
| | Diary | 0.625 |
| | Pamphlet | 0.620 |
| | Chapter | 0.617 |
| Computer | Computers | 0.688 |
| | Computing | 0.647 |
| | Programmer | 0.614 |
| | Console | 0.604 |
| | Hardware | 0.603 |

### 3.2  FBoW

BoW suffers extreme sparsity when representing short texts since short texts only contain very few words. FBoW adopts a fuzzy mapping mechanism to solve this problem. A document in FBoW model is represented by $\mathbf{z} = [z_1, z_2 \ldots, z_l]$, where the $z_i$ is the weight of the $i$th basis term, and is computed as:

$$z_i = c_i \sum_{w_j \in \mathsf{w}} A_{t_i}(w_j) x_j \tag{2}$$

where $c_i$ is a controlling parameter, w is the set of all the words occurred in the document, $x_j$ is term frequency of word $w_j$ in current document, and $A_{t_i}(w_j)$ is the semantically similarity between the $i$th basis term $t_i$ and word $w_j$, which is computed as:

$$A_{t_i}(w_j) = \begin{cases} \cos(\mathbf{W}[t_i], \mathbf{W}[w_j]), & \text{if } \cos(\mathbf{W}[t_i], \mathbf{W}[w_j]) > 0 \\ 0, & \text{otherwise} \end{cases} \tag{3}$$

where $\mathbf{W}[w_j]$ denotes word embedding of word $w_j$.

In addition, Zhao et al. also proposed Fuzzy Bag-of-Words Cluster model (FBoWC) to alleviate the high dimensional disaster in FBoW. The basic idea of FBoWC is to use word clusters instead of words as basis terms, and word clusters is obtained by clustering words in basis terms. To obtain the similarity between word $w_i$ and a word cluster $\mathbf{t}_i$, similarities between $w_i$ and every word in $\mathbf{t}_i$ should be considered, and among them the mean, maximum, or minimum value can be defined as similarity between $w_i$ and $\mathbf{t}_i$.

### 3.3  Fuzzy Bag-of-Topics Model

K-means clustering algorithm is applied in FBoWC to obtain word clusters. Thus the performance of FBoWC model can be influenced by K-means clustering algorithm. However, the result of K-means is instable. Without appropriate initialization, K-means may work badly, resulting in incoherence word clusters.

In previous researches, some researchers utilize auxiliary external knowledge to overcome the sparsity in short texts in some text mining tasks [12,13] and achieve success. Inspired by their works, we proposed our FBoT model based on FBoWC.

The core idea of FBoT model is to utilize word communities as basis features in document vector. In order to obtain word communities, firstly, we run LDA on an external corpus to obtain some topics. Then, from each topic, we acquire the top words to form a word community, i.e. the $i$th word community is defined as the set:$\{word \mid p_i(word) > \lambda\}$, where $p_i(\cdot)$ is the probability of the word in the $i$th topic, and $\lambda$ is the threshold. Only top high-probability words are selected to form word communities. If no word meet the condition, the highest-probability word would be selected. In FBoT model, a document is represented by $\mathbf{z} = [z_1, z_2 \ldots, z_K]$, which is calculated as:

$$\mathbf{z} = \mathbf{xH} \tag{4}$$

where

$$\mathbf{x} = [x_1, x_2, \ldots, x_v] \tag{5}$$

where $v$ is the number of words in vocabulary, $x_i$ denotes the number of occurrence of word $w_i$ in the document.

$$\mathbf{H} = \begin{bmatrix} \mathrm{sim}(w_1, \mathbf{t}_1) & \mathrm{sim}(w_1, \mathbf{t}_2) & \cdots & \mathrm{sim}(w_1, \mathbf{t}_K) \\ \mathrm{sim}(w_2, \mathbf{t}_1) & \mathrm{sim}(w_2, \mathbf{t}_2) & \cdots & \mathrm{sim}(w_2, \mathbf{t}_K) \\ \vdots & \vdots & & \vdots \\ \mathrm{sim}(w_v, \mathbf{t}_1) & \mathrm{sim}(w_v, \mathbf{t}_2) & \cdots & \mathrm{sim}(w_v, \mathbf{t}_K) \end{bmatrix} \tag{6}$$

where $K$ is the number of word communities predefined and $\mathrm{sim}(w_j, \mathbf{t}_i)$ is the similarity between word $w_j$ and word community $\mathbf{t}_i$, which is computed as:

$$\mathrm{sim}(w_j, \mathbf{t}_i) = \frac{1}{n_i} \sum_{w_k \in \mathbf{t}_i} \cos(\mathbf{W}[w_j], \mathbf{W}[w_k]) \tag{7}$$

where $\mathbf{W}[w]$ is the word embedding of $w$, and $n_i$ is the number of words in the $i$th word community.

The procedure of FBoT model is presented in Algorithm 1.

---

**Algorithm 1.** FBoT Framework

---

**Input:** A text corpus with N documents; the vocabulary D and its corresponding word embedding matrix $W \in \mathbb{R}^{v \times d}$, where $v$ is the vocabulary size and $d$ is the dimensionality of word embeddings

1: Based on the vocabulary D, obtain document-word matrix X, where $x_{ij}$ is term frequency of $w_j$ in the $i$th document
2: Calculate H according to (6)
3: Calculate Z by Z = XH
4: Return Z

**Output:** Learned document vectors Z for the corpus.

---

## 4    Experiments

To validate the effectiveness of our proposed FBoT model, we conduct experiments of text classification based on different text representation methods.

### 4.1    Datasets

Google Snippets [13]: This dataset contains 8 categories, as shown in Table 2, including business, computers, health, and so on. The training dataset consists of 10060 labeled snippets and the test dataset consists of 2280 snippets. On average, each snippet has 18.07 words.

Amazon Reviews[1]: This dataset contains 10000 labeled reviews belonging to 2 categories in a ratio of 5097 to 4903. The average length of reviews is 79.6. Five-fold cross validation was conducted on this dataset, and the ratio of train set to test set is 4:1.

### 4.2    Experimental Setup

In order to verify the effectiveness of our proposed FBoT model, we conduct short text classification experiments by different text representation methods. Our proposed FBoT model are compared with the following methods:

(1) BoW: Bag-of-Words model.
(2) LSA: Latent Semantic Analysis model in [9].
(3) LDA: Latent Dirichlet Allocation model in [8].
(4) AE: Average Embeddings, the mean word embedding vector of all the words occurred in a document.

---

[1] https://gist.github.com/kunalj101.

**Table 2.** Data distribution of Google Snippets

| Labels | Training | Test |
|---|---|---|
| Business | 1200 | 300 |
| Computers | 1200 | 300 |
| Culture-arts-entertainment | 1880 | 330 |
| Education-Science | 2360 | 300 |
| Engineering | 220 | 150 |
| Health | 880 | 300 |
| Politics-Society | 1200 | 300 |
| Sports | 1120 | 300 |
| Total | 10060 | 2280 |

(5) PV: Paragraph Vector model in [11].
(6) FBoW: Fuzzy Bag-of-Words model in [2].
(7) FBoWC: Fuzzy Bag-of-Words Cluster model in [2].

For BoW and FBoW model, only top 1000 high-frequency words were considered because they are more meaningful. For LSA, LDA, and BoT, the number of latent topics were all set to 300. The dimensionality of the pre-trained word embeddings was set to 300, thus text vector calculated by AE also has a dimensionality of 300. For PV, dimensionality was set to 300. And for FBoWC, the number of word clusters was also set to 300. The mapping bound in FBoW and FBoWC was set to zero, since FBoW and FBoWC perform best according to [2]. $\lambda$ was set to 0.15 in FBoT model. Thus, the derived dimensionality of short text vector from LSA, LDA, AE, PV, FBoWC and FBoT are all 300.

In our experiments, word2vec [3] was utilized to generate word embeddings, which were trained on a Google News corpus (over 100 billion words). Word communities were obtained from a news corpus which contains nearly 20000 news by LDA [8]. Support Vector Machine (SVM) was utilized as classification algorithm in our experiments, which was implemented by sklearn[2]. And the implementations of BoW, LSA, LDA are based on gensim[3].

### 4.3   Classification Results

Table 3 shows the results of short text classification. It can be easily observed that FBoT achieved the highest accuracies on two datasets, which means that our method is effective. The results of FBoWC was slightly lower than FBoT. This mainly because word communities in FBoT are more coherent and meaningful than word clusters in FBoWC. FBoWC performed better than FBoW, which might be because word clusters alleviate the noises and redundancies in basis

---

[2] http://scikit-learn.org.
[3] https://radimrehurek.com/gensim.

**Table 3.** Classification accuracies (%) for compared methods on two datasets.

|        | Google Snippets | Amazon Reviews |
|--------|-----------------|----------------|
| BoW    | 69.0            | $67 \pm 0.2$   |
| LSA    | 70.0            | $72 \pm 0.3$   |
| LDA    | 46.8            | $65.5 \pm 0.6$ |
| AE     | 80.3            | $73.0 \pm 0.1$ |
| PV     | 67.4            | $76.8 \pm 0.3$ |
| FBoW   | 78.6            | $73.2 \pm 0.3$ |
| FBoWC  | 82.2            | $75.1 \pm 0.2$ |
| FBoT   | **84.7**        | **$78.0 \pm 0.3$** |

terms. BoW and LDA, however, performed really bad due to the sparsity of short texts. We can also find that LDA performed better in Amazon Reviews than in Google Snippets due to the increases of documents length.

### 4.4 Analysis of Parameter

In our experiments, we also analyze the influence of the value of $\lambda$ in our proposed FBoT model, which is defined as the threshold to select words to form word communities. In the experiment, we set $\lambda$ in the range of 0.05 to 0.5 to observe its effect. The results on Google Snippets dataset are illustrated in Fig. 1 (Results were similar in the other dataset.). As the figure shows, when $\lambda$ is lower than

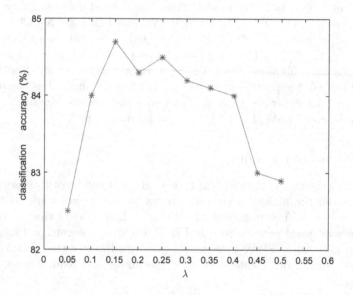

**Fig. 1.** Performance of FBoT for different $\lambda$.

0.1, the classification accuracy boosts with $\lambda$ grows. When $\lambda$ increases over 0.4, the accuracy degrades gradually. And when $\lambda$ is set in the range from 0.1 to 0.4, the accuracy fluctuates within a narrow range. Thus it is rational to set $\lambda$ in this interval. Most word communities would have only one word if $\lambda$ is too large. And too small $\lambda$ may introduce noise.

## 5  Conclusion

We have proposed a new model, FBoT, for short text representation learning. Word communities were defined as basis terms in document vector. And we obtained word communities by run topic model on external corpus. Experimental results show that word communities in FBoT are more coherent and meaningful than word clusters in FBoWC. In experiments of short text classification, text vectors represented by FBoT achieved the highest accuracies compared with 7 other methods, which verified the effectiveness of our proposed FBoT model. However, a limitation of FBoT is that the generalization capacity of the model depends on the corpus which word communities extracted from. In the future, we will explore ways to employ better paradigms, e.g. lifelong learning [14], to generate more generalized and meaningful word communities by utilizing a large amount of corpus.

**Acknowledgments.** This work was supported in part by Shanghai Innovation Action Plan Project under the grant No. 16511101200.

## References

1. Spärck Jones, K.: A statistical interpretation of term specificity and its application to retrieval. J. Doc. **28**, 11–21 (1972)
2. Zhao, R., Mao, K.: Fuzzy bag-of-words model for document representation. IEEE Trans. Fuzzy Syst. **26**, 794–804 (2018)
3. Mikolov, T., Chen, K., Corrado, G., Dean, J.: Efficient Estimation of Word Representations in Vector Space. http://arxiv.org/abs/1301.3781
4. Banerjee, S., Ramanathan, K., Gupta, A.: Clustering short texts using Wikipedia. In: 30th International ACM SIGIR Conference on Research and Development in Information Retrieval, pp. 787–788. ACM, Amsterdam (2007)
5. Hu, X., Sun, N., Zhang, C., Chua, T.S., et al.: Exploring internal and external semantics for the clustering of short texts using world knowledge. In: 18th ACM Conference on Information and Knowledge Management, pp. 919–928. ACM, Hong Kong (2009)
6. Wang, L., Jia, Y., Han, W.: Instant message clustering based on extended vector space model. In: Kang, L., Liu, Y., Zeng, S. (eds.) ISICA 2007. LNCS, vol. 4683, pp. 435–443. Springer, Heidelberg (2007). https://doi.org/10.1007/978-3-540-74581-5_48
7. Sahami, M., Heilman, T.D.: A web-based kernel function for measuring the similarity of short text snippets. In: 15th International Conference on World Wide Web, pp. 377–386. ACM, Edinburgh (2006)

8. Blei, D.M., Ng, A.Y., Jordan, M.I.: Latent Dirichlet allocation. J. Mach. Learn. Res. **3**, 993–1022 (2003)
9. Dumain S.: Latent Semantic Indexing (LSI): TREC-3 Report. In: Harman, M. (ed.) The Third Text REtrieval Conference, vol. 500, no. 226, pp. 219–230. NIST Special Publication, Gaithersburg (1995)
10. Hofmann, T.: Unsupervised learning by probabilistic latent semantic analysis. Mach. Learn. **42**, 177–196 (2001)
11. Le, Q., Mikolov, T.: Distributed Representations of Sentences and Documents. http://arxiv.org/abs/1405.4053
12. Phan, X.H., Nguyen, C.T., Le, D.T., Nguyen, L.M., Horiguchi, S., Ha, Q.T.: A hidden topic-based framework towards building applications with short web documents. Trans. KDE **23**, 961–976 (2011)
13. Phan, X.H., Nguyen, L.M., Horiguchi, S.: Learning to classify short and sparse text and web with hidden topics from large-scale data collections. In: 17th International World Wide Web Conference, pp. 91–100. ACM, Beijing (2008)
14. Daniel, L.S., Yang, Q., Li, L.: Lifelong machine learning systems: beyond learning algorithms. In: Proceedings of the AAAI Spring Symposium on Lifelong Machine Learning, pp. 49–55. AAAI, Palo Alto (2013)

# W-Net: One-Shot Arbitrary-Style Chinese Character Generation with Deep Neural Networks

Haochuan Jiang[1], Guanyu Yang[1], Kaizhu Huang[1(✉)], and Rui Zhang[2]

[1] Department of EEE, Xi'an Jiaotong - Liverpool University,
No. 111 Ren'ai Road, Suzhou, Jiangsu, People's Republic of China
Kaizhu.huang@xjtlu.edu.cn
[2] Department of MS, Xi'an Jiaotong - Liverpool University,
No. 111 Ren'ai Road, Suzhou, Jiangsu, People's Republic of China

**Abstract.** Due to the huge category number, the sophisticated combinations of various strokes and radicals, and the free writing or printing styles, generating Chinese characters with diverse styles is always considered as a difficult task. In this paper, an efficient and generalized deep framework, namely, the *W-Net*, is introduced for the one-shot arbitrary-style Chinese character generation task. Specifically, given a single character (one-shot) with a specific style (e.g., a printed font or hand-writing style), the proposed *W-Net* model is capable of learning and generating any arbitrary characters sharing the style similar to the given single character. Such appealing property was rarely seen in the literature. We have compared the proposed *W-Net* framework to many other competitive methods. Experimental results showed the proposed method is significantly superior in the one-shot setting.

## 1 Introduction

Chinese is a special language with both messaging functions and artistic values. On the other hand, Chinese contains thousands of different categories or over 10,000 different characters among which 3,755 characters, defined as level-1 characters, are commonly used. Given a limited number of Chinese characters or even one single character with a specific style (e.g., a personalized hand-writing calligraphy or a stylistic printing font), it is interesting to mimic automatically many other characters with the same specific style. This topic is very difficult and rarely studied simply because of the large category number of different Chinese characters with various styles. This problem is even harder due to the unique nature of Chinese characters among which each is a combination of various strokes and radicals with diverse interactive structures.

Despite these challenges, there are recently a few proposals relevant to the above-mentioned generation task. For example, in [13], strokes are represented by time-series writing evenly-thick trajectories. Then it is sent to the Recurrent Neural Network based generator. In [6], font feature reconstruction for standardized character extraction is achieved based on an additional network to assist the

© Springer Nature Switzerland AG 2018
L. Cheng et al. (Eds.): ICONIP 2018, LNCS 11305, pp. 483–493, 2018.
https://doi.org/10.1007/978-3-030-04221-9_43

one-to-one image-to-image translation framework. Over 700 pre-selected training images are needed in this framework. In the *Zi2Zi* [12] model, a one-to-many mapping is achieved with only a single model by the fixed Gaussian-noise based categorical embedding with over 2,000 training examples per style.

There are several main limitations in the above approaches. On one hand, the performance of these methods usually relies heavily on a large number of samples with a specific style. In the case of a few-shot or even one-shot generation, these methods would fail to work. On the other hand, these methods may not be able to transfer to a new style which has not been seen during training. Such drawbacks may hence present them from being used practically.

(a) printing font          (b) hand-writing style

**Fig. 1.** Generated traditional characters by the proposed *W-Net* model **with one single sample available** (the right-bottom character with red boxes). (Color figure online)

In this paper, aiming to generate Chinese characters when even given one shot sample with a specific arbitrary style (seen or unseen in training), we propose a novel deep model named *W-Net* as a generalized style transformation framework. This framework better solves the above-mentioned drawbacks and could be easily used in practice. Particularly, inherent from the *U-Net* framework [9] for the one-to-one image-to-image translation task [4], the proposed *W-Net* employs two parallel convolution-based encoders to extract style and content information respectively. The generated image will be obtained by the deconvolution-based decoder by using the encoded information. Short-cut connections [9] and multiple residual blocks [2] are set to deal with the gradient vanishing problem and balance information from both encoders to the decoder. The training of the *W-Net* follows an adversarial manner. Inspired by the recently proposed Wasserstein Generative Adversarial Network (W-GAN) framework with gradient penalty [1], an independent discriminator[1] ($D$) are employed to assist the *W-Net* ($G$) learning.

---

[1] The discriminator actually attaches an auxiliary classifier proposed in [8].

As a methodological guidance, only one-shot arbitrary-style Chinese character generation is demonstrated in this paper, as examples given in Fig. 1. However, the *W-Net* framework can be extended to a variety of related topics on one-shot arbitrary-style image generation. With such a proposal, the data synthesizing tasks with few samples available can be fulfilled much more readily and effectively than previous approaches in the literature.

## 2   Model Definition

### 2.1   Preliminary

Denote $X$ be a Chinese character dataset, consisting of $J$ different characters with in total $I$ different fonts. Let $x_j^i$ be a specific sample in $X$, regarded as the **real target**. Following [3,5], the superscript $i \in [0, 1, 2, ..., I]$ represents $i$-th style, while the subscript $j \in [1, 2, ..., J]$ denotes $j$-th example.

Specifically, during the training, when $i = 0$, $x_j^0$ denotes the image of the $j$-th character with a standardized style information, named as the **prototype content**. Meanwhile, $x_k^i, k \in [1, 2, ..., J]$ is defined as a **style reference** equipping with the $i$-th style information, the same as $x_j^i$. Be noted that commonly, $j$ and $k$ are different. In the proposed model, each $x_j^i$ is assumed to be combined with information from the prototype $x_j^0$ and the $i$-th writing style learned from $x_k^i$. The proposed *W-Net* model will then produce the **generated target** $G(x_j^0, x_k^i)$ which is similar to $x_j^i$ by taking both $x_j^0$ and $x_k^i$ simultaneously.

Be noted that only single style character is required to produce the generated target. It is defined as the **One-Shot Arbitrary-Style Character Generation** task. Specifically, the given single sample (E.g., $x_p^m$, where $m$ can be any arbitrary style, meanwhile $p$ could be any single character. Both $m$ and $p$ can be irrelevant to $[1, 2, ..., I]$ and $[1, 2, ..., J]$ respectively) is seen as the **one-shot** style reference. The task can be readily fulfilled by feeding any content prototype $(x_q^0)$ of the desired $q$-th character on condition of those relevant outputs of the $Enc_r$ (to be connected to the $Dec$ with both shortcut or residual block connections, as will be demonstrated in Sect. 2.2) to produce $G(x_q^0, x_p^m)$ given the single style example $(x_p^m)$. In such the setting, alternating $q$ will lead to synthesizing different characters. Simultaneously, all the generated examples are expected to imitate the $m$-th style information given by $x_p^m$. Similarly, $q$ could also be out of $[1, 2, ..., J]$ as well.

### 2.2   *W-Net* Architecture

Figure 2 illustrates the basic structure of the proposed *W-Net* model. It consists of the content prototype encoder ($Enc_p$, the blue part), the style reference encoder ($Enc_r$, the green part), and the decoder ($Dec$, the red part).

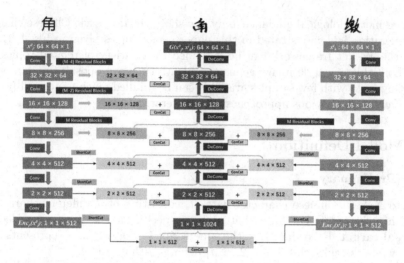

**Fig. 2.** The *W-Net* (better viewed in colors), where the blue part represents $Enc_p$, green for $Enc_r$, and red for $Dec$. Conv: $5 \times 5$ convolution; DeConv: $5 \times 5$ deconvolution. Fixed stride 2 and ReLU are applied to both Conv and DeConv. ConCat: Feature concatenation on the channel; ShortCut: Feature Shortcut.

The $Enc_p$ and $Enc_r$ are constructed as sequences of convolutional layers, where $5 \times 5$ filters with fixed stride 2 and ReLU function are implemented. By this setting, $64 \times 64$ prototype and reference images $x_j^0$ and $x_j^k$ will be mapped into $1 \times 512$ feature vector, denoted as $Enc_p(x_j^0)$ and $Enc_r(x_j^k)$ respectively.

Identical to the decoder in the *U-Net* framework [9], $Dec$ is designed as a deconvolutional progress layer-wisely connected with $Enc_p$ and $Enc_r$. It produces a generated image, the size of which is consistent with all the input images of both encoders. Specifically, for higher-level features between the decoder and both the encoders, connections are achieved by simple feature shortcut. For lower-level layers of $Enc_p$, a series of residual blocks[2] [2] are applied and connected to the $Dec$. The number of blocks is controlled by a super parameter $M$. On the contrast, as the writing style is a kind of high-level deep feature, there is only one residual block connection (with $M$ blocks) between $Enc_r$ and $Dec$, omitting lower-level feature concatenation at the same time.

## 2.3 Optimization Strategy and Losses

The proposed *W-Net* is trained adversarially based on the Wasserstein Generative Adversarial Network (W-GAN) framework, regarded as the generator $G$. Specifically, it takes the content prototype and the style reference, and then returns generated target as $G(x_j^0, x_k^i) = Dec(Enc_p(x_j^0), Enc_r(x_k^i))$ closed to $x_j^i$. $G$ is optimized by taking advantages of the adversarial network $D$ as well as several optimization losses defined as follows.

---

[2] The structure of the residual block follows the setting in [6].

**Training Strategy:** The learning of the *W-Net* follows the adversarial train-ing scheme. In each learning iteration, there are two independent procedures, including the $G$ training and the $D$ training respectively. The $G$ and the $D$ are trained to optimize Eqs. (1) and (2) respectively.

$$\mathbb{L}_G = -\alpha\mathbb{L}_{adv-G} + \beta_d\mathbb{L}_{dac} + \beta_p\mathbb{L}_{enc-p-cls} + \beta_r\mathbb{L}_{enc-r-cls}$$
$$+ \lambda_{l1}\mathbb{L}_1 + \lambda_\phi\mathbb{L}_\phi + \psi_p\mathbb{L}_{Const_p} + +\psi_r\mathbb{L}_{Const_r} \tag{1}$$

$$\mathbb{L}_D = \alpha\mathbb{L}_{adv-D} + \alpha_{GP}\mathbb{L}_{adv-GP} + \beta_d\mathbb{L}_{dac} + \beta_p\mathbb{L}_{enc-p-cls} + \beta_r\mathbb{L}_{enc-r-cls} \tag{2}$$

**Adversarial Loss:** $G$ optimizes $\mathbb{L}_{adv-G} = D(x_j^0, G(x_j^0, x_k^i), x_k^i)$, while $D$ mini-mizes $\mathbb{L}_{adv-D} = D(x_j^0, x_j^i, x_k^i) - D(x_j^0, G(x_j^0, x_k^i), x_k^i)$. Be noted that a gradient penalty is set as $\mathbb{L}_{adv-GP} = ||\nabla_{\widehat{x}}D(x_j^0, \widehat{x}, x_k^i) - 1||_2$ [1], where $\widehat{x}$ is uniformly interpolated along the line between $x_j^i$ and $G(x_j^0, x_k^i)$.

**Categorical Loss of the Discriminator Auxiliary Classifier:** $\mathbb{L}_{dac} = \left[\log C_{dac}(i|x_j^0, x_j^i, x_k^i)\right] + \left[\log C_{dac}(i|x_j^0, G(x_j^0, x_k^i), x_k^i)\right]$, inspired by [8].

**Reconstruction Losses** consists the pixel-level difference ($\mathbb{L}_1 = ||(x_j^i - G(x_j^0, x_k^i))||_1$) and the high-level feature variation ($\mathbb{L}_\phi = \sqrt{\sum_\phi \left[\phi(x_j^i) - \phi(G(x_j^0, x_k^i))\right]^2}$). $\phi(.)$ represents a specific deep feature. The VGG-16 network [10] trained with multiple character styles is employed here. In this optimization, in total five convolutional features including $\phi_{1-2}$, $\phi_{2-2}$, $\phi_{3-3}$, $\phi_{4-3}$, $\phi_{5-3}$ are involved.

**Constant Losses of the Encoders:** The constant losses [11] are also employed for both encoders. They are given by $\mathbb{L}_{Const_p} = ||Enc_p(x_j^0) - Enc_p(G(x_j^0, x_k^i))||^2$ and $\mathbb{L}_{Const_r} = ||Enc_r(x_k^i) - Enc_r(G(x_j^0, x_k^i))||^2$ respectively for $Enc_p$ and $Enc_r$.

**Categorical Losses on Both Encoders:** To ensure the specific functional-ities of the two encoders, we forced the content and style features extracted by them to be equipped with the corresponding commonality separately for the same kind. It is designated by adding a fully-connecting to implement the category classification task, which leads to that both encoders learn their own representative features, simultaneously over-fitting is avoided thereby. $\theta_p$ and $\theta_r$ are used to denote the fully-connecting and softmax functions together for both output feature vectors of encoders respectively, while the classifications are noted as $C_{encp}$ and $C_{encr}$. The upon-mentioned cross entropy losses of both classifications are given as $\mathbb{L}_{enc-p-cls} = \left[\log C_{encp}(j|\theta_p(Enc_p(x_j^0)))\right]$ and $\mathbb{L}_{enc-p-cls} = \left[\log C_{encr}(i|\theta_r(Enc_r(x_k^i)))\right]$ respectively. Be noted that $i$ and $j$ rep-resent the specific style and the character labels.

## 3   Experiment

A series of experiments have been conducted to verify the effectiveness of the proposed *W-Net* network. Both printed and hand-writing fonts are evaluated. Several relevant baselines are also referred for the comparison as well.

### 3.1  Experiment Setting

80 fonts are specially chosen in standard Chinese printed font database. 50 of them, each containing 3,755 level-1 simplified Chinese characters, are involved in the training set. The offline version of both CASIA-HWDB-1.1 (for simplified isolated characters) and the CASIA-HWDB-2.1 (for simplified cursive characters) [7] are involved as the hand-writing data set. Characters written by 50 writers (No. 1,101 to 1,150) are selected as the training set, resulting in total 249,066 samples (4,980 examples per writer averagely). For both sets, the testing data are chosen due to different evaluation purposes. *HeiTi* (boldface font) is used as the prototype font for both the sets, as examples given in Fig. 3(a).

Baseline models include two upgraded version of the *Zi2Zi* [12] framework which were modified for the few-shot new-coming style synthesization task. One utilizes a fine-tuning strategy (noted as *Zi2Zi-V1*), where the style information is assumed to be the linear combination of multiple known styles represented by the fixed Gaussian-noise based categorical embedding; The other (*Zi2Zi-V2*) discards the categorical embedding by introducing the final softmax output of a pre-trained VGG-16 network (embedder network), identical to the one employed in Sect. 2.3. All the other network architecture and training settings of these baselines are all the same by following [12]. Characters from both databases are represented by $64 \times 64$ gray-scale images, after which they are then binarized. One thing to be particularly noted is that both the proposed *W-Net* and the *Zi2Zi-V2* follow the **one-shot** setting, where only single style example ($x_p^m$) is referred during the evaluation process. However, the *Zi2Zi-V1* employ the few-shot (32 references) scheme in order to obtain a valid fine-tuning performance.

Hyper-parameters during *W-Net* training are tuned empirically and set as follows: residual block number is $M = 5$; relevant penalties are: $\alpha = 3$, $\alpha_{GP} = 10$, $\beta_d = 1$, $\beta_p = \beta_r = 0.2$, $\lambda_{l1} = 50$, $\lambda_\phi = 75$, $\psi_p = 3$ and $\psi_r = 5$.

The Adam optimizer with $\beta_1 = 0.5$ and $\beta_2 = 0.999$ is implemented, while the initial learning rate is set to be 0.0005 and decayed exponentially after each training epoch. The architecture of $D$ follows the setting of the *Zi2Zi* framework [12] with the W-GAN framework. For the sake of speeding up and stabilizing the training progress, the batch normalization is applied several layers to the $G$ network, while the layer normalization is selected for $D$. Drop out trick is also applied to both $G$ and $D$ to improve the generalization performance. Weight decay is also engaged to avoid the over-fitting issue. The proposed *W-Net* framework and other baselines are implemented with the Tensorflow (r1.5).

### 3.2  Model Reasonableness Evaluation

The *W-Net* model is verified by setting $p = q$ for content $x_q^0$ and the style $x_p^m$ in this section. Hereby, the reference is exactly the real target ($x_p^m = x_q^m$). For each evaluation, as previously instructed, only single style reference ($x_p^m$, characters of 2nd rows in (b)–(e) of Fig. 3) is engaged. The generated image is seen to follow the style tendency of the one-shot reference if the proposed *W-Net* is capable of reconstructing the extracted style information in the reference image $x_p^m$.

**Fig. 3.** Several examples of generated data of unseen printing and hand-writing styles. (a) The input content prototypes; (b)–(e) 1st row: generated characters; 2nd row: corresponding style references (ground truth characters).

Figure 3 illustrates several examples of the comparison result for synthesizing unseen styles during training. It can be observed that styles of both printed and hand-writing types are learned and transferred to the prototypes by the *W-Net* model with the proper performance by maintaining the style consistency.

### 3.3   Model Effectiveness Evaluation

The effectiveness of *W-Net* is tested by generating commonly-used Chinese characters (simplified and traditional) with alternative styles. In this setting, $x_p^m$ are randomly selected one-shot character with the $m$-th style information to imitate the real application scenario, while $q$ are referred to the desired content prototypes to be generated. Commonly, $p \neq q$.

Figures 4 and 5 list several examples of the generated images by *W-Net* and two baselines for seen and unseen styles during training respectively. Particularly, only simplified Chinese characters are accessible during training, as seen in the left four columns of each subfigure of those two illustrations. There is no ground truth data in both the involved databases of those traditional images in other remaining columns.

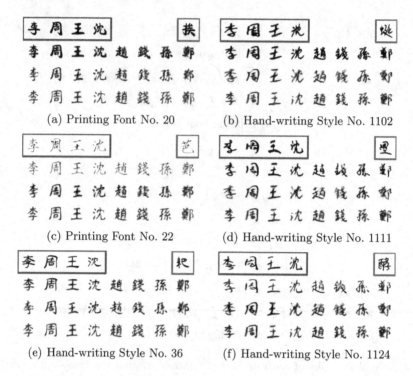

(a) Printing Font No. 20

(b) Hand-writing Style No. 1102

(c) Printing Font No. 22

(d) Hand-writing Style No. 1111

(e) Hand-writing Style No. 36

(f) Hand-writing Style No. 1124

**Fig. 4.** Several examples of generated characters of seen styles. (a), (c) and (e) are printing fonts; (b), (d) and (f) are hand-writing styles. In each figure, 1st row: ground truth characters (with blue boxes) and the one-shot style reference (with red boxes); 2nd: *W-Net* generated characters; 3rd row: *Zi2Zi-V1* performance; 4th row: *Zi2Zi-V2* performance. (Color figure online)

When generating characters with a specific seen style during training, it can be intuitively observed in Fig. 4 that even given one-shot style reference, the generated fonts by *W-Net* look very similar to the corresponding real targets. Differently, under the few-shot setting, the *Zi2Zi-V1* still produces blurred images, while *Zi2Zi-V2* seems to synthesize characters with the averaged style. The proposed *W-Net* outperforms others by producing characters with both desired contents and consistent styles with only one-shot style reference available.

Simultaneously, acceptable generations can still be obtained from the Fig. 5 by the proposed scheme when constructing unseen styles with one-shot style reference as well. Though the generated samples are not similar enough as that in previous examples, a clear stylistic tendency can still be clearly observed. On the contrast, *Zi2Zi-V1* failed to produce high-quality images even 32 references are given for the fine-tuning due to the over-fitting issue. At the same time, the *Zi2Zi-V2* failed to generate distinguishable styles since it is only capable of learning styles from the original basis provided by the embedder network (VGG-16).

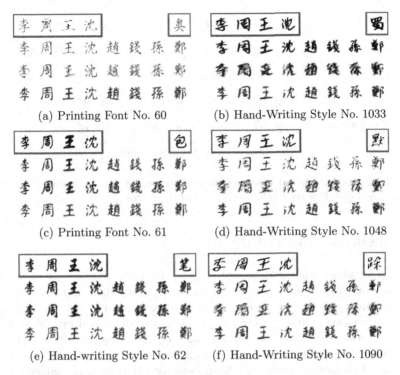

(a) Printing Font No. 60          (b) Hand-Writing Style No. 1033

(c) Printing Font No. 61          (d) Hand-Writing Style No. 1048

(e) Hand-writing Style No. 62     (f) Hand-Writing Style No. 1090

**Fig. 5.** Several examples of generated characters of unseen styles. (a), (c) and (e) are printing fonts; (b), (d) and (f) are hand-writing styles. In each figure, 1st row: ground truth characters (with blue boxes) and the one-shot style reference (with red boxes); 2nd: *W-Net* generated characters; 3rd row: *Zi2Zi-V1* performance; 4th row: *Zi2Zi-V2* performance. (Color figure online)

### 3.4 Analysis on Failure Examples

The proposed model would sometimes fail to capture the style information when it is over far away from the prototype font. For example, some cursive writing may play a negative role in the generation process since input contents are all isolated characters. Some failed generated characters are given in Fig. 6, of which the 2nd row lists the corresponding one-shot style references.

Upon the proposed *W-Net*, each target is regarded as a non-linear style transformation from a reference to a prototype. However, when the style is too different from the content font, the model fails to learn this complicated mapping relationship. In such extreme circumstances, the provided single prototype font in this paper might be an inappropriate choice. In this case, it could be a good idea to learn additional mappings which can transform the original prototype to suitable latent features so as to better handle free writing styles in real scenarios.

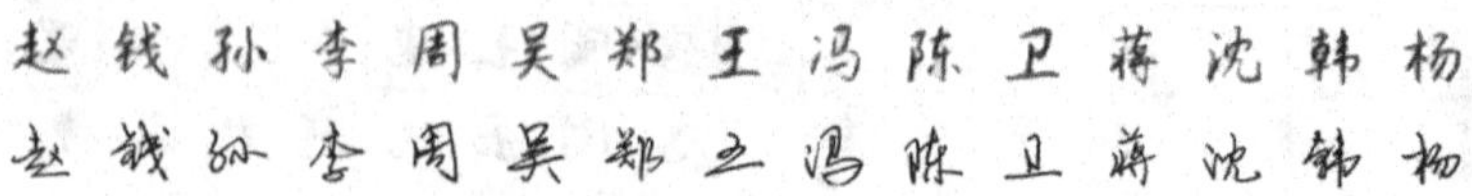

赵 钱 孙 李 周 吴 郑 王 冯 陈 卫 蒋 沈 韩 杨
赵 钱 孙 李 周 吴 郑 之 冯 陈 卫 蒋 沈 韩 杨

(a) Printing Font No. 71

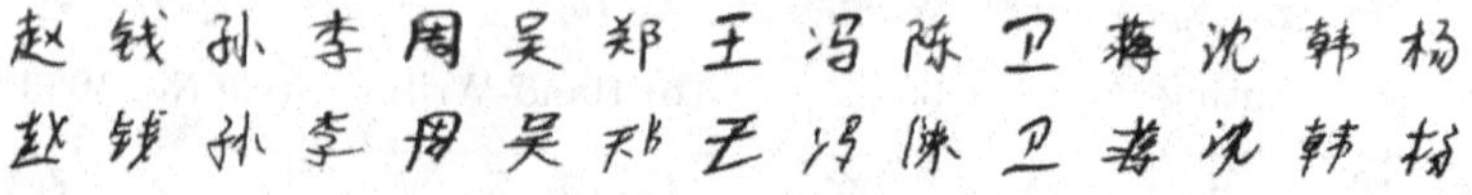

赵 钱 孙 李 周 吴 郑 王 冯 陈 卫 蒋 沈 韩 杨
赵 钱 孙 李 周 吴 郑 王 冯 陈 卫 蒋 沈 韩 杨

(b) Hand-Writing Style No. 1289

**Fig. 6.** Unsatisfied generated examples. In each figure: 1st row: generated characters; 2nd row: corresponding style references (ground truth characters)

## 4   Conclusion and Future Work

A novel generalized framework named *W-Net* is introduced in this paper in order to achieve one-shot arbitrary-style Chinese character generation task. Specifically, the proposed model, composing of two encoders and one decoder with several layer-wised connections, is trained adversarially based on the W-GAN scheme. It enables synthesizing any arbitrary stylistic character by transferring the learned style information from one single reference to the input content prototype. Extensive experiments have demonstrated the reasonableness and effectiveness of the proposed *W-Net* model in the one-shot setting.

Extensions to more proper mapping architectures for image reconstruction will be studied in the future so as to capture sufficiently complicated and free writing styles. Meanwhile, practical applications are to be developed not only restricted in the character generation domain, but also in other relevant arbitrary-style image generation tasks.

**Acknowledgements.** The work was partially supported by the following: National Natural Science Foundation of China under no. 61473236 and 61876155; Natural Science Fund for Colleges and Universities in Jiangsu Province under no. 17KJD520010; Suzhou Science and Technology Program under no. SYG201712, SZS201613; Jiangsu University Natural Science Research Programme under grant no. 17KJB-520041; Key Program Special Fund in XJTLU under no. KSF-A-01 and KSF-P-02.

## References

1. Gulrajani, I., Ahmed, F., Arjovsky, M., Dumoulin, V., Courville, A.C.: Improved training of Wasserstein GANs, pp. 5769–5779 (2017)
2. He, K., Zhang, X., Ren, S., Sun, J.: Identity mappings in deep residual networks. In: Leibe, B., Matas, J., Sebe, N., Welling, M. (eds.) ECCV 2016. LNCS, vol. 9908, pp. 630–645. Springer, Cham (2016). https://doi.org/10.1007/978-3-319-46493-0_38
3. Huang, K., Jiang, H., Zhang, X.Y.: Field support vector machines. IEEE Trans. Emerg. Top. Comput. Intell. **1**(6), 454–463 (2017)

4. Isola, P., Zhu, J.Y., Zhou, T., Efros, A.A.: Image-to-image translation with conditional adversarial networks. arXiv preprint (2017)
5. Jiang, H., Huang, K., Zhang, R.: Field support vector regression. In: Liu, D., Xie, S., Li, Y., Zhao, D., El-Alfy, E.S. (eds.) ICONIP 2017. LNCS, vol. 10634, pp. 699–708. Springer, Heidelberg (2017). https://doi.org/10.1007/978-3-319-70087-8_72
6. Jiang, Y., Lian, Z., Tang, Y., Xiao, J.: DCFont: an end-to-end deep Chinese font generation system. In: SIGGRAPH Asia 2017 Technical Briefs, p. 22. ACM (2017)
7. Liu, C.L., Yin, F., Wang, D.H., Wang, Q.F.: Casia online and offline Chinese handwriting databases, p. 37–41 (2011)
8. Odena, A., Olah, C., Shlens, J.: Conditional image synthesis with auxiliary classifier GANs. arXiv preprint arXiv:1610.09585 (2016)
9. Ronneberger, O., Fischer, P., Brox, T.: U-Net: convolutional networks for biomedical image segmentation. In: Navab, N., Hornegger, J., Wells, W.M., Frangi, A.F. (eds.) MICCAI 2015. LNCS, vol. 9351, pp. 234–241. Springer, Cham (2015). https://doi.org/10.1007/978-3-319-24574-4_28
10. Simonyan, K., Zisserman, A.: Very deep convolutional networks for large-scale image recognition. arXiv preprint arXiv:1409.1556 (2014)
11. Taigman, Y., Polyak, A., Wolf, L.: Unsupervised cross-domain image generation. arXiv preprint arXiv:1611.02200 (2016)
12. Tian, Y.: zi2zi: master Chinese calligraphy with conditional adversarial networks (2017). https://github.com/kaonashi-tyc/zi2zi/
13. Zhang, X.Y., Yin, F., Zhang, Y.M., Liu, C.L., Bengio, Y.: Drawing and recognizing Chinese characters with recurrent neural network. IEEE Trans. pattern Anal. Mach. Intell. **40**(4), 849–862 (2017)

# Automatic Grammatical Error Correction Based on Edit Operations Information

Quanbin Wang and Ying Tan(✉)

Key Laboratory of Machine Perception (Ministry of Education)
and Department of Machine Intelligence,
School of Electronics Engineering and Computer Science, Peking University,
Beijing 100871, People's Republic of China
{qbwang362,ytan}@pku.edu.cn

**Abstract.** For second language learners, a reliable and effective Grammatical Error Correction (GEC) system is imperative, since it can be used as an auxiliary assistant for errors correction and helps learners improve their writing ability. Researchers have paid more emphasis on this task with deep learning methods. Better results were achieved on the standard benchmark datasets compared to traditional rule based approaches. We treat GEC as a special translation problem which translates wrong sentences into correct ones like other former works. In this paper, we propose a novel correction system based on sequence to sequence (Seq2Seq) architecture with residual connection and semantically conditioned LSTM (SC-LSTM), incorporating edit operations as special semantic information. Our model further improves the performance of neural machine translation model for GEC and achieves state-of-the-art $F_{0.5}$-score on standard test data named CoNLL-2014 compared with other methods that without any re-rank approach.

**Keywords:** Grammatical error correction · Edit operations
Natural language processing · Semantically conditioned LSTM
Sequence to sequence

## 1 Introduction

With the development of globalization, the number of second language learners is growing rapidly. Errors, including grammar, misspelling and collocation (for simplicity, we call all these types of errors grammatical errors) are inevitable for freshman who just begin to learn a new language. In order to help those learners to avoid making errors in their learning process and improve their skills both for writing and speaking, an automatic grammatical error correction (GEC) system is necessary.

Specifically, GEC for English has attracted much attention as an important natural language processing (NLP) task since 1980s. Macdonald et al. developed a GEC tool named Writer's Workbench based on some rules in 1982 [22], this

© Springer Nature Switzerland AG 2018
L. Cheng et al. (Eds.): ICONIP 2018, LNCS 11305, pp. 494–505, 2018.
https://doi.org/10.1007/978-3-030-04221-9_44

work leads the research in this field. Rule based error correction methods can achieve high precision but with lower recall because the lack of generalization. Some learning based approaches had been adopted to alleviate this drawback, such as learning correction rules with corpora and machine learning algorithms with N-grams features. Mangu et al. proposed a method which learned rules for misspelling correction from a data set called Brown [13]. In addition, [29] used N-grams and language model (LM) to cope with GEC problem.

From a common accepted perspective, researchers always treat GEC as a special translation task which translates text with errors to correct one. On account of this, many machine translation methods are utilized to rectify errors. Statistical machine translation (SMT) as one of the most effective approach, was first adopted to GEC in 2006 [3], they used SMT based model to correct 14 kinds of noun number ($Nn$) errors and achieved much better performance than rules based systems. Compared to traditional rules based and learning based methods, machine translation based approaches only need corpora with pairwise sentences. What is more, they have no limit to specific error types and can construct general correction model for all kinds of errors. Whereas the main drawback of SMT based GEC is that it handles each word or phrase independently, which results in ignoring global context information and relationship between each entity. With the aim of making up this deficiency, researchers attempted to take the advantages of some neural encoder-decoder architectures such as sequence to sequence (Seq2Seq) [27] with recurrent neural networks (RNN), since these models considered the global source text and all the preceding words when decoding. Xie et al. proposed a neural machine translation (NMT) model based GEC system in 2014 [31], which was the first attempt to combine encoder-decoder architecture with attention mechanism like the work in NMT [1]. They used character level embedding and gated recurrent unit (GRU) [5] to correct all kinds of errors and obtained a result on pair with state-of-the-art at that time.

In this paper, we further exploit neural encoder-decoder architecture with RNN and attention mechanism which is similar as commonly used in NMT. In addition, we utilize residual connection as in ResNet but with RNN [16] between every two layers to make training process stable and effective. Different from [31], we adopt long short term memory (LSTM) [17] in both encoder and decoder steps with special semantic information called edit operations. We conclude 3 kinds of different edit operations in correction process as "Delete, Insert and Substitute", which can also be considered as 3 kinds of simple error types as "Unnecessary, Missing and Replacement" as defined in [4,11]. For the purpose of using these edit operations information, semantically conditioned LSTM (SC-LSTM) [30] is applied to our RNN based Seq2Seq model. In view of only a small part of the whole text need to be corrected, we use a gate for those edit operations. Through the results of our experiments, the gate is very useful to improve the performance of SC-LSTN in GEC task. Since whether opening the gate or not in a decoding step is mainly depends on all the words had been generated until now, and there exists a clear distinction between training process and inference, the model may give error gate information because of some mistakes made in former steps. With the aim of alleviating this drawback, we take the advantage of the scheduled sampling technique [2]. With all of these methods, our automatic GEC system with edit operations information achieves 48.67% $F_{0.5}$-score on the benchmark CoNLL-2014 test set [23]. It is state-of-the-art performance compared to other approaches without the help of large language model and other tricks to re-rank candidate corrections.

## 2   Related Work

Researchers in the field of NLP had paid much emphasis on GEC task since 2013, with the organization of CoNLL-2013 and 2014 shared tasks [23,24], of which were competitions to cope with grammatical error correction problem of essays written by second language learners. The test set in 2014 shared task had been used as a standard benchmark since then and many works were made to perform well on it.

The most commonly used methods in recent years are all related to machine translation including statistical and neural models. All the top-ranking teams in CoNLL shared tasks are used SMT based approaches to correct grammatical errors, such as CAMB [12] and AMU [19]. Susanto et al. proposed a system which combined SMT based method with a classification model and got a better result [26]. The most effective technique which purely based on SMT was put forward by Chollampatt et al. [6], they designed some sparse and dense features manually and incorporated some tricks, such as LM, spelling checker and neural network joint model (NNJMs) [8], to further improve their model's performance which was similar to [20].

In spite of the success of SMT based model for GEC task, those kinds of methods suffer from ignoring global context information and lacking of smooth representation which resulted in lower generalization and unnatural correction. To address these issues, several correction systems which adopted neural encoder-decoder framework have been presented. *RNNSearch* [1]was the first NMT model be utilized to correct grammatical errors by Yuan et al. [32]. They additionally applied an unsupervised word alignment technique and a word level SMT for unknown words replacing. However, their work were conducted with Cambridge Learner Corpus (CLC) which is non-public. Xie et al. [31] used a model with similar architecture, but they chose character level granularity to avoid unknown words problem effectively. They trained their model on two publicly available corpora called NUCLE [10] and Lang-8 [28]. For supplementary, they synthesized examples with frequent errors by some rules. A N-gram LM and edit classifier were incorporated to choose solutions. Ji et al. also proposed a RNN based Seq2Seq model with hybrid word and character level embedding and attention for known and unknown words respectively [18]. Except NUCLE and Lang-8, they employed non-public CLC dataset like [32] for training. What is more, they further improved the performance of their correction system by a candidates rescoring LM based on a very large scale corpora. Researchers have investigated the effectiveness of convolutional neural networks (CNN) for encoder-decoder architecture to cope with GEC task. Chollampatt et al. proposed a Seq2Seq model fully based on multi-layer CNN [7], they adopted the famous model in [14] with BPE-based sub-word units embedding. In order to select the best correction, they trained a resoring model with edit operations and LM as features explicitly.

The most valid correction system until now was put forward by Grundkiewicz et al. [15], they combined NMT and SMT model together and used corrections from the best SMT as the inputs of NMT model, incorporating with SMT based spelling checker and RNN based LM, they achieved state-of-the-art performance on CoNLL-2014 test set. Moreover, a most related work was proposed by Schmaltz et al. in 2017 [25]. Different from [7] and our work, they used edit operations as special tags in the target sentences and predicted those tags as atomic tokens in decoding.

## 3    GEC Based on Edit Operations

In the following sections, we will describe our work in details, including the corpora we used, our model architecture, experimental settings and results. At last, a results' analysis was presented.

### 3.1    Datasets

As general, we collected two publicly available corpora as talked above, NUCLE [10] and Lang-8 [28]. The details of this two data sets are shown in Table 1.

**Table 1.** Corpora statistical information

| Corpora | Class | Max-Len | Min-Len | Avg-Len | Words-Num | Chars-Num |
|---------|-------|---------|---------|---------|-----------|-----------|
| NUCLE | Source | 222 | 3 | 20.89 | 33805 | 115 |
| | Target | 222 | 3 | 20.68 | 33258 | 114 |
| Lang-8 | Source | 448 | 3 | 12.35 | 126667 | 94 |
| | Target | 494 | 3 | 12.6 | 109537 | 94 |
| CoNLL-2014 test set | | 227 | 1 | 22.96 | 3143 | 75 |

Since NUCLE corpora is homologous with CoNLL-2014 test set but in a small amount compared with Lang-8, we adopt a simple up-sampling technique that using these samples twice for training. In data preprocessing step, we discard samples with more than 200 characters despite in source or target, in addition, we only use parallel samples that the difference of length between source text and target one are less than 50. Moreover, some samples' correct target texts are with all words been removed, we throw away all these kinds of data directly. After those processing steps, we split the whole corpora into training and validation sets randomly and results in over 0.9M training samples and nearly 10 K for validation. For model's performance comparison, we choose CoNLL-2104 test set [23] which has 1312 samples as commonly used in this task.

## 3.2   Model Architecture

The main architecture of our GEC system is the commonly used Seq2Seq [27] framework but with a soft attention mechanism in decoder which is similar as [1]. The simplified version of our model architecture with 3 layers is shown in Fig. 1. Our model is constituted by 4 layers encoder and 4 layers decoder with residual connection between each 2 layers and attention mechanism is adopted in the last decoder layer. The bottom-left corner represents the encoder of our model which encodes source text in character level including space symbol. The bottom layer is a bi-directional RNN with half layer size and traditional LSTM cell compared to upper layers, and process embedding data forward and backward respectively to make sure the encoder can obtain contextual information of the source text.

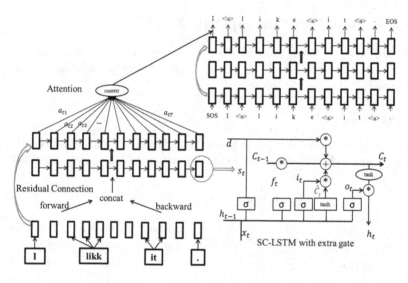

**Fig. 1.** The architecture of our GEC system with residual connection, attention mechanism and SC-LSTM with extra gate.

Upper layers are all in forward style and with SC-LSTM [30] which is very similar with traditional LSTM but with a semantical vector **d** that represents the semantical information of the text, in our model, it represents the edit operations needed for this error text. Since not all tokens need to be changed, we add a semantical gate to control the information flow of this vector. The SC-LSTM which illustrated in the bottom-right corner of Fig. 1 is defined by the following equations with main difference in Eq. 6.

$$\mathbf{i}_t = \sigma(\mathbf{W}_{wi}\mathbf{w}_t + \mathbf{W}_{hi}\mathbf{h}_{t-1}) \tag{1}$$

$$\mathbf{f}_t = \sigma(\mathbf{W}_{wf}\mathbf{w}_t + \mathbf{W}_{hf}\mathbf{h}_{t-1}) \tag{2}$$

$$o_t = \sigma(\mathbf{W}_{wo}\mathbf{w}_t + \mathbf{W}_{ho}\mathbf{h}_{t-1}) \tag{3}$$

$$s_t = \sigma(\mathbf{W}_{ws}\mathbf{w}_t + \mathbf{W}_{hs}\mathbf{h}_{t-1}) \tag{4}$$

$$\hat{\mathbf{c}}_t = \tanh(\mathbf{W}_{ws}\mathbf{w}_t + \mathbf{W}_{hs}\mathbf{h}_{t-1}) \tag{5}$$

$$\mathbf{c}_t = \mathbf{f}_t \odot \mathbf{c}_{t-1} + \mathbf{i}_t \odot \hat{\mathbf{c}}_t + \mathbf{s}_t \odot \tanh(\mathbf{W}_{dc}\mathbf{d}) \tag{6}$$

$$\mathbf{h}_t = \mathbf{o}_t \odot \tanh(\mathbf{c}_t) \tag{7}$$

To avoid gradient vanishing and make training process stable, we adopt residual connection both in encoder and decoder which is represented by red-curved arrow. It changes the inputs of middle layers, of which can be defined by following equations. $\mathbf{I}_t$ indicates the inputs of time $t$ and $i$ means $i$th layer, $\mathbf{x}$ represents the source or target text with word embedding and $\mathbf{h}$ is the hidden states of RNN cells.

$$\mathbf{I}_t = \begin{cases} \mathbf{x}_t & i = 0 \\ \mathbf{h}_t^{i-1} & i = 1 \\ \mathbf{h}_t^{i-1} + \mathbf{h}_t^{i-2} & i > 1 \end{cases} \tag{8}$$

Another important component of our model is attention mechanism as used in [1] which is shown in the top-left corner of Fig. 1. We use weighted sum of encoder outputs as context vector in the last decoder layer for generates characters. The weight $a_{tk}$ is computed as defined in Eqs. 9–11 where $t$ indicates the decoding step that from 1 to $T_t$, and $e_k$ represents the $k$th encoder output. $k$ and $j$ both range from 1 to $T_s$. $\phi_1$ and $\phi_2$ are two feedforward affine transforms, $T_s$ and $T_t$ represent the length of source error text and target right one respectively. $\mathbf{h}_t^L$ is the $t$th hidden state of the last decoder layer and $C_t$ means of context vector computed by weights and encoder outputs for decoding at step $t$.

$$u_{tk} = \phi_1(\mathbf{h}_t^L)^T \phi_2(\mathbf{e}_k) \tag{9}$$

$$a_{tk} = \frac{u_{tk}}{\displaystyle\sum_{j=1}^{T_s} u_{tj}} \tag{10}$$

$$\mathbf{C}_t = \sum_{j=1}^{T_s} a_{tj}\mathbf{e}_j \tag{11}$$

### 3.3  Experiments

For experiments, we use the model described above with character level operations. In view of the correction of misspelling, we represent each sample in character style with a vocabulary constituted by 99 unique characters. The embedding dimension of each character is 256 and maximum sentence length is limited to 200.

The most important part of our method is the edit operations information $\mathbf{d}$ used in SC-LSTM which are extracted by a ERRor ANnotation Toolkit (ERRANT) [4,11]. The toolkit is designed to automatically annotate parallel English sentences with rule based error type information, all errors are grouped into 3 kinds of edit operations named "Unnecessary, Missing and Replacement". They are determined by whether tokens are deleted, inserted or substituted respectively. We use this toolkit to extract all edit operations, and represent them with a 3 dimensional one-hot vector **d** to indicate whether the operations are needed or not for a specific error sentence.

**Training.** The model is trained using negative log-likelihood loss function as defined in Eq. 12, where $N$ is the number of pairwise samples in a batch and $T_t^i$ is the number of characters in the $i$th target right sentence, **x** and **d** indicate the source error text and edit operations vector respectively $y_{i,j}$ represents the $j$th token in the correction for the $i$th instance.

$$Loss = -\frac{1}{N} \sum_i^N \frac{1}{T_t^i} \sum_{j=1}^{T_t^i} \log(p(y_{i,j}|y_{i,1}, ..., y_{i,j-1}, \mathbf{x}, \mathbf{d})) \tag{12}$$

The parameters are optimized by Adaptive Moment Estimation (Adam) [21] with learning rate set as 0.0003.

Another useful technique we adopt in our experiments is scheduled sampling [2]. On account of the computation of gate for edit operations information relies heavily on the preceding tokens. The different usage of target sentence between training and inference affects the accuracy of computing semantical gates greatly. In order to alleviate the influence of this distinction, we utilize scheduled sampling with linear decay on some randomly chosen samples to bridge the gap between training and inference.

**Inference.** For inference and testing, the edit operations we use in training are unavailable since we do not know the corrections of samples in test set. We take a simple traversal approach which means we consider all possible combinations of edit operations. This method results in 8 kinds of different cases. We do correction for each of them using beam search technique with same beam size. 24 candidates are obtained and sorted by the cumulative probability of each token, the top one is regarded as the best correction.

## 3.4   Results and Analysis

**Experimental Results.** We compare the loss on validation set for 3 different conditions as shown in Fig. 2, the green-triangle one represents experiments without edit operations information with traditional LSTM and orange-star one shows the loss without scheduled sampling technique in training. The blue-dot one is the performance of our final model.

More concretely, the MaxMatch ($M^2$) [9] scores computed by standard evaluation metric on CoNLL-2014 test set for those three different experimental settings are shown in Table 2.

In Table 2, the top 5 lines are some baselines of previous works by other researchers. The bottom 3 lines show the results of our model in which EOI

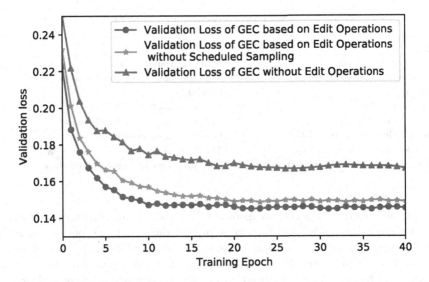

**Fig. 2.** The loss comparison of three different conditions on test set

**Table 2.** $M^2$ score comparison on CoNLL-2014 test set among our model and other previous work without the help of re-rank technique

| Model | Parallel train data | P | R | $F_{0.5}$ |
|---|---|---|---|---|
| **Baseline** | | | | |
| SMT of [6] | Lang-8,NUCLE | 58.24 | 24.84 | 45.90 |
| SMT of [20] | Lang-8,NUCLE | 57.99 | 25.11 | 45.95 |
| NMT of [18] | Lang-8,NUCLE,CLC | - | - | 41.53 |
| NMT of [31] | Lang-8,NUCLE | 45.86 | 26.40 | 39.97 |
| MLConv [7] | Lang-8,NUCLE | 59.68 | 23.25 | 45.36 |
| MLConv(4 ens.) [7] | Lang-8,NUCLE | **67.06** | 22.52 | 48.05 |
| **Ours** | | | | |
| GEC w/o EOI | Lang-8,NUCLE | 60.43 | 20.61 | 43.58 |
| GEC_EOI w/o SS | Lang-8,NUCLE | 54.55 | 30.16 | 46.95 |
| Best GEC w/ EOI | Lang-8,NUCLE | 55.34 | **32.83** | **48.67** |

is the representation of Edit Operations Information and SS means Scheduled Sampling. In addition, some correction examples are show in Table 3.

**Analysis.** To be fair, all the baselines are without the help of re-rank or rescoring methods such as large scale LM since all of our experiments are conducted without any those kinds of techniques. From the results, we can conclude that our method obtain the best overall performance and edit operations are very effective for grammatical error correction. Of which some previous work also

**Table 3.** Some examples corrected by our EO_GEC

| Source error sentence | Target right correction |
|---|---|
| It's **heavy rain** today | It **rained heavily** today |
| Everyone wants to be **success** | Everyone wants to be **successful** |
| I **likk** it | I **like** it |
| I **has a** apple | I **have an** apple |
| **I start** to learning English again | **I'm starting** to learn English again |
| I am very **interes on** the book | I am very **interested in** the book |
| The poor man needs a house to live | The poor man needs a house to live **in** |
| We must return **back** to school this afternoon | We must return to school this afternoon |

had proved in other aspects, for example, [7] used edit operations information to train rescoring model and further improved their system's performance. In detail, compared with other approaches, our model achieves much higher recall but with lower precision, the main reason is that edit operations bring more information to correct errors. In addition, our straightforward traversal skill in inference is more likely to do more corrections which further results in higher recall but may lose precision.

## 4 Conclusion

In conclusion, we propose a neural sequence to sequence grammatical error correction system which utilizes edit operations information in encoder and decoder directly, the model with SC-LSTM achieves state-of-the-art performance on standard benchmark compared to other former effective approaches with fair conditions. To our knowledge, it is the first attempt to exploit edit operations as semantic information to control the correction process. The usage of character level representation, residual connection and scheduled sampling further improve our method's robustness and effectiveness. The traversal technique for edit operations in inference is intuitive but very valid. We can further enhance its capacity by some kinds of selection tricks to avoid unnecessary modification and result in promotion of precision. We will explore further in this direction in the future. What's more, direct utilization of error types information may be more effective but with many difficulties since there are more categories of errors, but it is a valuable research work.

**Acknowledgments.** This work was supported by the Natural Science Foundation of China (NSFC) under grant no. 61673025 and 61375119 and Supported by Beijing Natural Science Foundation (4162029), and partially supported by National Key Basic Research Development Plan (973 Plan) Project of China under grant no. 2015CB352302.

# References

1. Bahdanau, D., Cho, K., Bengio, Y.: Neural machine translation by jointly learning to align and translate. arXiv preprint arXiv:1409.0473 (2014)
2. Bengio, S., Vinyals, O., Jaitly, N., Shazeer, N.: Scheduled sampling for sequence prediction with recurrent neural networks. In: Advances in Neural Information Processing Systems 28, Annual Conference on Neural Information Processing Systems 2015, 7–12 December 2015, Montreal, Quebec, Canada, pp. 1171–1179 (2015)
3. Brockett, C., Dolan, W.B., Gamon, M.: Correcting ESL errors using phrasal SMT techniques. In: Proceedings of the 21st International Conference on Computational Linguistics and 44th Annual Meeting of the Association for Computational Linguistics, Sydney, Australia. Association for Computational Linguistics, pp. 249–256 (2006)
4. Bryant, C., Felice, M., Briscoe, T.: Automatic annotation and evaluation of error types for grammatical error correction. In: Proceedings of the 55th Annual Meeting of the Association for Computational Linguistics, ACL 2017, Vancouver, Canada, 30 July–4 August, Volume 1: Long Papers, pp. 793–805 (2017). https://doi.org/10.18653/v1/P17-1074
5. Cho, K., et al.: Learning phrase representations using RNN encoder-decoder for statistical machine translation. arXiv preprint arXiv:1406.1078 (2014)
6. Chollampatt, S., Ng, H.T.: Connecting the dots: towards human-level grammatical error correction. In: Proceedings of the 12th Workshop on Innovative Use of NLP for Building Educational Applications, BEA@EMNLP 2017, Copenhagen, Denmark, 8 September 2017, pp. 327–333 (2017). https://aclanthology.info/papers/W17-5037/w17-5037
7. Chollampatt, S., Ng, H.T.: A multilayer convolutional encoder-decoder neural network for grammatical error correction. In: Proceedings of the Thirty-Second AAAI Conference on Artificial Intelligence, New Orleans, Louisiana, USA, 2–7 February 2018 (2018). https://www.aaai.org/ocs/index.php/AAAI/AAAI18/paper/view/17308
8. Chollampatt, S., Taghipour, K., Ng, H.T.: Neural network translation models for grammatical error correction. In: Proceedings of the Twenty-Fifth International Joint Conference on Artificial Intelligence, IJCAI 2016, New York, NY, USA, 9–15 July 2016, pp. 2768–2774 (2016). http://www.ijcai.org/Abstract/16/393
9. Dahlmeier, D., Ng, H.T.: Better evaluation for grammatical error correction. In: Human Language Technologies, Conference of the North American Chapter of the Association of Computational Linguistics, Proceedings, 3–8 June 2012, Montréal, Canada, pp. 568–572 (2012). http://www.aclweb.org/anthology/N12-1067
10. Dahlmeier, D., Ng, H.T., Wu, S.M.: Building a large annotated corpus of learner English: the NUS corpus of learner English. In: Proceedings of the Eighth Workshop on Innovative Use of NLP for Building Educational Applications, BEA@NAACL-HLT 2013, 13 June 2013, Atlanta, Georgia, USA, pp. 22–31 (2013). http://aclweb.org/anthology/W/W13/W13-1703.pdf
11. Felice, M., Bryant, C., Briscoe, T.: Automatic extraction of learner errors in ESL sentences using linguistically enhanced alignments. In: COLING 2016, 26th International Conference on Computational Linguistics, Proceedings of the Conference: Technical Papers, 11–16 December 2016, Osaka, Japan, pp. 825–835 (2016). http://aclweb.org/anthology/C/C16/C16-1079.pdf

12. Felice, M., Yuan, Z., Andersen, Ø.E., Yannakoudakis, H., Kochmar, E.: Grammatical error correction using hybrid systems and type filtering. In: Proceedings of the Eighteenth Conference on Computational Natural Language Learning: Shared Task, CoNLL 2014, Baltimore, Maryland, USA, 26–27 June 2014, pp. 15–24 (2014). http://aclweb.org/anthology/W/W14/W14-1702.pdf

13. Francis, W.N., Kucera, H.: The brown corpus: a standard corpus of present-day edited American English. Department of Linguistics, Brown University [producer and distributor], Providence, RI (1979)

14. Gehring, J., Auli, M., Grangier, D., Yarats, D., Dauphin, Y.N.: Convolutional sequence to sequence learning. In: Proceedings of the 34th International Conference on Machine Learning, ICML 2017, Sydney, NSW, Australia, 6–11 August 2017, pp. 1243–1252 (2017). http://proceedings.mlr.press/v70/gehring17a.html

15. Grundkiewicz, R., Junczys-Dowmunt, M.: Near human-level performance in grammatical error correction with hybrid machine translation. In: Proceedings of the 2018 Conference of the North American Chapter of the Association for Computational Linguistics: Human Language Technologies, NAACL-HLT, New Orleans, Louisiana, USA, 1–6 June 2018, Volume 2 (Short Papers), pp. 284–290 (2018). https://aclanthology.info/papers/N18-2046/n18-2046

16. He, K., Zhang, X., Ren, S., Sun, J.: Deep residual learning for image recognition. CoRR abs/1512.03385 (2015). http://arxiv.org/abs/1512.03385

17. Hochreiter, S., Schmidhuber, J.: Long short-term memory. Neural Comput. 9(8), 1735–1780 (1997)

18. Ji, J., Wang, Q., Toutanova, K., Gong, Y., Truong, S., Gao, J.: A nested attention neural hybrid model for grammatical error correction. In: Proceedings of the 55th Annual Meeting of the Association for Computational Linguistics, ACL 2017, Vancouver, Canada, 30 July 30–4 August, Volume 1: Long Papers, pp. 753–762 (2017). https://doi.org/10.18653/v1/P17-1070

19. Junczys-Dowmunt, M., Grundkiewicz, R.: The AMU system in the CoNLL-2014 shared task: grammatical error correction by data-intensive and feature-rich statistical machine translation. In: Proceedings of the Eighteenth Conference on Computational Natural Language Learning: Shared Task, CoNLL 2014, Baltimore, Maryland, USA, 26–27 June 2014, pp. 25–33 (2014). http://aclweb.org/anthology/W/W14/W14-1703.pdf

20. Junczys-Dowmunt, M., Grundkiewicz, R.: Phrase-based machine translation is state-of-the-art for automatic grammatical error correction. In: Proceedings of the 2016 Conference on Empirical Methods in Natural Language Processing, EMNLP 2016, Austin, Texas, USA, 1–4 November 2016, pp. 1546–1556 (2016). http://aclweb.org/anthology/D/D16/D16-1161.pdf

21. Kingma, D.P., Ba, J.: Adam: a method for stochastic optimization. CoRR abs/1412.6980 (2014). http://arxiv.org/abs/1412.6980

22. Macdonald, N., Frase, L., Gingrich, P., Keenan, S.: The writer's workbench: computer aids for text analysis. IEEE Trans. Commun. 30(1), 105–110 (1982)

23. Ng, H.T., Wu, S.M., Briscoe, T., Hadiwinoto, C., Susanto, R.H., Bryant, C.: The CoNLL-2014 shared task on grammatical error correction. In: Proceedings of the Eighteenth Conference on Computational Natural Language Learning: Shared Task, CoNLL 2014, Baltimore, Maryland, USA, 26–27 June 2014, pp. 1–14 (2014). http://aclweb.org/anthology/W/W14/W14-1701.pdf

24. Ng, H.T., Wu, S.M., Wu, Y., Hadiwinoto, C., Tetreault, J.R.: The CoNLL-2013 shared task on grammatical error correction. In: Proceedings of the Seventeenth Conference on Computational Natural Language Learning: Shared Task, CoNLL 2013, Sofia, Bulgaria, 8–9 August 2013, pp. 1–12 (2013). http://aclweb.org/anthology/W/W13/W13-3601.pdf

25. Schmaltz, A., Kim, Y., Rush, A.M., Shieber, S.M.: Adapting sequence models for sentence correction. In: Proceedings of the 2017 Conference on Empirical Methods in Natural Language Processing, EMNLP 2017, Copenhagen, Denmark, 9–11 September 2017, pp. 2807–2813 (2017). https://aclanthology.info/papers/D17-1298/d17-1298

26. Susanto, R.H., Phandi, P., Ng, H.T.: System combination for grammatical error correction. In: Proceedings of the 2014 Conference on Empirical Methods in Natural Language Processing, EMNLP 2014, 25–29 October 2014, Doha, Qatar. A meeting of SIGDAT, a Special Interest Group of the ACL, pp. 951–962 (2014). http://aclweb.org/anthology/D/D14/D14-1102.pdf

27. Sutskever, I., Vinyals, O., Le, Q.V.: Sequence to sequence learning with neural networks. In: Advances in Neural Information Processing Systems, pp. 3104–3112 (2014)

28. Tajiri, T., Komachi, M., Matsumoto, Y.: Tense and aspect error correction for ESL learners using global context. In: The 50th Annual Meeting of the Association for Computational Linguistics, Proceedings of the Conference, 8–14 July 2012, Jeju Island, Korea - Volume 2: Short Papers, pp. 198–202 (2012). http://www.aclweb.org/anthology/P12-2039

29. Zhang, K.L., Wang, H.F.: A unified framework for grammar error correction. In: CoNLL-2014, pp. 96–102 (2014)

30. Wen, T., Gasic, M., Mrksic, N., Su, P., Vandyke, D., Young, S.J.: Semantically conditioned LSTM-based natural language generation for spoken dialogue systems. In: Proceedings of the 2015 Conference on Empirical Methods in Natural Language Processing, EMNLP 2015, Lisbon, Portugal, 17–21 September 2015, pp. 1711–1721 (2015). http://aclweb.org/anthology/D/D15/D15-1199.pdf

31. Xie, Z., Avati, A., Arivazhagan, N., Jurafsky, D., Ng, A.Y.: Neural language correction with character-based attention. arXiv preprint arXiv:1603.09727 (2016)

32. Yuan, Z., Briscoe, T.: Grammatical error correction using neural machine translation. In: NAACL HLT 2016, The 2016 Conference of the North American Chapter of the Association for Computational Linguistics: Human Language Technologies, San Diego California, USA, 12–17 June 2016, pp. 380–386 (2016). http://aclweb.org/anthology/N/N16/N16-1042.pdf

# An Online Handwritten Numerals Segmentation Algorithm Based on Spectral Clustering

Renrong Shao[1], Cheng Chen[2], and Jun Guo[1(✉)]

[1] Computer Center, East China Normal University, 3663 Zhong Shan Rd. N.,
Shanghai, China
jguo@cc.ecnu.edu.cn

[2] iQIYI Innovation Building, No. 365 Linhong Road, Shanghai, China

**Abstract.** In our previous work, without considering the stroke information, a method based on spectral clustering (SC) for solving handwritten touching numerals segmentation was proposed and obtained very good performance. In this paper, we extend the algorithm to an online system, and propose an improved method where the stroke information is involved. First, the features of the numerals image are extracted by a sliding window. Second, the obtained feature vectors are trained by support vector machine to generate an affinity matrix. Thereafter, the stroke information of original images is used to generate another affinity matrix. Finally, these two affinity matrices are added and trained by SC. Experimental results show that the proposed method can further improve the accuracy of segmentation.

**Keywords:** Stroke information · Spectral clustering
Handwritten numerals segmentation · Affinity matrix · Sliding window

## 1   Introduction

The segmentation of handwritten touching numerals has been a research focus in the OCR field, because it is the foundation of handwritten numeral string and has great influences on the effect of recognition [12]. In the usual recognition of numerals, it is common practice to divide a numeral string image into single numeral image and then classified by the recognizer and obtain the classification result. It's easy to segment printed numerals. However, for handwritten numerals, due to the connection of handwritten numeral as shown in Fig. 1, these types of numerals will make the segmentation become difficult.

In the traditional algorithms, it is usually based on the contour [3,8,18], the skeleton [2,7,13], the reservoir [15] or the combination of contour, profile and other morphological features to detect the type of adhesions by connection points [14,16,20]. All of these algorithms were proposed to solve some problems of connected numbers, but there are still some drawbacks of these algorithms.

© Springer Nature Switzerland AG 2018
L. Cheng et al. (Eds.): ICONIP 2018, LNCS 11305, pp. 506–516, 2018.
https://doi.org/10.1007/978-3-030-04221-9_45

With the application of machine learning algorithms in related fields, handwritten single number recognition has become perfect. The recognition accuracy in some excellent recognition systems can reach more than 97% [5], which is close to the level of human. However, the accuracy of handwriting touching numerals recognition have not been able to achieve the desired effect. As far as we know that in ICDAR 2013 (HDRC) competition the single numeral recognition achieved the accuracy of 97.74% [5], and in ICFHR 2014 competition the best accuracy of handwritten numerals string recognition is 85.3% [6]. Both competitions use samples from the CVL database, in which the single numeral samples are extracted from numerals string samples. In recent years, some state-of-the-art algorithms have been proposed to see numerals string as a word, but we still emphasize the importance of segmentation as mentioned in our previous paper [1,9].

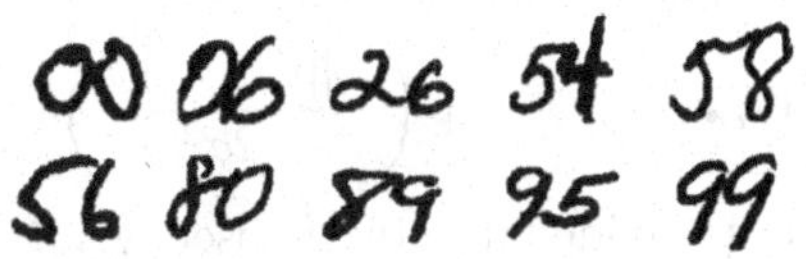

**Fig. 1.** Different types of touching numerals samples in NIST dataset.

In our previous study, we proposed an approach to extract features by a sliding window which is similar to convolution [1] and use support vector machine (SVM) to predict the affinity matrix of spectral clustering (SC). The matrix predicted in this way need to rely on prior knowledge and large scale samples trained by supervised learning. It only considers the position information of left and right. However, In numerals strings sequence, the left and right information can be replaced by the writing sequence, which means that we can also use strokes sequence instead of left and right information.

Inspired by method in [4], we also can get the gesture and stroke information of our online system. As stroke can be more complete representation of numerals information. Therefore, based on the previous experiments, we use numerals' stroke information to construct a matrix $S_m$ and then add it to the affinity matrix $S$ to obtain an enhanced affinity matrix $W^*$. Such design can further reflect the internal association of the different numerals, and can more completely response the internal information of the numerals, thus theoretically it can improve the effect of clustering. Experiments show that the improved algorithm has a better effect on the segmentation and further reduces the error rate on the original dataset.

The structure of this article will be introduced as follows: In Sect. 2, the related theoretical algorithms will be introduced. In Sect. 3, we will introduce the improved method and the whole experimental process. In Sect. 4, the experimental results will be shown the comparison effects of our method and previous method on different datasets. Finally, in Sect. 5, we will present our conclusion and future work.

## 2    Brief of Spectral Clustering

Spectral Clustering is a kind of clustering method based on graph theory. It divides a weighted undirected graph into two or more optimal subgraphs by the distance between nodes in the graph. In graph theory, each graph can be represented as $G = (V, E)$, $V$ is defined as a set of nodes in the graph $\{v_1, v_2...v_n\} \in V$. $E$ is defined as a set of any two nodes connected by the edge, $E(v_i, v_j) \in E$, where $i, j \in n$. Each edge has a weight $w_{ij}$. As is defined in graph theory, we can construct a matrix of degree of $n * n$ dimensions. However, in practice, the weight $w_{ij}$ of the sample does not actually exist, so it is necessary for us to construct such weights according to certain rules. The common practice in SC is to construct a the affinity matrix $S$ by calculating the distance between any two nodes in the sample. The construction of adjacency matrix $W$ is depended on $S$. The usual way to calculate distance between two points in space based on their Euclidean distance. The formula is as follows:

$$W_{ij} = S_{ij} = \exp\left(-\frac{||v_i - v_j||^2}{2\sigma^2}\right).$$

If a graph's points set is divided into two subsets $\{A, B\}$, then you can define their $cut$ as follows:

$$cut(A, B) = \sum_{i \in A, j \in B} w_{ij},$$
$$s.t. \quad A \cup B = V, \ A \cap B = \emptyset.$$

In spectral clustering, the $cut(A, B)$ gives a measure of association between the two subgraphs. If the graph is divided in an optimal way, then the corresponding $cut$ value often is a local minimum. However, in practical problems, a graph with $K$ subgraphs: $A_1$, $A_2$, ...$A_k$, there exists multiple local minimum and we do not know which one to use. If we just simply find the global minimum, we will get some unbalanced results, which often leads to isolated points in the cut. In order to avoid the poor result caused by the minimum cut, we need to make a scale to limit each subgraph. Therefore, only by solving the optimization problem: $\min_A Ncut(A_i, A_2, ...A_k)$, we can get the optimal partition of a graph. Although this optimization problem is NP-complete, we still can solve it by introducing slack variables as follow:

$$(D - W)y = \lambda Dy,$$

where $D_i = \sum_{j=1}^{n} w_{ij}$. From this formula, we know that the key to SC is to construct affinity matrix $W$ and its degree matrix $D$. Based on this theory, our previous method was proposed to construct such a affinity matrix based on pixel features, and now we further strengthen such matrices by adding stroke information.

## 3    The Proposed Method

In our method, Firstly, sliding window was used to extract features. The design of sliding window refers to the approach of convolution neural network (CNN), which makes the feature extraction not only limited to the neighborhood calculation of the original image pixels, but can take a deeper step to utilize the left and right position information that hidden inside the numerals. So the relationship between the image pixels can be excavated to make the meaning of the image feature richer. SVM which is a classical machine learning algorithm has a good theoretical basis and extensive application. In addition, we add the stroke information to the sample, and construct the affinity matrix by the stroke information, which can completely reflect the internal correlation of numerals to make the overall effect of the experiment further improved.

### 3.1   Feature Extraction by Sliding Window

The key to the application of machine learning in field of computer vision (CV) is to completely extract effective features of the image and represent the feature data correctly. In our method, we refer to the convolution operation and use the convolution-like sliding window to extract the features of the samples, which can represent the relation of multi-dimensional data in the image more completely. Each image size in our samples is $32 * 16$ dimensions and sliding window size is $16 * 16$. The main process is to traverse each foreground pixel of the handwritten numerals image. When it encounters the foreground pixel $x_i$, taking $x_i$ as the center and using the sliding window to extract the pixels around $x_i$. If the central target pixel is closer to the edge of the image, the part of the sliding window in the periphery of the image is considered as '0' in the region. If the target pixel in the region of window we mark the blank of window as '1'. If there is not any target pixel in the window we define the blank of window as '0'. So we will get a $16 * 16$ dimensional feature data when a slide window skate over each target pixel. For a sample containing $m$ target pixels, there are $m * 16 * 16$ dimensional feature data. Sliding window to extract the feature data is shown in Fig. 2.

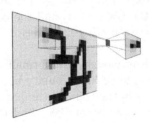

**Fig. 2.** The $5 * 5$ dimensional sliding window to extract the feature data around the target pixel.

### 3.2   Construction of Affinity Matrix by SVM

In the part of Sect. 2, the theory shows that using SC to achieve the numerals segmentation based on affinity matrix $S$. But the key problem is how to construct this matrix. In a sample, the target pixels is composed of the pixels of the stroke. Usually, the similarity of the SC is to directly calculate the Euclidean distance between any points then to construct the adjacency matrix. Although this method is straightforward and quick, for highly conglutinated numerals, the segmentation effect is still poor. It neglects some of the hidden features and global information of the sample, such as left or right information and stroke information. These features can be used to measure weights between different numerals pixel pairs, so the main task is to extract the features implicit in the foreground pixels and use these hidden features to construct an affinity matrix about the foreground pixels. We can regard the numerals foreground pixels as a one-dimensional matrix, which means it can be defined as $V = \{i_1, i_2, ...i_N\}$ where $N$ represents the number of target pixels, and $i_1, i_2, ...i_N$ represent the target pixel sequence. In the construction of affinity matrix, we can extract the features of $16 * 16$ dimensions around each foreground pixel through the sliding window.

The purpose of the experiment is to get a two-class clustering result by using the SC. Therefore, we hope to get the classification of '0' and '1' which can represent the left-right position of the image through the training of the extracted features. But how to judge such '0' and '1', we can consider it as a dichotomous question. In previous work, we proposed to obtain the value of similarity between two points by SVM prediction [1], So as to obtain a $m * m$ dimensional affinity matrix $S$ of the target pixel. Therefore, the similarity between any two points in a pair of images can be expressed as follows:

$$S_{ij} = P_{svm}(Mbox(i, j)),$$

where $i, j \in N$.

Here $P_{svm}$ represents SVM for predicting the similarity of two points, and $Mbox(i, j)$ represents the use of sliding window $Mbox$ to extract the features of $N*N$ dimensions ($N$ represents arbitrary dimension of matrix, In our experiment is 16) around the center pixel $i, j$. The correlation between the two pixel pairs is predicted by calculating the features around the center point. This not only limited to the calculation between adjacent pixels, but can make full use of the image features.

In previous work of training models, We normalized the training samples of the data set such that each sample has a binary image of $M$ ($M$ is arbitrary) pixels. The touching numerals in the training sample set are composed by single numeral, and before the synthesis, we marked the left and right numeral as '$L$' or '$R$' to represent their location information. Similarly, Using a sliding window to capture a $16 * 16$ dimensional feature of each target pixel, if the central pixel pair has the same label ('$L$' '$L$' or '$R$' '$R$'), the corresponding training sample is marked as positive, otherwise marked as negative.

So if a sample with $N$ foreground pixels, we can get a total of $N * (N - 1)$ data with label, except for the comparison of the pixels themselves (the diagonal of affinity matrix is 0). All these feature data will be used as the input for SVM to train a recognizer model and it will be used to predict the wight of pixel pair in one image (Fig. 3).

**Fig. 3.** Example of handwritten touching numerals '23'. (a) is origin sample, the red dots on same side in (b) (d) represent positive samples, the red dots on different side in (c) represent negative samples. (Color figure online)

### 3.3   Construction of Affinity Matrix by Stroke

In practical online application, such as tablet or smart phone, we can get stroke of writing. When the nip of pen touches the pad and leaves the pad, we can record the coordinates of the gliding path. We record each pixel of the numerals to constructs an affinity matrix based on stroke. For the strokes of the handwritten numerals, the pixels in the one stroke are more relevant than the pixels in another strokes. So the correlation coefficient in the same stroke is stronger than other strokes, that is our idea of clustering. The following is our concrete approach. For the one image sample, the stroke of the target pixels can be defined as a set $R$. If the numeral is composed by two strokes, we define $\{R_1, R_2\} \in R$ for it. Since the numeral '4' and '5' are both composed of two strokes, so in practical writing connected numerals will appear three or four strokes. The composition of the three stroke information can be represented as $\{R_1, R_2, R_3\} \in R$. For handwritten touching numerals was composed by two numbers, there will not be more than four stroke information in the combination. Maybe someone questions that how we deal with the cluster of numerals more than two strokes. Here, we solve this problem by add the matrix based on our previous method. Because the previous method has divided the numeral string into two categories, and now the numeral with three or four stokes will still be divided into two categories. We extract the target pixel to be a one-dimensional matrix $A[n]$, where $n$ is the length of a one-dimensional matrix. $n$ is also the size of the target pixel while corresponds to the length of affinity matrix. In order to facilitate the calculation, we set two cursors $i, j$, where $i$ represents the current pixel, and $j$ represents the other target pixels. The calculation of algorithm as follows:

$$S_m = A_{ij} \odot A_{ji}{}^T,$$
$$s.t. \;\; i, j < n.$$
$$W^* = S + S_m,$$

Here $S_m$ represents affinity matrix of the strokes. $\odot$ represents calculation of XNOR gate or equivalence gate. From the above calculation of the matrix, we will get a affinity matrix of strokes. The matrix $S_m$ contains the relevant information of the stroke and adds it to the affinity matrix predicted by the SVM before, so as to obtain a new affinity matrix $W^*$. The matrix $W^*$ is used as the input of SC algorithm. Experiments show that the method with stroke information added performs better than the previous approach. The segmentation effect is improved significantly. The effect of segmentation comparison is shown as below Fig. 4.

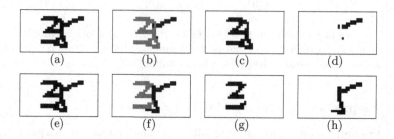

(a)    (b)    (c)    (d)

(e)    (f)    (g)    (h)

**Fig. 4.** Example of handwritten numerals '35'. (a)–(d) is the segmentation effect of handwritten numerals by SC algorithm in previous work. (e)–(h) is the segmentation effect of handwritten touching numerals by SC algorithm with stroke information in this paper.

### 3.4   Segmentation Verification

From the above, the numerals will be split into two single characters. The following work is to identify the numeral of segmentation image and calculate the accuracy of the segmentation. The common practice is to use the origin single numeral images with prior knowledge as training samples to obtain a recognizer and calculate the accuracy through cross-validation. As our previous test data set was synthesized by single numeral images, so we can train these single numeral data set to get a common recognizer for our segmentation numerals. We choose SVM to do the classification and cross-validation, this part also can be replaced by multi-layer perception (MLP) or back propagation (BP). All the origin image was classified to '0' to '9' and marked the label, then the training samples from '0' to '9' were cross-validated by SVM to get the segmentation accuracy. This approach reflects the advantages of the idea of segmentation identification. If we recognized the numerals from '00' to '99' directly, we need to make 100 classifiers of these numerals respectively. If we do like that, it will increases the complexity of classifier and reduces the reusability. In order to make the whole system more reliable, the accuracy of each numeral' recognizer over 99% on our original sample will be accepted.

## 4   Experimental Results

The work in this paper is based on the real part of OCR products. Our experimental data was collected from a such a tablet, and data set was supplemented before the experiment. We collected 8, 654 single numeral added to the previous data set to train the recognizer. We also collected 9, 896 handwritten connected numerals with stroke information and the size of each sample is 32 * 16. The

**Table 1.** Comparison of segmentation effects on our data set and NIST SD-19 data set.

| Data sets | Our online data set | | NIST SD-19 | |
|---|---|---|---|---|
| Type of sample | Previous algorithm (%) | Improved algorithm (%) | Previous algorithm (%) | Improved algorithm (%) |
| 0[0−9] | 97.68 | 98.22 | 99.33 | 99.46 |
| 1[0−9] | 96.43 | 97.54 | 97.84 | 98.31 |
| 2[0−9] | 97.32 | 98.47 | 96.15 | 97.56 |
| 3[0−9] | 95.34 | 98.64 | 98.65 | 99.23 |
| 4[0−9] | 94.16 | 96.73 | 95.34 | 98.39 |
| 5[0−9] | 92.67 | 95.44 | 96.02 | 98.34 |
| 6[0−9] | 96.59 | 97.83 | 97.03 | 98.72 |
| 7[0−9] | 97.48 | 98.92 | 98.92 | 99.22 |
| 8[0−9] | 93.08 | 95.71 | 95.88 | 97.06 |
| 9[0−9] | 96.08 | 97.43 | 98.58 | 99.05 |

**Table 2.** Comparison of different segmentation algorithms using NIST SD-19 data set.

| Ref. | Primitives | Ligatures | Pre-class | ≥ 2 | OverSeg | Approach | Size | Perf.(%) |
|---|---|---|---|---|---|---|---|---|
| Pre. approach [1] | Affinity matrix | Yes | Yes | No | Yes | Seg-Rec-Seg | 2000 | 97.65 |
| [21] | Contour, Concavities | No | No | Yes | No | Seg-then-Rec | 3287 | 94.8 |
| [11] | Contour, Concavities | No | Yes | No | No | Rec-Based | 3500 | 92.5 |
| [12] | Contour | No | Yes | Yes | Yes | Rec-Based | 3359 | 97.72 |
| [19] | Skeleton | Yes | No | No | No | Seg-then-Rec | 2000 | 88.7 |
| [17] | Skeleton | Yes | Yes | Yes | Yes | Rec-Based | 5000 | 96.5 |
| New approach | Improved affinity matrix | Yes | Yes | No | Yes | Seg-then-Rec | 2000 | 98.53 |

number of each numerals is about 100 ranging from '00' to '99'. In the experiment, the sample picture and the binary pixel text with the stroke information are used as input for the experiment. The whole process of collecting spent a lot of time and resources and the purpose is to ensure the quantity and quality of the sample, because this part of work has an important role on the following feature extraction, and affects the entire experiment results.

In the part of feature extraction, we still use the previous experiment method. In the part of algorithm, we added an affinity matrix based on the stroke information. In the part of recognition and verification, the data set is equally divided into 10 subsets to use 10-fold cross-validation. For each fold, we use nine of the subsets to build the training sets and the rest of subsets is used as test data for verification. All the algorithms are mainly implemented by python and MATLAB. The prediction of affinity matrix and cross-validation part use the open source framework LIBSVM. In our experiment, we only focus on the accuracy

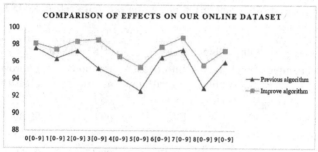

(a) the effects on our online data set

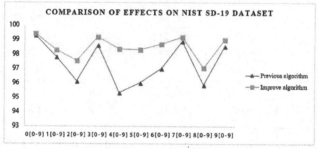

(b) the effects on NIST data set

**Fig. 5.** Comparison of experimental results on different data sets

of segmentation, so there is no comparison of calculation time, nor the specific operating conditions associated with specific experiments.

The experiment is tested on our collected online data set and the standard data set (NIST SD-19) [10]. In the our dataset, the previous method without adding the strokes information to obtains an accuracy of 95.68%. Afterwards, when we add the strokes information as input, the whole experiment result has been improved to 97.49%. In order to fully demonstrate the credibility and accuracy of our algorithm, we also validate it on another common data set, NIST SD-19. Since NIST SD-19 is an offline data set, it does not provide strokes information. To bring the experimental data closer to our online data set, we perform some preprocessing on the SD-19 data set, manually marking the strokes for samples that have not been divided by the previous method. By comparing with previous method, we found that the segmentation results of the experiment have been significantly improved after adding the stroke information to samples. The segmentation effect of numerals with '3' or '5' is obviously improved. The whole accuracy of the experimental results on the NIST SD-19 data set can reach 98.53%, which is about 0.9% higher than before. This verifies our previous assumption. The details of comparison of specific experiments is shown in Tables 1, 2 and Fig. 5.

# 5   Conclusion and Outlook

In this paper, we propose an algorithm that merging the stroke information to the construction of affinity matrix, which improves the correlation of matrix elements and effect of handwritten touching numerals segmentation algorithm. Different from traditional algorithms, our approach divides the connected numerals image that is based on all pixels of graph, so that it can overcome the disadvantage of the traditional cutting which only calculates neighborhood of images. In our online system, we make full use of the stroke information to build an affinity matrix to improve the SC. These strokes often are ignored by some applications. Our approach has been tested on our collected data set and NIST SD-19 data set, which achieves a good result. The accuracy of segmentation has all improved to 97%, which further proves that online stroke information has good generalization in numerals segmentation. At present, our experiments are mainly used in online writing systems to solve the problem of double-numerals. In the following work, we will continue to study the segmentation problem of numerals string and characters based on this research.

# References

1. Chen, C., Guo, J.: A general approach for handwritten digits segmentation using spectral clustering. In: International Conference on Document Analysis and Recognition (IAPR), pp. 547–552 (2018)
2. Chen, Y.K., Wang, J.F.: Segmentation of single or multiple-touching handwritten numeral string using background and foreground analysis. IEEE Pattern Anal. **1**, 1304–1317 (2000)
3. Congedo, G., Dimauro, G., Impedovo, S., Pirlo, G.: Segmentation of numeric strings. In: International Conference on Document Analysis and Recognition, vol. 2, pp. 1038–1041 (1995)
4. Corr, P.J., Silvestre, G.C., Bleakley, C.J.: Open source dataset and deep learning models for online digit gesture recognition on touchscreens. In: Irish Machine Vision and Image Processing Conference (IMVIP) (2017)
5. Diem, M., Fiel, S., Garz, A., Keglevic, M., Kleber, F., Sablatnig, R.: ICDAR 2013 competition on handwritten digit recognition (HDRC 2013). In: International Conference on Document Analysis and Recognition, pp. 1422–1427 (2013)
6. Diem, M., et al.: ICFHR 2014 competition on handwritten digit string recognition in challenging datasets (HDSRC 2014). In: International Conference on Frontiers in Handwriting Recognition, pp. 779–784 (2014)
7. Elnagar, A., Alhajj, R.: Segmentation of connected handwritten numeral strings. Pattern Recogn. **36**(3), 625–634 (2003)
8. Fujisawa, H., Nakano, Y., Kurino, K.: Segmentation methods for character recognition: from segmentation to document structure analysis. Proc. IEEE **80**(7), 1079–1092 (1992)
9. Graves, A.: Offline arabic handwriting recognition with multidimensional recurrent neural networks. In: Advances in Neural Information Processing Systems, pp. 549–558 (2008)
10. Grother, P.J.: Nist special database 19, Handprinted forms and characters database. National Institute of Standards and Technology (NIST) (1995)

11. Kim, K.K., Jin, H.K., Suen, C.Y.: Segmentation-based recognition of handwritten touching pairs of digits using structural features. Pattern Recogn. Lett. **23**, 13–24 (2002)
12. Lei, Y., Liu, C.S., Ding, X.Q., Fu, Q.: A recognition based system for segmentation of touching handwritten numeral strings. In: International Workshop on Frontiers in Handwriting Recognition, pp. 294–299. IEEE (2004)
13. Lu, Z., Chi, Z., Siu, W.C., Shi, P.: A background-thinning-based approach for separating and recognizing connected handwritten digit strings. Pattern Recogn. **32**(6), 921–933 (1999)
14. Oliveira, L.S., Lethelier, E., Bortolozzi, F.: A new approach to segment handwritten digits. In: Proceedings of 7th International Workshop on Frontiers in Handwriting Recognition, pp. 577–582 (2000)
15. Pal, U., Choisy, C.: Touching numeral segmentation using water reservoir concept. Pattern Recogn. Lett. **24**(1), 261–272 (2003)
16. Ribas, F.C., Oliveira, L.S., Britto, A.S., Sabourin, R.: Handwritten digit segmentation: a comparative study. Int. J. Doc. Anal. Recogn. **16**(2), 127–137 (2013)
17. Sadri, J., Suen, C.Y., Bui, T.D.: A genetic framework using contextual knowledge for segmentation and recognition of handwritten numeral strings. Pattern Recogn. **40**(3), 898–919 (2007)
18. Shi, Z., Govindaraju, V.: Segmentation and recognition of connected handwritten numeral strings. Pattern Recogn. **30**(9), 1501–1504 (1997)
19. Suwa, M., Naoi, S.: Segmentation of handwritten numerals by graph representation. In: International Workshop on Frontiers in Handwriting Recognition, vol. 2, pp. 334–339 (2004)
20. Vellasques, E., Oliveira, L.S., Koerich, A.L., Sabourin, R.: Filtering segmentation cuts for digit string recognition. Pattern Recogn. **41**(10), 3044–3053 (2008)
21. Yu, D., Yan, H.: Separation of touching handwritten multi-numeral strings based on morphological structural features. Pattern Recogn. **34**(3), 587–599 (2001)

# Analysis, Classification and Marker Discovery of Gene Expression Data with Evolving Spiking Neural Networks

Gautam Kishore Shahi[1(✉)], Imanol Bilbao[3], Elisa Capecci[2], Durgesh Nandini[1], Maria Choukri[4], and Nikola Kasabov[2]

[1] Dipartimento di Ingegneria e Scienza dell'Informazione (DISI), University of Trento, via Sommarive 9, 38100 Povo, Trento, TN, Italy
gautamshahi16@gmail.com
[2] Knowledge Engineering and Discovery Research Institute (KEDRI), Auckland University of Technology (AUT), AUT Tower, Level 7, cnr Rutland and Wakefield Street, Auckland 1010, New Zealand
[3] University of the Basque Country, Bilbao, Spain
[4] Ara Institute of Canterbury, Christchurch, New Zealand

**Abstract.** The paper presents a methodology to assess the problems behind static gene expression data modelling and analysis with machine learning techniques. As a case study, transcriptomic data collected during a longitudinal study on the effects of diet on the expression of oxidative phosphorylation genes was used. Data were collected from 60 abdominally overweight men and women after an observation period of eight weeks, whilst they were following three different diets. Real-valued static gene expression data were encoded into spike trains using Gaussian receptive fields for multinomial classification using an evolving spiking neural network (eSNN) model. Results demonstrated that the proposed method can be used for predictive modelling of static gene expression data and future works are proposed regarding the application of eSNNs for personalised modelling.

**Keywords:** Evolving spiking neural networks
Gaussian receptive fields · Static data · Gene expression · Microarray
Transcriptome data analysis

## 1 Introduction

Biomedical research is one of the most significant areas of investigation in data science. This area produces a tremendous amount of data and provides an opportunity for extensive exploration and application in several domains, such as personalised medicine. Personalised modelling is now a trend and is considered beneficial for diagnosis, treatments, and advance in medical science. The prospect of personalised treatments are now promising [1,2].

© Springer Nature Switzerland AG 2018
L. Cheng et al. (Eds.): ICONIP 2018, LNCS 11305, pp. 517–527, 2018.
https://doi.org/10.1007/978-3-030-04221-9_46

The current size of bioinformatics data collected for this purpose was estimated around 75 petabytes in 2015, and it is expected to grow tremendously [3]. Data scientists are continuously proposing novel methods for the analysis of such data and cost-effective data modelling is the hot spot and a future trend. In this respect, Artificial Neural Networks (ANN) have demonstrated their potential for the analysis of biological data and the application in personalised medicine in a cost-effective way [4].

There are different sources of bioinformatics data, one of them is gene expression data. Modelling of gene expression data is challenging in terms of time-costly and reliable results [5]. Several researchers have tried to tackle these problems by developing new methodologies for gene data modelling and analysis, e.g. [6]. In particular, advancements have been made by using ANN techniques, some of them proving their ability to deal with complex, multidimensional data by using Evolving Spiking Neural Networks (eSNN) [7–11].

The eSNN architecture was first proposed by Kasabov [7] as an extension to the evolving connectionist systems principle [12]. This architecture can change both connection weights and structure during training to adapt to the input information encoded. This behavior is based on biological neurons, which are able to evolve and adapt according to their input [13]. The eSNN architecture is divided into three layers: an encoding layer that binarizes the input data into spike trains; a hidden layer for supervised learning; and an evolving output neural model for data classification.

The eSNN demonstrate their potential to handle different types of data (e.g. [9,11,14]), yet they have never been applied for static gene expression data modelling and classification. This research work proposes a computational model for static gene expression data analysis and predictive modelling using eSNN techniques. The computational model is applied to static gene expression data to study a nutrigenomics problem.

Some of the research questions that have been covered in this research work are:

- Can we replace traditional ANN techniques with a promising computational model based on eSNN for gene expression data analysis?
- What are the benefits of using eSNN versus traditional ANN?
- Can we discriminate different types of diets by using the proposed computational model and a set of signature genes selected from the multitude of features available?
- Can the computational model proposed be applied to nutrigenomics data to build a predictive model for the identification of weight-related genetic issues?
- Using the above-proposed model, can weight-related issues be detected at an early stage?

The next section, Sect. 2, describes the material and methods used to carry out the study, Sect. 3 presents a case study on static gene expression data modelling with eSNN, Sect. 4 discusses the results, and Sect. 5 draws the conclusion and future work.

# 2    Materials and Methods

## 2.1    Gaussian Receptive Fields

Population encoding combined with arrays of receptive fields makes it possible to represent continuous input value with graded and overlapping Gaussian activation functions [15] miming the behaviour of biological sensory neurons [11,13,15]. To encode static values into temporal patterns, it is sufficient to divide the range of possible values into several overlapping Gaussian segments, relating one neuron to each segment. Stimulation of the neuron will be greater when the static value belongs to its segment or nearby segments. Then, assigning early spikes to those neurons that are highly stimulated and later (or non) spikes to those neurons that are less stimulated, the extension to temporal coding will be straightforward. Temporal encoding of static values achieves both sparse codings, by assigning firing times to significantly activated neurons only, and effective event-based network simulation [16]. Additionally, it achieves a coarse coding by encoding each variable independently, as every input is encoded by an optimal number of neurons [15]. Gaussian receptive fields permit the encoding of static inputs by applying a number of neurons with overlapping sensitivity functions, where each input data are binarized by $N$ dimensional receptive fields.

For an input variable a range of minimum and maximum value is defined as $I_{min}^n$ and $I_{max}^n$ respectively. For neuron $j$, the Gaussian receptive field can be defined by its centre $C_j$, calculated as:

$$C_j = I_{min}^n + \frac{(2j-3)}{2} * \frac{(I_{max}^n - I_{min}^n)}{(N-2)} \tag{1}$$

and its width $W_j$, calculated as:

$$W_j = \frac{1}{\beta} * \frac{(I_{max}^n - I_{min}^n)}{(N-2)} \tag{2}$$

with $1 \leq \beta \leq 2$. The parameter $\beta$ directly regulates the width of every Gaussian. Classification accuracy of the model can be improved by adjusting the centre, the width and the number of receptive fields.

## 2.2    Evolving Spiking Neural Networks

As shown in Fig. 1, the architecture of an eSNN model can be divided into three layers: an encoding layer; a hidden layer for supervised learning; and an evolving output neural model for data classification. In the input layer, the input data is presented to the network, where the static gene data is encoded into spike events before learning. Here, a number of overlapping Gaussian receptive functions is implemented to transform the input into spike sequences to represent the first layer of neurons. Equations 1 and 2 are used to compute the centre and width of the Gaussian, which influence the firing intensity of the receptive function. In

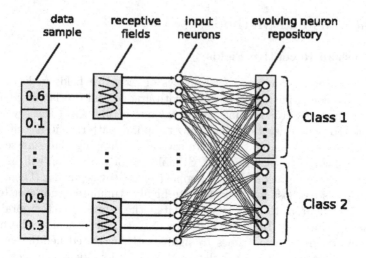

**Fig. 1.** Overall of architecture of eSNN [11].

the hidden layer, the encoded value is presented to a pre-synaptic neuron and the receptive field with the highest related value generates the first spike.

$$\tau_i = [T(1 - Output_j)] \tag{3}$$

where $T$ is the simulation or spike time interval. The neurons of the hidden and output layer are completely connected. In the evolving output neural model, spiking neurons evolve during training to represent the input spike trains that correspond to the same class. Connection weights change during learning. The eSNN architecture is based on Thorpe's model [17], as this uses the timing of the incoming spike to adjust neural connection weights. More specifically, earlier spikes denote stronger connection weights as opposed to later spikes. A neuron spikes just if its post-synaptic potential (PSP) reaches a threshold as per Eq. 4.

$$PSP_i = \begin{cases} 0, & \text{if fired} \\ \sum w_{ji} * mod^{order(j)}, & \text{otherwise} \end{cases} \tag{4}$$

where $w_{ji}$ represents the weight of a synaptic connection between pre-synaptic neuron j to the output neuron i, $mod$ is the modulation factor ($0 \leq mod \leq 1$) and function $order(j)$ is the rank of the spike emitted by neurons j. Single pass-feedforward learning is applied in the output neurons. Each input data point is assigned to an output neuron, which has a weight and threshold value associated and stored in the neuron repository. This weight is compared to the weight of other neurons in the network to check for similarity. If the similarity is greater than a defined threshold, then this weight will be updated with the weight and threshold of the most similar neuron, otherwise, it evolves a new one. This is computed by calculating the average between the new output neuron weight vector and the merged neuron weight vector, and the average between the new output neuron threshold and the merged neuron threshold respectively. After merging the trained vector is discarded and not stored in the repository. After learning, testing is performed by passing the test sample spikes to all the trained output neurons. The output neuron that fires first is associated to a class label defined by the test sample.

**Parameters of the eSNN Model.** The performance of eSNN depends on several parameters [7,11], some of the most important parameters of the eSNN are:

- Number of Gaussian receptive fields - Increasing the number of receptive fields, a priori, will help to distinguish between data samples, but it would require a high computational cost. Lowering the number of receptive fields will make the process faster, but it could decrease the accuracy.
- Width of the receptive field - By decreasing the width of the Gaussian fields, we could make the response more localized. This, in turn, will increase the number of neurons in the network.
- Modulation factor - Modulation factor controls the initials weights, which affects the role of each spike and therefore the PSP, The modulation factor can take a value between 0 and 1.
- Spiking threshold - When a pre-synaptic neuron reaches a set threshold this causes a spike to be emitted by the post-synaptic neuron. The threshold is calculated as a fraction of the total PSP gathered during the presentation of the input pattern.
- Similarity threshold - This parameter sets a threshold value by which output layer neurons are generated or updated.

## 3    A Case Study on Transcriptomics Data Modelling with eSNN

### 3.1    Data Description

Gene expression data is collected using microarrays techniques [18]. Microarray data represents matrices where the expression level of thousands of genes are measured simultaneously. For our case study, gene expression data was obtained from the publicly available Gene Expression Omnibus (GEO) repository of functional genomics data (NCBI GEO [19] accession GSE30509; [20]). A Norwegian scientist published the data as an extension of his research demonstrating that saturated fatty acids contribute to obesity [21]. The data describes three types of diet: Western Diet with Saturated Fatty Acid (SFA), Western Diet with monounsaturated fat (MUFA) and Mediterranean Diet (MED). The Mediterranean (MED) diet is believed to be health-promoting due to its high content of MUFA and polyphenols [19]. These bioactive compounds can affect gene expression, and therefore they can regulate pathways and proteins related to cardiovascular disease interference. This research identified the effects of a MED-type diet, and the replacement of SFA with MUFA in a Western-type diet, on peripheral blood mononuclear cell (PBMC) gene expression and plasma proteins. Overweight men and women with the abdominal distribution of fat were allocated to an eight-week, completely controlled SFA diet (19% daily energy as SFA), a MUFA diet (20% daily energy MUFA), or a MED diet (21% daily energy MUFA). Concentrations of 124 plasma proteins and PBMC whole-genome transcriptional profiles were assessed. Results demonstrated that consumption of MUFA and MED diets decreased the expression of oxidative phosphorylation (OXPHOS) genes, plasma connective tissue growth factor, and apo-B concentrations when compared with the SFA diet. Moreover, the MUFA diet changed the expression of genes involved in B-cell receptor signaling and endocytosis signaling, when compared with the MED and SFA diets. Participants who consumed the MED diet had lower concentrations of pro-inflammatory proteins at eight weeks compared with baseline.

### 3.2 Gene Expression Data Encoding with Gaussian Receptive Fields

Each instance of a sample is composed of real value features. Population encoding algorithm maps a real input value into a series of spikes over time using an array of Gaussian receptive fields. Then, a rank order learning is applied over those spikes.

### 3.3 Learning and Classification with eSNN

To build our computational model, the eSNN architecture has been applied to classify and analyse the gene expression data. This is the first time that an eSNN model is applied for the study of gene expression data to address a nutrigenomics problem. The focus of the work is to study and model gene expression data, and the proposed computational model could be use as benchmarks for gene expression data analysis for personalised modelling. Figure 2 shows a graphical representation of the experimental procedure designed for the eSNN system for static gene expression data analysis. Our proposed procedure is summarised below:

- Dataset selection - Gene expression data is selected for the experiment.
- Data preprocessing - Preprocessing of data includes data cleaning, modification, and noise removal. In our experiments, the genePattern portal [22] provided by the Broad Institute is being used for the analysis of the selected gene expression data.
- Feature selection - Feature selection plays a key role in machine learning, especially in the case of high-dimensional gene expression data, which consists of thousands of features with only a few numbers of samples. In our case study, there were 23941 features and only 49 samples before preprocessing. The high number of features leads to the curse of dimensionality problem [18].

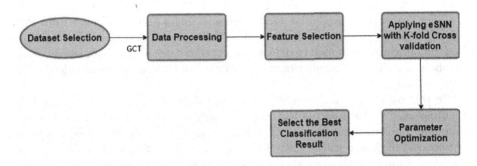

**Fig. 2.** Shows a graphical representation of the experimental procedure designed

To solve the problem of high dimensionality, first, a set of 29 signature genes were selected, as indicated by the literature [20].

Then, a number of feature selection algorithms were applied to the original 23941 genes. For our work, we have used the Maximum Relevance Minimum Redundancy (mRMR) [23] and Multi-task Lasso (MT-LASSO) [24] algorithms. The top 50 features from both the algorithms were selected.

Finally, the intersection of all the three sets of genes was performed. This resulted in a final set of 26 genes.

- Cross-validation - K-fold cross validation is applied to have an unbiased result from the computational model. In this case, K = 10 was considered for the experiments.
- Parameter optimization - A genetic algorithm was applied for parameter optimization. Here, the number of receptive fields and the width of the Gaussian receptive functions were optimized to maximize the accuracy of results.
- Classification results evaluation - By using evolutionary computation algorithms, the highest classification accuracy is calculated with respect to the optimised combination of parameters.

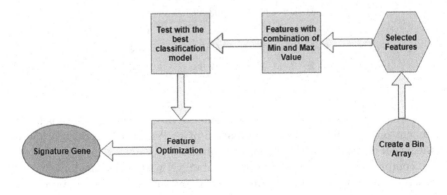

**Fig. 3.** Signatory gene selection

Then, we build a predictive model, by selecting a subset of genes from the already selected 26 features. This subset of features can in turn be used to construct a personalised modell to detect weight related issues. This can be achieved by observing the least number of genes from peripheral blood mononuclear cells (PBMC) data.

In this paper, we created a Bin array containing combinations of (0,1) and then passed it to the selected features. The output was a list of features having minimum and maximum values. The minimum and maximum valued features obtained were passed to the best classification model. Optimisation of features was done using genetic algorithm. By selecting relevant genes, we will decrease the computational cost of the model and make it faster and more accurate. These genes could be used for decision support on diet-related issues. A diagram is used in Fig. 3 to summarise this procedure.

## 4    Results and Discussion

### 4.1    Classification

K-Fold cross-validation is used to estimate the accuracy of the model. In the computational model, only two parameters (the number of the receptive fields and the parameter $\beta$) were optimised. Table 1 reports the accuracy and F1-score obtained with the combination of several parameters for the case study gene expression data.

**Table 1.** Classification accuracy and F1 score obtained for the case study data with different combinations of parameters.

| Parameter combination | Accuracy % | F1 score |
|---|---|---|
| $N = 3$ & $\beta = 1.4$ | 57.14 | 0.70 |
| $N = 4$ & $\beta = 1.2$ | 52.46 | 0.82 |
| $N = 5$ & $\beta = 1.1$ | 50.0 | 0.63 |
| $N = 6$ & $\beta = 1.6$ | 63.59 | 0.70 |
| $N = 7$ & $\beta = 1.4$ | 79.46 | 0.91 |
| $N = 8$ & $\beta = 1.6$ | 50.0 | 0.67 |
| $N = 9$ & $\beta = 1.7$ | 47.68 | 0.60 |

The above results prove that the performance of the eSNN network is largely dependent on parameters. The combination of $\beta$: 1.4, $N = 7$ produced the best classification accuracy for the given test data.

To compare the classification accuracy of eSNN with other machine learning algorithms, further experiments were carried out using multilayer perceptron (MLP) and support vector machines (SVM), The highest classification accuracy for MLP was obtained with 4 hidden layers and 12 neurons at each hidden layer.

The best result for SVM was obtained with polynomial kernels and a degree of 2. These results are reported in Table 2. Based on the Signature Gene matrix, each possible combination of genes were tested in our computational model and the classification accuracy was observed. A set of five genes achieved an accuracy of around 79%. Hence, these genes can be used to determine the type of diet for a person and what are their current status and weight. The list of Signatures genes are shown in Table 3.

**Table 2.** Comparative analysis of eSNN, SVM and MLP classification algorithms.

| Algorithm | Accuracy | F1 score |
|-----------|----------|----------|
| eSNN      | 79.46    | 0.91     |
| SVM       | 70.96    | 0.76     |
| MLP       | 73.24    | 0.82     |

**Table 3.** Signatory genes.

| Gene    | Description |
|---------|-------------|
| NDUFB2  | NADH: Ubiquinone Oxidoreductase Subunit B2 |
| NDUFS7  | NADH: Ubiquinone Oxidoreductase Core Subunit S7 |
| TNNI2   | Troponin I2, Fast Skeletal Type |
| COX8A   | Cytochrome C Oxidase Subunit 8A |
| ATP6V1A | ATPase H+ Transporting V1 Subunit A |

## 5    Conclusion and Future Work

In this research, a computational model is built for the analysis of static gene expression data. Additionally, an algorithm is proposed for the selection of signature genes. Our results demonstrated that our computational model can help predict weight related issues by assessing gene expression data. For the first time, Gaussian receptive fields are used to encode gene expression data in an eSNN. The method has proven its ability to classify the selected signature genes with an higher classification accuracy, when compared with traditional classification methods. This constitutes a machine based automatic technique for fast and highest classification results, which could find its application in the area of personalised modelling.

A natural extension of this project would be to combine the eSNN architecture with a suitable evolutionary algorithm for optimising the remaining parameters of the model (i.e. the modulation factor, spiking and similarity threshold). This would help improving the classification accuracy.

For a proper comparison of the eSNN model with other machine learning algorithms, like the NeuCube architecture [25,26], the hyperparameters of all the compared algorithms should also be optimized. In this way, a comparative analysis could be done to study eSNN application to different kind of gene expression data.

**Acknowledgments.** The presented study was a collaboration between the Knowledge Engineering and Discovery Research Institute (KEDRI, https://kedri.aut.ac.nz/) funded by the Auckland University of Technology of New Zealand and the University of Trento in Italy. Several people have contributed to the research that resulted in this paper, especially: Y. Chen, J. Hu, E. Tu and L. Zhou.

# References

1. Cornetta, K., Brown, C.G.: Perspective: balancing personalized medicine and personalized care. Acad. Med.: J. Assoc. Am. Med. Coll. **88**(3), 309 (2013)
2. Dunn, M.C., Bourne, P.E.: Building the biomedical data science workforce. PLoS Biol. **15**(7), e2003082 (2017)
3. Greene, C.S., Tan, J., Ung, M., Moore, J.H., Cheng, C.: Big data bioinformatics. J. Cell. Physiol. **229**(12), 1896–1900 (2014)
4. Lancashire, L.J., Lemetre, C., Ball, G.R.: An introduction to artificial neural networks in bioinformatics-application to complex microarray and mass spectrometry datasets in cancer studies. Brief. Bioinform. **10**(3), 315–329 (2009). https://doi. org/10.1093/bib/bbp012
5. Ramasamy, A., Mondry, A., Holmes, C.C., Altman, D.G.: Key issues in conducting a meta-analysis of gene expression microarray datasets. PLoS Med. **5**(9), e184 (2008)
6. Ay, A., Arnosti, D.N.: Mathematical modeling of gene expression: a guide for the perplexed biologist. Crit. Rev. Biochem. Mol. Biol. **46**(2), 137–151 (2011)
7. Kasabov, N.K.: Evolving Connectionist Systems: The Knowledge Engineering Approach. Springer, London (2007). https://doi.org/10.1007/978-1-84628-347-5
8. Schliebs, S., Defoin-Platel, M., Worner, S., Kasabov, N.: Integrated feature and parameter optimization for an evolving spiking neural network: exploring heterogeneous probabilistic models. Neural Netw. **22**(5), 623–632 (2009)
9. Soltic, S., Kasabov, N.: Knowledge extraction from evolving spiking neural networks with a rank order population coding. Int. J. Neural Syst. **20**(06), 437–445 (2010)
10. Wysoski, S.G., Benuskova, L., Kasabov, N.: Evolving spiking neural networks for audiovisual information processing. Neural Netw. **23**(7), 819–835 (2010)
11. Schliebs, S., Kasabov, N.: Evolving spiking neural networks: a survey. Evol. Syst. **4**(2), 87–98 (2013)
12. Kasabov, N.: Global, local and personalised modeling and pattern discovery in bioinformatics: an integrated approach. Pattern Recognit. Lett. **28**(6), 673–685 (2007)
13. Bohte, S.M.: The evidence for neural information processing with precise spike-times: a survey. Natural Comput. **3**(2), 195–206 (2004)
14. Schliebs, S., Defoin-Platel, M., Kasabov, N.: Integrated feature and parameter optimization for an evolving spiking neural network. In: Köppen, M., Kasabov, N., Coghill, G. (eds.) ICONIP 2008. LNCS, vol. 5506, pp. 1229–1236. Springer, Heidelberg (2009). https://doi.org/10.1007/978-3-642-02490-0_149

15. Bohtea, S.M., Koka, J.N., La Poutr, H.: Error-backpropagation in temporally encoded networks of spiking neurons. Neurocomputing **48**(1), 17–37 (2002)

16. Delorme, A., Gautrais, J., Van Rullen, R., Thorpe, S.: Spikenet: a simulator for modeling large networks of integrate and fire neurons. Neurocomputing **26**, 989–996 (1999)

17. Thorpe, S.J., Gautrais, J.: Rapid visual processing using spike asynchrony. In: Advances in Neural Information Processing Systems, pp. 901–907 (1997)

18. Keogh, E., Mueen, A.: Curse of dimensionality. In: Sammut, C., Webb, G.I. (eds.) Encyclopedia of Machine Learning and Data Mining, pp. 314–315. Springer, Boston (2017). https://doi.org/10.1007/978-1-4899-7687-1_192

19. Barrett, T., et al.: NCBI GEO: archive for functional genomics data setsupdate. Nucleic Acids Res. **41**(D1), D991–D995 (2012)

20. van Dijk, S.J., et al.: Consumption of a high monounsaturated fat diet reduces oxidative phosphorylation gene expression in peripheral blood mononuclear cells of abdominally overweight men and women–4. J. Nutr. **142**(7), 1219–1225 (2012)

21. van Dijk, S.J., et al.: A saturated fatty acid-rich diet induces an obesity-linked proinflammatory gene expression profile in adipose tissue of subjects at risk of metabolic syndrome–. Am. J. Clin. Nutr. **90**(6), 1656–1664 (2009)

22. Reich, M., Liefeld, T., Gould, J., Lerner, J., Tamayo, P., Mesirov, J.P.: Genepattern 2.0. Nat. Genet. **38**(5), 500 (2006)

23. Peng, H., Long, F., Ding, C.: Feature selection based on mutual information criteria of max-dependency, max-relevance, and min-redundancy. IEEE Trans. Pattern Anal. Mach. Intell. **27**(8), 1226–1238 (2005)

24. Argyriou, A., Evgeniou, T., Pontil, M.: Multi-task feature learning. In: Advances in Neural Information Processing Systems, pp. 41–48 (2007)

25. Kasabov, N., Dhoble, K., Nuntalid, N., Indiveri, G.: Dynamic evolving spiking neural networks for on-line spatio-and spectro-temporal pattern recognition. Neural Netw. **41**, 188–201 (2013)

26. Schliebs, S., Kasabov, N.: Computational modeling with spiking neural networks. In: Kasabov, N. (ed.) Springer Handbook of Bio-/Neuroinformatics, pp. 625–646. Springer, Heidelberg (2014). https://doi.org/10.1007/978-3-642-30574-0_37

# Improving Off-Line Handwritten Chinese Character Recognition with Semantic Information

Hongjian Zhan, Shujing Lyu, and Yue Lu[✉]

Shanghai Key Laboratory of Multidimensional Information Processing,
Department of Computer Science and Technology, East China Normal University,
Shanghai 200062, China
hjzhan@stu.ecnu.edu.cn, {sjlv,ylu}@cs.ecnu.edu.cn

**Abstract.** Off-line handwritten Chinese character recognition (HCCR) is a well-developed area in computer vision. However, existing methods only discuss the image-level information. Chinese character is a kind of ideograph, which means it is not only a symbol indicating the pronunciation but also has semantic information in its structure. Many Chinese characters are similar in writing but different in semantics. In this paper, we add semantic information into a two-level recognition system. First we use a residual network to extract image features and make a premier prediction, then transform the image features into a semantic space to conduct a second prediction if the confidence of the previous prediction is lower than a threshold. To the best of our knowledge, we are the first to introduce semantic information into Chinese handwritten character recognition task. The results on ICDAR-2013 off-line HCCR competition dataset show that it is meaningful to add semantic information to HCCR.

**Keywords:** Handwritten Chinese character recognition
Semantic information · Character embedding

## 1 Introduction

Information in the real world comes through multiple input channels. Different channels typically carry different kinds of information. Useful representations can be learned about these information by fusing them into a joint representation that captures the real world "concept" that those information corresponds to [1]. For example, given a concept "apple", people may describe it in their own ways, referring to its shape, color, taste and so on. We may not recognize apple with just one feature, but we can make the right decision with two or more characteristics.

Combinations of mulit-channel information appear in many work. Karpathy et al. [2] introduced a model for bidirectional retrieval of images and sentences through a multi-modal embedding of visual and natural language data.

© Springer Nature Switzerland AG 2018
L. Cheng et al. (Eds.): ICONIP 2018, LNCS 11305, pp. 528–536, 2018.
https://doi.org/10.1007/978-3-030-04221-9_47

Ma et al. [3] proposed m-CNNs for matching image and sentence. The m-CNN provides an end-to-end framework with convolutional architectures to exploit image representation, word composition, and the matching relations between the two modalities. Kiela et al. [4] first combined linguistic and auditory information into multi-modal representations.

For HCCR task, besides the fantastic neural network architectures are applied, the combinations of traditional features and deep learning methods also play important roles. Zhong et al. [5] added Gabor feature to GoogleNet for HCCR. Zhang et al. [6] applied DirectMap and convolution networks to this task and introduced a new benchmark. These literature indicated that multi-channel information is beneficial to final performance. But these works only focus on image-level information.

Chinese character is a kind of ideograph and has a long history. It has undergone several big changes since it was created. Figure 1 shows some significant nodes in the evolution progress of Chinese character. Each line shows the same Chinese character. The character images are selected from HanDian[1]. The first column is Oracle bone script. It is derived from the real world directly and appears no later than Shang Dynasty (16th-11th B.C.). The second column is small seal script, which is the official script of Qin Dynasty (2th B.C.). The third column is traditional Chinese. The last column shows the simplified Chinese, which is the official script of Chinese character now and the processing target of HCCR. The abstractness of Chinese character increases while its development.

**Fig. 1.** Some samples of the evolution of Chinese character. (a) Oracle bone script, which is the original form of Chinese character; (b) small seal script; (c) traditional Chinese; (d) shows samples of simplified Chinese, which is used in Chinese mainland and the official script of China.

According to the experience, many Chinese characters are similar in image-level. Figure 2 shows some groups of such kind of characters. In this paper, we present a second classification to these image-level similar characters. If the image classifier like residual network is not very confident to the prediction result, we will classify it in semantic space. The main contribution of this work

---

[1] http://www.zdic.net/.

is to introduce semantic information into handwritten Chinese character recognition. With the two-level classifier, we improve the performance of HCCR on the common used dataset.

**Fig. 2.** Each line shows a group of image-level similar Chinese characters.

This paper is organized as follows. In Sect. 2 we describe the proposed method. Then, the details of our experiments are presented in Sect. 3. Section 4 concludes this paper and discusses the future work.

## 2 Methods

The outputs of softmax function can be treated as confidences of input image belongs to the classes. If the maximum output is small, it always give a wrong prediction. Furthermore, we find that the right class may always appears in second or third highest output. For HCCR, the top-3 predictions are always image-level similar characters. So we try to transfer the low-confidence samples into semantic space to assist making the right prediction.

Our method has two steps. The flow chart is shown in Fig. 3. First, we train a residual network to make the first prediction and create character embeddings. Then we train a transform network from image features to the corresponding character embedding. If the confidence of predicted class is lower than a threshold, we will project the image features by the transform matrix. Then in semantic space, we will make the second prediction, which is the final result.

### 2.1 Residual Neural Network

As convolutional neural networks are successful in most computer vision tasks, we are going to keep and make the most of its advantages.

Generally speaking, deeper CNNs mean better performance. However, when the networks going deeper, a degradation problem has been exposed, which

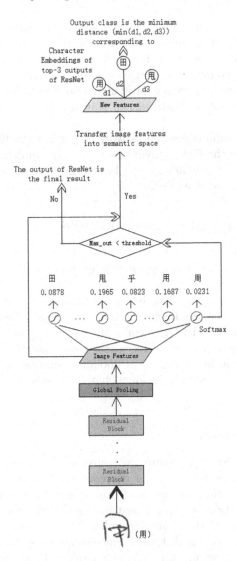

**Fig. 3.** The flow chart of our two-level recognition system.

makes the network to be more difficult to train and slower to converge. In order to address this issue, He et al. [7] introduced a deep residual learning framework, i.e. the ResNet. ResNet is composed of a number of stacked residual blocks, and each block contains direct links between the lower layer outputs and the higher layer layer inputs.

The residual block (described in Fig. 4) is defined as:

$$\mathbf{y} = \mathcal{F}(\mathbf{x}, \mathbf{W}_i) + \mathbf{x}, \tag{1}$$

where $\mathbf{x}$ and $\mathbf{y}$ are the input and output of the layers considered, and $\mathcal{F}$ is the stacked nonlinear layers mapping function. Note that identity shortcut connections of $\mathbf{x}$ do not add extra parameters and computational complexity. With the presence of residual connections, ResNet can improve the convergence speed in training and gains accuracy from greatly increased depth. With residual convolutional layers, we can obtain more efficient feature representations than previous architectures [8,9].

There are many famous residual network. We follow the ResNet-50 proposed in [7,10]. We reduce the feature map size to fit our input image (64 * 64), and apply global pooling at the last pooling layer. The residual network used in our experiment is also donated as ResNet-50.

## 2.2 Image Feature and Character Embedding

The output of each layer can be treated as the feature representation of the input image. We use the output of last pooling layer, i.e., the global pooling layer, as the feature vector.

We train character embedding by word2vec toolkit[2] with different dimensions. The corpus is the Wikimedia dumps[3] and Sogou News [11], which will be described in Sect. 3.1. To build embedding features, we first use jieba tokenizer[4] to tokenize each location name where a user has visited before. Then we simply adopt the min, max, and mean pooling operations on all tokens in the location names. We use 200-dimensional character embeddings in the following experiments.

## 2.3 Transform Network

We utilize a four layers fully-connected neural network to model the relations between image features and character embeddings. Number of Neurons in each layer is 1000. It is a one-way mapping from image to semantic. After we get the vector representation of image in semantic space, we calculate the distances between this vector and the character embedding of top-k characters. We use cosine distance to matric the distances. Since the range of elements in character embeddings is $(-1,1)$, we use tanh as the activation function.

Given an input image $I$, the final prediction label is $Y$. Donate the top-3 outputs of ResNet-50 as $O_1$, $O_2$, $O_3$ with the corresponding labels are $L_1$, $L_2$, $L_3$. The character embeddings of $L_1$, $L_2$, $L_3$ are donated as $E_1$, $E_2$, $E_3$. Y can be calculated like:

if $\max(O_1, O_2, O_3) > threshold$:

$$Y = L_i, i = \underset{i=1,2,3}{\arg\max}(O_i) \tag{2}$$

---

[2] http://code.google.com/archive/p/word2vec.
[3] https://archive.org/details/zhwiki-20160501.
[4] github.com/fxsjy/jieba.

else:

$$Y = L_i, i = \arg\min_{i=1,2,3} cosdi(F, E_i) \tag{3}$$

where $cosdi(F, E_i)$ is the cosine distance between $F$ and $E_i$, and $F$ is the vector representation of $I$.

In this paper, we empirically set threshold $= 0.5$.

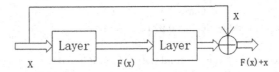

**Fig. 4.** The structure of residual block.

**Table 1.** The statistics for off-line HCCR datasets.

| DB name | #writers | #samples |
|---|---|---|
| HWDB1.0 | 420 | 1,556,675 |
| HWDB1.1 | 300 | 1,121,749 |
| HWDB1.2 | 300 | 4,463 |
| Off-line ICDAR2013 | 60 | 224,419 |

## 3    Experiments

### 3.1    Datasets

There are two categories of dataset used in our experiments. For image, we use the off-line handwritten character datasets CASIA-HWDB1.1, which contains $1,121,749$ images written by 300 writers. CASIA-HWDB is a series datasets created by Institute of Automation, Chinese Academy of Sciences (CASIA), which now is the most widely used handwritten Chinese character dataset. All samples in CASIA-HWDB1.1 are used as training set, and there is no validation set while training ResNet-50.

For character embedding, we use two corpuses. One is Wikimedia dumps Chinese corpus, which contains 876,239 Web pages. Contents of these Web pages are extracted by Wikipedia Extractor and a total of 3,736,800 sentences are collected after preprocessing. The other is SogouCA[5], which selects eighteen channels for Sohu news.

---

[5] http://www.sogou.com/labs/.

**Table 2.** Recognition rates of different models on ICDAR-2013 off-line HCCR competition dataset.

| Method | Ref. | Top_1 Acc | Training set | Distortion | Writer info |
|---|---|---|---|---|---|
| ICDAR-2013 Winner | [12] | 0.9477 | 1.1 | Yes | No |
| HCCR-Gabor-GoogLeNet | [5] | 0.9635 | 1.0 + 1.1 | No | No |
| CNN-Single | [13] | 0.9658 | 1.0 + 1.1 + 1.2 | Yes | No |
| DirectMap + Conv | [6] | 0.9695 | 1.0 + 1.1 | No | No |
| DirectMap + Conv + Adap | [6] | **0.9737** | 1.0 + 1.1 | No | Yes |
| ResNet-50 | Ours | 0.9688 | 1.1 | No | No |
| ResNet-50 + Char-embedding | Ours | **0.9713** | 1.1 | No | No |

**Table 3.** Top-k accuracy of our residual network on ICDAR-2013 competition dataset.

| top_k | top_1 | top_2 | top_3 | top_4 | top_5 |
|---|---|---|---|---|---|
| Accuracy | 0.9688 | 0.9895 | 0.9936 | 0.9954 | 0.9964 |

The test data of HCCR is the ICDAR-2013 off-line competition datasets. It contains 224419 characters that written by 60 writers. The number of character classes in all datasets is 3755, which is the level-1 set of **GB2312 − 80** standard. The statistical data of HCCR datasets are shown in Table 1.

## 3.2    Experimental Results

We first train the residual network with CASIA-HWDB1.1. The results on ICDAR-2013 competition dataset are shown in Table 3. We find that after $k = 3$ for top-k, the accuracy increases a little. In order to reduce the computation complexity in semantic space, we only calculate the semantic distance with the top-3 recognition results.

To train the transform network, we first extract image features from CASIA-HWDB1.1. 80% of the image features and character embeddings are used as training set and reset for validation.

Table 2 shows our final results on ICDAR-2013. The proposed method gives a very competitive result. Methods in [6] applied writer adaptation layer. It needs the writer information of the character image, which is a strong priori knowledge. In general situations, it is almost impossible to provide such information. So we also make our result bold to show the best result without harsh conditions.

We think there are several reasons for the success of our method. Firstly it benefits from the powerful ability of residual network. With ResNet-50 we stand a high starting point. Secondly the ideographic nature of Chinese character also helps a lot. They make it possible to find a relationship between the image features and character embeddings. Another important reason we think is that we bring the top_k information into second-level classification. It not only reduces the search space, but also increases the accuracy by removing most of distractions.

## 3.3  Experimental Cost

Our experiments are performed on a SuperMicro server. The CPU is Intel Xeon E5-2630 with 2.2 GHz and the GPU is NVIDIA GeForce 1080TI. The software is the latest version of Caffe [14] and Tensorflow [15] with cuDNN V5 accelerated on Ubuntu 14.04 LTS system. The average testing time is 3ms per image. We train residual network with Caffe, and apply Tensorflow and python interface of Caffe to implement the remainder programs.

# 4  Conclusion and Future Work

Because of the grounding-problem [4], we can not account for the fact that human semantic knowledge is grounded in the perceptual system.

In this paper, we try to find the relationship between multi-channel information that is beneficial to handwritten Chinese character recognition. There are three parts in our method, image features, character embeddings and the transform network. With residual network, we can get a high-performance feature representation. For semantic area, we use a common method to create character embedding, and at last we use these features to train a transform network. But now it is not an end-to-end trainable architecture. In the future, on one hand we will introduce more efficient Chinese character embedding such as association to the structure of Chinese character, on the other hand, we can apply more advanced methods to model the relations between two feature spaces.

# References

1. Srivastava, M., Salakhutdinov, R.: Multimodal learning with deep Boltzmann machines. J. Mach. Learn. Res. **15**, 2949–2980 (2014)
2. Karpathy, A., Joulin, A., Li, F.: Deep fragment embeddings for bidirectional image sentence mapping. In: 27th Advances in Neural Information Processing Systems, pp. 1889–1897 (2014)
3. Ma, L., Lu, Z., Shang, L., Li, H.: Multimodal convolutional neural networks for matching image and sentence. In: 14th IEEE International Conference on Computer Vision, pp. 2623–2631 (2015)
4. Kiela, D., Clark, S.: Multi- and cross-modal semantics beyond vision: grounding in auditory perception. In: 20th Conference on Empirical Methods in Natural Language Processing, pp. 2461–2470 (2015)
5. Zhong, Z., Jin, L., Xie, Z.: High performance offline handwritten Chinese character recognition using GoogleNet and directional feature maps. In: 13th International Conference on Document Analysis and Recognition, pp. 846–850 (2015)
6. Zhang, X., Bengio, Y., Liu, C.: Online and offline handwritten chinese character recognition: a comprehensive study and new benchmark. Pattern Recogn. **61**, 348–360 (2017)
7. He, K., Zhang, X., Ren, S., Sun, J.: Deep residual learning for image recognition. In: 29th IEEE Conference on Computer Vision and Pattern Recognition, pp. 770–778 (2016)

8. Simonyan, K., Zisserman, A.: Very deep convolutional networks for large-scale image recognition. arXiv preprint arXiv:1409.1556 (2014)
9. Szegedy, C., et al.: Going deeper with convolutions. In: 28th IEEE Conference on Computer Vision and Pattern Recognition, pp. 1–9 (2015)
10. Simon, M., Rodner, E., Denzler, J.: ImageNet pre-trained models with batch normalization. arXiv preprint arXiv:1612.01452 (2016)
11. Wang, C., Zhang, M., Ma, S., Ru, L.: Automatic online news issue construction in web environment. In: 17th International Conference on World Wide Web, pp. 457–466 (2008)
12. Yin, F., Wang, Q., Zhang, X., Liu, C.: ICDAR 2013 Chinese handwriting recognition competition. In: 12th International Conference on Document Analysis and Recognition, pp. 1464–1470 (2013)
13. Chen, L., Wang, S., Fan, W., Sun, J., Naoi, S.: Beyond human recognition: a CNN-based framework for handwritten character recognition. In: 3rd Asian Conference on Pattern Recognition, pp. 695–699 (2015)
14. Jia, Y., et al.: Caffe: convolutional architecture for fast feature embedding. arXiv preprint arXiv:1408.5093 (2014)
15. Abadi, M., et al.: TensorFlow: a system for large-scale machine learning. In: 12th USENIX Symposium on Operating Systems Design and Implementation, pp. 265–283 (2016)

# Text Simplification with Self-Attention-Based Pointer-Generator Networks

Tianyu Li, Yun Li$^{(\boxtimes)}$, Jipeng Qiang$^{(\boxtimes)}$, and Yun-Hao Yuan

School of Information Engineering, Yangzhou University, Yangzhou 225137, China
{liyun,jpqiang}@yzu.edu.cn

**Abstract.** Text Simplification aims to reduce semantic complexity of text, while still retaining the semantic meaning. Recent work has started exploring neural text simplification (NTS) using the Sequence-to-sequence (Seq2seq) attentional model which achieves success in many text generation tasks. However, dealing with long-range dependencies and out-of-vocabulary (OOV) words remain the challenge of Text Simplification task. In this paper, in order to solve these problems, we propose a text simplification model that incorporates self-attention mechanism and pointer-generator network. Our experiments on Wikipedia and Simple Wikipedia aligned datasets demonstrate that our model is outperforms the baseline systems.

**Keywords:** Text simplification · Seq2seq · Self-attention
Pointer-generator networks

## 1 Introduction

The goal of text simplification is to convert complex sentences into simpler sentences without significantly altering the original meaning. Text simplification can reduce reading complexity and make complex sentences suitable for people with limited linguistic skills [23], such as children, non-native speakers and patients with linguistic and cognitive disabilities [3].

Previous work, such as PBMT-R model [11], treats the simplification process as a monolingual text-to-text generation task, borrowing the idea of machine translation. It uses a simplified text corpora, such as Simple English Wikipedia [2,26], and adopts statistical machine translation models to text simplification. Recently, Neural Machine Translation (NMT) [1] based on the Seq2seq attentional model [17] shows more powerful effect than traditional statistical machine translation models. Inspired by the success of NMT, recent works have started exploring neural text simplification (NTS) [12,20–22] using the Seq2seq attentional model.

Although NTS model has achieved better result than previous work, there are still many problems to be solved. The first one is long sentences problem.

© Springer Nature Switzerland AG 2018
L. Cheng et al. (Eds.): ICONIP 2018, LNCS 11305, pp. 537–545, 2018.
https://doi.org/10.1007/978-3-030-04221-9_48

In NTS model, the number of operations required to relate signals from two arbitrary input or output positions grows in the distance between positions. This makes it more difficult to learn dependencies between distant positions [8]. The second one is out-of-vocabulary (OOV) words problem. Text simplification dataset exist a large number of out-of-vocabulary (OOV) words, but when an OOV word appears in input sequence, NTS model does not know how to handle this word. Inspired by recent NMT models, we propose self-attention-based pointer-generator networks. It replaces Transformer [18] model with the original Seq2seq model in Pointer-generator network [15].

In this paper, our contributions are as follows: We combine the self-attention mechanism with pointer-generator networks and apply it to neural text simplification. This model is easier to capture the characteristics of the long-range dependencies in the sentence and can solve OOV problems very well. When the OOV words appear in the input sentence, pointer-generator network can choose to copy the OOV word directly from the input sentence as the output word.

We evaluate our proposed model on English Wikipedia and Simple English Wikipedia datasets. The experimental results indicate that our model achieves better results than a series of baseline algorithms in terms of BLEU score and SARI score.

## 2    Related Work

In previous studies, researchers of text simplification mostly address the simplification task as a monolingual machine translation problem. Recent progress in deep learning with neural networks brings great opportunities for the development of stronger NLP systems such as NMT. Neural text simplification [12,20,28] is inspired by the success of NMT and gradually became the main research direction. It is based on the Seq2seq model [1,17]. The common Seq2seq models use Long Short-Term Memory (LSTM) [9] or Gated Recurrent Unit (GRU) [5] for the encoder and decoder [12], and use attention mechanism [1] align the words in the encoder and decoder. Nisioi et al. [12] implemented a standard Seq2seq model and found that they outperform previous models. Zhang et al. [29]implement a novel constrained neural generation model to simplify sentences given simplified words. Zhang et al. [28] viewed the encoder-decoder model as an agent and employed a deep reinforcement learning framework [24]. Vu et al. [20] use Neural Semantic Encoders to augmented memory capacities. These methods are aim to make the NTS model perform better, but the NTS model still has many problems to solve.

Self-attention mechanism is a special case of the attention mechanism. Self-attention can learn the internal word dependencies of the sentence and capture the internal structure of the sentence. Self-attention has been successfully used for a variety of tasks, including machine translation, reading comprehension, summarization [4,14]. Transformer model [18] based on self-attention mechanism [10], and is the first Seq2seq model based entirely on attention, achieving state-of-the-art performance on machine translation task.

Pointer-generator network [15] has been successfully applied to abstractive summarization, which is similar to CopyNet [7]. It is a hybrid between seq2seq model and pointer network [19]. The pointer network is a Seq2seq model that uses the attention distribution to produce an output sequence consisting of elements from the input sequence. When the OOV words appears in the input sentence, pointer-generator network can choose to copy the OOV word directly from the input sentence as the output word.

Overall, both Self-attention and Pointer-generator network have their own advantages. In this paper, We replace Transformer model with the original Seq2seq model in Pointer-generator network to achieve the goal of combining these advantages.

# 3    Proposed Model

## 3.1    Model Overview

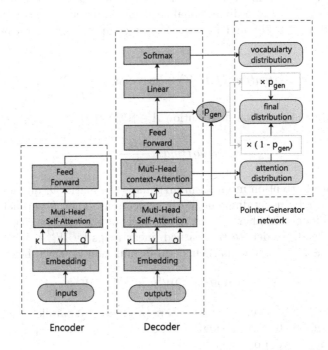

**Fig. 1.** Self-attention-based pointer-generator network model architecture.

Our model is the combination of Transformer and Pointer-generator network, and it follows encoder-decoder architecture. Encoder and Decoder shown in the left and middle of Fig. 1 respectively. Encoder layer consist of a multi-head self-attention layer and a feed-forward network. In addition to the two layers in encoder, the decoder inserts a Multi-head context-attention layer which imitate the typical encoder-decoder attention mechanisms in Seq2seq model. Multi-head context-attention layer can obtain information of attention distribution. Finally, the Decoder outputs a vocabulary distribution.

The right of graph is Pointer-generator network. It will generate the last final distribution based on vocabulary distribution and attention distribution. Final distribution contains the information that choose whether to copy the word directly from the input sentence as the output word.

## 3.2  Transformer

We adopt the Transformer model because it is the first Seq2seq model based on the self-attention mechanism, and achieving state-of-the-art performance on machine translation task. We adopt the multi-head attention [18] formulation to calculate attention value. The core of the formulation is the scaled dot-product attention, which is a variant of dot-product attention. We compute the attention function on a set of d-dimensional queries simultaneously, packed together into a matrix $Q \in \mathbb{R}^{n \times d}$. The d-dimensional keys and values are also packed together into matrices $K \in \mathbb{R}^{m \times d}$ and $V \in \mathbb{R}^{m \times d}$. We compute the matrix of attention value as:

$$Attention(Q, K, V) = softmax(\frac{QK^T}{\sqrt{d}})V \tag{1}$$

The matrix of the attention distribution $A \in \mathbb{R}^{n \times m}$ is:

$$A = softmax(\frac{QK^T}{\sqrt{d}}) \tag{2}$$

The multi-head attention mechanism first maps the matrix of queries, keys and values matrices by using different linear projections. Then $h$ parallel heads are employed to focus on different part of channels of the value vectors. Formally, for the $i$-th head, we denote the learned linear maps by $W_i^Q \in \mathbb{R}^{d \times d/h}, W_i^K \in \mathbb{R}^{d \times d/h}, W_i^V \in \mathbb{R}^{d \times d/h}$. The mathematical formulation is shown below:

$$M_i = Attention(QW_i^Q, KW_i^K, VW_i^V) \tag{3}$$

where $M_i \in \mathbb{R}^{n \times d/h}$. Then, all the vectors produced by parallel heads are concatenated together to form a single vector. Again, a linear map is used to mix different channels from different heads:

$$M = Concat(M_1, \ldots, M_h)W \tag{4}$$

where $M \in \mathbb{R}^{n \times d}$, $W \in \mathbb{R}^{d \times d}$. Then, the feed-forward layer is the following formula:

$$F(X) = ReLU(XW_1)W_2 \tag{5}$$

where $W_1$ and $W_2$ are trainable matrices. Given $M_o$ as attention value of context-attention layer, the output of the feed-forward for the decoder $O \in \mathbb{R}^{n \times d}$ is:

$$O = F(M_o) \tag{6}$$

Finally, Given $d_v$ as the output vocabulary size, the decoder outputs a vocabulary distribution $P_{vacab}$:

$$P_{vocab} = softmax(V^{'}(OV + b) + b^{'}) \tag{7}$$

where $V \in \mathbb{R}^{d_v \times n}$, $V^{'} \in \mathbb{R}^{d \times 1}$, $b$ and $b'$ are learnable parameters. $P_{vocab}$ is a probability distribution over all words in the vocabulary.

## 3.3 Pointer-Generator Network

Compared to the basic seq2seq model, Pointer-Generator network calculates the generation probability $p_{gen}$ after calculating attention distribution and vocabulary distribution, which is a scalar value between 0 and 1. This represents the probability of generating a word from the vocabulary, versus copying a word from the input sentence. On each timestep $t$ decoder calculates $p_{gen}$ from $O$ and $Q^*$, where $Q^*$ is the output of muti-head self-attention layer for decoder:

$$p_{gen} = \sigma(w_o{}^T o_t + w_q{}^T q_t + b) \tag{8}$$

where vectors $w_o$, $w_q$ and scalar b are learnable parameters, and $o_t$ is the $t$-th row vector of matrix $O$ and $q_t$ is the $t$-th row vector of matrix $Q^*$ and $\sigma$ is the sigmoid function. The generation probability $p_{gen}$ is used to weight and combine the vocabulary distribution $P_{vocab}$ and the attention distribution $\alpha_t$, where $\alpha_t$ is the $t$-th row vector of attention distribution matrix $A$. The final distribution $P_f inal$ via the following formula:

$$P_{final}(w) = p_{gen} P_{voacb}(w) + (1 - p_{gen}) \sum_{i:w_i=w} \alpha_{ti} \tag{9}$$

This formula just says that the probability of producing word w is equal to the probability of generating it from the vocabulary plus the probability of pointing to it anywhere it appears in the source text. When the OOV words appears in the input sentence, the value of $p_{gen}$ will be close to 0 and final distribution equals attention distribution. So, it can choose to copy the OOV word directly from the input sentence as the output word.

During training, the loss for timestep t is the negative log likelihood of the target word $w_t$ for that timestep and the overall loss for the whole output sequence is

$$loss = \frac{1}{T} \sum_{t=0}^{T} -log P(w_t) \tag{10}$$

where T is length of output sentence.

## 4  Experiments

### 4.1  Experimental Setting

**Datasets.** We conducted experiments on two simplification datasets. WikiSmall [2] is a parallel corpus that has been widely used as a benchmark for evaluating text simplification systems. It contains automatically aligned complex-simple sentences pairs from the complex and simple English Wikipedias. The dataset has 88,837/205/100 pairs for train/dev/test. WikiLarge [28] has 296,402/2,000/359 pairs for train/dev/test. It is a large corpus and the training set is a mixture of three Wikipedia datasets [25].

**Training Details.** We trained our models on an Nvidia GPU card. For all experiments, our models have 512-dimensional hidden states and 512-dimensional word embeddings. The number of network layers is 6 and init parameters with Xavier initialization [6]. Batch size is set to 2048. We used dropout [16] for regularization with a dropout rate of 0.2.

**Evaluation.** We used BLEU [13] that scores the output by counting n-gram matches with the reference, or SARI [27], which evaluates the quality of the output by comparing it against the source and reference simplifications. Both measures are commonly used to automatically evaluate the quality of simplification output.

**Comparison Systems.** We compare our model with several systems previously proposed in the literature. PBMT-R [11] is a monolingual phrase-based statistical machine translation system. DRESS [28] is a deep reinforcement learning model for Text Simplification. EncDecA [12] is the basic attentional encoder-decoder model with two LSTM layers experimented with beam sizes of 5. Compare with our model, TSSP-Self is a seq2seq model that uses only Transformer but not Pointer-Generator network.

### 4.2  Results

The result of the automatic evaluation on WikiSmall and WikiLarge are displayed in Table 1. In WikiSmall Dataset, Our model achieved best BLEU score (50.73) and best SARI score(35.36). All neural models obtain higher SARI compared to PBMT-R. In WikiLarge Dataset, PBMT-R achieved best SARI(38.56), but its BLEU score is lower than seq2seq models. Our model achieved best BLEU(92.10). Overall, our model has a good potential for BELU and SARI scores.

Example model outputs on WikiLarge are provided in Table 2. The first example sentence is a long sentence, and it also contains an OOV word 'Gregorian Calendar'. From the simplified sentence of our model, it is seen that the simplified sentence omit some of the original sentence, and the overall semantics of the sentence have not changed. So we can see that our model performs well for long sentences. The second sentence shows that our model still retains the overall semantics of the sentence and that the OOV word 'Soviet Union' is well copied to the output sentence.

**Table 1.** Model performance using automatic evaluation measures (BLEU and SARI).

| Dataset | Model | BLEU | SARI |
|---------|-------|------|------|
| WikiSmall | PBMT-R | 46.31 | 15.97 |
| | DRESS | 34.53 | 27.48 |
| | EncDecA | 47.93 | 13.61 |
| | TSSP-Self | 40.32 | 35.22 |
| | TSSP(our proposal) | **50.73** | **35.36** |
| Dataset | Model | BLEU | SARI |
| WikiLarge | PBMT-R | 81.11 | **38.56** |
| | DRESS | 77.81 | 37.08 |
| | EncDecA | 88.85 | 35.66 |
| | TSSP-Self | 70.40 | 34.04 |
| | TSSP(our proposal) | **92.10** | 30.49 |

**Table 2.** Example model outputs on wikiLarge. Substitutions are shown in bold

| System | Output |
|--------|--------|
| Complex | December is the twelfth and last month of the year in the Gregorian Calendar and one of seven Gregorian months with the length of 31 days |
| Reference | December is the twelfth and last month of the year, with 31 days |
| PBMT-R | December is the twelfth and last month of the year **with 31 days** |
| DRESS | December is the twelfth and last month of the year |
| EncDecA | **November** is the twelfth and last month of the year in the Gregorian Calendar and one of seven Gregorian months with the length of 31 days |
| TSSP-Self | December is the twelfth and last month of the year |
| TSSP | December is the twelfth and last month of the year **with 31 days** |
| System | Output |
| Complex | Restoration of independence In 1991, the Soviet Union broke apart and Armenia re-established its independence |
| Reference | Armenia received its independence from the Soviet Union in 1991 |
| PBMT-R | Restoration of independence In 1991, the Soviet Union broke apart and Armenia **set up** its independence |
| DRESS | **In 1991**, the Soviet Union broke apart and Armenia |
| EncDecA | **In 1991**, the Soviet Union broke apart and Armenia **reorganised** its independence |
| TSSP-Self | **In 1991**, the Soviet Union broke apart and Armenia re-established its independence |
| TSSP | **In 1991**, the Soviet Union broke apart and Armenia **made** its independence |

## 5  Conclusions

In this paper, we propose a text simplification model that incorporates self-attention mechanism and a pointer generator network, which can directly capture the relationships between two tokens regardless of their distance and copy words from the source text via pointing to aids accurate reproduction of information. We run experiments on the parallel datasets of WikiSmall and WikiLarge. The results show that our method outperforms the baseline and performs well in simplifying long sentences and handling OOV words.

**Acknowledgments.** This research is partially supported by the National Natural Science Foundation of China under grants (61703362, 61402203), the Natural Science Foundation of Jiangsu Province of China under grants (BK20170513, BK20161338), the Natural Science Foundation of the Higher Education Institutions of Jiangsu Province of China under grant 17KJB520045, and the Science and Technology Planning Project of Yangzhou of China under grant YZ2016238.

## References

1. Bahdanau, D., Cho, K., Bengio, Y.: Neural machine translation by jointly learning to align and translate. arXiv preprint arXiv:1409.0473 (2014)
2. Bernhard, D., Gurevych, I.: A monolingual tree-based translation model for sentence simplification. In: The 23rd International Conference on Computational Linguistics Proceedings of the Main Conference, vol. 2, pp. 1353–1361 (2010)
3. Carroll, J., Minnen, G., Pearce, D., Canning, Y., Devlin, S., Tait, J.: Simplifying text for language-impaired readers. In: Ninth Conference of the European Chapter of the Association for Computational Linguistics (1999)
4. Cheng, J., Dong, L., Lapata, M.: Long short-term memory-networks for machine reading. arXiv preprint arXiv:1601.06733 (2016)
5. Cho, K., et al.: Learning phrase representations using RNN encoder-decoder for statistical machine translation. arXiv preprint arXiv:1406.1078 (2014)
6. Glorot, X., Bengio, Y.: Understanding the difficulty of training deep feedforward neural networks. In: Proceedings of the Thirteenth International Conference on Artificial Intelligence and Statistics, pp. 249–256 (2010)
7. Gu, J., Lu, Z., Li, H., Li, V.O.: Incorporating copying mechanism in sequence-to-sequence learning. arXiv preprint arXiv:1603.06393 (2016)
8. Hochreiter, S., Bengio, Y., Frasconi, P., Schmidhuber, J., et al.: Gradient flow in recurrent nets: the difficulty of learning long-term dependencies (2001)
9. Hochreiter, S., Schmidhuber, J.: Long short-term memory. Neural Comput. **9**(8), 1735–1780 (1997)
10. Lin, Z., et al.: A structured self-attentive sentence embedding. arXiv preprint arXiv:1703.03130 (2017)
11. Narayan, S., Gardent, C.: Hybrid simplification using deep semantics and machine translation. In: Proceedings of the 52nd Annual Meeting of the Association for Computational Linguistics (Volume 1: Long Papers). vol. 1, pp. 435–445 (2014)
12. Nisioi, S., Štajner, S., Ponzetto, S.P., Dinu, L.P.: Exploring neural text simplification models. In: Proceedings of the 55th Annual Meeting of the Association for Computational Linguistics (Volume 2: Short Papers), vol. 2, pp. 85–91 (2017)

13. Papineni, K., Roukos, S., Ward, T., Zhu, W.J.: BLEU: a method for automatic evaluation of machine translation. In: Proceedings of the 40th Annual Meeting on Association for Computational Linguistics, pp. 311–318. Association for Computational Linguistics (2002)
14. Parikh, A.P., Täckström, O., Das, D., Uszkoreit, J.: A decomposable attention model for natural language inference. arXiv preprint arXiv:1606.01933 (2016)
15. See, A., Liu, P.J., Manning, C.D.: Get to the point: summarization with pointer-generator networks. arXiv preprint arXiv:1704.04368 (2017)
16. Srivastava, N., Hinton, G., Krizhevsky, A., Sutskever, I., Salakhutdinov, R.: Dropout: a simple way to prevent neural networks from overfitting. J. Mach. Learn. Res. **15**(1), 1929–1958 (2014)
17. Sutskever, I., Vinyals, O., Le, Q.V.: Sequence to sequence learning with neural networks. In: Advances in Neural Information Processing Systems, pp. 3104–3112 (2014)
18. Vaswani, A., et al.: Attention is all you need. In: Advances in Neural Information Processing Systems, pp. 6000–6010 (2017)
19. Vinyals, O., Fortunato, M., Jaitly, N.: Pointer networks. In: Advances in Neural Information Processing Systems, pp. 2692–2700 (2015)
20. Vu, T., Hu, B., Munkhdalai, T., Yu, H.: Sentence simplification with memory-augmented neural networks. arXiv preprint arXiv:1804.07445 (2018)
21. Wang, T., Chen, P., Amaral, K., Qiang, J.: An experimental study of LSTM encoder-decoder model for text simplification. arXiv preprint arXiv:1609.03663 (2016)
22. Wang, T., Chen, P., Rochford, J., Qiang, J.: Text simplification using neural machine translation. In: AAAI, pp. 4270–4271 (2016)
23. Watanabe, W.M., Junior, A.C., Uzêda, V.R., Fortes, R.P.d.M., Pardo, T.A.S., Aluísio, S.M.: Facilita: reading assistance for low-literacy readers. In: Proceedings of the 27th ACM International Conference on Design of Communication, pp. 29–36. ACM (2009)
24. Williams, R.J.: Simple statistical gradient-following algorithms for connectionist reinforcement learning. In: Sutton, R.S. (ed.) Reinforcement Learning. The Springer International Series in Engineering and Computer Science (Knowledge Representation, Learning and Expert Systems), vol. 173. Springer, Boston (1992). https://doi.org/10.1007/978-1-4615-3618-5_2
25. Woodsend, K., Lapata, M.: Learning to simplify sentences with quasi-synchronous grammar and integer programming. In: Proceedings of the Conference on Empirical Methods in Natural Language Processing, pp. 409–420. Association for Computational Linguistics (2011)
26. Xu, W., Callison-Burch, C., Napoles, C.: Problems in current text simplification research: new data can help. Trans. Assoc. Comput. Linguist. **3**(1), 283–297 (2015)
27. Xu, W., Napoles, C., Pavlick, E., Chen, Q., Callison-Burch, C.: Optimizing statistical machine translation for text simplification. Trans. Assoc. Comput. Linguist. **4**, 401–415 (2016)
28. Zhang, X., Lapata, M.: Sentence simplification with deep reinforcement learning. arXiv preprint arXiv:1703.10931 (2017)
29. Zhang, Y., Ye, Z., Feng, Y., Zhao, D., Yan, R.: A constrained sequence-to-sequence neural model for sentence simplification. arXiv preprint arXiv:1704.02312 (2017)

# Integrating Topic Information into VAE for Text Semantic Similarity

Xiangdong Su[1,2], Rong Yan[1,2(✉)], Zheng Gong[1,2], Yujiao Fu[1], and Heng Xu[1]

[1] College of Computer Science, Inner Mongolia University, Hohhot 010021, China
{cssxd, csyanr}@imu.edu.cn
[2] Inner Mongolia Key Laboratory of Mongolian Information Processing Technology, Hohhot 010021, China

**Abstract.** Representation learning is an essential process in the text similarity task. The methods based on neural variational inference first learn the semantic representation of the texts, and then measure the similar degree of these texts by calculating the cosine of their representations. However, it is not generally desirable that using the neural network simply to learn semantic representation as it cannot capture the rich semantic information completely. Considering that the similarity of context information reflects the similarity of text pairs in most cases, we integrate the topic information into a stacked variational autoencoder in process of text representation learning. The improved text representations are used in text similarity calculation. Experiment shows that our approach obtains the state-of-art performance.

**Keywords:** Representation learning · Variational autoencoder
Topic information · Semantic similarity

## 1 Introduction

In text similarity task, previous statistical methods generally regard actual text as bag of words and ignore the text structure information. The bag-of-words model cannot distinguish the semantic ambiguity of natural language and further affected the accuracy of document similarities. Recently, word, sentence and text representations become hot areas with the rapid development of deep learning [1–3]. Deep neural network based methods have obtained great progress on representation learning tasks. Many methods use text representation resulting from deep neural network models instead of word vector in natural language processing. Our goal in this paper is integrating the topic information into variation autoencoder (VAE) to learn an improved text representation for text semantic similarity calculation.

As a generative model, VAE induces more possibilities in text processing [4], in which we consider that the textual data can be observed is generated by some hidden variables that we cannot see. The hidden variables are explicit representation of the original document with lower dimensions. It can be seen that the acquisition of these hidden variables is crucial to obtain the semantic information of these texts. Bowman

© Springer Nature Switzerland AG 2018
L. Cheng et al. (Eds.): ICONIP 2018, LNCS 11305, pp. 546–557, 2018.
https://doi.org/10.1007/978-3-030-04221-9_49

et al. [5] employed VAE to get the hidden representation of the whole sentence. Samples from the prior over these sentence representations remarkably produce diverse and well-formed sentences through simple deterministic decoding. Miao et al. [6] constructed a neural variational framework, proposed an unsupervised model and a supervised model based on this for document modeling tasks and answer filtering tasks, respectively. Inspired by the above work, this paper explores the resulted text representation (hidden variables) from VAE in the semantic similarity task.

In fact, only using the text as input of VAE, the resulting text representation can capture the semantic content of the text, but the effect of contextual information is ignored. It was proved that the context of text can provide significance information for accurate semantic understanding and comprehensive representation learning. At the same time, the similarity of context information can reflect the similarity of texts. For this reason, we refer to the way proposed in [7] to integrate context information into VAE.

Topic information is one of the very important context information. It has an intimate relationship with text semantics, and is used most in natural language processing. This paper integrates the topic information into the process of text representation learning together with each text. Non-Negative Matrix Factorization (NMF) is used to extract the topic information in the preprocessing stage.

The innovations of our work are as follows: (1) Integrating topic information into VAE; (2) Stacking VAE with topics to form a deep model, the latent representations get from this model are used to compute semantic similarity. We investigate the utility of context information in two semantic similarity tasks: word-pair task on the SCWS dataset provided by Huang et al. [8] and sentence-pair task on the SICK dataset provided by Marelli et al. [9]. Compared with the recent works, our model outperforms in both tasks.

The remainder of the paper is organized as follows: Sect. 2 provides background on VAE and works that integrate topic information into deep learning models. Section 3 describes the basic autoencoder, VAE and our model TVAE. Section 4 presents the results of experiments on SCWS and SICK. Finally, we conclude and provide future research directions in Sect. 5.

## 2  Related Works

In this paper, we focus our attention on VAE by analyzing the current status of research. VAE, an extension of AE, has spawned a renaissance in latent variable models. Bowman et al. [5] proposed a RNN-based variational autoencoder language model to model holistic properties of sentences, then got diverse and well-formed latent representations of entire sentences through simple decode. Xu et al. [10] proposed a conditional variational autoencoder and used latent variables to represent multiple possible trajectories. It predicted events in a wide variety of scenes successfully in vision task. In addition, VAE are also used for dialogue generating [11] and sentence compressing [12], etc. VHRED model presented by Serban et al. [11] was improved based on HRED model, an extension of the RNNLM, at the utterance level with a stochastic latent variable. It helped to produce long and diverse responses and maintain dialogue state in the task of dialogue response generation. Unlike most works about VAE, Miao and Blunsom [12] modelled language as a discrete latent variable. They proposed a generative model based on a variational auto-encoding framework, and applied it to the task of compressing sentences. Currently, most works about VAE stay in single stochastic layer. Sønderby et al. [13] for the first time trained deep variational autoencoder up to five stochastic layers. However, model with multiple stochastic layers is scarce in NLP.

Topic information provides a convenient way for people to obtain important information. Modeling topics for textual data by topic model enables access to potential topic information in large amounts of documents. In recent researches on presentation learning, many works combined the topic model with Recurrent Neural Networks [14–17]. Mikolov and Zweig [14] improved the performance of RNNLM with the vector which used to convey contextual information about the sentence. They achieved the vector by Latent Dirichlet Allocation (LDA) and proposed a topic-conditioned RNNLM. Ji et al. [15] proposed three document-context language models combined local and global information in language modeling. They obtained the contextual information representation from the hidden states of the previous sentence by RNN language model, and provided various approaches to integrate the contextual information into the language model of the current sentence. CLSTM presented by Ghosh et al. [16] incorporated contextual features, such as topics of text segments, into the recurrent neural network LSTM model, and verified was beneficial for different NLP tasks. Dieng et al. [17] combined latent topic models with RNNs and proposed a TopicRNN model, then captured both semantic dependencies and syntactic dependencies by using it. Certainly, there are also researches combined with other deep learning models. Amiri et al. [7] integrated context information into DAE and proposed a pairwise context-sensitive Autoencoder to calculated the similarity of text pairs. It mapped the input and the hidden representation for the given context vector into a context-sensitive representation. In addition, Xing et al. [18] also incorporated topic model into the encoder-decoder structure network. The TAJA-Seq2Seq model they proposed regarded topic information as prior knowledge, and merged it into Seq2Seq.

# 3   Models

## 3.1   Variational Autoencoder

Basic autoencoder gets latent representation from the input data by weighting and mapping, then inverts maps and minimizes the error function by repeated iteration to reconstructs the outputs and obtain the approximate values of original contents. A single autoencoder learn a characteristic variation $z = f_\theta(x)$ through a three layers structure: $x \rightarrow z \rightarrow \hat{x}$. The latent vector z obtained after training phase can be regarded as a new input to train a new autoencoder and stacks in this way to form a stacked deep autoencoder.

Variational autoencoder (Fig. 1) is a generative model based on variational Bayesian inference. Unlike traditional autoencoder, the VAE replaces the deterministic function $z = f_\theta(x)$ with a posterior recognition model $p(x|z)$, and replaces the deterministic function $x = f'_\emptyset(z)$ with an inference model $q(\sim z|x)$. In other words, VAE not learn an arbitrary deterministic function but a set of parameters from a probability distribution of the input, and reconstruct the input by sampling in the probability distribution. A generative process of the VAE takes two steps: step one is generating a set of latent variables $z$ from the prior distribution $p_\theta(z)$, and step two is reconstructing the data $x$ based on $z$ by the generative distribution $p_\theta(x|z)$. $p_\theta(z), p_\theta(x|z)$ are parameterized probability distribution functions. The posterior distribution $p_\theta(z|x)$ is intractable and thus approximate inference should be applied. VAE estimates parameters by variational inference. It introduces an inference model $q_\emptyset(z|x)$ which approximated by a standard Gaussian distribution to converge to the true posterior probability $p_\theta(z|x)$.

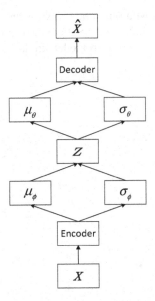

**Fig. 1.** VAE model. $Z$ is the latent variable sampled from the approximate posterior distribution with mean and variances parameterized using neural networks.

## 3.2   Deep Autoencoder Integrating Topic Information

A continuous hidden variable $z$ generated from VAE carries a rich text of its own semantic information, but need the supports of context information in text similarity task. We propose an improved VAE model which integrates topic information in it, called topic variational autoencoder (TVAE). The latent variable $z$ is generated by a mixture of input x and topic information $t$, which similar to the method for conditional multimodal by CMMA in [19].

Our aim is to find the parameters so as maximize the joint log-likelihood of $x$ and $t$ for the given sequence, the joint log-likelihood can be written as:

$$\log p_\theta(x|t) = KL\big[q_\phi(z|x)p_\theta(z|x,t)\big] + \mathrm{E}_{q_\phi(z|x,t)}\log\frac{p_\theta(x,z|t)}{q_\phi(z|x)}$$

$$\geq \mathrm{E}_{q_\phi(z|x,t)}\log\frac{p_\theta(x,z|t)}{q_\phi(z|x)}$$

(1)

Here, the variational lower bound is:

$$\ell_{TVAE}(x,t;\theta,\phi) = \mathrm{E}_{q_\phi(z|x,t)}\log\frac{p_\theta(x|z)}{q_\phi(z|x,t)} + \mathrm{E}_{q_\phi(z|x,t)}\log p_\theta(x|z)$$

$$= -KL\big[q_\phi(z|x,t)p_\theta(z|t)\big] + \mathrm{E}_{q_\phi(z|x,t)}\log p_\theta(x|z)$$

(2)

However, shallow model with only one or two layers of randomly latent variables limits the flexibility of the latent representations. To improve this, we stacked VAE joined topics information to form a deep model similar to the stacked autoencoder, as shown in Fig. 2.

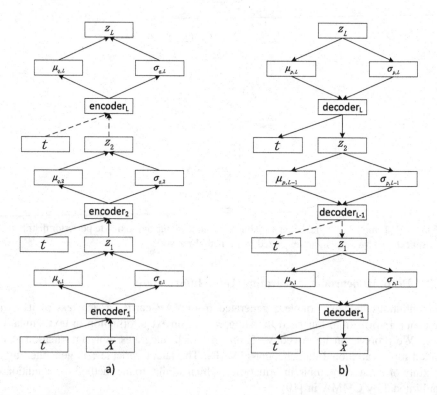

**Fig. 2.** TVAE model. (a) inference (or encoder) model of TVAE, (b) generative (or decoder) model of TVAE. The $z_n$ are latent variables sampled from the approximate posterior distribution with x and topic information t.

In inference model, $z_i$ is built on the output of the $z_{i-1}$ and topic information $t$ and so on until form a deep network with $L$ layers. There are $L$ latent variables in it, the hierarchical specification allows the lower layers of the latent variables to be highly correlated but still maintain the computational efficiency of fully factorized models. Each layer in the inference model $q_\phi(z|x)$ is specified using a Gaussian distribution:

$$\begin{cases} q_\phi(z_1|x,t) = \mathrm{N}\left(z|\mu_{q,1}(x,t), \sigma_{q,1}^2(x,t)\right) \\ q_\phi(z_i|z_{i-1},t) = \mathrm{N}\left(z_i|\mu_{q,i}(z_{i-1},t), \sigma_{q,1}^2(z_{i-1},t)\right) \end{cases} \tag{3}$$

$$q_\phi(z|t) = q_\phi(z_L|t) \prod_{i=1}^{L} q_\phi(z_i|z_{i-1},t) \tag{4}$$

In generative model, random variable $z_i$ is treated as a Gaussian distribution conditioned on $z_{i+1}$ and topic information $t$:

$$\begin{cases} p_\theta(z_i|z_{i+1},t) = \mathrm{N}\left(z_i|\mu_{p,i}(z_{i+1},t), \sigma_{p,1}^2(z_{i+1},t)\right) \\ p_\theta(x|z_1,t) = \mathrm{N}\left(x|\mu_{p,0}(z_1,t), \sigma_{p,0}^2(z_1,t)\right) \end{cases} \tag{5}$$

$$p_\theta(z|t) = p_\theta(z_L|t) \prod_{i=1}^{L-1} p_\theta(z_i|z_{i+1},t) \tag{6}$$

We train the generative and inference parameters $\theta$ and $\emptyset$ by optimizing Eq. (2) using stochastic gradient descent.

### 3.3    Topic Information

Currently, adding topic representation as additional information into deep learning model can often provide significant improvements to the results. Ghos et al. [16] evaluated their CLSTM in three NLP tasks. The gains are all quite significant especially in the next sentence selection task, it improved about 20% on accuracy. Extracting topic information by topic models such as LDA, LSI and NMF is popular in NLP. LDA is a popular algorithm for topic-modelling, the co-occurrence information of terms is used to find the topic structure of text. NMF is often regarded as LDA with fixed parameters and can obtain sparse solutions. Although NMF model is less flexible than the LDA model, it can handle short text data sets better than LDA. LSI based on the SVD method to get the topics of the text. The algorithm principle is simple: a singular value decomposition to get the topic model. But the SVD is very time-consuming in text processing due to the huge quantity of words and documents. For this limitation, NMF is more advantageous in computation speed. Furthermore, it is more reasonable and easily interpretable because its non-negativity condition can provide semantically meaningful representation.

In this paper, we use NMF with sparseness constraint [20] to extract topic information, and each text will be represented as a weighted combination of topics. For a desired dimension $r$ (the number of topics in our work), NMF approximately decompose a non-negative matrix $V$ into $W$ and H: $V_{n \times m} = W_{n \times r} \times H_{r \times m}$. Input dataset is represented as the column of $V$, where $n$ is the number of texts, $m$ is the number of vocabulary. NMF finds matrix $W$ and matrix H by minimizing the objective function:

$$\frac{1}{2} \sum_i^n \sum_j^m \left( V_{ij} - (WH)_{ij} \right)^2 + \mu ||H||_1 \tag{7}$$

where $\mu$ is a penalty parameter used to control the sparseness. The result of NMF can be viewed as topic modeling results directly due to the non-negativity constraints [21]. Each column of $H$ is the sparse representation of this text over all topics which we aim to obtain.

## 4 Experiments

### 4.1 Datasets and Context Information

Most previous work has reported results on two datasets "SCWS" and "SICK", so we evaluate our method on "SCWS" and "SICK" for word similarity and text pair semantic similarity tasks. We use common metric the Spearman's $\rho$ correlation to evaluate the performance. We will explain in detail on Spearman's $\rho$ correlation later.

The dataset SCWS is a word similarity dataset with ground-truth labels on similarity of pairs of target words in sentential context from [8]. It consists of 2003 words pairs and their sentential contexts, including 1328 noun-noun pairs, 399 verb-verb pairs, 140 verb-noun, 97 adjective-adjective, 30 noun-adjective, 9 verb-adjective, and 241 same word pairs.

The dataset SICK consists of about 10,000 English sentence pairs, generated starting from two existing sets: the 8K ImageFlickr data set and the SemEval 2012 STS MSR-Video Description data set.

For each word in SCWS, topic information is obtained from its surrounding context sentences. Average word embedding is used to create context vectors for target words. We use NMF, the same way in [7] to create context global context vectors, which are used as the input $t$ of our TVAE model.

### 4.2 Parameter Setting

We use pre-trained word vectors from GloVe [22] to convert natural language word into vector representation, which is used as the input $x$ of our TVAE model. For the SCWS and SICK task, we use 100-dimensional, 200-dimensional, 300-dimensional, 400-dimensional word embeddings to represent target word respectively, with 50-dimensional context vectors performing best from [8]. In the experiment, our model performs best by tuning the parameters when we set the weight parameter $\lambda = 0.5$, masking noise $\eta = 0$, depth of our model $n = 3$.

### 4.3    Word Similarity

In order to demonstrate the effectiveness of topic information for our model in the word similarity task, we use two cases to measure: **SimS** and **SimC**. SimS uses one representation per word to compute Similarities by our model in different dimension ignoring information from the context; SimC calculates the similarity with context information. The Spearman's $\rho$ correlation between each model's similarity judgement and the human judgement in context is used.

Table 1 shows the performance of different models [7, 8, 23–26] on the SCWS dataset. TVAE outperforms the other models in general. Especially, the result performs better when the word embedding is 300-dimension. So, the word embedding's dimension also affects the performance of the model.

**Table 1.** Spearman's $\rho$ performance of different models on the SCWS dataset (the numbers are $\rho \times 100$). Fields marked with "-" indicate that the results were not available for assessment.

| Model | | SimS | SimC |
|---|---|---|---|
| Huang et al. [8] | | 58.6 | 65.7 |
| Chen et al. [28] | | 64.2 | 68.9 |
| Neelakantan et al. [29] | | 65.5 | 69.3 |
| Rothe and Schütze [23] | | – | 69.8 |
| Amiri et al. [7] | | 61.1 | 70.9 |
| Zheng et al. [24] | | – | 69.9 |
| Our model | TVAE-100d | 63.5 | 68.2 |
| | TVAE-200d | 64.6 | 69.3 |
| | TVAE-300d | **66.6** | **71.3** |
| | TVAE-400d | 64.3 | 68.8 |

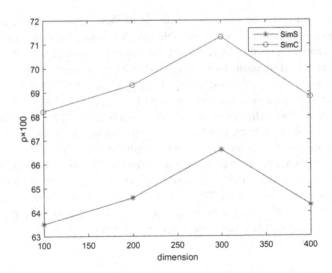

**Fig. 3.** Performance of the different dimension word embedding of our TVAE model. We only extract the 100-dimension, 200-dimension, 300-dimension, 400-dimension word embeddings for the TVAE model.

Figure 3 shows clearly the impact of the dimension on the result. The 300-dimension word embedding outperforms other dimension word embedding. Cosine similarity calculation does not handle High-dimensional vectors well while it performs well for Low-dimensional vector. As is shown in Fig. 4, the SimC method performs better than the SimS method. Topic information improves the performance of our model.

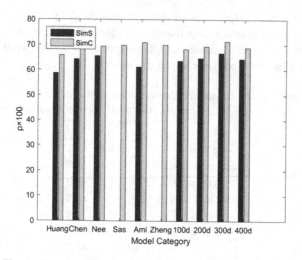

**Fig. 4.** Spearman's $\rho \times 100$ performance of different models.

## 4.4  Text Pair Semantic Similarity

For a pair of input texts, we measure the degree of their semantic similarity by computing the cosine similarity between their hidden representations from our model. We test text representation on the SICK dataset. Representing sentences by the average of their constituent word representations is proved to be surprisingly effective. That is, a sentence $s$ can be represented as: $v(s) = 1/|s| \sum_{w \in s} v(w)$, which has been shown effective in encoding the semantic information of sentences in [19].

Table 2 shows the performance of different models [25–27] for text pair semantic similarity on the SICK dataset. Our model outperforms the other models. The size of word embedding also affects the experimental results. As shown in Table 2 and Fig. 5, the result is best when the size of word embedding is 300-dimension. In this experiment, we use the topic information captured from sentences representation in the whole process. Comparing with others, our model is still a bit improved, but does not noticeably perform well. The reason is that our topic information is helpful, but sentence itself does not contain rich contextual information contrast to surrounding text which absent in this dataset.

**Table 2.** Experimental results in the SICK task. The numbers are Spearman's correlation $\rho \times 100$ between each model's similarity judgments and the human judgments.

| Model | | $\rho \times 100$ |
|---|---|---|
| SDAE | | 46 |
| Hill et al. [25] | | 60 |
| Sent2Vect | | 62 |
| Pagliardini et al. [26] | | 60 |
| Li et al. [27] | | 61 |
| Deep VAE | | 60 |
| Our model | TVAE-100d | 59.3 |
| | TVAE-200d | 61.2 |
| | TVAE-300d | **63.1** |
| | TVAE-400d | 62.3 |

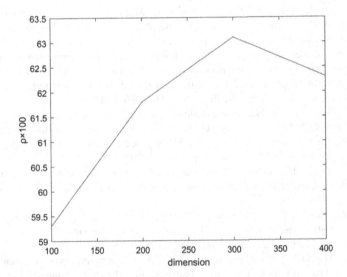

**Fig. 5.** The performance of the different word embedding size of our model.

## 5   Conclusions

This paper introduces the use of deep topic variational autoencoder (TVAE) to model natural language text, and calculate semantic similarity of text pair on this basis.

We integrate topic information of sentence or document as additional information into deep VAE. Our topic representation is obtained from the NMF with sparseness constraint. We compared with our model and the model without topic information on SCWS and SICK dataset, and find that the results with topic information perform better. This demonstrates that such integration is effective in semantic similarity tasks because it contains rich contextual information. In addition, our TVAE stakes the VAE with topic information up to three layers. The latent representations learned from our model are more flexible than the shallow network. We verify our model in word semantic similarity task and sentence semantic similarity task, and find that different dimension exert a great influence in both tasks.

**Acknowledgements.** This work was funded by National Natural Science Foundation of China (Grant No. 61762069), Natural Science Foundation of Inner Mongolia Autonomous Region (Grant No. 2017BS0601, Grant No. 2018MS06025) and program of higher-level talents of Inner Mongolia University (Grant No. 21500-5165161).

# References

1. Lin, R., Liu, S., Yang, M., Li, M., Zhou, M., Li, S.: Hierarchical recurrent neural network for document modeling. In: Conference on Empirical Methods in Natural Language Processing, Lisbon, pp. 899–907 (2015)
2. Liu, P., Qiu, X., Chen, X., Wu, S., Huang, X.: Multi-timescale long short-term memory neural network for modelling sentences and documents. In: Conference on Empirical Methods in Natural Language Processing, Lisbon, pp. 2326–2335 (2015)
3. Tai, K.S., Socher, R., Manning, C.D.: Improved semantic representations from tree-structured long short-term memory networks. In: The 53rd Annual Meeting of the Association for Computational Linguistics and the 7th International Joint Conference on Natural Language Processing, Beijing, pp. 1556–1566 (2015)
4. Kingma, D.P., Welling, M.: Auto-encoding variational bayes. arXiv:1312.6114 (2014)
5. Bowman, S.R., Vilnis, L., Vinyals, O., Dai, A.M., Jozefowicz, R., Bengio, S.: Generating sentences from a continuous space. In: SIGNLL Conference on Computational Natural Language Learning (CONLL), Berlin, pp. 10–21 (2016)
6. Miao, Y., Yu, L., Blunsom, P.: Neural variational inference for text processing. In: the 33rd International Conference on International Conference on Machine Learning, New York, pp. 1727–1736 (2016)
7. Amiri, H., Resnik, P., Boyd-Graber, J., Daumé III, H.: Learning text pair similarity with context-sensitive autoencoders. In: The 54th Annual Meeting of the Association for Computational Linguistics, Berlin, pp. 1882–1892 (2016)
8. Huang, E.H., Socher, R., Manning, C.D., Ng, A.Y.: Improving word representations via global context and multiple word prototypes. In: The 50th Annual Meeting of the Association for Computational Linguistics, Jeju, pp. 873–882 (2012)
9. Marelli, M., Menini, S., Baroni, M., Bentivogli, L., Bernardi, R., Zamparelli, R.: A SICK cure for the evaluation of compositional distributional semantic models. In: The Ninth International Conference on Language Resources and Evaluation (2014)
10. Xu, W., Sun, H., Deng, C., Tan, Y.: Variational autoencoders for semi-supervised text classification. In: The Thirty-First AAAI Conference on Artificial Intelligence, San Francisco, pp. 3358–3364 (2017)
11. Serban, I.V., et al.: A hierarchical latent variable encoder-decoder model for generating dialogues. In: Thirty-First AAAI Conference on Artificial Intelligence, San Francisco (2017)
12. Miao, Y., Blunsom, P.: Language as a latent variable: discrete generative models for sentence compression. In: The 2016 Conference on Empirical Methods in Natural Language Processing, Austin, pp. 319–328 (2016)

13. Sønderby, C.K., Raiko, T., Maaløe, L., Sønderby, S.K., Winther, O.: How to train deep variational autoencoders and probabilistic ladder networks. In: The 33rd International Conference on Machine Learning, New York (2016)
14. Mikolov, T., Zweig, G.: Context dependent recurrent neural network language model. In: Spoken Language Technology Workshop, Miami, pp. 234–239 (2013)
15. Ji, Y., Cohn, T., Kong, L., Dyer, C., Eisenstein, J.: Document context language models. In: ICLR 2016 Workshop Track, San Juan, pp. 1–10 (2016)
16. Ghosh, S., Vinyals, O., Strope, B., Roy, S., Dean, T., Heck, L.: Contextual LSTM (CLSTM) models for Large scale NLP tasks. arXiv:1602.06291 (2016)
17. Dieng, A.B., Wang, C., Gao, J., Paisley, J.: TopicRNN: a recurrent neural network with long-range semantic dependency. arXiv:1611.01702 (2016)
18. Xing, C., et al.: Topic augmented neural response generation with a joint attention mechanism. arXiv:1606.08340 (2016)
19. Wieting, J., Bansal, M., Gimpel, K., Livescu, K.: From paraphrase database to compositional paraphrase model and back. J. Trans. Assoc. Comput. Linguist. 3, 345–358 (2015)
20. Rezende, D.J., Mohamed, S., Wierstra, D.: Stochastic backpropagation and approximate inference in deep generative models. In: The 31st International Conference on Machine Learning, PMLR, vol. 32, no. 2, pp. 1278–1286 (2014)
21. Chung, J., Kastner, K., Dinh, L., Goel, K., Courville, A., Bengio, Y.: A recurrent latent variable model for sequential data. In: The 28th International Conference on Neural Information Processing Systems, Montreal, vol. 2, pp. 2980–2988 (2015)
22. Pennington, J., Socher, R., Manning, C.: Glove: global vectors for word representation. In: Conference on Empirical Methods in Natural Language Processing, Doha, pp. 1532–1543 (2014)
23. Rothe, S., Schütze, H.: Autoextend: extending word embeddings to embeddings for synsets and lexemes. In: the 53rd Annual Meeting of the Association for Computational Linguistics and the 7th International Joint Conference on Natural Language Processing, Beijing, pp. 1793–1803 (2015)
24. Zheng, X., Feng, J., Chen, Y., Peng, H., Zhang, W.: Learning context-specific word/character embeddings. In: The Thirty-First AAAI Conference on Artificial Intelligence, San Francisco, pp. 3393–3399 (2017)
25. Hill, F., Cho, K., Korhonen, A.: Learning distributed representations of sentences from unlabelled data. In: NAACL-HLT, San Diego, pp. 1367–1377 (2016)
26. Pagliardini, M., Gupta, P., Jaggi, M.: Unsupervised learning of sentence embeddings using compositional n-gram features. arXiv:1703.02507 (2017)
27. Li, B.F., Liu, T., Zhao, Z., Wang, P.W., Du, X.Y.: Neural bag-of-ngrams. In: The Thirty-First AAAI Conference on Artificial Intelligence, San Francisco, pp. 3067–3074 (2017)
28. Chen, X., Liu, Z., Sun, M.: A unified model for word sense representation and disambiguation. In: Conference on Empirical Methods in Natural Language Processing, pp. 1025–1035 (2014)
29. Neelakantan, A., Shankar, J., Passos, A., Mccallum, A.: Efficient non-parametric estimation of multiple embeddings per word in vector space. In: Conference on Empirical Methods in Natural Language Processing, pp. 1059–1069 (2014)

# Mongolian Word Segmentation Based on Three Character Level Seq2Seq Models

Na Liu[1,2,3], Xiangdong Su[1,2(✉)], Guanglai Gao[1,2], and Feilong Bao[1,2]

[1] College of Computer Science, Inner Mongolia University, Hohhot, China
cssxd@imu.edu.cn
[2] Inner Mongolia Key Laboratory of Mongolian Information Processing Technology, Hohhot, China
[3] Department of Science, Hetao University, Bayannur, China

**Abstract.** Mongolian word segmentation is splitting the Mongolian words into roots and suffixes. It plays an important role in Mongolian related natural language processing tasks. To improve performance and avoid the tedious work of rule-making and statistics over large-scale corpus in early methods, this work takes a Seq2Seq framework to realize Mongolian word segmentation. Since each Mongolian word consisted of several sequential characters, we map Mongolian word segmentation to character-level Seq2Seq task, and further propose three different models from three different prospective to achieve the segmentation goal. The three character-level Seq2Seq models are (1) *translation model*, (2) *true and pseudo mapping model*, (3) *binary choice model*. The main differences of these three models are the output sequences and the architectures of the RNNs in segmentation. We employ an improved beam search to optimize the second segmentation model and boost the segmentation process. All the models are trained on a limited dataset, and the second model achieved the state-of-the-art accuracy.

**Keywords:** Mongolian · Word segmentation · Seq2Seq
Limited search strategy · LSTM

## 1 Introduction

Mongolian is an agglutinative language which normally ranks as a member of the Altaic language family, a family whose principal members are Turkish, Mongolian and Manchu (with Korean and Japanese listed as possible relations). Mongolian words are formed by attaching suffixes to roots. The suffix falls into two groups: derivational suffix and inflection suffix. Derivational suffix is also called the word-building suffix. They are added to the root and give the original words new meanings. The root adding one or more derivation suffixes is called a stem. Inflection suffix is also called word-changing suffix. They are added to the stems and give the original words grammatical meanings. These suffixes serve to integrate a word into sentence. In some special cases, there are nearly seventy suffixes following a root. Therefore, the number of various words is theoretically numerous [1]. Since the role difference between derivation suffix and inflection suffix, many NLP tasks pay more attention to the suffix processing. And

© Springer Nature Switzerland AG 2018
L. Cheng et al. (Eds.): ICONIP 2018, LNCS 11305, pp. 558–569, 2018.
https://doi.org/10.1007/978-3-030-04221-9_50

thus the Mongolian word segmentation becomes the requisite step in these NLP tasks, which segment the Mongolian word into roots and suffixes.

Automatic segmentation of Mongolian words is of utmost importance for development of more sophisticated NLP systems such as information retrieval [2], Machine Translation [3], Name Entity Recognition [4], Speech Synthesis [5] and so on. On the contrary, erroneous segmentation (including overcutting or undercutting) will degrade the overall performance of these systems. Due to the sparse labeled dataset and rich suffix variation, limited resources segmentation approaches were proposed to facilitate the Mongolian word segmentation. These approaches usually depend on three sources, including root dictionary, suffix dictionary and segmenting rules. The methods given in [6, 7] first match roots in the root dictionary, and then compare the remaining part with the suffix dictionary. These methods could give the segmentation results automatically, but ambiguity also follows. That is, there may be more than one root or suffix matching the substring of the original word and which one is reasonable is difficult to decide automatically. Another work [8] tries to solve this problem by referring to POS file. It strongly relies on the lexical information in the corpus. Furthermore, these methods still could not deal with Out of Vocabulary (OOV) very well.

To improve performance and avoid the tedious work of rule-making and statistics over large-scale corpus in early methods, this work takes a sequence-to-sequence (Seq2Seq) framework for Mongolian word segmentation. Since each Mongolian word consisted of several sequential characters, we mapped Mongolian word segmentation into a character-level Seq2Seq task, and further propose three different models from three different prospective to achieve the segmentation goal. The first model is named as translation model, in which the segmentation process is treated as a translation process from the character sequence of the original word to the target sequence which consists of the characters of the root, suffixes, and the character "-" between them. The second model is named as true and pseudo mapping model, in which each character in the original word is mapped to itself or a pseudo character. Here the pseudo character represents the input character plus the segmentation symbol "-". We change the rules of beam search to optimize this model and boost the segmentation process. The third model is called binary choice model, representing the fact that we make a decision that whether the segmentation is needed after each character in the original word. All the models are trained on a limited dataset, and the second model achieved the state-of-the-art accuracy. In the experiment, we exploit the base LSTM following [9], bidirectional LSTM (Bi-LSTM), and their variants with attention mechanism (LSTM+A, Bi-LSTM+A).

The main contributions of the work can be summarized as follows:

1. To the best of our knowledge, however, we are the first group to map Mongolian word segmentation to a Seq2Seq task without resorting to dictionaries, rules, and large training corpus.
2. We propose three different Seq2Seq models from three different prospective to achieve the segmentation goal. Meanwhile, we compare the performance of various LSTMs in Seq2Seq process.
3. We improved the beam search strategy to optimize the true and pseudo mapping model and boost the segmentation process.

## 2  Related Works

Early approaches for Mongolian word segmentation usually depend on three sources, including root dictionary, suffix dictionary and segmenting rules. The methods given in [6, 7] match roots and suffixes in the positive or negative direction using the root dictionary and suffix dictionary. When there is more than one root or suffix matching the substring of the original word, an ambiguity occurs. The segmentation rules are then used to deal with the ambiguity and make the final decision. Another work [8] tries to solve this problem by referring to POS file, in which the word needs to be labeled the POS tag in advance.

In this paper, we take a Seq2Seq to implement Mongolian word segmentation. Seq2Seq learning has made astounding progress in various natural language processing (NLP) tasks, including but not limited to Machine Translation [10–14], speech to text [15, 16], dialogue systems [17] and text summarization [18, 19]. The Seq2Seq model using deep neural networks mainly has two parts: encoder and decoder. The former reads input data in a sequence style and generates the hidden unite. The latter uses the output generated by encoder and produces the sequence of outputs. Recurrent neural networks (RNN), are always used to implement encoder and decoder, such as LSTMs [9, 10, 20, 21]. A common LSTM unit is composed of a cell, an input gate, an output gate and a forget gate. The cell is responsible for "remembering" values over arbitrary time intervals; hence the word "memory" in LSTM. Each of the three gates can be thought of as a "conventional" artificial neuron, as in a multi-layer (or feedforward) neural network. LSTMs were developed to deal with the exploding and vanishing gradient problem when training traditional RNNs.

An important extension of the Seq2Seq model is by adding an attention mechanism [10, 13, 15, 22]. It is approved that attention mechanism is able to focus on the most effective information (word or phonological) adaptive to each token to maximize the information gain. In order to amplify the contribution of important elements in the final segmentation results we use an attention mechanism following [10], that aggregates all the hidden states using their relative importance.

## 3  Approach

As mentioned above, Mongolian words are formed by adding derivational suffix and inflection suffix to roots. If we treat the original word as character sequence, and transform the target root and suffixes into a new sequence, the segmentation process is a character-level Seq2Seq task. From three different prospective, we propose three different models to achieve the segmentation goal, including translation model, true and pseudo mapping model, binary choice model.

### 3.1  Translation Model

Mongolian word segmentation is splitting word into root and suffixes. Translation model treats segmentation as a Seq2Seq translation task. Take the Mongolian word "ᠠᠷᠠᠳ/ᠨ" for example, its root is "ᠠᠷᠠᠳ" and suffixes are " ᠵ " and "ᠨ". We connect the roots and the suffixes with the character "-" in order, (the character "-" is not a Mongolian character), and obtains a new sequence is "ᠪᠠᠨ _ ᠳᠤ ᠴᠢ" which we called target sequence. If we can translate the character sequence of the original word to the target sequence with the Seq2Seq framework, then we can obtain the root and suffixes by dividing the target sequence by character "-". From this point of view, the segmentation is a character-level translation process. That is, we translate the original character sequence to the target sequence. Thus, we employ a character-level translation model to do Mongolian word segmentation. Figure 1 demonstrates the translation model. Each Mongolian word also consists of characters. For simplicity, we used the Latin character to represent Mongolian character in some examples and figures.

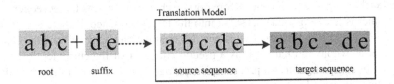

Translation Model

root      suffix          source sequence      target sequence

**Fig. 1.** Translation model in Mongolian word segmentation

This paper employs recurrent neural network (RNN) to implement the translation model for Mongolian word segmentation task. In Seq2Seq learning task, RNN (specifically LSTM) involves an encoder-decoder architecture, which have demonstrated state-of-the-art performance. Let $X = (x_1, x_2, \cdots, x_I)$ and $Y = (y_1, y_2, \cdots, y_J)$ be the input/output sequence, with $I$ and $J$ respectively being the input/output lengths. Seq2Seq learning can be expressed finding the most probable output sequence given the input:

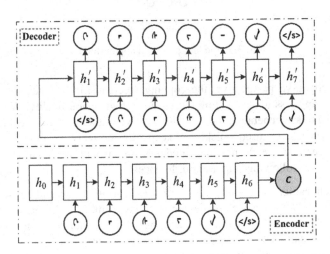

**Fig. 2.** The encoder-decoder framework. An encoder converts character-level input sequence "ᠴ ᠷ ᠠ ᠷ ᠳ" (Mongolian word "ᠴᠠᠷᠳ") into a fixed length vector $c$ which is passed through a decoder to produce the segmentation "ᠴ ᠷ ᠠ ᠷ _ ᠳ".

$$argmax_{Y \in YP}(\boldsymbol{Y}|\boldsymbol{X}) \tag{1}$$

Where $Y$ is the set of all possible sequences. Figure 2 gives a schematic of this simple framework for Mongolian word segmentation problem.

## 3.2   True and Pseudo Mapping Model

In translation model, we convert the character sequence of the original word into the target sequence. We assume the translation process is completely correct. That is we can obtain the expected root and suffixes. Now, we make a comparison between the original word and the target sequence, we find that two character operations happened in the translation process: (1) one is copying the character into the target sequence, and (2) the other is inserting a character "-" into the target sequence. If there is a character "c" and a "-" after it in the target sequence, we treat the character "c" and the character "-" after it as whole and named it as a pseudo character of the character "c".

Then the segmentation is a process of mapping each character in the original word to itself or to its pseudo character. This is also one-to-one mapping, and we call it true and pseudo mapping model. For simplicity, we use Latin characters to demonstrate this in Fig. 3.

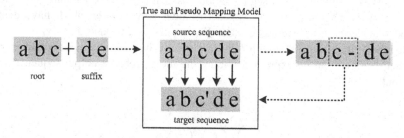

**Fig. 3.** True and pseudo mapping model.

From a cognitive perspective, the copying mechanism is related to rote memo-rization, requiring less understanding but ensuring high literal fidelity. We expect that when the results are generated, we can filter and view them, and if we can use some strategy, which can be used quickly to keep high copying loyalty. From the data perspective, true and pseudo mapping model is similar with translation model above, with however the following important differences:

- **Pseudo character:** For the input word sequence "ᠣ ᠷ ᠡ ᠷ ᠣ" and its segmentation sequence "ᠣ ᠷ ᠡ ᠷ ᠆ ᠣ", we combine the dividing symbol " -" and its nearest left neighbor character "ᠷ" into a new abstract pseudo character "ᠷ". Therefore, the size of the target dictionary is about 2 times that of the previous.
- **Output rewritten:** We rewrite the output sequence based on the above way of pseudo character formation.

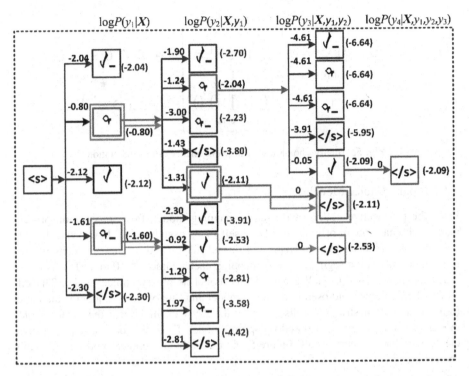

**Fig. 4.** A simple of beam ($n = 2$) search example. The red boxes and arrows are the search paths of standard beam search. The green boxes and arrows are the search paths of limited search strategy. (Color figure online)

- **Limited search strategy:** To improve the performance and accelerate the segmentation process, we limit the output target in advance and change the standard n-best beam search strategy [23] to 2-best search strategy, called limited search strategy (LSS). To explain LSS, we take word "᠊ᠣ᠊ᠸ᠊" (input "᠊ᠣ᠊ᠸ᠊") for a simple example in the following paragraph (see Fig. 4).

We also employ the RNN to implement the true and pseudo mapping model, in which beam search is responsible for finding the best output sequence. Beam size is critical to the decoding speed and performance, and is set to larger than 2 in general. However, in the true and pseudo mapping model, every possible output in each decoding step is either the input character or its pseudo character. Based on this rule, we set the beam size to 2. From the first step to the end, at each step, we keep the top 2-best conditional hypotheses by the input character and its abstract pseudo character. Since in this case, the conditional hypotheses are restricted in relatively very small areas, leading to the improvement in the efficiency and accuracy of the algorithm.

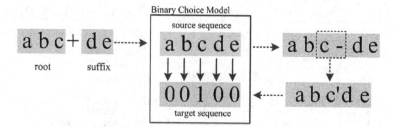

**Fig. 5.** Binary choice model in Mongolian word segmentation

### 3.3    Binary Choice Model

From the prospective of true and pseudo mapping model, the segmentation process maps each character in original word to itself or its pseudo form. If we consider that mapping the character to itself represent 0, and mapping the character to its pseudo form represents 1, the segmentation is mapping each character to 0 or 1. This is also a sequence to sequence task. We named this model as binary choice model. The difference between true and pseudo mapping model and binary choice model is the target form or the output string. We also employ the RNN to implement the binary choice model as we do in true and pseudo mapping model. From this decoding perspective, the search space is very small. Figure 5 shows the binary choice model using Latin characters.

## 4    Experiments

### 4.1    Dataset

Nowadays, there is no public annotated corpus about Mongolian word segmentation. In this paper, we have collated and expanded the laboratory corpus. The dataset we used has been annotated and reviewed manually by a group of Mongolian native speakers, including 60,000 Mongolian words. We split it into training dataset (42,000 words, 70%), developing dataset (6,000 words, 10%) and testing (12,000 words, 20%). It is important to note that our test samples and training samples are completely different. That means the testing words we evaluated are all out-of-vocabulary (OOV) words, which is more challenging.

### 4.2    Evaluation Metrics

Quantitative evaluation of the segmentation systems is performed using Precision ($P$), Recall($R$), $F1$, and Word precision ($Wp$). This is the same as those used in [6–8]. We defined each segment as one unit after Mongolian word segmentation. For example, the word "ᠬᠥᠸ" is segmented into two units "ᠬᠥ" and "ᠸ". The following are the formulas for evaluation metrics, where the MU is the total number of the units in manually annotated testing samples, RU is the number of resulting units from the testing samples after segmentation, CU is the number of correct units among resulting units of the testing sample, CW represents the number of the word who is correctly segmented in the testing dataset, TW represents the number of the word in testing dataset.

$$P = \frac{CU}{RU} \times 100\% \tag{2}$$

$$R = \frac{CU}{MU} \times 100\% \tag{3}$$

$$F1 = \frac{2 \times P \times R}{P + R} \tag{4}$$

$$Wp = \frac{CW}{TW} \times 100\% \tag{5}$$

### 4.3  Experimental Setting

In experiment, we use the naïve LSTM, Bi-LSTM and their variants with attention mechanism (LSTM+A, Bi-LSTM+A) in the Seq2Seq process. The input and output are designed according to the three models (translation model, true and pseudo mapping model and binary choice model). Our Bi-LSTM models have 2 layers. The core of the experiments involved training a large LSTM. As described in [9], we use greedy algorithm to separate corpus into two buckets based on the length of each Mongolian word during training. It is important to note that the search strategy used in our other LSTMs is $n$-best search strategy, where $n = 5$. We initialized all of the LSTM's parameters with the uniform distribution between $-0.1$ and $0.1$. We used stochastic gradient descent without momentum, with a fixed learning rate of 0.8. After 5 epochs, we begin to half the learning rate every epoch. We compare the evaluation metrics by using three models. And we also compare the evaluation metrics by true and pseudo mapping model with limited search strategy (LSS). Other super parameters are selected according to the performance on the developing dataset.

## 5  Results and Discussion

### 5.1  The Effect of Different LSTMs

To validate the effectiveness of these different LSTMs on the proposed three models, we evaluate their performance on the testing dataset and the scores are reported in Table 1. For true and pseudo mapping model, we list the score of the different LSTMs with the LSS. According to the results, the following conclusions were obtained:

- For translation model and true and pseudo model, the Bi-LSTM performs better than naïve LSTM, because it provides more temporal context than the unidirectional LSTM.

**Table 1.** Performance of different LSTMs

| Model | Seq2Seq | P(%) | R(%) | F1(%) | Wp(%) |
|---|---|---|---|---|---|
| Translation model | LSTM | 92.62 | 92.67 | 92.65 | 88.00 |
| | Bi-LSTM | 93.82 | 93.62 | 93.72 | 88.99 |
| | LSTM+A | 95.52 | 95.37 | 95.44 | 92.36 |
| | Bi-LSTM+A | 95.87 | 95.64 | 95.75 | 92.72 |
| True and pseudo mapping model | LSTM+LSS | 92.77 | 92.94 | 92.86 | 89.83 |
| | Bi-LSTM+LSS | 94.56 | 93.98 | 94.27 | 90.79 |
| | LSTM+A+LSS | 95.59 | 95.87 | 95.73 | 93.92 |
| | Bi-LSTM+A+LSS | 96.37 | 96.01 | 96.19 | 94.02 |
| Binary choice model | LSTM | 93.64 | 93.39 | 93.52 | 91.45 |
| | Bi-LSTM | 94.43 | 92.92 | 93.67 | 92.04 |
| | LSTM+A | 93.97 | 93.68 | 93.83 | 93.22 |
| | Bi-LSTM+A | 94.65 | 94.18 | 94.41 | 93.26 |

- Each segmentation model with attention mechanism achieves better result than that without this mechanism, because the attention mechanism is good at finding important information and focusing on it.

## 5.2  Three Model Comparison

This section compares with the best results of the above three segmentation models. And the same test set is used to validate the segmentation system described in literature [6] (abbr. Hou [6]). Of course, we manually tagged the POS before running.

As shown in Table 2, all of our models perform better than the traditional model Hou [6] without any feature engineering or dictionary. Considering these evaluate metrics, the true and pseudo mapping model with Bi-LSTM+A+LSS is the best of all. It establishes performance of 96.19 $F1$ and 94.02 $Wp$, outperforming the model Hou [6] 9.95 $F1$ and 11.49 $Wp$. The results indicate that our models can better extract the adhesion characteristics of Mongolian words. That is, the internal high cohesion and the external low-coupling of Mongolian morphemes (roots, suffixes).

**Table 2.** Performance of our models and Hou [6]

| Model | P(%) | R(%) | F1(%) | Wp(%) |
|---|---|---|---|---|
| Hou [6] | 85.90 | 86.59 | 86.24 | 82.53 |
| Translation model | 95.87 | 95.64 | 95.75 | 92.72 |
| True and pseudo mapping model | 96.37 | 96.01 | 96.19 | 94.02 |
| Binary choice model | 94.65 | 94.18 | 94.41 | 93.26 |

## 5.3  Evaluate the Limited Search Strategy

This section investigates the results by true and pseudo mapping model with and without Limited Search Strategy (LSS) to examine the effect of LSS. We add an extra metric speed (characters per second) to evaluate the decoding process of each model.

At the top and bottom of Table 3, we report the segmentation results obtained by true and pseudo mapping model with standard $n$-best search($n = 5$) and with LSS individually. It is obvious that LSS improves the performance of true and pseudo mapping model, no matter which seq2seq framework is used.

As shown in Table 3, our model with LSS achieves a speed more than 5 times faster the same model with standard $n$-best ($n = 5$) search. Because our limited search strategy reduces the search space and the impact of data sparseness. The results are consistent with our intuition.

**Table 3.** Performance of true and pseudo mapping model

| Seq2Seq | P(%) | R(%) | F1(%) | Wp(%) | Speed |
|---|---|---|---|---|---|
| LSTM | 78.45 | 79.17 | 78.81 | 68.43 | 85.71 |
| Bi-LSTM | 82.64 | 82.76 | 82.70 | 72.48 | 79.12 |
| LSTM+A | 91.81 | 92.21 | 92.01 | 86.74 | 68.99 |
| Bi-LSTM+A | 92.77 | 92.50 | 92.64 | 87.72 | 78.30 |
| LSTM+LSS | 92.77 | 92.94 | 92.86 | 89.83 | 486.62 |
| Bi-LSTM+LSS | 94.56 | 93.98 | 94.27 | 90.79 | 481.44 |
| LSTM+A+LSS | 95.59 | 95.87 | 95.73 | 93.92 | 383.52 |
| Bi-LSTM+A+LSS | 96.37 | 96.01 | 96.19 | 94.02 | 367.93 |

In addition, the model with attention mechanism achieves better result than that without such mechanism. The framework Bi-LSTM generally obtains better result than the naïve LSTM. Comparing with the translation model listed in Table 2, we find that the true and pseudo mapping model without LSS performs worse than the translation model, due to the fact that when pseudo-characters are added, the problem of sparse data is highlighted compared to translation model.

## 6  Conclusion

In this paper, we have proposed three character-level Seq2Seq models for Mongolian word segmentation without resorting to dictionaries, rules, and large training corpus. The three models are translation model, true and pseudo mapping model, and binary choice model. The key idea of our approaches is mapping the segmentation task into a Seq2Seq task by treating the original word as character sequence and transforming the target root and suffixes into a new sequence. The main differences of these three models are the output sequences and the architectures of the RNNs in segmentation.

Our experiments show that these models are able to obtain quite competitive results compared to early method. The true and pseudo mapping model achieved the state-of-the-art accuracy when Bi-LSTM framework plus attention mechanism and limited search strategy (LSS) is used. LSS is an improved beam search strategy, which not only improves the accuracy metrics of Mongolian word segmentation, but also boosts the segmentation speed. That indicates our models can better extract the adhesion characteristics of Mongolian words.

Furthermore, each segmentation model with attention mechanism achieves better result than that without this mechanism. For translation model and true and pseudo model, the Bi-LSTM performs better than the naïve LSTM, because it provides more temporal context than the unidirectional LSTM.

**Acknowledgements.** This work was funded by National Natural Science Foundation of China (Grant No. 61762069), Natural Science Foundation of Inner Mongolia Autonomous Region (Grant No. 2017BS0601), Research program of science and technology at Universities of Inner Mongolia Autonomous Region (Grant No. NJZY18237).

# References

1. Hankamer, J.: Finite state morphology and left to right phonology. In: Proceedings of the West Coast Conference on Formal Linguistics, vol. 5, pp. 41–52 (1986)
2. Na, L., Junyi, W., Guiping, L.: Query expansion based on mongolian semantics. In: Third World Congress on Software Engineering IEEE Computer Society, pp. 25–28. IEEE, Wuhan (2012)
3. Jing, W., Hou, H., Bao, F., Jiang, Y.: Template-based model for Mongolian - Chinese machine translation. In: Technologies and Applications of Artificial Intelligence, pp. 352–357. IEEE, Tainan (2016)
4. Weihua, W., Feilong, B., Guanglai, G.: Mongolian named entity recognition with bidirectional recurrent neural networks. In: IEEE International Conference on TOOLS with Artificial Intelligence, pp. 495–500. IEEE, San Jose (2017)
5. Liu, R., Bao, F., Gao, G., Wang, Y.: Mongolian text-to-speech system based on deep neural network. In: Tao, J., Zheng, T.F., Bao, C., Wang, D., Li, Y. (eds.) NCMMSC 2017. CCIS, vol. 807, pp. 99–108. Springer, Singapore (2018). https://doi.org/10.1007/978-981-10-8111-8_10
6. Hongxu, H., Liu, Q., Nasanurtu, M.: Mongolian word segmentation based on statistical language model. Pattern Recognit. Artif. Intell. **22**(1), 108–112 (2009)
7. Ming, Y., Hongxu, H.: Researching of Mongolian word segmentation system based on dictionary, rules and language model. M.S. Thesis, Inner Mongolia University, Hohhot, Inner Mongolia, China (2011)
8. Jianguo, S., Hongxu, H., Bao, F.: Research on Slavic Mongolian word segmentation based on dictionary and rule. J. Chin. Inf. Process. **29**(1), 197–202 (2015)
9. Hochreiter, S., Schmidhuber, J.: Long short-term memory. Neural Comput. **9**(8), 1735–1780 (1997)
10. Luong, T., Pham, H., Manning, C.D.: Effective approaches to attention-based neural machine translation. In: Conference on Empirical Methods in Natural Language Processing, EMNLP 2015, pp. 1412–1421 (2015)

11. Wu, Y., Schuster, M., Chen, Z., Le, Q.V., Norouzi, M.: Google's neural machine translation system: bridging the gap between human and machine translation. arXiv preprint arXiv: 1609.08144 (2016)
12. Vaswani, A., et al.: Tensor2tensor for neural machine translation. arXiv preprint arXiv:1803. 07416 (2018)
13. Bahdanau, D., Cho, K., Bengio, Y.: Neural machine translation by jointly learning to align and translate. Computer Science. arXiv preprint arXiv:1409.0473(2014)
14. Shiqi, S., Yong, C., Zhongjun, H., Wei, H., et al.: Minimum risk training for neural machine translation. In: ACL 2016: Annual Meeting of the Association for Computational Linguistics, vol. 1, pp. 1683–1692. ACL, Berlin (2016)
15. Bahdanau, D., Chorowski, J., Serdyuk, D., Brakel, P., Bengio, Y.: End-to-end attention-based large vocabulary speech recognition. Computer Science, pp. 4945–4949 (2016)
16. Arık, S.O., Chrzanowski, M., Coates, A., Diamos, G., Gibiansky, A.: Deep voice: realtime neural text-to-speech. In: International Conference on Machine Learning and Computing, ICMLC 2017, pp. 195–2049. ACM, Singapore (2017)
17. Asri, L.E., He, J., Suleman, K.: A sequence-to-sequence model for user simulation in spoken dialogue systems. In: Conference of the International Speech Communication Association, Interspeech, pp. 1151–1155. IEEE, San Francisco (2016)
18. Nallapati, R., Zhou, B., Santos, C., Gulcehre, C., Xiang, B.: Abstractive text summarization using sequence-to-sequence RNNs and beyond. In: Proceedings of the 20th SIGNLL Conference on Computational Natural Language Learning, CoNLL 2016, pp. 280–290. ACL, Berlin (2016)
19. See, A., Liu, P.J., Manning, C.D.: Get to the point: summarization with pointer-generator networks. In: ACL 2017: Annual Meeting of the Association for Computational Linguistics, vol. 1, pp. 1073–1083. ACL, Vancouver (2017)
20. Sutskever, I., Vinyals, O., Le, Q.V.: Sequence to sequence learning with neural networks. In: Advances in Neural Information Processing Systems 27 (NIPS 2014), vol. 4, pp. 3104–3112. MIT Press Cambridge, Montréal (2014)
21. Hakkani-Tür, D., Tur, G., Celikyilmaz, A., Chen, Y.N., Gao, J., Deng, L.: Multi-domain joint semantic frame parsing using bi-directional RNN-LSTM. In: Conference of the International Speech Communication Association, Interspeech, pp. 715–719. IEEE, San Francisco (2016)
22. Mnih, V., Heess, N., Graves, A., Kavukcuoglu, K.: Recurrent models of visual attention. In: Advances in Neural Information Processing Systems 27 (NIPS 2014), vol. 3, pp. 2204–2212. MIT Press Cambridge, Montréal (2014)
23. Neubig, G.: Neural machine translation and sequence-to-sequence models: a tutorial, pp. 41– 43. arXiv preprint arXiv: 1703.01619v1 (2017)

# Author Index